Lecture Notes in Physics

Edited by H. Araki, Kyoto, J. Ehlers, München, K. Hepp, Zürich
R. L. Jaffe, Cambridge, MA, R. Kippenhahn, München, D. Ruelle, Bures-sur-Yvette
H. A. Weidenmüller, Heidelberg, J. Wess, Karlsruhe and J. Zittartz, Köln

375

C. Bartocci U. Bruzzo
R. Cianci (Eds.)

Differential Geometric Methods in Theoretical Physics

Proceedings of the 19th International Conference
Held in Rapallo, Italy, 19–24 June 1990

Springer-Verlag
Berlin Heidelberg GmbH

Editors

Claudio Bartocci
Ugo Bruzzo
Roberto Cianci
Dipartimento di Matematica, Università di Genova
Via L. B. Alberti 4, I-16132 Genova, Italy

ISBN 978-3-662-13866-3 ISBN 978-3-540-47090-8 (eBook)
DOI 10.1007/978-3-540-47090-8

Originally published by Springer-Verlag Berlin Heidelberg New York in 1991
Softcover reprint of the hardcover 1st edition 1991

2153/3140-543210 – Printed on acid-free paper

Preface

The 19th International Conference on Differential Geometric Methods in Theoretical Physics was held in Rapallo (Italy) from June 19th to 24th, 1990. Since its inception in 1971 in Bonn, this series of conferences has focused, broadly speaking, on those mathematical methods in physics which are of a differential-geometric nature. Among the various topics that formed the programmes of the last conferences, four themes seem to have played a central role, namely, non-commutative differential geometry, quantum groups, superalgebras and supermanifolds, and a complex of arguments embracing conformal field theory, integrable systems, and statistical mechanics. It was therefore decided to devote this 19th conference to these topics. The papers included in this proceedings volume are correspondingly divided into the following four sections:

— non-commutative differential geometry;
— quantum groups and integrable systems;
— conformal field theory and related topics;
— superalgebras and supermanifolds.

A fifth section (problems in quantum field theory) contains two papers which do not really fit into any of the above divisions.

The conference programme included some thirty-five invited lectures. Since this 1990 meeting was intended to be particularly specialized, some "crash courses" covering basic materials were planned; they each consisted of two one-hour lectures, and were held by D. Kastler (non-commutative differential geometry), C. De Concini (quantum groups), and J. Fröhlich (low-dimensional quantum field theory). To our regret, not all the lecturers submitted the text of their lectures for publication in these proceedings, so that only thirty-one papers have been included. In addition to these, eight short articles — reflecting the contents of some of the posters presented during the conference — make up a separate section.

A major feature of this conference was that, for the first time in this series, a fair number of the participants (about 30 out of a total of 160) came from Eastern Europe (including the German Democratic Republic and the USSR). This is of course another consequence of the remarkable political changes that have occurred in those countries in recent years. We hope this is a good omen for the future and that a closer and closer worldwide collaboration among physicists and mathematicians will take place.

A conference like this involves a considerable financial effort. It is therefore a pleasure to acknowledge financial help from those institutions which made the conference possible. These were the Committee for Mathematics of the Italian Research Council (CNR), the National Group for Mathematical Physics of CNR, the National Institute for Nuclear Physics (INFN), the City of Rapallo administration, the Regione Liguria administration, the University of Genoa, and the "Cassa di Risparmio di Genova e Imperia." We also acknowledge clerical help from the

Department of Mathematics of the University of Genoa, the City of Rapallo, and the Tourist Office of Rapallo.

The conference was opened by some welcome speeches, which were delivered by the Rector of the University of Genoa, Prof. Enrico Beltrametti, by the Mayor of the City of Rapallo, Dr. Mauro Cordano, by the Chairman of the Department of Mathematics of the University of Genoa, Prof. Giovanni Pistone, by the coordinator of the National Research Project "Metodi geometrici in relatività e teorie di campo" of the Italian Ministry for Universities, Research and Technology (MURST), Prof. Mauro Francaviglia, and, finally, by Prof. Konrad Bleuler. We would like to thank all these distinguished personalities for their contributions to the conference. Special thanks are due to Professor Bleuler for his exceptional and long-standing commitment to this series of conferences on the differential-geometric methods in theoretical physics and for the enthusiasm he is able to communicate to all of us.

Genoa, November 1990 The Editors

Foreword by Professor K. Bleuler

Institute for Theoretical Nuclear Physics
University of Bonn

Recent years have witnessed most impressive and far-reaching developments in mathematical physics. This is to a large extent due to a renewed and deep-rooted discussion between mathematicians and physicists. In both domains there have been, in completely different directions, important developments (i.e. topology on the one hand and particles on the other) leading to a breathtaking confrontation. We are therefore reminded of that great earlier period of the 1920s with the simultaneous creation and development of general relativity and quantum theory; that decisive step in the history of physics could not be conceived of without the contributions of the greatest mathematicians of that time, e.g. Hilbert, Weyl, Poincaré, and others. Their feeling of that profound mystery of "finding, or better, discovering basic mathematical structures hidden far behind empirical physical facts" has remained the leading idea for understanding nature ever since.

In the course of recent years such "idealistic" guidelines have determined the research projects to a large extent, too: generalized geometrical, topological and group theoretical principles have become the decisive tools for the interpretation and understanding of the vast mass of data resulting from the enormous experimental efforts of our time. This situation had, in a way, been foreseen in Plato's philosophy: his famous words, as emphasized and adapted by Heisenberg, "symmetries are more basic than particles" appear in fact to have been realized in a literal way by modern gauge theory.

Within the huge body of empirical data in hadron physics there are a practically infinite number of different heavy fermions and bosons originally assumed to be elementary; one of the aims of the gauge-theoretical approach to strong interactions is to reinterpret these different mass values as eigenvalues of the quark-gluon system which is singled out by the Yang-Mills local gauge principle and by a special choice of the gauge group, i.e. the group $SU(3)$. Thus the invariances or symmetries of this system, in analogy to Einstein's gravitation theory and Salam and Weinberg's theory of electroweak processes, should determine all empirical masses.

In a rather problematic additional step the remaining elementary particles, i.e. leptons and quarks, were to be interpreted as the quantum states of a new and enlarged geometric structure: the so-called string. This "heroic" attempt, called "string theory," with its enormous hopes (it has been called "the theory of everything") and its great disappointments, led, in any case, to an unprecedented impetus to mathematical research related to this geometric structure. The result was a far-reaching development of already known and new methods, even leading to the creation of previously unknown domains of pure mathematics. These in turn allowed the discovery and understanding of interesting interrelations among various conventional structures and physical problems of practical importance, such as statistical mechanics, superconductivity and the quantum Hall effect. As more general examples I might cite the sequence

knots and links ⟶ Jones polynomials ⟶ Yang-Baxter equations ⟶ statistical mechanics ⟶ conformal field theory

and the wonderful revival of Hopf's classical work triggered off by the present-day concept of "quantum groups".

The main consequence of this undertaking is the insight that the conventional concept of space-time must undergo — according to Riemann's very first suggestion — a thourough revision for the case of smallest dimensions. This led to the development of a "p-adic geometry," and to the "non-commutative differential geometry" of A. Connes and D. Kastler. With this proposal a completely new and basic chapter in the history of physics has, in fact, been opened: according to a personal "message" of W. Pauli the most disturbing mathematical difficulties encountered in relativistic quantum field theory (the basis of theoretical physics for half a century) will appear in a new light and might lead to a new concept of elementary particles — perhaps in a certain way analogous to the one suggested by string theory.

Thus, a novel and exciting stage of research has been initiated: new and extended geometric structures cover simultaneously and in a most successful way very different domains of physics, from the lowest to the highest energies. This prompts an exchange between so far completely separate domains in physics and calls for a deeper understanding of the abstract but nonetheless intuitive structures in modern mathematics that are inherent in physical laws.

Enlarging this quest for human understanding and exchange to all nations of our world we are immediately reminded of the beautiful location of this meeting: Rapallo was, in fact, the place of the very first contacts and handshakes between the enemies of the first world war. It might thus in our days contribute to a deeper understanding and a real friendship between East and West. For this reason I heartily welcome our friends and participants from various eastern countries. In this connection I should not forget to express our great appreciation to the mayor of this city, Dr. Mauro Cordano, for this extremely kind hospitality, as well as — speaking on behalf of us all — to convey our heartiest thanks and greetings to the dedicated organizers of our meeting from the Department of Mathematics of the University of Genoa, Ugo Bruzzo, Claudio Bartocci and Roberto Cianci: you did really a wonderful job, very much in the "spirit of Rapallo."

Bonn, October 1990

Foreword by Professor M. Francaviglia

Istituto di Fisica Matematica "J.-L. Lagrange"
University of Turin

It is a great honour for me to be here in Rapallo at the official opening of this important and beautiful international conference and to have the opportunity of welcoming in Italy all the friends and colleagues who have come from all over the world to participate.

My present task is twofold. On the one hand I have the pleasant duty of sitting here and representing one of the major sponsors who made this conference possible. I offer you a warm welcome from CNR (the Italian National Research Council); on behalf of its Scientific Committee I bring you in particular best wishes for a fruitful stay from GNFM (the National Group of Mathematical Physics). GNFM supports a large part of the scientific activity in Italy in the domain of mathematical physics, through a Visiting Professorship program and through the sponsorship of a limited number of conferences, among which this 19th DGM will certainly be one of the major events of 1990. On the other hand I also have the great honour of welcoming all participants on behalf of the National Research Project "Geometria e Fisica", of our MURST (Ministry of Universities, Research and Technology). This project was started in Italy ten years ago, with the aim of promoting and coordinating Italian research and international collaboration in the fields of interaction between physics and geometry. Besides being a relevant source of research funding in these years, this project has extensively helped the Italian scientific community to develop a number of coherent lines of research in this beautiful domain. Nowadays the project comprises over a hundred Italian investigators, including a large number of young researchers, belonging to 16 Universities scattered through the whole country. As the national coordinator of this project, it is my greatest pleasure to be able to participate in such an important task. This conference, which came into being entirely due to the active and strong will of the local group of this project working at the University of Genoa, will surely represent a milestone in the life of the project itself.

Although I can see many colleagues in the audience who would be much better than me in this job, I will nonetheless try here to stress in a few words the importance of the subject we shall be discussing in Rapallo for these six days. The relations between geometry and physics were already hidden in the celebrated treatise "Méchanique Analitique", written by Joseph-Louis Lagrange in 1788, where the basis of the modern approach to theoretical physics was laid down. These intimate relations were subtly envisaged as a means of understanding the very structure of our universe by the genius of Bernhard Riemann (1854) and fully developed by Albert Einstein in his famous and fundamental theory of general relativity

(1916). The ideas and the methods embraced by Einstein in fact constituted one of the leading themes in the development of "classical" (non-quantum) theoretical physics, both in the direction of gravitational theories and, more recently, in the interpretation of gauge field theories in terms of principal connections. Soon after the formulation of general relativity, physics and mathematics seemed to have come to live in perpetual harmony, and for a few years it was believed that geometry would forever be the unifying language of physics. In the 1920s, however, the revolution of quantum mechanics began. With the advent of quantum mechanics, and later on with the extensive development of quantum field theory, physicists increasingly regarded analysis as the natural language for their discipline. Thus, theoretical physics and geometry, the latter oriented in those years towards the creation of more abstract conceptions, experienced a sort of repulsive force. Geometry, which in fact plays a fundamental role in the local formulation of classical physical laws, was long believed to be incapable of shedding light on quantum phenomena and hence relegated to be the language of just the "old" physics. Perhaps one of the reasons why a satisfactory solution to the problem of a coherent unification of gravity and quantum physics has so far remained elusive should be looked for in this divorce. This apparent dichotomy lasted for almost fifty years. However, the last two decades have seen a profound internal unification, both in mathematics and in physics, where deep-seated relations between seemingly unrelated fields have been discovered. Along with these internal revolutions, a renewed and stronger interaction between geometry and physics has taken place, and today we sense a great excitement as large portions of both disciplines are coming together. While until a few years ago the interaction of geometry and physics was mainly limited to the domains of differential geometry and to "classical" field theories, in the recent past other fundamental branches of geometry have found their way into physics and stimulated its development, often giving an enormous impetus to the investigations concerning the global behaviour of physical fields and their quantum properties, the structure of continua with their defects, the fascinating world of solitons and completely dynamical and quantum integrable systems. While the physicist of 20 years ago just spoke of groups, manifolds, tensors and perhaps principal connections, today we see the growing importance of subjects which perhaps are less familiar to physicists, like cohomology, supermanifolds, algebraic geometry, deformations, and even non-commutative geometry.

The meetings on Differential Geometrical Methods in Theoretical Physics have in these past 20 years been a fundamental forum where mathematicians and physicists could come together and work for the reconstruction of this important symbiosis. Looking at the impressive program of this 19th conference we can be sure that this tradition is still alive and able to mantain the rapid evolution of the subject. Let us enjoy this conference and thank once more the organizers for their magnificent work!

Turin, October 1990

Contents

1. Non-commutative differential geometry

2. Quantum groups and integrable systems

3. Conformal field theory and related topics

4. Superalgebras and supermanifolds

5. Problems in quantum field theory

6. Short contributions

List of Participants

I. Aref'eva, Steklov Mathematical Institute, Ul. Vavilova 42, Moscow, GSP-1 117966 U.S.S.R.

I. Avramidi, Universität Karlsruhe, Institut für Theoretische Physik, Kaiserstraße 12, D-7500 Karlsruhe, F. R. Germany. E-Mail: BE12@DKAUNI2.BITNET

H. Bacry, C.N.R.S. — Centre de Physique Théorique, Case 907, F-13288 Marseille Luminy Cedex 8, France

C. Bartocci, Università di Genova, Dipartimento di Matematica, Via L.B. Alberti 4, I-16132 Genova, Italia.

M. Batchelor, Cambridge University, D.A.M.M.S., Mill Lane, Cambridge CB3 9EW, United Kingdom

C.M. Becchi, Università di Genova, Dipartimento di Fisica, Via Dodecaneso 33, I-16146 Genova, Italia

R. Bekhechi, King's College, Dept. of Mathematics, Strand, London WC2R 2LS, United Kingdom. E-Mail: RBEKHECHI@OAK.CC.KCL.AC.UK

L. Bettge, Universität Dortmund, FB Physik, Postfach 500500, D-4600 Dortmund 50, F. R. Germany

L.C. Biedenharn, Duke University, Dept. of Physics, Durham, NC 27706, U.S.A. E-Mail: JEC@PHY.DUKE.EDU

W. Bischoff, Albert-Ludwig-Universität, Fakultät für Physik, Hermann-Herder-Straße 3, D-7800 Freiburg, F. R. Germany

A. Blasi, Università di Genova, Dipartimento di Fisica, Via Dodecaneso 33, I-16146 Genova, Italia

K. Bleuler, Institut für Theoretische Kernphysik der Universität Bonn, Nussallee 14-16, D-5300 Bonn, F. R. Germany

F.J. Bloore, The University of Liverpool, D.A.M.T.P., P.O. Box 147, Liverpool L69 3BX, United Kingdom. E-Mail: SX35@LIVERPOOL.AC.UK

L. Bonora, S.I.S.S.A., Strada Costiera 11, I-34014 Miramare-Grignano TS, Italia. E-Mail: BONORA@ITSSISSA.BITNET

V. Bonservizi, S.I.S.S.A., Strada Costiera 11, I-34014 Miramare-Grignano TS, Italia. E-Mail: BONSERVI@ITSSISSA.BITNET

L.J. Boya, University of Texas at Austin, Physics Department, Austin, TX 78712 U.S.A. E-Mail: LUISJO@UTAPHY.BITNET

U. Bruzzo, Università di Genova, Dipartimento di Matematica, Via L.B. Alberti 4, I-16132 Genova, Italia. E-Mail: BRUZZO@IGECUNIV.BITNET

K. Bugajska, York University, Dept. of Mathematics and Statistics, North York, Ontario, Canada MJ3 1P3. E-Mail: BUGAJSKA@YORKVM1.BITNET

R.K. Bullough, U.M.I.S.T., Dept. of Mathematics, Sackville Street, P.O. Box 88, Manchester M6O 1QD United Kingdom. E-Mail: RKBULLOUGH@UMIST.AC.UK

N. Burroughs, Cambridge University, D.A.M.T.P., Silver Street, Cambridge CB3 9EW, United Kingdom. E-Mail: NJB16@PHX.CAM.AC.UK

C. Buzzanca, Università di Palermo, Dipartimento di Matematica e Applicazioni, Via Archirafi 34, I-90123 Palermo, Italia.

E. Caccese, Dipartimento di Matematica, Università della Basilicata, Potenza, Italia

S. Carillo, Università di Roma "La Sapienza", Dipart. Metodi e Modelli Matematici, Via A. Scarpa 10, I-00161 Roma, Italia. E-Mail: CARILLO@ROMA1.INFN.IT
U. Carow-Watamura, Universität Karlsruhe, Institut für Theoretische Physik, Kaiserstraße 12, D-7500 Karlsruhe, F. R. Germany. E-Mail: BE05@DKAUNI2.-BITNET
R. Catenacci, Università di Trieste, Dipartimento di Matematica, P.le Europa 1, I-34017 Trieste, Italia. E-Mail: CATENACCI@PAVIA.INFN.IT
S. Catto, The City University of New York, Baruch College, Physics. Dept., 17 Lexington Ave., New York, NY 10010 U.S.A. E-Mail: SCBBB@CUNYVM.BITNET
R. Cianci, Università di Genova, Dipartimento di Matematica, Via L.B. Alberti 4, I-16132 Genova, Italia. E-Mail: CIANCI@IGECUNIV.BITNET
R. Collina, Università di Genova, Dipartimento di Fisica, Via Dodecaneso 33, I-16146 Genova, Italia
C. De Concini, Scuola Normale Superiore, Piazza dei Cavalieri 7, I-56100 Pisa, Italia
A. De Pantz, Corticella Leoni 4, I-37121 Verona, Italia
H.J. De Vega, Université de Paris VII, Labo. Phys. Theor. H. Energies, 4 Pl. Jussieu, Tour 16, 1er Et., F-75252 Paris Cedex 5, France.
G. Dell'Antonio, Università di Roma "La Sapienza", Dipartimento di Matematica, P.le A. Moro 2, I-00185 Roma, Italia. E-Mail: GIANFA@IRMUNISA.BITNET
J.A. Dixon, University of Texas at Austin, Theory Group, Physics Department, Austin, TX 78712 U.S.A. E-Mail: DIXON@UTAPHY.BITNET
M. Djurdjevič, University of Belgrade, Dept. of Physics, P.O. Box 550, 19001 Belgrade, Yugoslavia. E-Mail: YUBGSS21@EPMFF41.BITNET
V.K. Dobrev, Bulgarian Academy of Sciences, Institute of Nuclear Research, 72 Boul. Lenin, 1784 Sofia, Bulgaria.
J.-A. Domínguez Pérez, Universidad de Salamanca, Depto. de Matemáticas, Plaza de la Merced 1-4, E-37008 Salamanca, España.
B. Drabant, Universität Karlsruhe, Institut für Theoretische Physik, Kaiserstraße 12, D-7500 Karlsruhe, F. R. Germany. E-Mail: BE08@DKAUNI2.BITNET
M. Dubois-Violette, Université de Paris-Sud, Labo. Phys. Theor. H. Energies, Bat. 211, F-91405 Orsay, France. E-Mail: MADORE@FRCPN11.BITNET
I.L. Egusquiza, University of Cambridge, D.A.M.T.P., Silver St., Cambridge CB3 9EW, United Kingdom. E-Mail: ILE10@PHX.CAM.AC.UK
C. Emmrich, Albert-Ludwig-Universität, Fakultät für Physik, Hermann-Herder-Straße 3, D-7800 Freiburg, F. R. Germany. E-Mail: CEMM@DFRRUF1.BITNET
O. Eyal, Universität Karlsruhe, Institut für Theoretische Physik, Kaiserstraße 12, D-7500 Karlsruhe, F. R. Germany. E-Mail: BE01@DKAUNI2.BITNET
G. Falqui, S.I.S.S.A., Strada Costiera 11, Miramare-Grignano, I-34014 Trieste. E-Mail: FALQUI@ITSSISSA.BITNET
A. Fernández Martínez, Universidad de Salamanca, Depto. de Matemáticas, Plaza de la Merced 1-4, E-37008 Salamanca, España.
F. Ferrari, Universität Wien, Institut für Theoretische Physik, Boltzmanngasse 5, A-1090 Wien, Austria. E-Mail: FERRARI@AWIRAP.BITNET
M. Ferraris, Università di Cagliari, Dipartimento di Matematica, Via Ospedale 72, I-09100 Cagliari, Italia. E-Mail: FERRARIS@VAXCA2.INFN.IT

T. Filk, Universität Freiburg, Fakultät für Physik, Hermann-Herder-Straße 3, D-7800 Freiburg, F. R. Germany. E-Mail: FILK@DFRRUF1.BITNET

A. Floer, Fakultät für Mathematik, Ruhr-Universität, Universitätstraße 150 NA 6/27, 4630 Bochum, F. R. Germany.

A. Folacci, Université de Corse, Fac. des Sciences, 15 Quartier des 4 Fontaines, F-20250 Corte, France.

M. Forger, Universität Freiburg, Fakultät für Physik, Hermann-Herder-Straße 3, D-7800 Freiburg, F. R. Germany. E-Mail: FORGER@DFRRUF1.BITNET

M. Francaviglia, Università di Torino, Istituto di Fisica Matematica, Via Carlo Alberto 10, I-10123 Torino, Italia. E-Mail: FRANCAVIGLIA@ASTRTO.INFN.IT

J. Fröhlich, ETH-Hönggerberg, Theoretical Physics, CH-8093 Zürich, Schweiz. E-Mail: FROEHLICH@CZHETH5A.BITNET

A. Ganchev, Bulgarian Academy of Science, Institute for Nuclear Research, Boul. Lenin 72, 1784 Sofia, Bulgaria.

A. Gavrilik, Institute for Theoretical Physics, 252130 Kiev 130, U.S.S.R.

F. Ghaboussi, Universität Konstanz, Fakultät für Physik, Postfach 5560, D-7750 Konstanz, F. R. Germany

R. Giachetti, Università di Firenze, Istituto di Matematica Applicata, Via S. Marta 3, I-50139 Firenze, Italia

C. Gómez, Universidad de Salamanca, Depto. de Física Teórica, Pl. de los Caidos, E-37008 Salamanca, España

M. González León, Universidad de Salamanca, Depto. de Matemáticas, Plaza de la Merced 1-4, E-37008 Salamanca, España

S. Gotzes, Universität Dortmund, FB Physik, Postfach 500500, D-4600 Dortmund 50, F. R. Germany. E-Mail: UPH418@DDOHRZ11.BITNET

J.M. Guilarte, Universidad de Salamanca, Depto. de Física Teórica, Pl. de los Caidos, E-37008 Salamanca, España. E-Mail: ESANZ@USAL.ES

P. Hajac, Mathematical Institute, 24-29 St Giles', Oxford OX1 3LB, United Kingdom. E-Mail: PMH@VAX.OX.AC.UK

K.M. Happle, Universität Freiburg, Fakultät für Physik, Hermann-Herder-Straße 3, D-7800 Freiburg, F. R. Germany. E-Mail: KLHA@DFRRUF1.BITNET

M. Hayashi, Universität Karlsruhe, Institut für Theoretische Physik, Kaiserstraße 12, D-7500 Karlsruhe 1, F. R. Germany. E-Mail: BE08@DKAUNI2.BITNET

F. Hegenbarth, II Università di Roma, Dipartimento di Matematica, Via Fontanile di Carcaricola, I-00133 Roma, Italia

M. Hellmund, Karl-Marx-Universität, Sektion Physik, Karl-Marx-Platz, DDR-7010 Leipzig, German D. R.

D. Hernández Ruipérez, Universidad de Salamanca, Depto. de Matemáticas, Pl. de la Merced 1-4, E-37008 Salamanca, España

S. Huggett, Polytechnic South West, Dept. of Mathematics and Statistics, Plymouth PL4 8AA, United Kingdom. E-Mail: P07406@PA.PSW.AC.UK

C. Hull, Queen Mary College, Dept. of Physics, Mile End Road, London E1 4NS, United Kingdom. E-Mail: CMH@VI.PH.QMC.AC.UK

C. Itzykson, Service de Physique Théorique, CEN Saclay, F-91191 Gif-sur-Yvette Cedex, France.

B. Jensen, Université de Corse, Fac. des Sciences, B.P. 52, F-20250 Corte, France.

J. Kalkman, Rijksuniversiteit Utrecht, Mathematisch Instituut, Postbus 80 010, 3508 TA Utrecht, Nederland. E-Mail: KALKMAN@MATH.RUU.NL

D. Kastler, Université Aix–Marseille II, Centre de Physique Théorique, Case 907 - Luminy, 13288 Marseille Cedex 9, France.

A. Kellendonk, Physikalische Institut der Universität Bonn, 12 Nussallee, D-5300 Bonn 3, F. R. Germany

A. Kempf, Universität Karlsruhe, Institut für Theoretische Physik, Kaiserstraße 12, D-7500 Karlsruhe 1, F. R. Germany. E-Mail: BE04@DKAUNI2.BITNET

T. Kornhass, Universität Freiburg, Fakultät für Physik, Hermann-Herder-Straße 3, D-7800 Freiburg, F. R. Germany. E-Mail: KLHA@DFRRUF1.BITNET

D. Krupka, Masaryk University, Dept. of Mathematics, Janackovo Nam. 2A, 66295 Brno, Czechoslovakia

O. Krupkova, Masaryk University, Dept. of Mathematics, Janackovo Nam. 2A, 66295 Brno, Czechoslovakia

H.P. Künzle, University of Alberta, Dept. of Mathematics, Edmonton, Alberta, Canada T6G 2G1. E-Mail: HKUNZLE@UALTAVM.BITNET

J. Laartz, Albert-Ludwig-Universität, Fakultät für Physik, Hermann-Herder-Straße 3, D-7800 Freiburg, F. R. Germany

N.P. Landsman, Cambridge University, D.A.M.T.P., Silver Street, Cambridge CB3 9EW, United Kingdom. E-Mail: NPL11@PHX.CAM.AC.UK

R.J. Lawrence, Harvard University, Mathematics Dept., 1 Oxford St., Cambridge, MA 02138, U.S.A. E-Mail: LAWRENCE@HUMA1

H.C. Lee, Chalk River Nuclear Lab, Theoretical Physics, Chalk River, Ontario, Canada K0J 1J0. E-Mail: 02153@AECLCR.BITNET

D. Leĭtes, University of Stockholm, Dept. of Mathematics, Box 6701, S-11385 Stockholm, Sweden. E-Mail: MLEITES@NADA.KTH.SE

A. Lichnerowicz, Collège de France, 3 Rue d'Ulm, F-75005 Paris, France

M. Lo Schiavo, Università di Roma "La Sapienza", Dip. Metodi Modelli Matematici, Via A. Scarpa 10, I-00161 Roma, Italia

R. Loll, Physikalische Institut der Universität Bonn, Nussallee 12, D-5300 Bonn 3, F. R. Germany. E-Mail: LOLL@DBNPIB5.BITNET

A. López Almorox, Universidad de Salamanca, Depto. de Matemáticas, Plaza de la Merced 1-4, E-37008 Salamanca, España

L. Lusanna, Sezione I.N.F.N. di Firenze, Largo E. Fermi 2, I-50125 Arcetri FI, Italia. E-Mail: LUSANNA@VAXFI.INFN.IT

G. Mackey, Harvard University, Mathematics Dept., 1 Oxford St., Cambridge, MA 02138, U.S.A.

G. Magnano, S.I.S.S.A., Strada Costiera 11, Miramare-Grignano, I-34014 Trieste. E-Mail: MAGNANO@ITSSISSA.BITNET

S. Majid, Cambridge University, D.A.M.T.P., Silver St., Cambridge CD3 9EW, United Kingdom. E-Mail: SHM10@PHX.CAM.AC.UK

M. Mamone Capria, Università di Perugia, Dipartimento di Matematica, Via Vanvitelli 1, I-06100 Perugia, Italia

K.B. Marathe, The City University of New York, Brooklyn College, Brooklyn, NY 11210, U.S.A. E-Mail: KBM@BKLYN.BITNET

P.A. Marchetti, Università di Padova, Dipartimento di Fisica, Via Marzolo 8, I-35131 Padova, Italia. E-Mail: MARCHETTI@VAXFPD.INFN.IT

M. Marinkovič, University of Belgrade, Dept. of Physics, P.O. Box 550, 19001 Belgrade, Yugoslavia. E-Mail: YUBGSS21@EPMFF41.BITNET

G. Marmo, Università di Napoli, Dipartimento di Fisica, Mostra d'Oltremare Pad. 19, I-80125 Napoli, Italia. E-Mail: GIMARMO@CLUSNA.INFN.IT

L. Martina, Università di Lecce, Dipartimento di Fisica, I-73100 Lecce, Italia. E-Mail: MARTINA@VAXLE.INFN.IT

V.B. Matveev, Institute for Aviation Instrumentation, Dept. of Mathematics, Gertzena 67, 190000 Leningrad, U.S.S.R.

M.E. Mayer, University of California, Dept. of Physics, Irvine, CA 92717 U.S.A. E-Mail: MMAYER@UCIVMSA.BITNET

G. Mendella, Università di Napoli, Dipartimento di Fisica, Mostra D'Oltremare Pad. 19, I-80125 Napoli, Italia

I. Mladenov, Bulgarian Acad. Sciences, Central Biophysics Laboratory, Ul. Acad. Boncnev 21, 1113 Sofia, Bulgaria

M. Modugno, Università di Firenze, Istituto di Matematica Applicata, Via S. Marta 3, I-50139 Firenze. E-Mail: MODUGNO@IFIIDG.BITNET

V. Molotkov, Bulgarian Academy of Sciences, Institute of Nuclear Research, 72 Boul. Lenin, 1784 Sofia, Bulgaria

K. R. Müller, Universität Karlsruhe, Institut für Logik, Postfach 6980, D-7500 Karlsruhe, F. R. Germany. E-Mail: KLAUS@IRA.UKA.DE

R. Myers, McGill University, Physics Department., Rutheford Bldg., 3600 University Str., Montreal, Canada H3A 2T8. E-Mail: RCM@PHYSICS.MCGILL.CA

Y. Ne'eman, Tel-Aviv University, Sackler Institute, Ramat-Aviv 69, 978 Tel Aviv, Israel. E-Mail: B21A@TAUNOS.BITNET

H. Ocampo, Universität Karlsruhe, Institut für Theoretische Physik, Kaiserstraße 12, D-7500 Karlsruhe, F. R. Germany. E-Mail: BE08@DKAUNI2.BITNET

O. Ogievetsky, Universität Karlsruhe, Institut für Theoretische Physik, Kaiserstraße 12, D-7500 Karlsruhe 1, Postfach 6980, F. R. Germany. E-Mail: BE12@DKAUNI2.BITNET

P. Orland, The City University of New York, Baruch College, Physics. Dept., 17 Lexington Ave., New York, NY 10010 U.S.A.

Z. Oziewicz, University of Wrocław, Institute of Theoretical Physics, Cybulskiego 36, Wrocław, 50205 Poland.

S. Pasquero, Dipartimento di Matematica, Via L.B. Alberti 4, I-16132 Genova, Italia E-Mail: DOTMAT@ICNUCEVM.BITNET

O. Pekonen, University of Jyvaskyla, Dept. of Mathematics, Seminaarinkatu 15, 40100 Jyvaskyla, Finland

I. Penkov, University of California Dept. of Mathematics Berkeley, CA 94720, U.S.A. E-Mail: PENKOV@MATH.BERKELEY.EDU

A. Perelomov, Institute Theor. Exp. Physics, Moscow, U.S.S.R.

E. Poletaeva, Pennsylvania State University, Dept. of Mathematics, 218 Mc Allister Bldg., University Park, PA 16802, U.S.A.

C. Procesi, Università di Roma "La Sapienza", Dipartimento di Matematica, P.le A. Moro 2, I-00185, Roma, Italia

J.M. Rabin, University of California at San Diego, Mathematics Dept., La Jolla, CA 92093 U.S.A. E-Mail: JRABIN@UCSD.EDU

O. Ragnisco, Università di Roma "La Sapienza", Dipartimento di Fisica, P.le A. Moro 2, I-00185 Roma, Italia. E-Mail: RAGNISCO@ROMA1.INFN.IT

P. Ramond, University of Florida, Dept. of Physics, Gainesville, FL 32611 U.S.A. E-Mail: RAMOND@UFPINE

J. Rawnsley, University of Warwick, Mathematics Institute, Coventry CV4 7AL, United Kingdom. E-Mail: JHR@MATHS.WARWICK.AC.UK

C. Reina, S.I.S.S.A., Strada Costiera 11, I-34014 Miramare TS, Italia. E-Mail: REINA@ITSSISSA.BITNET

L. Richardson, 44 Fosseway, Clevedon, Avon BS21 5EQ, United Kingdom.

S. Rodriguez-Romo, Universität Konstanz, Fakultät für Physik (LS Dehnen), Postfach 5560 7750, Konstanz, F. R. Germany. E-Mail: PHEBNER@DKNKURZ1.BITNET

A. Rogers, University of London, King's College, Strand, London WC2R 2LS, United Kingdom. E-Mail: UDAH039@OAK.CC.KCL.AC.UK

M. Rothstein, University of Georgia, Dept. of Mathematics, Athens, GA 30602, USA. E-Mail: ROTHSTEI@JOE.MATH.UGA.EDU

M. Ruiz-Altaba, Université de Geneve, Dept. de Physique Theorique, CH-1211 Genève 4, Suisse. E-Mail: RUIZALTB@CGEUGE52.BITNET

G. Sardanashvily, University of Moscow, Dept. of Theoretical Physics, Physics Faculty, Moscow 117234, U.S.S.R.

M. Saveliev, Institute for High Energy Physics, Theory Division, Protvino, Moscow Region 142284, U.S.S.R.

B. Sazdovic, Institute of Physics, P.O. Box 57, 11001 Belgrade, Yugoslavia. E-Mail: EIPH004@YUBGSS21.BITNET

U. Schäper, Universität Freiburg, Fakultät für Physik, Hermann-Herder-Straße 3, D-7800 Freiburg, F. R. Germany

A. Schirrmacher, Universität Karlsruhe, Institut für Theoretische Physik, Kaiserstraße 12, Karlsruhe D-7500, F. R. Germany. E-Mail: BE04@DKAUNI2.BITNET

M. Schlieker, Institut für Theoretische Physik, Kaiserstraße 12, D-7500 Karlsruhe, F. R. Germany. E-Mail: BE02@DKAUNI2.BITNET

T. Schmitt, Karl-Weierstrass-Institut für Mathematik, Mohrenstraße 39, Berlin 1086, Germany.

M. Schottenloher, Mathematische Institut der LMU, Theresienstraße 39, D-8000 München 2, F. R. Germany

J. Schwenk, Institut für Theoretische Physik, Kaiserstraße 12, D-7500 Karlsruhe, F. R. Germany. E-Mail: BE02@DKAUNI2.BITNET

S. Shnider, Ben Gurion University, Dept. of Mathematics and Computer Science, P.O. Box 653, 84105 Beer Sheva, Israel. E-Mail: SHNIDER@BIMACS.BIU.AC.IL

N. Sorace, Università di Firenze, Istituto di Matematica Applicata, Via S. Marta 3, I-50139 Firenze, Italia

M. Spera, Università di Padova, Dip. Metodi e Modelli Matematici, Via Belzoni 7, I-35131 Padova, Italia E-Mail: SPERA@PDMSA1.UNIPD.IT

A. Stahlhofen, Universität Stuttgart, Institut für Theoretische u. Ang. Physik, Pfeffenwaldring 57, D-7000 Stuttgart 80, F. R. Germany

O.T. Stoytchev, Institute for Nuclear Research, Elementary Particle Division, Boul. Lenin 72, 1184 Sofia, Bulgaria

R. Tammelo, Estonian Academy of Sciences, Institute of Physics, Riia 142, 202400 Tartu, Estonia, U.S.S.R.

P. Teofilatto, II Università di Roma, Dipartimento di Fisica, Via Orazio Raimondo, I-00173 Roma, Italia.

P. Truini, Università di Genova, Dipartimento di Fisica, Via Dodecaneso 33, I-16146 Genova, Italia. E-Mail: TRUINI@GENOVA.INFN.IT

V. Tsanov, University of Sofia, Dept. of Mathematics and Informatics, A. Ivanov Str. 5, Sofia 1126, Bulgaria

S.T. Tsou, Mathematical Institute, 24-29 St Giles', Oxford OX1 3LB, United Kingdom. E-Mail: TSOU@VAX.OX.AC.UK

A. Vaintrob, Ul. Akademika Tchelomeya 1 Kv. 328, Moscow 117630, U.S.S.R.

S. Vidussi, Università di Trieste, Dipartimento di Matematica, P.le Europa 1, I-34017 Trieste, Italia

G. Vilasi, Dipartimento di Fisica Teorica, Università di Salerno, Via S. Allende, 84081 Baronissi SA, Italy

P. Vitale, I.N.F.N Sezione di Napoli, Mostra D'Oltremare, 80125 Napoli, Italia

I. V. Volovich, Steklov Mathematical Institute, Ul. Vavilova 42, Moscow, GSP-1 117966 U.S.S.R.

S. Watamura, Institut für Theoretische Physik, Kaiserstraße 12, 7500 Karlsruhe, F. R. Germany. E-Mail: BE05@DKAUNI2.BITNET

S.L. Woronowicz, University of Warsaw, Dept. of Mathematical Methods in Physics, Hoza 74, 00682 Warsaw, Poland

C.S. Xiong, S.I.S.S.A., Strada Costiera 11, I-34014 Miramare TS, Italia. E-Mail: XIONG@ITSSISSA.BITNET

C.Z. Zha, Xinjiang University, Physics Dept., Urumqi, Xinjiang, China

D. Zwanziger, New York University, Department of Physics, Washington Sq., New York, NY 10003, U.S.A.

P. Zweydinger, Universität Karlsruhe, Institut für Theoretische Physik, Kaiserstraße 12, D-7500 Karlsruhe 1, Postfach 6980, F. R. Germany

1. Non-commutative differential geometry

HIGGS FIELDS AND SUPERCONNECTIONS

R. Coquereaux

Centre de Physique Théorique
CNRS Luminy - Case 907
F 13288 - Marseille Cedex 9 (France)

1. Introduction

We consider an extension of the formalism of gauge theories where the connection is no longer described by a Lie-algebra valued one-form on Space-Time, but incorporates both the Yang-Mills fields and the Higgs fields. This leads to gauge field models with symmetry breaking but with numerical constraints between the (otherwise free) parameters of the theory. From the physical point of view, Higgs fields are here associated to the gauging of discrete directions (an intuitive interpretation would involve a discussion of parallel universes). From the mathematical point of view, physical fields (gauge bosons and Higgs) can be interpreted as the components of a super-connection [5]. The present contribution is an overview of the paper [1] where more details can be found. Our approach shares several features with the work of [2] in the sense that some of the conclusions are identical; however the two approaches do not seem to be directly related. Another kind of "non-commutative connections" is also described in [4]. Our wish in the following is to give a very elementary account of the subject, for this reason, the only mathematical tools that we shall use will involve nothing else than matrices, differential forms and a few basic facts on Lie groups and Yang-Mills theory. Using more fancy mathematics, the whole approach could be summarised as follows. Step 1: Choose an algebra of complex matrices and give it the structure of a $(\mathbb{Z}_2)$-graded differential algebra. Step 2: Consider the space of differential forms on space-time as a $\mathbb{Z}_2$-graded differential algebra. Step 3: Build the graded differential algebra equal to the graded tensor product of algebras obtained in steps 1 and 2. Step 4: Consider connections as odd elements in step 3, compute their curvature and the norm-square of this curavture (to be the bosonic part of the Lagrangian defining the physical theory). Step 5: Add spinor fields. We shall first illustrate this method by starting (step 1) with an algebra of 2×2 matrices, this will lead to a $U(1) \times U(1)$ gauge theory with spontaneous symmetry (one massless photon and one massive Z) breaking exhibiting already all the interesting features. Then we shall illustrate the same method by starting (step 1) with an algebra of

3×3 antihermitian matrices, this will lead to a $SU(2)\times U(1)$ gauge theory very similar to the standard model of electro-weak interactions but with extra constraints on the value of the Weinberg angle and on the masses. In both examples, it can be seen that the generalized Yang Mills field transforms under a Lie super-algebra. The exhibited features –possibility of incorporatating the Higgs field in the connection, emergence of symmetry breaking and numerical constraints on coefficients– are very similar to those that appear in the case of dimensional reduction of the Yang-Mills action ([4] and references therein).

2. The $\mathbf{U(1)\times U(1)}$ model

Step 1.
Let a a 2×2 matrix, we decompose it into even and odd parts as follows: $a = a_e + a_o$ with

$$a_e = \begin{pmatrix} a_{11} & 0 \\ 0 & a_{22} \end{pmatrix}, a_O = \begin{pmatrix} 0 & a_{12} \\ a_{21} & 0 \end{pmatrix}$$

This defines a $\mathbb{Z}_2$ grading and we will set $\partial a = 0$ when a is even and $\partial a = 1$ when a is odd. Let us define

$$da = i \begin{pmatrix} a_{21} + a_{12} & a_{22} - a_{11} \\ a_{11} - a_{22} & a_{21} + a_{12} \end{pmatrix}$$

It is therefore obvious that $d^2 a = 0$ and that d satisfies the graded Leibnitz rule. It can be seen that da is nothing else that the graded commutator of a with the Pauli matrix τ_1, i.e. $da = i[\tau_1, a]_S = i(\tau_1 a - (-1)^{\partial a} a\tau_1$. Acually, replacing τ_1 by a matrix $\eta_\gamma = cos(\gamma)\tau_1 + sin(\gamma)\tau_2$. would lead to an equivalent theory. As we shall see later, chosing a particular η in this family amounts to choose the position of the vacuum on the circle of minima of the Higgs potential.

Step 2.
We already know how d acts on differential forms: $d = dx^\mu \frac{\partial}{\partial x^\mu}$. A differential form B of given degree b has a $\mathbb{Z}_2$ parity $\partial B = b \bmod 2$. Remember that

$$d(B \wedge C) = dB \wedge C + (-1)^{\partial B} B \wedge dC.$$

Step 3.
We now consider 2×2 matrices whose coefficients are differential forms on Space-Time, for instance $X = \begin{pmatrix} A & C \\ D & B \end{pmatrix}$. Here A, B, C and D are differential forms that are not necessarily of the same degree and not even of homogeneous degree. Multiplication of these matrices

is performed as follows. Assume first that A, B, C, D are forms of (homogeneous) degree a, b, c, d and therefore of $\mathbb{Z}_2$ parity $\partial A, \partial B, \partial C, \partial D$. Then, we set

$$\begin{pmatrix} A & C \\ D & B \end{pmatrix} \odot \begin{pmatrix} A' & C' \\ D' & B' \end{pmatrix} =$$
$$\begin{pmatrix} A \wedge A' + (-1)^{\partial C} C \wedge D' & C \wedge B' + (-1)^{\partial A} A \wedge C' \\ D \wedge A' + (-1)^{\partial B} B \wedge D' & B \wedge B' + (-1)^{\partial D} D \wedge C' \end{pmatrix}$$

The product of two arbitrary elements in this algebra is obtained from the previous rules by linearity. It is easy to check that this product is associative. For an arbitrary matrix

$$X = \begin{pmatrix} A & C \\ D & B \end{pmatrix}$$

we set

$$dX = i\left[\eta_\gamma, \begin{pmatrix} A & 0 \\ 0 & B \end{pmatrix}\right] + i\left[\eta_\gamma, \begin{pmatrix} 0 & C \\ D & 0 \end{pmatrix}\right]_+ + \begin{pmatrix} dA & -dC \\ -dD & dB \end{pmatrix}$$
$$= \begin{pmatrix} dA + i(e^{i\gamma}C + e^{-i\gamma}D) & -dC - ie^{-i\gamma}(A - B) \\ -dD + ie^{i\gamma}(A - B) & dB + i(e^{i\gamma}C + e^{-i\gamma}D) \end{pmatrix}$$

One can check that the previous operator d is a derivation ($d(X \odot Y) = dX \odot Y + (-1)^{\partial X} X \odot dY$) for the new product and for the total $\mathbb{Z}_2$-grading (∂X is the sum of the $\mathbb{Z}_2$-grading of X as a matrix –*i.e.* diagonal or antidiagonal– and as a form –*i.e.* of even or odd degree–). Moreover, one checks that

$$d^2 X = 0$$

Step 4.
We shall be interested mainly in the case of matrices where A and B are one-forms and C, D are scalar fields (zero-forms). The total $\mathbb{Z}_2$ grading of such matrices is 1 and they will be considered as generalized Yang-Mills potentials. The "generalized" gauge field is therefore described by the 2×2 anti-hermitian matrix

$$\mathcal{A} = \begin{pmatrix} A & i\mu^{-1}\phi \\ i\mu^{-1}\overline{\phi} & B \end{pmatrix}$$

Here A and B are abelian gauge fields but they could be (and will be) chosen as Lie algebra valued in an arbitrary Lie algebra without any essential modification. A and B are dimensionless so that A_μ and B_μ defined by $A = A_\mu dx^\mu$ and $B = B_\mu dx^\mu$ have dimension of a mass. We introduced a mass parameter μ to give the scalar field the dimension of a mass. The curvature will be, of course, defined as $\mathcal{F} = d\mathcal{A} + \mathcal{A} \odot \mathcal{A}$. When A and B are abelian, $A \wedge A = 0 = B \wedge B$ but we keep these terms for later purpose. In the same way, we do not use the fact that A,B and ϕ are, in the present case, commuting quantities. Notice that, even in this "abelian" case, the matrix $\mathcal{A} \odot \mathcal{A}$ does not vanish. Calling $F^A = dA + A \wedge A$ and $F^B = dB + B \wedge B$ the curvatures of the Yang-Mills fields A and B, we get the matrix

$$\mathcal{F} = \begin{pmatrix} \mathcal{F}_{11} & \mathcal{F}_{12} \\ \mathcal{F}_{21} & \mathcal{F}_{22} \end{pmatrix}$$

where

$$\begin{aligned}
\mathcal{F}_{11} &= F^A - \mu^{-2}(\mu(e^{i\gamma}\phi + \overline{e^{i\gamma}\phi}) + \phi\overline{\phi}) \\
\mathcal{F}_{12} &= -i\mu^{-1}(\nabla\phi + \mu e^{-i\gamma}(A - B)) \\
\mathcal{F}_{21} &= -i\mu^{-1}(\nabla\overline{\phi} - \mu e^{i\gamma}(A - B)) \\
\mathcal{F}_{22} &= F^B \; - \mu^{-2}(\mu(e^{i\gamma}\phi + \overline{e^{i\gamma}\phi}) + \overline{\phi}\phi)
\end{aligned}$$

with

$$\begin{aligned}
\nabla\phi &= d\phi + A\phi - \phi B \\
\nabla\overline{\phi} &= \overline{\nabla\phi} = d\overline{\phi} - \overline{\phi}A + B\overline{\phi}
\end{aligned}$$

We can calculate the norm square $\mathcal{L} = ||\mathcal{F}||^2 \doteq Tr < \overline{\mathcal{F}}, \mathcal{F} >= ||\mathcal{F}_{11}||^2 + ||\mathcal{F}_{12}||^2 + ||\mathcal{F}_{21}||^2 + ||\mathcal{F}_{22}||^2$ that will be our lagrangian for the bosonic fields. Here, $\overline{\mathcal{F}}$ denotes the hermitian conjugate of $\mathcal{F}$. To keep the dimensions right, one has also to insert the mass parameter μ in the right place when we perform the scalar products. In particular the scalar product of a p-form and of a q-form is zero when $p \neq q$, $< dx^\mu, dx^\nu >= \mu^{-2} g^{\mu\nu}$, and $< dx^\mu \wedge dx^\nu, dx^\rho \wedge dx^\sigma >= \frac{1}{2}\mu^{-4}(g^{\mu\rho}g^{\nu\sigma} - g^{\mu\sigma}g^{\nu\rho})$. The straighyforward calculation done below assumes an Euclidean signature or a Lorentzian signature $-+++$ (one gets corresponding results for the other signature $+---$ via an appropriate change of signs). In the non-abelian case, one has also to trace the right hand side of these expressions –*cf.* next section–. Setting $F = 1/2 F_{\mu\nu} dx^\mu \wedge dx^\nu$ for both fields A and B and using the fact that F is anti-hermitian, we have $< \overline{F}, F >= -1/4 F_{\mu\nu}F^{\mu\nu}$. Therefore

$$\mathcal{L} = -\frac{1}{4}((F^A_{\mu\nu})^2 + (F^B_{\mu\nu})^2) + 2\overline{D_\nu\phi}D^\nu\phi + 2(\mu(e^{i\gamma}\phi + \overline{e^{i\gamma}\phi}) + \phi\overline{\phi})^2$$

with

$$\begin{aligned}
D_\nu\phi &= \nabla_\nu\phi + \mu e^{-i\gamma}(A_\nu - B_\nu) \\
D_\nu\overline{\phi} &= \overline{D_\nu\phi} = \nabla_\nu\overline{\phi} - \mu e^{i\gamma}(A_\nu - B_\nu)
\end{aligned}$$

It is convenient to introduce a coupling constant g in the model, setting $\mathcal{L} = 1/g^2 \, ||\mathcal{F}||^2$. We then rescale the Yang-Mills fields by setting $iL = A/g$ and $iR = B/g$. We introduce a factor i so that L and R are hermitian (this choice is standard in physics). We also have to rescale the scalar field as follows

$$\chi = \phi\sqrt{2}/g$$

in order to get a conventional kinetic energy term in the Lagrangian (*i.e.* $\partial_\nu\overline{\chi}\partial^\nu\chi$ rather than $\frac{2}{g^2}\partial_\nu\overline{\phi}\partial^\nu\phi$). The Lagrangian can therefore be rewritten as

$$\mathcal{L} = YM(L,R) + KE(\chi) + V(\chi)$$

where

$$\begin{aligned}
YM(L,R) &= \frac{1}{4}(F^L_{\mu\nu})^2 + \frac{1}{4}(F^R_{\mu\nu})^2 \\
KE(\chi) &= D_\nu\overline{\chi}D^\nu\chi \\
&\quad \text{with} \quad D_\nu\chi = \nabla_\nu\chi + i\mu\sqrt{2}e^{-i\gamma}(L_\nu - R_\nu) \\
&\quad \text{and} \quad \nabla_\nu\chi = \partial_\nu\chi + ig(L_\nu\chi - \chi R_\nu) \\
V(\chi) &= \frac{1}{2}(\mu\sqrt{2}(e^{i\gamma}\chi + \overline{e^{i\gamma}\chi}) + g\chi\overline{\chi})^2 \\
&= \frac{2}{g^2}(\mid \frac{ge^{i\gamma}\chi}{\sqrt{2}} + \mu \mid^2 - \mu^2)^2
\end{aligned}$$

We therefore obtain the usual Yang-Mills action with a symmetry breaking Higgs potential. However the potential is already shifted onto an absolute minimum (this is a nice feature of the model : no further shift is necessary !).

The potential V is plotted as *Fig.1*. We call $x = Re\,(e^{i\gamma}\chi)$, $y = Im\,(e^{i\gamma}\chi)$. The circle of minima is $(x+\mu\sqrt{2}/g)^2+y^2 = 2\mu^2/g^2$. It is centered on the point $\Omega(x = -\mu\sqrt{2}/g, y = 0)$. V has a local maximum in Ω and $(V(\Omega) = 2\mu^4/g^2)$. The mass of the Higgs field is gotten from the potential.

$$M_\chi^2 = \frac{1}{2}\frac{\partial^2 V}{\partial\chi\partial\overline{\chi}}|_{\chi=0} = 4\mu^2$$

Therefore $M_\chi = 2\mu$. One can see easily on *Fig.1* that the freedom of choice for γ in the definition of the derivation d amounts to choose the position of the vacuum (the origin) on the circle of minima of V.

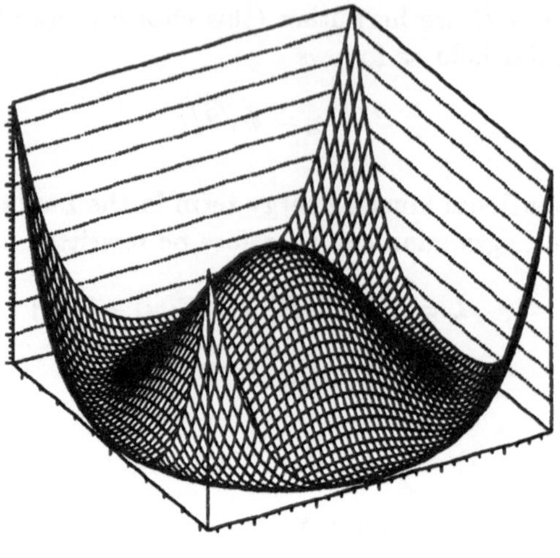

Figure 1

In order to get a kinetic term that is diagonal in the dynamical variables, we are led to redefine the fields as follows: $Z = (R - L)/\sqrt{2}$ and $P = (L + R)/\sqrt{2}$. Then

$$KE(\chi) = \{\nabla_\nu \overline{\chi} \nabla^\nu \chi - 2i\mu(e^{-i\gamma}\nabla_\nu \overline{\chi} - e^{i\gamma}\nabla_\nu \chi)Z^\nu + 4\mu^2 Z_\nu Z^\nu\}.$$

The abelian gauge field P_μ (the "photon") stays massless but the abelian gauge field Z_μ acquires a mass $\frac{1}{2}M_Z^2 = 4\mu^2$ so that $M_Z = 2\sqrt{2}\mu$.

Step 5.
We now want to add a fermionic sector to the theory. So we introduce a Dirac operator $\not{\mathcal{D}}$ coupled to the connection $\mathcal{A}$ in our "space" In the present paragraph, the signature of the metric has to be specified. We take it as purely euclidean. This explains in particular why there is an extra i factor in front of all the Dirac γ matrices in the expression of $\not{\mathcal{D}}$. Modulo the approriate insertion of -1 signs and i factors, calculations would be the same in the case of Lorentz signature. acting on the spinor field Ψ:

$$\not{\mathcal{D}} = \begin{pmatrix} \not\partial & \mu \\ \mu & \not\partial \end{pmatrix} + g \begin{pmatrix} i\gamma^\mu L_\mu & \chi/\sqrt{2} \\ \overline{\chi}/\sqrt{2} & i\gamma^\mu R_\mu \end{pmatrix}$$

$$\Psi = \begin{pmatrix} \psi_L \\ \psi_R \end{pmatrix} = \begin{pmatrix} (\frac{1-\gamma_5}{2})\psi \\ (\frac{1+\gamma_5}{2})\psi \end{pmatrix}$$

Here $\psi = \psi_L + \psi_R$ is a four component Dirac spinor. Therefore ψ_L and ψ_R denote respectively left and right spinors of opposite chiralities. The complete lagrangian is then $\mathcal{L} = \mathcal{L}_{Boson} + \mathcal{L}_{Fermion}$ with

$$\mathcal{L}_{Fermion} =$$
$$\overline{\psi}\gamma^\mu \partial_\mu \psi + i\frac{g}{\sqrt{2}}\overline{\psi}\not{P}\psi - i\frac{g}{\sqrt{2}}\overline{\psi}\gamma_5\not{Z}\psi + \mu\overline{\psi}\psi + \frac{g}{\sqrt{2}}(\overline{\psi}_L\chi\psi_R + \overline{\psi}_R\overline{\chi}\psi_L)$$

The reader may be puzzled by the appearance of the parameter μ in the expression of the Dirac operator. Indeed, in the standard model, fermions are massless before the shift in the Higgs scalar field and they acquire a mass only when this scalar field is shifted to a minimum. The reader should remember that, in the present formalism, the scalar field is already shifted to a minimum (*cf. Fig.1*). The situation here is therefore the same as usual but, in a sense, "chronologically reversed": if we shift the scalar field to its maximum, fermions become massless (as well as the Z field). This explain also why we inserted μ (rather than another scale) in the expression of the Dirac operator. From the physical (or intuitive) point of view, μ plays the rôle of a discrete (and constant) vector $\partial/\partial x^\mu$ in a direction "transverse" to Space-Time.

Conclusion

The model that we just discussed describes a massive spinor field of mass μ and a Higgs field of mass 2μ coupled to two abelian gauge fields. The model exhibits spontaneous symmetry breaking. One of these gauge fields (P) stays massless but the other (Z) acquires a mass $2\sqrt{2}\mu$. Analogous features will also appear in the more realistic model that we consider below. Notice that these mass relations come only from the normalization of the scalar product that we have defined in the space of all exterior forms. Here, we have chosen the "simplest" one in the sense that it only involves the parameter μ. Another normalization would have led to other constraints. The choice of this normalization seems to lie beyond the model itself. However, the value of the "Weinberg angle" in this toy model is fixed to 45 *deg*. Independently of the numerical "predictions", the advantage of this model is that it involves Higgs fields as components of a generalized Yang Mills field and that it leads naturally to symmetry breaking.

3. The $\mathbf{SU(2)} \times \mathbf{U(1)}$ model

Steps 1, 2, 3 and 4.

In order to take advantage of the previous calculations, it will be convenient to describe the generalized connection (*i.e.* Yang-Mills fields and Higgs scalars) by a 2×2 matrix whose elements (blocks) are themselves 2×2 matrices. Taking $\mathbf{A}$ and $\mathbf{B}$ antihermitian 2×2 matrices, we set :

$$\mathcal{A} = \begin{pmatrix} \mathbf{A} & i\mu^{-1}\Phi \\ i\mu^{-1}\overline{\Phi} & \mathbf{B} \end{pmatrix}$$

We first restrict our attention to a $U(2) \times U(1)$ theory by assuming that $\mathbf{B} = \begin{pmatrix} B & 0 \\ 0 & 0 \end{pmatrix}$. In order to get $SU(2)\times U(1)$, or better $S(U(2)\times U(1))$, we impose $Str\mathcal{A} = 0 = tr\mathbf{A} - tr\mathbf{B}$. Because of the choice that we made for $\mathbf{B}$, the matrix obtained by (charge) conjugating the right part of $\begin{pmatrix} \mathbf{A} & 0 \\ 0 & \mathbf{B} \end{pmatrix}$, namely $\begin{pmatrix} \mathbf{A} & 0 \\ 0 & -\mathbf{B} \end{pmatrix}$ has vanishing trace and belongs to $Lie[S(U(2) \times U(1))] \subset Lie[SU(3)]$. We can therefore decompose this traceless anti-Hermitian 3×3 matrix as

$$\begin{pmatrix} \mathbf{A} & 0 \\ 0 & -B \end{pmatrix} = \sum_{a \in \{1,2,3,8\}} \frac{i\lambda^a}{\sqrt{2}} W_a$$

where W_a are real 1-forms, and λ^a denote the Gell-Mann matrices. Using the fact that $\lambda_8 = \frac{1}{\sqrt{3}} diag(1, 1, -2)$, the generalized connection $\mathcal{A}$ can be rewritten as

$$\mathcal{A} = \begin{pmatrix} \frac{i}{\sqrt{2}} \vec{\tau} \vec{W} + \frac{i}{\sqrt{6}} \mathbf{1} W_8 & \frac{i}{\mu} \begin{pmatrix} \phi_0 & 0 \\ \phi_+ & 0 \end{pmatrix} \\ \frac{i}{\mu} \begin{pmatrix} \overline{\phi}_0 & \overline{\phi}_- \\ 0 & 0 \end{pmatrix} & \begin{pmatrix} +\frac{2i}{\sqrt{6}} W_8 & 0 \\ 0 & 0 \end{pmatrix} \end{pmatrix}$$

Notice the sign of the component $\mathcal{A}_{33}$, which is such that $Str(\mathcal{A}) = 0$. In the present case, we could actually remove the last line and the last column of $\mathcal{A}$ since they are only filled with zeros. Calculations (steps 1, 2 ,3 and 4) are exactly the same as in the previous section modulo the fact that the d operator acting on a 3×3 matrix will build a 4×4 matrix. One has therefore to introduce a projection matrix $p = diag(1, 1, 1, 0)$ to reproject. This introduces a constant $pdpdp$ in the extression of the curvature. Again, it is convenient to introduce a (unique) coupling constant g by multiplying the Lagrangian by $\frac{1}{g^2}$ and to rescale the (anti-hermitian) fields $\mathbf{A}$ and $\mathbf{B}$. We also set $\chi = \Phi\sqrt{2}/g$. We could also multiply the gauge fields by a factor i, in order to make them hermitian, as we did in the case of the $U(1) \times U(1)$ model.

The lagrangian is

$$\mathcal{L} = -\frac{1}{4} Tr(F^{\mathbf{A}}_{\mu\nu})^2 - \frac{1}{4} Tr(F^{\mathbf{B}}_{\mu\nu})^2 + Tr(D_\nu \overline{\chi} D^\nu \chi) + V(\chi)$$

where

$$D_\nu \chi = \partial_\nu \chi + g(\mathbf{A}_\nu \chi - \chi \mathbf{B}_\nu) + \mu\sqrt{2}(\mathbf{A}_\nu - \mathbf{B}_\nu)$$
$$D_\nu \overline{\chi} = \overline{D_\nu \chi}$$
$$V(\chi) = \frac{1}{2} Tr[\mu\sqrt{2}(\chi + \overline{\chi}) + g\chi\overline{\chi}]^2 + \mu^4/g^2$$

Here we removed the factor $e^{i\gamma}$ describing the gauge freedom in the choice of d (i.e. we set $\gamma = 0$. Here it describes an arbitrary element of $SU(2)$.

As usual, the $U(1)$-symmetry corresponding to weak-hypercharge is spontaneously broken, but not the one corresponding to electric charge.

The mass term for the gauge bosons is given by

$$\begin{aligned} -2\mu^2 Tr(\mathbf{B}-\mathbf{A})^2 &= 2\mu^2 Tr[\frac{1}{\sqrt{2}}\vec{\tau}\vec{W} + \frac{\mathbb{1}}{\sqrt{6}}W_8 - \frac{2}{\sqrt{6}}\begin{pmatrix}1 & 0\\ 0 & 0\end{pmatrix}W_8]^2 \\ &= 2\mu^2(W_1^2 + W_2^2 + (W_3 - \frac{1}{\sqrt{3}}W_8)^2) \end{aligned}$$

In order to diagonalize the kinetic term for the gauge fields, one has to set

$$\begin{aligned} Z &= -\cos\theta\, W_3 + \sin\theta\, W_8 \\ P &= \sin\theta\, W_3 + \cos\theta\, W_8 \\ \theta &= \frac{\pi}{6} \end{aligned}$$

and one can also define as usual $W_\pm = \frac{1}{\sqrt{2}}(W_1 \pm iW_2)$. The Yang-Mills term of the Lagrangian for the photon P and the Z is then normalized and becomes $1/4[(F^Z_{\mu\nu})^2 + (F^P_{\mu\nu})^2]$, and the mass term for the gauge bosons is equal to $2\mu^2(2W_+W_- + \frac{4}{3}Z^2)$. The mass of the W is then $M_W = 2\mu$ and the mass of the Z is $M_Z = M_W/cos\theta = \frac{4}{\sqrt{3}}\mu$. The photon P stays massless. The χ_0 is not coupled to the photon (it is the usual neutral Higgs particle), whereas the χ_+ is coupled to all the gauge bosons. Both of these particles get the same mass as in the $U(1)\times U(1)$ model that we analysed previously, namely $M_\chi = 2\mu$.

The particular prediction $sin^2\theta = 0.25$ of the model comes from the fact that we required $\mathcal{A}$ to be of vanishing supertrace. In the present case, $SU(2)\times U(1)$ gauge invariance alone would allow more freedom. This extra hypothesis of vanishing supertrace, not only fixes the Weinberg angle to $\theta = 30°$ but also, as we shall see below fixes the couplings of the leptons (the hypercharge matrix). The comments made at the end of section 3 are still valid here. If we change the normalization of the scalar product in the space of all exterior forms, the mass relations are modified but the prediction for $sin\theta$ remains.

Step 5.

In the usual approach to the standard model, the hypercharge content of multiplets is gotten from experiment (namely from the known experimental value of the electric charge for elementary particles). But here, the fact of setting *a priori* $Str\mathcal{A} = 0$ implies that leptonic matter described by a left doublet and a right singlet has to be coupled to the abelian gauge field B via a hypercharge matrix proportional to $\lambda_8^c = \frac{1}{\sqrt{3}}diag(1,1,+2)$. If , for instance, the left doublet is chosen as $(neutrino_{Left}, electron_{Left})$ and if the right singlet is $(electron_{Right})$, we see that our requirement is fulfilled since the hypercharge content is $diag(-1,-1,-2)$. In an equivalent way, we could couple the $U(1)$ gauge field to a left doublet and to the charge conjugated of a right singlet with a hypercharge matrix $diag(-1,-1,+2)$, which is proportional to λ_8. The present model can therefore describe leptonic families. The hypercharge content of quarks being very different, the present model has to be improved to incorporate them. We now describe one leptonic family. We introduce the Dirac operator $\not{\mathcal{D}}$, acting on a spinor $\Psi = \begin{pmatrix} \psi_L \\ \psi_R \end{pmatrix}$:

$$\not{\mathcal{D}} = \begin{pmatrix} \gamma^\mu \partial_\mu & 0 & \mu & 0 \\ 0 & \gamma^\mu \partial_\mu & 0 & 0 \\ \mu & 0 & \gamma^\mu \partial_\mu & 0 \\ 0 & 0 & 0 & 0 \end{pmatrix} + \begin{pmatrix} \gamma^\mu \mathbf{A}_\mu & \Phi \\ \overline{\Phi} & \gamma^\mu \mathbf{B}_\mu \end{pmatrix}$$

with $\psi_L = \begin{pmatrix} l_L \\ \nu_L \end{pmatrix}$ and $\psi_R = \begin{pmatrix} l_R \\ 0 \end{pmatrix}$. The leptonic part of the lagrangian is $\overline{\Psi} \not{\mathcal{D}}\, \Psi$. Calculations are similar to those made in the previous section. In particular, the mass term is $\mu(\overline{l}_L l_L + \overline{l}_R l_R) = \mu \overline{l} l$ where $l = l_L + l_R$ is the corresponding Dirac spinor. It acquires therefore a mass $m = \mu$. The other spinor (ν_L) stays massless. Here again, the value of m could be changed if we introduce an new constant in the definition of the scalar product of spinors. The fact that the parameter μ enters the expression of $\not{\mathcal{D}}$ itself should not be surprising since an appropriate shift in the potential can make the lepton l massless.

Independently of the numerical values for the masses (they depend more on a choice for normalization of scalar products that on the model itself), we believe that the main interest of the above approach is to present a new theoretical framework that allow us to discuss gauge theories with symmetry breaking in a rather natural way. From the phenomenological point of view the model itself has to be improved.

REFERENCES

[1] R. Coquereaux, G. Esposito-Farese, G. Vaillant, Higgs fields as Yang-Mills fields and discrete symmertries. C.P.T. preprint, (1990) /P. 2407.

[2] A. Connes, J. Lott, Particle models and Non-commutative geometry , I.H.E.S. preprint, 1990.

[3] R. Coquereaux, A. Jadczyk : Symmetries of Einstein-Yang-Mills fields, Commun. Math. Phys. 98, 1985.

[4] M. Dubois-Violette, R. Kerner, J. Madore , Non-commutative differential geometry and new models of gauge theory, J. Math. Phys. 31, 1990.

[5] D.Quillen, V. Matthai, Superconnections, Thom classes, and equivariant differential forms, Topology 25, 1985.

Noncommutative Differential Geometry, Quantum Mechanics and Gauge Theory

Michel Dubois–Violette

Laboratoire de Physique Théorique et Hautes Energies, Bâtiment 211, Université Paris XI, 91405 Orsay, France

Abstract: We describe a noncommutative differential calculus, introduced in [1], which generalizes the differential calculus of differential forms of E. Cartan. We show that besides the classical (commutative) situation, this differential calculus is well suited to deal with ordinary quantum mechanics. That is quantum mechanics falls in the framework of a noncommutative symplectic geometry. We then introduce the simplest corresponding gauge theories. We show that these theories describe ordinary gauge theories but with multivacua structures which provide a sort of alternative to the Higgs mechanism. Most of this lecture is based on a joint work with R. Kerner and J. Madore [2], [3], [4], [5].

1. Preliminaries

1.1 Origin

The idea of noncommutative geometry or quantum geometry comes from quantum mechanics [6], [7], [8]. It was realized at the very beginning of quantum mechanics that it is included in the framework of a noncommutative generalization of the notion of Poisson manifold [9]. We shall show here that, by introducing the appropriate generalization of the notion of differential form, quantum mechanics is included in the framework of a noncommutative generalization of symplectic geometry. Furthermore, we shall use this generalization of differential forms to construct new models of gauge theory.

1.2 The Role of Derivations

Formally, to pass to noncommutative geometry, one first replaces the space by the algebra of complex functions on it, considered as an abstract commutative $*$–algebra and then, one "forgets" commutativity by considering noncommutative $*$–algebras (as analog of "functions" on "noncommutative spaces"). To understand this one must first realize that, given a smooth manifold V, one can really study it by using the algebra $C^\infty(V)$ of smooth complex functions on V considered as an abstract $*$–algebra. At this level, it is worth noticing that the choice of smooth functions here is not innocent. Indeed, if one uses instead say the algebra $C^0(V)$ of continuous functions one loses all the information about the differentiable structure of V. Now, at the algebraic level, what distinguishes $C^\infty(V)$ from $C^0(V)$ is the fact that $C^\infty(V)$ has "many" derivations; these are the vector fields over V. Here, we take the point of view that the noncommutative generalization of the notion of vector field is that of derivation and that the analog of the differentiable structure is encoded in the Lie algebra of derivations. This point of view is of course not original, but it implies a natural corresponding noncommutative generalization of the notion of differential form. This is clearly very well suited for quantum mechanics where what replaces the hamiltonian vector fields are the derivations. In his famous article [9] Dirac refers to derivations as "quantum differentiations".

1.3 Modules

As usual [10] the noncommutative generalization of the notion of vector bundle which we take is that of finite projective module. Indeed given a smooth complex vector bundle of finite rank E over the smooth manifold V, one can replace it by the set $\Gamma(E)$ of its smooth sections. $\Gamma(E)$ is a finite projective module over $C^\infty(V)$ and the correspondence $E \mapsto \Gamma(E)$ is an equivalence of categories. Similarly, the notion of hermitian vector bundle generalizes in a notion of hermitian module for a $*$–algebra [11]. Finally once one has a noncommutative generalization Ω of the differential algebra of differential forms there is a natural notion of connection on modules which generalizes the notion of connection on vector bundles [10]. Here we take for Ω our noncommutative generalization of the algebra of differential forms.

2. The Graded Differential Algebra $\Omega_D(\mathcal{A})$.

2.1 The Differential Algebra $C(\mathrm{Der}(\mathcal{A}),\mathcal{A})$.

In the following, $\mathcal{A}$ denotes an associative algebra over $\mathbb{C}$ and $\mathrm{Der}(\mathcal{A})$ is the Lie algebra of all derivations of $\mathcal{A}$, i.e. $\mathrm{Der}\mathcal{A} = \{X \in \mathrm{End}(\mathcal{A}) | X(AB) = X(A)B + AX(B), \forall A, B \in \mathcal{A}\}$ and the Lie bracket is the commutator in $\mathrm{End}(\mathcal{A})$.Let $C^n(\mathrm{Der}(\mathcal{A}),\mathcal{A})$ be the space of antisymmetric multilinear mappings of $[\mathrm{Der}(\mathcal{A})]^n$ in $\mathcal{A}$ and let $C(\mathrm{Der}(\mathcal{A}),\mathcal{A})$ be the graded vector space

$$C(\mathrm{Der}(\mathcal{A}),\mathcal{A}) = \bigoplus_n C^n(\mathrm{Der}(\mathcal{A}),\mathcal{A}).$$

There is a natural product on $C(\mathrm{Der}(\mathcal{A}),\mathcal{A})$ obtained by using the product of $\mathcal{A}$ and antisymmetrisation in the arguments in $\mathrm{Der}(\mathcal{A})$. With this product, $C(\mathrm{Der}(\mathcal{A}),\mathcal{A})$ is a graded algebra with $\mathcal{A}$ as subalgebra of elements of degree zero. Define

$$d : C(\mathrm{Der}(\mathcal{A}),\mathcal{A}) \to C(\mathrm{Der}(\mathcal{A}),\mathcal{A})$$

with $dC^n(\mathrm{Der}(\mathcal{A}),\mathcal{A}) \subset C^{n+1}(\mathrm{Der}(\mathcal{A}),\mathcal{A})$ by:

$$d\omega(X_0,\dots,X_n)=\sum_k(-1)^k X_k\,\omega(X_0,\overset{k}{\vee}\!\!\dots,X_n)$$
$$+\sum_{r<s}(-1)^{r+s}\omega([X_r,X_s],X_0,\overset{r}{\vee}\!\dots\overset{s}{\vee}\!\dots,X_n),$$

$\forall X_k\in \mathrm{Der}(\mathcal{A})$, where $\overset{k}{\vee}$ means omission of X_k. One checks that d is an antiderivation of degree one of $C(\mathrm{Der}(\mathcal{A}),\mathcal{A})$ and that $d^2=0$; thus $C(\mathrm{Der}(\mathcal{A}),\mathcal{A})$ is a graded differential algebra.

2.2 The Differential Algebra $\Omega_D(\mathcal{A})$

Let $\Omega_D(\mathcal{A})$ be the smallest differential subalgebra of $C(\mathrm{Der}(\mathcal{A}),\mathcal{A})$ which contains $\mathcal{A}$. Each element of $\Omega_D(\mathcal{A})$ is a finite sum of elements of the form $A_0dA_1dA_2\dots dA_k$ with $A_i\in\mathcal{A}, k\in\mathbb{N}$, [1]. Thus if $\mathcal{A}=C^\infty(V)$ where V is a "good" smooth manifold (say with the topology of a CW complex) $\Omega_D(C^\infty(V))$ is just the graded differential algebra $\Omega(V)$ of differential forms on V. The graded differential algebra $\Omega_D(\mathcal{A})$ is the noncommutative generalization of the algebra of differential forms that will be used here. Occasionally we shall need a completion $\hat{\Omega}_D(\mathcal{A})$ of $\Omega_D(\mathcal{A})$ defined by the following. The $\hat{\Omega}_D(\mathcal{A})$ is the set of elements $\omega\in C(\mathrm{Der}(\mathcal{A}),\mathcal{A})$ such that for any finite dimensional subspace $\mathcal{F}$ of $\mathrm{Der}(\mathcal{A})$ there is a $\omega_\mathcal{F}\in\Omega_D(\mathcal{A})$ such that $\omega(X_1,\dots,X_n)=\omega_\mathcal{F}(X_1,\dots,X_n)$ for $X_i\in\mathcal{F}$. In the case $\mathcal{A}=C^\infty(V)$ where V is as above a "'good" smooth manifold, one has $\hat{\Omega}_D(C^\infty(V))=\Omega_D(C^\infty(V))$. We shall not make the distinction and we shall write $\Omega_D(\mathcal{A})$ for $\hat{\Omega}_D(\mathcal{A})$. In fact the only example that we shall meet where $\hat{\Omega}_D(\mathcal{A})$ is bigger than $\Omega_D(\mathcal{A})$ is the one of §3.4. $\hat{\Omega}_D(\mathcal{A})$ is of course also a graded differential subalgebra of $C(\mathrm{Der}(\mathcal{A}),\mathcal{A})$.

2.3 The Operation of $\mathrm{Der}(\mathcal{A})$ in $\Omega_D(\mathcal{A})$.

Let $X\in\mathrm{Der}(\mathcal{A})$ and define $i_X : C^n(\mathrm{Der}(\mathcal{A}),\mathcal{A})\to C^{n-1}(\mathrm{Der}(\mathcal{A}),\mathcal{A})$ by $i_X\omega(X_1,\dots,X_{n-1})=\omega(X,X_1,\dots,X_{n-1})$ for $\omega\in C^n(\mathrm{Der}(\mathcal{A}),\mathcal{A})$ and $X_i\in\mathrm{Der}(\mathcal{A})$. Then i_X is an antiderivation of degree -1 and if one defines the derivation L_X by $L_X=i_Xd+di_X$ one has: $i_{X_1}i_{X_2}+i_{X_2}i_{X_1}=0$, $L_{X_1}i_{X_2}-i_{X_2}L_{X_1}=i_{[X_1,X_2]}$ and $L_{X_1}L_{X_2}-L_{X_2}L_{X_1}=L_{[X_1,X_2]}$, i.e. one has an operation of the Lie algebra $\mathrm{Der}(\mathcal{A})$ in $C(\mathrm{Der}(\mathcal{A}),\mathcal{A})$ in the sense of H. Cartan [12]. The i_X restrict to $\Omega_D(\mathcal{A})$ (and to $\hat{\Omega}_D(\mathcal{A})$) so we have an operation of $\mathrm{Der}(\mathcal{A})$ in $\Omega_D(\mathcal{A})$, [1]. One has $L_XA=X(A)$ for $A\in\mathcal{A}=\Omega^0_D(\mathcal{A})$. An element $\omega\in\Omega_D(\mathcal{A})$ will be called invariant if one has $L_X\omega=0$ for any $X\in\mathrm{Der}(\mathcal{A})$. i_X is the generalization of inner derivation of forms by vector field and L_X is the generalization of Lie derivative.

2.4 Reality

Assume now that $\mathcal{A}$ is a $*$–algebra. One defines an antilinear involution $X \mapsto X^*$ on $\mathrm{Der}(\mathcal{A})$ by setting $X^*(A) = X(A^*)^*$ for $X \in \mathrm{Der}(\mathcal{A})$ and $A \in \mathcal{A}$. The derivations X satisfying $X = X^*$ will be called real, they form a real Lie-subalgebra $\mathrm{Der}_{\mathbb{R}}(\mathcal{A})$ of $\mathrm{Der}(\mathcal{A})$. One then extends the involution of $\mathcal{A}$ to an antilinear involution of $\Omega_D(\mathcal{A})$ by setting $\omega^*(X_1, \dots, X_k) = \omega(X_1^*, \dots, X_k^*)^*$ for $\omega \in \Omega_D^k(\mathcal{A})$ and $X_i \in \mathrm{Der}(\mathcal{A})$. With this involution $\Omega_D(\mathcal{A})$ becomes a differential graded $*$–algebra in the sense that one has $d(\omega^*) = (d\omega)^*$ for $\omega \in \Omega_D(\mathcal{A})$ and $(\alpha\beta)^* = (-1)^{k\ell}\beta^*\alpha^*$ for $\alpha \in \Omega_D^k(\mathcal{A})$ and $\beta \in \Omega_D^\ell(\mathcal{A})$. The elements ω of $\Omega_D(\mathcal{A})$ satisfying $\omega = \omega^*$ will be called real.

2.5 Examples

In the case $\mathcal{A} = C^\infty(V)$, we know that $\Omega_D(C^\infty(V))$ is the graded differential algebra of differential forms on V, that $\mathrm{Der}(C^\infty(V))$ is the Lie algebra of vector fields on V and that all what we introduced so far reduces to classical notions of differential geometry. Let us describe the situation for $\mathcal{A} = M_n(\mathbb{C})$, [1],[2]. The derivations of $M_n(\mathbb{C})$ are all inner so the complex Lie algebra $\mathrm{Der}(M_n(\mathbb{C}))$ reduces to $\mathfrak{sl}(n)$ and the real Lie algebra $\mathrm{Der}_{\mathbb{R}}(M_n(\mathbb{C}))$ reduces to $\mathfrak{su}(n)$. On the other hand one has $\Omega_D(M_n(\mathbb{C})) = C(\mathrm{Der}M_n(\mathbb{C}), M_n(\mathbb{C})) = C(\mathfrak{sl}(n), M_n(\mathbb{C}))$ as can be shown directly [1] and as also follows from formulas below [2]. It follows that the cohomology $H_D(M_n(\mathbb{C}))$ of $\Omega_D(M_n(\mathbb{C}))$ reduces to the Lie algebra cohomology $H^\bullet(\mathfrak{sl}(n))$. This implies in particular that $H_D^1(M_n(\mathbb{C})) = H_D^2(M_n(\mathbb{C})) = 0$ so every closed element of $\Omega_D^1(M_n(\mathbb{C}))$ or $\Omega_D^2(M_n(\mathbb{C}))$ is exact. Let $E_k, k \in \{1, 2, \dots, n^2-1\}$ be a base of self–adjoint traceless $n \times n$–matrices. The $\partial_k = \mathrm{ad}(iE_k)$ form a basis of real derivations i.e. a basis of $\mathrm{Der}_{\mathbb{R}}(M_n(\mathbb{C})) = \mathfrak{su}(n)$. One has $[\partial_k, \partial_\ell] = C_{k\ell}^m \partial_m$, the $C_{k\ell}^m$ are the corresponding structure constants of $\mathfrak{su}(n)$, (or $\mathfrak{sl}(n)$). Define $\theta^k \in \Omega_D^1(M_n(\mathbb{C}))$ by $\theta^k(\partial_\ell) = \delta_\ell^k \mathbb{1}$.The following formula (1) to (5) give a presentation of the graded differential algebra $\Omega_D(M_n(\mathbb{C}))$ [2]:

$$E_k E_\ell = g_{k\ell}\mathbb{1} + (S_{k\ell}^m - \frac{i}{2}C_{k\ell}^m)E_m \tag{1}$$

$$E_k\theta^\ell = \theta^\ell E_k \tag{2}$$

$$\theta^k\theta^\ell = -\theta^\ell\theta^k \tag{3}$$

$$dE_k = -C_{k\ell}^m E_m \theta^\ell \tag{4}$$

$$d\theta^k = -\frac{1}{2}C_{\ell m}^k \theta^\ell\theta^m \tag{5}$$

where $g_{k\ell} = g_{\ell k}$, $S_{k\ell}^m = S_{\ell k}^m$ are real, $g_{k\ell}$ are the components of the Killing form of $\mathfrak{su}(n)$ and $C_{k\ell}^m = -C_{\ell k}^m$ are as above the (real) structure constants of $\mathfrak{su}(n)$.Formula (4) can be inverted and one has [3] $\theta^k = -\frac{i}{n^2} g^{\ell m} g^{kr} E_\ell E_r dE_m$ where $g^{k\ell}$ are the components of the inverse matrix of $(g_{k\ell})$. The element $\theta = E_k\theta^k$ of $\Omega_D^1(M_n(\mathbb{C}))$ is real, $\theta = \theta^*$, and independent of the choice of the E_k, in fact $\theta(\mathrm{ad}(iA)) = A - \frac{1}{n}\mathrm{tr}(A)$, [5]. Furthermore θ is invariant, $L_X\theta = 0$, and any invariant element of $\Omega_D^1(M_n(\mathbb{C}))$ is a scalar multiple of θ. We call θ the ***canonical invariant element*** of $\Omega_D^1(M_n(\mathbb{C}))$. Using it, (4) and (5) can be rewritten in the form

$$dM = i[\theta, M], \ \forall M \in M_n(\mathbb{C}) \tag{6}$$

$$d(-i\theta) + (-i\theta)^2 = 0. \tag{7}$$

Finally one can study the case $\mathcal{A} = C^\infty(V) \otimes M_n(\mathbb{C})$ which mixes the above examples. It can be shown that one has [3]

$$\Omega_D(C^\infty(V) \otimes M_n(\mathbb{C})) = \Omega_D(C^\infty(V)) \otimes \Omega_D(M_n(\mathbb{C}))$$

where the last tensor product is the usual (twisted) tensor product of graded differential algebras.

3. Noncommutative Symplectic Structures

3.1 Definition

An element ω of $\Omega_D^2(\mathcal{A})$ will be called a ***symplectic structure for*** $\mathcal{A}$ [2] if it satisfies the following conditions a) and b)

a) For any $H \in \mathcal{A}$, there is a derivation $\mathrm{ham}(H) \in \mathrm{Der}(\mathcal{A})$ such that

$$\omega(X, \mathrm{ham}(H)) = X(H) \qquad \text{for any} \qquad X \in \mathrm{Der}(\mathcal{A}).$$

b) ω is closed, i.e. $d\omega = 0$.

Notice that a) implies that $\mathrm{ham}(H) \in \mathrm{Der}(\mathcal{A})$ is unique for a given $H \in \mathcal{A}$, i.e. one has a linear mapping $\mathrm{ham} : \mathcal{A} \to \mathrm{Der}(\mathcal{A})$. Notice also that in the commutative case $\mathcal{A} = C^\infty(V)$, a) means that ω is a non–degenerate differential 2–form on V and therefore ω is a symplectic structure for $C^\infty(V)$ if and only if (V, ω) is a symplectic manifold. This is our first example of a symplectic structure.

3.2 Poisson Brackets

Let ω be a symplectic structure for $\mathcal{A}$.We define the corresponding ***Poisson bracket*** $\{A, B\}$ of $A, B \in \mathcal{A}$ by

$$\{A, B\} = \omega(\mathrm{ham}(A), \mathrm{ham}(B)).$$

Then one has $\{A, B\} = -\{B, A\}$ and the mapping $B \mapsto \{A, B\}$ is a derivation of $\mathcal{A}$ which is precisely $\mathrm{ham}(A)$. Furthermore the condition $d\omega = 0$ is then equivalent to the Jacobi identity $\{A, \{B, C\}\} + \{B, \{C, A\}\} + \{C, \{A, B\}\} = 0$ and $[\mathrm{ham}(A), \mathrm{ham}(B)] = \mathrm{ham}(\{A, B\})$, [2]. Thus everything works as in the commutative case $\mathcal{A} = C^\infty(V)$, i.e. as in the classical situation which is our first example.

3.3 Second Example: The Case $\mathcal{A} = M_n(\mathbb{C})$.

The case $\mathcal{A} = M_n(\mathbb{C})$ is interesting because it is simple, purely noncommutative and corresponds to the typical quantum system of a spin $s = \frac{n-1}{2}$. Suppose that there is a symplectic structure for $M_n(\mathbb{C})$. Then its Poisson bracket must be proportional to the commutator since it is a derivation in each variable which is antisymmetric and since all derivations are inner. Thus one must have $\{A, B\} = \frac{i}{\hbar}[A, B]$, $\forall A, B \in M_n(\mathbb{C})$, where $\hbar$ is some number. On the other hand, since $\Omega_D^2(M_n(\mathbb{C})) = C^2(Der(M_n(\mathbb{C})), M_n(\mathbb{C}))$, one defines an element ω of $\Omega_D^2(M_n(\mathbb{C}))$ by setting

$$\omega(\mathrm{ad}(\frac{i}{\hbar}A), \mathrm{ad}(\frac{i}{\hbar}B)) = \frac{i}{\hbar}[A, B] \; (= \mathrm{ad}(\frac{i}{\hbar}A)B), \; \forall A, B \in M_n(\mathbb{C}),$$

which implies immediately that ω is a symplectic structure and that $\mathrm{ham}(H) = \mathrm{ad}(\frac{i}{\hbar}H)$ for $H \in M_n(\mathbb{C})$ and that furthermore the corresponding Poisson bracket is given by $\{A, B\} = \frac{i}{\hbar}[A, B]$. The above $\omega \in \Omega_D^2(M_n(\mathbb{C}))$ is invariant and closed so it is exact. In fact, one has $\omega = d(\hbar\,\theta)$ where θ is the canonical invariant element of $\Omega_D^1(M_n(\mathbb{C}))$ defined by $\theta(\mathrm{ad}(iA)) = A - \frac{1}{n}\mathrm{tr}(A)$. Thus we have just shown that the quantum mechanics of a spin $s = \frac{n-1}{2}$ is included in the framework of noncommutative symplectic geometry of $M_n(\mathbb{C})$, [2], (here the converse is also true up to the normalization factor $\hbar$ since there are no other symplectic structure for $M_n(\mathbb{C})$ as we saw above).

3.4 Third Example: The Case of the Heisenberg Algebra $\mathcal{A}_\hbar$

Define the Heisenberg algebra $\mathcal{A}_\hbar$ as the $*$–algebra with unit generated by two hermitian elements p and q with relation

$$[q, p] = i\hbar\,\mathbb{1}.$$

As it is well known this algebra is well suited for the description of the quantized version of a mechanics system with one degree of freedom, i.e. it corresponds to *the quantum phase space*. Here we consider only one degree of freedom for notational convenience but the extension to a finite number of degrees of freedom is straightforward. It is easy to show that all derivations of $\mathcal{A}_\hbar$ are inner derivations so again, if there is a symplectic structure for $\mathcal{A}_\hbar$, the corresponding Poisson bracket must be proportional to the commutator. In fact,[9] what replaces the Poisson bracket here is given by

$$\{A, B\} = \frac{i}{\hbar}[A, B] \quad \text{for} \quad A, B \in \mathcal{A}_\hbar.$$

On the other hand, one defines an element ω of $C^2(\mathrm{Der}(\mathcal{A}_\hbar), \mathcal{A}_\hbar)$ by setting $\omega(\mathrm{ad}(\frac{i}{\hbar}A), \mathrm{ad}(\frac{i}{\hbar}B)) = \frac{i}{\hbar}[A, B]$. Now, one verifies by direct computation that ω is also given by

$$\omega = \sum_n \frac{1}{(i\hbar)^n (n+1)!} [\dots[dp, \underbrace{p], \dots, p}_{n}][\dots[dq, \underbrace{q,], \dots, q}_{n}]$$

and thus ω is in $\Omega^2_D(\mathcal{A}_\hbar)$, (more precisely in $\hat{\Omega}^2_D(\mathcal{A}_\hbar)$), and the properties of the commutator imply that ω is indeed a symplectic structure. One has $\mathrm{ham}(H) = \mathrm{ad}(\frac{i}{\hbar}H)$ and the corresponding Poisson bracket is of course $\{A,B\} = \frac{i}{\hbar}[A,B]$. We have therefore shown that elementary quantum mechanics of spin systems and quantum systems of finite numbers of degrees of freedom is included in the framework of the above noncommutative generalization of symplectic geometry. This shows the relevance of $\Omega_D(\mathcal{A})$ as noncommutative generalization of the differential algebra of differential forms. We end this section by mentioning that we do not know if the above symplectic structure for $\mathcal{A}_\hbar$ is exact. We guess that it is not because there is no trace on $\mathcal{A}_\hbar$. Notice also that since $[\dots[dp,p],\dots,p][\dots[dq,q],\dots,q] = o(\hbar^{2n})$ the "formal classical limit" of the symplectic structure ω of $\mathcal{A}_\hbar$ is $dp.dq$ as expected.

4. Connections and Curvatures

4.1 Connections and Modules

Let $\mathcal{M}$ be a right $\mathcal{A}$–module. We shall use the notion of Ω– connection of A. Connes [10] but with $\Omega = \Omega_D(\mathcal{A})$. So we define a ***connection on*** $\mathcal{M}$ to be a linear mapping $\nabla : \mathcal{M} \to \mathcal{M} \otimes_\mathcal{A} \Omega^1_D(\mathcal{A})$ which satifies

$$\nabla(\Phi A) = (\nabla\Phi)A + \Phi \otimes dA \tag{8}$$

for any $\Phi \in \mathcal{M}$ and $A \in \mathcal{A}$.One extends ∇, as usual [10], as a linear mapping of $\mathcal{M} \otimes_\mathcal{A} \Omega_D(\mathcal{A})$ into itself by setting $\nabla(\Phi\omega) = (\nabla\Phi)\omega + \Phi d\omega$, for $\Phi \in \mathcal{M}$ and $\omega \in \Omega_D(\mathcal{A})$. Equation (8) implies that $\nabla^2(\Phi A) = (\nabla^2\Phi)A$, for $\Phi \in \mathcal{M}$ and $A \in \mathcal{A}$. Thus ∇^2 is a right–module homomorphism of $\mathcal{M}$ in $\mathcal{M} \otimes_\mathcal{A} \Omega^2_D(\mathcal{A})$ which is called the ***curvature*** of the connection ∇. The difference $\nabla - \nabla'$ of two connections is also a module homomorphism in view of [8].Connections always exist on a finite projective module [10]. Recall that $\mathcal{M}$ is a finite projective module if there is another module $\mathcal{N}$ such that the direct sum $\mathcal{M} \oplus \mathcal{N}$ is a free module of finite rank. For $\mathcal{A} = C^\infty(V)$, a finite projective module is the module of smooth sections of a smooth vector bundle over V and a connection on such a module is a connection on the corresponding vector bundle in the usual sense.

4.2 Hermitian Modules and Hermitian Connections

In this paragraph we again assume that $\mathcal{A}$ is a $*$–algebra. An element of $\mathcal{A}$ is said to be ***positive*** if it is a finite sum of elements of the form A^*A with $A \in \mathcal{A}$. The set $\mathcal{A}^+$ of all positive elements of $\mathcal{A}$ is a convex cone which we assume to be ***strict*** in the sense that $\mathcal{A}^+ \cap (-\mathcal{A}^+) = \{0\}$. This property is typically satisfied for $*$–algebras of operators in Hilbert spaces. Following [11], let us introduce the following definitions. A ***hermitian structure*** on the right $\mathcal{A}$–module $\mathcal{M}$ is a sesquilinear mapping $h : \mathcal{M} \times \mathcal{M} \to \mathcal{A}$ which satisfies

a) $h(\Phi A, \Psi B) = A^* h(\Phi,\Psi)B,\ \forall\Phi,\Psi \in \mathcal{M},\ \forall A,B \in \mathcal{A}$

b) $h(\Phi,\Phi) \in \mathcal{A}^+$, $\forall \Phi \in \mathcal{M}$, and $h(\Phi,\Phi) = 0 \Rightarrow \Phi = 0$.
A right module $\mathcal{M}$ equipped with a hermitian structure h will be called a ***hermitian module***. A ***hermitian connection*** on such a hermitian module is a connection ∇ in the above sense which satisfies

$$dh(\Phi,\Psi) = h(\nabla\Phi,\Psi) + h(\Phi,\nabla\Psi),\ \forall\Phi,\Psi \in \mathcal{M}.$$

Hermitian connections always exist on finite projective hermitian modules. For $\mathcal{A} = C^\infty(V)$ a finite projective hermitian module is the module of smooth sections of a smooth hermitian vector bundle and a hermitian connection is a hermitian connection on the corresponding hermitian vector bundle in the usual sense.

4.3 Gauge Transformations

Let $\mathcal{M}$ be a right $\mathcal{A}$–module. The group Aut($\mathcal{M}$) of all module automorphisms of $\mathcal{M}$ acts on the affine space of all connections on $\mathcal{M}$ via $\nabla \mapsto \nabla^U$ with $\nabla^U\Phi = U^{-1}\nabla(U\Phi)$, for $\Phi \in \mathcal{M}$, $U \in$ Aut($\mathcal{M}$). If $\mathcal{A}$ is a $*$–algebra as above and $(\mathcal{M},h)$ is a hermitian module the group Aut($\mathcal{M},h$) is the group of all module automorphisms U of $\mathcal{M}$ which preserve h, i.e. $h(U\Phi,U\Psi) = h(\Phi,\Psi)$, $\forall\Phi,\Psi \in \mathcal{A}$. The group Aut($\mathcal{M},h$) acts then, (with the same formula), on the space of hermitian connections. We shall refer to Aut($\mathcal{M},h$) as the ***gauge group*** and to its elements as ***gauge transformations***. The set of connections (resp. hermitian connections) ∇ with zero curvature ($\nabla^2 = 0$) is invariant by Aut($\mathcal{M}$) (resp. Aut($\mathcal{M},h$)). In the case $\mathcal{A} = C^\infty(V)$ with V simply connected, there is at most one orbit of connections with zero curvature on a finite projective module. The next example shows that this is not generally true for a noncommutative algebra $\mathcal{A}$.

4.4 Connections on $M_n(\mathbb{C})$–Modules

The $*$–algebra $M_n(\mathbb{C})$ is simple with only one irreductible representation in $\mathbb{C}^n$. A general finite right–module (which is projective) is the space $M_{Kn}(\mathbb{C})$ of $K \times n$–matrices with right action of $M_n(\mathbb{C})$. One has Aut($M_{Kn}(\mathbb{C})$) $= GL(K)$ with left matrix multiplication. The module $M_{Kn}(\mathbb{C})$ is naturally hermitian with $h(\Phi,\Psi) = \Phi^*\Psi$ where Φ^* is the $n \times K$ matrix hermitian conjugate to Φ. The gauge group is then the unitary group $U(K)(\subset GL(K))$. Here, there is a natural origin $\overset{0}{\nabla}$ in the space of connections given by $\overset{0}{\nabla}\Phi = -i\Phi\theta$ where $\Phi \in M_{Kn}(\mathbb{C})$ and where θ is the canonical invariant element of $\Omega^1_D(M_n(\mathbb{C}))$. The fact that this defines a connection follows from (6) and (8). This connection is hermitian and its follows from (7) that its curvature vanishes, i.e. $(\overset{0}{\nabla})^2 = 0$. Any connection ∇ is of the form $\nabla\Phi = \overset{0}{\nabla}\Phi + A\Phi$ where $A = A_k\theta^k$ with $A_k \in M_K(\mathbb{C})$ and $A\Phi$ means $A_k\Phi\otimes\theta^k$. The connection ∇ is hermitian if and only if the A_k are antihermitian i.e. $A_k^* = -A_k$. The curvature of ∇ is given by $\nabla^2\Phi = F\Phi$ $(= F_{k\ell}\Phi \otimes \theta^k\theta^\ell)$ with

$$F = \frac{1}{2}([A_k, A_\ell] - C^m_{k\ell}A_m)\theta^k\theta^\ell. \tag{9}$$

Thus $\nabla^2 = 0$ if and only if the A_k form a representation of the Lie algebra $sl(n)$ in $\mathbb{C}^K$ and two such connections are in the same $\mathrm{Aut}(M_{Kn}(\mathbb{C}))$–orbit if and only if the corresponding representations of $sl(n)$ are equivalent. This implies that ***the gauge orbits of flat*** **(**$\nabla^2 = 0$**)** ***hermitian connections are in one–to–one correspondence with the unitary classes of representations of*** $su(n)$ ***in*** $\mathbb{C}^K$. For instance if $n = 2$, these orbits are labelled by the number of partitions of the integer K. (See reference [2] for more details).

5. Models of Gauge Theory

5.1 Connections on $C^\infty(\mathbb{R}^{s+1}) \otimes M_n(\mathbb{C})$–Modules

Let x^μ, $\mu \in \{0,1,\dots,s\}$, be the canonical coordinates of $\mathbb{R}^{s+1}$. One has $\Omega_D(C^\infty(\mathbb{R}^{s+1}) \otimes M_n(\mathbb{C})) = \Omega_D(C^\infty(\mathbb{R}^{s+1})) \otimes \Omega_D(M_n(\mathbb{C}))$ so one can split the differential as $d = d' + d''$ where d' is the differential along $\mathbb{R}^{s+1}$ and d'' is the differential of $\Omega_D(M_n(\mathbb{C}))$. A typical finite projective right module is $C^\infty(\mathbb{R}^{s+1}) \otimes M_{Kn}(\mathbb{C})$. This is an hermitian module with hermitian structure given by $h(\Phi,\Psi)(x) = \Phi(x)^*\Psi(x)$, $(x \in \mathbb{R}^{s+1})$. As a $C^\infty(\mathbb{R}^{s+1})$–module, this module is free (of rank $K.n$), so $d'\Phi$ is well defined for $\Phi \in C^\infty(\mathbb{R}^{s+1}) \otimes M_{Kn}(\mathbb{C})$. In fact, $d'\Phi(x) = \frac{\partial \Phi}{\partial x^\mu}(x)dx^\mu$. A connection on the $C^\infty(\mathbb{R}^{s+1}) \otimes M_n(\mathbb{C})$–module $C^\infty(\mathbb{R}^{s+1}) \otimes M_{Kn}(\mathbb{C})$ is of the form $\nabla\Phi = d'\Phi - i\Phi\theta + A\Phi$ with $A = A_\mu dx^\mu + A_k\theta^k$, where the A_μ and the A_k are $K \times K$ matrix valued functions on $\mathbb{R}^{s+1}$ (i.e. elements of $C^\infty(\mathbb{R}^{s+1}) \otimes M_K(\mathbb{C})$) and where $A\Phi(x) = A_\mu(x)\Phi(x)dx^\mu + A_k(x)\Phi(x)\theta^k$. Such a connection is hermitian iff the $A_\mu(x)$ and the $A_k(x)$ are antihermitian, $\forall x \in \mathbb{R}^{s+1}$. The curvature of ∇ is given by $\nabla^2\Phi = F\Phi$ where

$$\begin{aligned} F = \frac{1}{2}(\partial_\mu A_\nu - \partial_\nu A_\mu + [A_\mu, A_\nu])dx^\mu dx^\nu \\ + (\partial_\mu A_k + [A_\mu, A_k])dx^\mu \theta^k \\ + \frac{1}{2}([A_k, A_\ell] - C^m_{k\ell} A_m)\theta^k \theta^\ell \end{aligned} \tag{10}$$

The connection ∇ is flat (i.e. $\nabla^2 = 0$) if and only if each term of (10) vanishes, which implies that ∇ is gauge equivalent to a connection for which one has $A_\mu = 0$, $\partial_\mu A_k = 0$ and $[A_k, A_\ell] = C^m_{k\ell}A$. Furthermore two such connections are equivalent if and only if the corresponding representations of $su(n)$ in $\mathbb{C}^K$ (given by the constant $K \times K$–matrices A_ℓ) are equivalent. So again, ***the gauge orbits of flat hermitian connections are in one to one correspondence with the unitary classes of (antihermitian) representations of*** $su(n)$ ***in*** $\mathbb{C}^K$. Again, in the case n=2, the number of such orbits is the number of partitions of the integer$K$ i.e.

$$\mathrm{card}\{(n_r) | \sum_r n_r.r = K\}.$$

5.2 Actions

We consider $\mathbb{R}^{s+1}$ as the $(s+1)$–dimensional space–time and we replace the algebra of smooth functions on $\mathbb{R}^{s+1}$ by $C^\infty(\mathbb{R}^{s+1}) \otimes M_n(\mathbb{C})$ which we interpret as the algebra of "smooth functions on a noncommutative generalized space–time". It is clear, from (10), that the generalization of the (euclidean) Yang–Mills action for a hermitian connection ∇ as in 5.1 on $C^\infty(\mathbb{R}^{s+1}) \otimes M_{Kn}(\mathbb{C})$ is

$$\begin{aligned} \|F\|^2 = \int d^{s+1}x \, \mathrm{tr}\Big\{ & \frac{1}{4}\sum(\partial_\mu A_\nu - \partial_\nu A_\mu + [A_\mu, A_\nu])^2 \\ & + \frac{1}{2}\sum(\partial_\mu A_k + [A_\mu, A_k])^2 + \frac{1}{4}\sum([A_k, A_\ell] - C^m_{k\ell}A_m)^2\Big\} \end{aligned} \tag{11}$$

where the metric of space-time is $g_{\mu\nu} = \delta_{\mu\nu}$ and where the basis E_k of hermitian traceless $n \times n$–matrices of §2.5 is chosen in such a way that $g_{k\ell} = \delta_{k\ell}$, i.e. $\mathrm{tr}(E_k E_\ell) = n\delta_{k\ell}$. This can be more deeply justified by introducing the analog of the Hodge involution on $\Omega_D(M_n(\mathbb{C}))$, the analog of the integration of elements of $\Omega_D^{n^2-1}(M_n(\mathbb{C}))$ (essentially the trace) and by combining these operations with the corresponding one on $\mathbb{R}^{s+1}$ to obtain a scalar product on $\Omega_D(C^\infty(\mathbb{R}^{s+1}) \otimes M_n(\mathbb{C}))$ etc. See in [2] and [3] for more details.

5.3 Discussion

The action (11) is the Yang–Mills action on the noncommutative space corresponding to $C^\infty(\mathbb{R}^{s+1}) \otimes M_n(\mathbb{C})$. However it can be interpreted as the action of a field theory on the $(s+1)$–dimensional space–time $\mathbb{R}^{s+1}$. At first sight, this field theory consists of a $U(n)$–Yang–Mills potential $A_\mu(x)$ minimally coupled with scalar fields $A_k(x)$ with values in the adjoint representation which interact among themselves through a quartic potential. The action (11) is positive and vanishes for $A_\mu = 0$ and $A_k = 0$, but is also vanishes on other gauge orbits. Indeed $\|F\|^2 = 0$ is equivalent to $F = 0$, so the gauge orbits on which the action vanishes are labelled by unitary classes of representations of $su(n)$ in $\mathbb{C}^K$. By the standard semi–heuristic argument, these gauge orbits are interpreted as different vacua for the corresponding quantum theory. To specify a quantum theory, one has to choose one such vacuum and to translate the fields in order that the zero of these translated fields corresponds to the vacuum (i.e. is the corresponding zero of the action). The variables A_μ, A_k in (11) are thus adapted to the specific vacuum φ_0 corresponding to the trivial representation $A_k = 0$ of $su(n)$. If one chooses the vacuum φ_α corresponding to a representation $\overset{\alpha}{R}_k$ of $su(n)$, (i.e. one has $[\overset{\alpha}{R}_k, \overset{\alpha}{R}_\ell] = C^m_{k\ell}\overset{\alpha}{R}_m$), one must instead use the variables A_μ and $\overset{\alpha}{B}_k = A_k - \overset{\alpha}{R}_k$. Making this change of variable one observes that components A_μ become massive and that the $\overset{\alpha}{B}_k$ have different masses; the whole mass spectrum depends on α. This is very analogous to the Higgs mechanism. Here however the gauge invariance is not broken, the non–invariance of the mass–terms of the A_μ is compensated by the fact that the gauge transformation of the $\overset{\alpha}{B}_k$ becomes inhomogeneous (they are components of a connection). When $\mathbb{R}^{s+1}$ is interpreted as space–time the x^μ have the dimension of a length. To write down the action (11) we have implicitly set a length or, equivalently, a mass m equal to one. One can recover the dimension by writing $m x^\mu$ instead of x^μ in (11), which gives the mass scale of the theory. The fields must be correspondingly renormalized to give its right dimension to the action,[3].

5.4 Models

The simplest models with the above features are obtained for $n = 2$ by taking as algebra the algebra $C^\infty(\mathbb{R}^{s+1}) \otimes M_2(\mathbb{C})$ of 2×2–matrix valued functions. The simplest one for which there are several vacua (in fact two) is obtained by taking the free hermitian module of rank one i.e. $C^\infty(\mathbb{R}^{s+1}) \otimes M_2(\mathbb{C})$ considered as a right module. This is the analog of the free Maxwell theory, ($C^\infty(\mathbb{R}^{s+1}) \otimes M_2(\mathbb{C})$ replacing $C^\infty(\mathbb{R}^{s+1})$). In this case there are two vacua φ_0 and $\varphi_{1/2}$ corresponding to the two inequivalent representations of $su(2)$ in $\mathbb{C}^2$, the trivial one $\{0\} \oplus \{0\}$ and the representation $\{\frac{1}{2}\}$ of spin 1/2. For the vacuum φ_0 the variables of (11) are the good ones and there is not much to say. The interesting vacuum is $\varphi_{1/2}$. One then has to make the translation in the A_k as explained in 5.3; the traceless part of the A_μ becomes massive and there is a mass spectrum for the whole theory which is described in [3]. This model with vacuum $\varphi_{1/2}$ is interesting but not very realistic. Indeed it looks a little like the bosonic sector for the Weinberg–Salam model with the $U(1) \times SU(2)$ group, but one must identify the $U(1)$–gauge potential with $\mathrm{tr}(A_\mu)\mathbb{1}$ so it is not coupled with the other fields so, for instance, there is no Weinberg angle etc. In order to obtain more realistic models one must look at other modules. The next simplest $C^\infty(\mathbb{R}^{s+1}) \otimes M_2(\mathbb{C})$ right hermitian module is $C^\infty(\mathbb{R}^{s+1}) \otimes M_{3\,2}(\mathbb{C})$. In this case there are three vacua, φ_0, $\varphi_{1/2}$ and φ_1 corresponding to the three inequivalent representations of $su(2)$ in $\mathbb{C}^3$, $\{0\}\oplus\{0\}\oplus\{0\}$, $\{\frac{1}{2}\}\oplus\{0\}$ and $\{1\}$. Using the vacuum $\varphi_{1/2}$ corresponding to the representation $\{\frac{1}{2}\}\oplus\{0\}$, one obtains a model close to the Weinberg–Salam model in the bosonic sector by identifying appropriately the $U(1)$ part of the $U(1)\times SU(2)$ gauge potential (and by making the field translations corresponding to $\varphi_{1/2}$). A defect of this model is that there are too many bosonic fields. There is first a $U(1)$–gauge field which is completely decoupled and may be probably eliminated by introducing a generalization of a fiber volume for the module. Secondly there are two identical pairs of $W^\pm$ fields and two Z fields. It may be that this can be cured by adding some structure on the module to be conserved by the connections. It may also be that this is not a real defect. In any case what is missing is the fermionic sector and for that one has to define a noncommutative analog of the notion of spinor.

6. Conclusion

We did not define noncommutative generalizations of linear connections. The reason is that Der($\mathcal{A}$) is not a module on $\mathcal{A}$ but only a module over the center of $\mathcal{A}$. So in general one cannot use the notion of connection of section 4 for Der($\mathcal{A}$). For similar reasons, we have not for the moment a natural generalization of spinors. Finally, we remark that although to study noncommutative symplectic geometry we used explicitely the details of the structure of $\Omega_D(\mathcal{A})$, (i.e. the operation of Der($\mathcal{A}$) in $\Omega_D(\mathcal{A})$), to discuss models of gauge theory we need much less: Only in fact the $\mathbb{Z}_2$-grading of $\Omega_D(\mathcal{A})$ and the existence of the differential d. Replacing $\Omega_D(\mathcal{A})$ by more general $\mathbb{Z}_2$-graded differential algebras containing $\mathcal{A}$, one arrives at models of gauge theory such that the one proposed by A. Connes and J. Lott [13], or the one described by R. Coquereaux in this conference [14].

References

[1] M. DUBOIS-VIOLETTE,"Dérivations et calcul différentiel non-commutatif", *C.R. Acad. Sci. Paris* **307**, Série I, 403, (1988).

[2] M. DUBOIS-VIOLETTE, R. KERNER, J. MADORE,"Noncommutative differential geometry of matrix algebras", Preprint Orsay 1988, SLAC PPF 88-45, *J. Math. Phys.* **31**, 316, (1990).

[3] M. DUBOIS-VIOLETTE, R. KERNER, J. MADORE, "Noncommutative differential geometry and new models of gauge theory", Preprint Orsay 1988, SLAC PPF 88-49,*J. Math. Phys.* **31**, 323, (1990).

[4] M. DUBOIS-VIOLETTE, J. MADORE, R. KERNER, "Gauge bosons in a noncommutative geometry", *Phys. Lett.* **B217**, 485, (1989).

[5] M. DUBOIS-VIOLETTE, R. KERNER, J. MADORE, "Classical bosons in a noncommutative geometry", *Class. Quantum Grav.* **6**, 1709, (1989).

[6] W. HEISENBERG, "Über quantentheoretische Umdeutung kinematischer und mechanischer Beziehungen", *Zs. f. Phys.* **33**, 879, (1925).

[7] M. BORN, P. JORDAN, "Zur Quantenmechanik", *Zs. f. Phys.* **34**, 858, (1925).

[8] M. BORN, W. HEISENBERG, P. JORDAN, "Zur Quantenmechanik II",*Zs. f. Phys.* **35**, 557, (1926).

[9] P.A.M. DIRAC, "The fundamental equations of quantum mechanics", *Proc. Roy. Soc.* **A109**, 642, (1926).

[10] A. CONNES, "Noncommutative differential geometry", *Publi. I.H.E.S.,* **62**, 257, (1986).

[11] A. CONNES,"C*-algèbres et géométrie différentielle", *C.R. Acad. Sci. Paris* **290**, Série A, 599, (1980).

[12] H. CARTAN, in "Colloque de Topologie", Bruxelles 1950, Masson, Paris 1951.

[13] A. CONNES, J. LOTT, "Particle models and noncommutative geometry", Preprint I.H.E.S. 1989.

[14] R. COQUEREAUX, G. ESPOSITO-FARESE, G. VAILLANT, "Higgs fields as Yang-Mills fields and discrete symmetries", Preprint Marseille 1990.

Introduction to non-commutative geometry and Yang-Mills model-building

Daniel KASTLER

Centre de Physique Théorique, CNRS Luminy,
Case 907, 13288 Marseille Cedex 9, France

I have been asked to give two lectures of introduction to Alain Connes's non-commutative differential geometry. This is a futuristic (in fact quantum-) type of mathematics, designed amongst other things (and perhaps primarily) for general use in quantum physics, with the far aim of recasting quantum field theory (ideally in terms of quantized space).

Now, Alain Connes has recently advanced an astonishing proposal incorporating the standard model to his non-commutative geometry [1],[2],[3] following a general scheme of derivation of the usual Yang-Mills action from a "quantum Yang-Mills action" [4] applied to the classical frame - with, in this case, a "doubling of the space" automatically producing the Higg particle, whose mass is predicted to be $\sqrt{2}$ times that of the top (at the tree approximation level).

Since our gathering is a physics conference, I feel that I should also try to sketch this latest, hopefully germinal, development. Therefore I will devote lecture I to a mathematical introduction to non-commutative differential geometry, and lecture II to Yang-Mills model building. Needless to say, this is a bit of a challenge in the imparted time . Therefore you should not expect more than a sketch. In part I, I shall entirely skip the (strategically important) cyclic cohomology, concentrating on geometry-analysis[1]. For more details I refer to the original papers [1] [2] [4] [5] , to Alain Connes's recent book [3], and, for a more elementary treatment, to the forthcoming book [7].

I . THE BASIC OBJECTS .

As we shall see Alain Connes's theory provides substitutes for the usual items of differential geometry (resp. analysis and differential topology), thereby allowing to develop these disciplines in the generalised quantum context. In this lecture we shall proceed as follows: first review the main items of classical differential geometry: namely: (i): the basic algebra; (ii): the basic differential algebra; (iii): vector bundles and covariant derivatives; (iv): elliptic operators (particularly the basic Dirac operator), focussing on the aspects which lend themselves to non-commutative generalisation: the "functional" or "algebraic" (as contrasted to

---

[1] See, e.g., my talk at the former Lake Tahoe Conference for a sketch of cyclic cohomology and a general philosophical orientation.

"spatial") features. I thereby hope to make the subsequent description of the corresponding "quantum" objects more natural and easier to memorize.

[I.1] ***REVIEW OF CLASSICAL DIFFERENTIAL GEOMETRY*** (of a compact, $spin^c$, d- dimensional manifold **M**.)

(i): The basic algebra is the *-algebra $C^\infty(M)=A$ of smooth complex functions on **M** (with the pointwise product of functions, and their complex conjugation as the *-operation). The algebraic viewpoint consists in addressing **A** instead of **M** (whereby **M** can be reconstructed from **A** as its *spectrum* - set of *characters*).

(ii): The basic differential algebra is the set $\Omega(M)$ of smooth differential forms on **M** (= the *De Rham complex of* **M**). $\Omega(M)$ is a graded-commutative $\mathbb{N}$-graded differential algebra having **A** as its zero-grade subalgebra (this under the wedge product of differential forms as the product, the order of differential forms as the $\mathbb{N}$-grading, and the exterior derivative as the differential). We recall that a $\mathbb{N}$-*graded differential algebra* $(\Omega=\oplus_{n\in\mathbb{N}}\Omega^n,d)$ is a (complex associative) algebra Ω fullfilling $\Omega^n\Omega^m\subset\Omega^{n+m}$, $n,m\in\mathbb{N}$, moreover equipped with a *differential* d of grade one (i.e. a linear operator of Ω s.t. $d\Omega^n\subset\Omega^{n+1}$, $n\in\mathbb{N}$, enjoying the derivation property $d(\omega\omega')=\omega d\omega'+(-1)^n(d\omega)\omega'$, $\omega\in\Omega^n$, $\omega'\in\Omega$, and with vanishing square: $d^2=0$. These latter features carry over to the non-commutative frame, as well as the fact that the zero-grade part coincides with **A**. The graded commutative property, however, will have to be abandoned (see comment below).

(iii): To get an algebraic description of vector bundles over **M**, we consider instead their *sets of smooth sections* (a generalisation of the replacement of spaces by their algebras of functions). Given a (finite rank) smooth vector bundle V over **M**, its set E of smooth sections is a (finite projective) **A**-module, with module addition resp. multiplication arising from the (fiberwise) sum of sections, resp. the multiplication of sections by functions on the base -and this is a characterization of the modules of vector bundle-sections.

A *hermitean structure of* V then corresponds to an **A**-valued, **A**-linear definite-positive scalar product on E (archetype: the **A**-valued scalar product on the tangent bundle T_M embodying the riemannian structure).

The algebraic interpretation of *connections* (covariant derivatives) of V is obtained as follows: given a *covariant derivative* ∇_ξ in the direction of the tangent vector-field ξ (=infinitesimal parallel transport along ξ, in coordinates: $(\nabla_\xi\eta)^i=\xi^\mu[\partial_\mu\eta^i+\omega_\mu{}^i{}_k\eta^k)$, the reinterpretation:

$$\nabla_\xi\eta=\nabla(\eta,\xi)=(\nabla\eta)(\xi) \qquad ,\ \eta\in E, \tag{I,1}$$

yields a $\mathbb{C}$-linear map $\nabla: E\to E\otimes_A\Omega^1(M)$ fulfilling[1]:

---

[1] where ηa denotes the product of $\eta\in E$ by $a\in A$ - we consider E as a right **A**-module. Right modules are more natural,particularly in connection with "dyadic calculus", see (01,11) below.

(I,2) $$\nabla(\eta a)=(\nabla\eta)a+\eta\otimes da \qquad ,\eta\in E, a\in A.$$

This property, of the type of a "Leibnitz rule", becomes a bona-fide (graded) module-derivation property if we extend the module E to that of E-*valued exterior forms*:

(I,3) $$E_\Omega=E\otimes_A\Omega(M),$$

(with $\mathbb{N}$-grading induced by that of $\Omega(M)$, including E as its zero-grade part from which it inherits an obvious hermitean structure): ∇ then actually uniquely extends to a graded d-*derivation of the right module* E_Ω i.e. a map $\nabla:E_\Omega\to E_\Omega$ fulfilling[2]:

(I,4) $$\nabla(Xa)=(\nabla X)a+X\otimes da \qquad ,X\in E_\Omega, a\in A.$$

The *curvature* θ is then elegantly defined as the square:

(I,5) $$\theta=\nabla^2,$$

in fact an endomorphism of E_Ω ($\Omega(M)$-linear, as the square of a graded d-derivation)[2]:

(I,6) $$\theta(Xa)=(\theta X)a \qquad ,X\in E_\Omega, a\in \mathbb{A}.$$

The *hermitean connexions* (covariant derivatives) are then singled out by the requirement:

(I,7) $$(\nabla\eta,\eta')+(\eta,\nabla\eta')=d(\eta,\eta') \qquad ,\eta,\eta'\in E,$$

entailing for the exterior covariant derivative and for the curvature the properties:

(I,8) $$(\nabla X,X')+(X,\nabla X')=d(X,X') \qquad ,X,X'\in E_\Omega,$$

(I,9) $$(\theta X,X')=(X,\theta X'). \qquad ,X,X'\in E_\Omega.$$

We mentioned earlier that the A-modules of smooth sections of the smooth vector bundles of M are *finite projective*: in fact this property characterizes them as A-modules. We recall (see **[01,2] (iii)** below for more details) that *finite projective modules* are characterized by the existence of a finite number of generators $e_i\in E$ and $\varepsilon^i\in E^*$ (E^* the dual module), i=1...n, fulfilling the completeness relation:

---

2 this extension to the V-valued exterior forms is the **exterior covariant derivative** .

Note that the graded commutator $[\nabla_1,\nabla_2]$ of a d_1-derivation ∇_1 and d_2-derivation ∇_2 is a $[d_1,d_2]$-derivation: thus $\nabla^2=1/2[\nabla,\nabla]$ is a 0-derivation, i.e. an endomorphism.

(I,10) $e_i\, \varepsilon^i = id_E$ (summation over i),

where the "*dyad*" $u\phi$,$u\in E$, $\phi\in E^*$, denotes the following map : $E\rightarrow E$:

(I,11) $(u\phi)v = u(\phi v)$,$v\in E$.

(iv): We shall consider the elliptic operators as acting on (Sobolev-)Hilbert spaces. We concentrate on the Dirac operator D, which plays a fundamental role, e.g. because it embodies the whole structure of **M**, allowing one to reconstruct the geodesic distance d (and therefore the whole structure of **M**) in the following simple way : one has, with $\|\cdot\|$ the operator norm of bounded operators on $L^2(S_M)$, the L^2-space of the spin bundle (=Hilbert space of square-integrable spinor fields), and $\underline{a}$ the multiplication operator by the function $a\in \mathbf{A}$:

(01,12) $d(x,y)=\text{Inf}\ \{|a(x)-a(y)|\ ;\ a\in \mathbf{A},\ \|[D,\underline{a}]\| \leq 1\}$,$x,y\in \mathbf{M}$),

The essential structural properties of D are the following: first $L^2(S_M)$ is a Z/2-graded Hilbert space (half spinors,the physical helicity!), on which D acts as an odd self adjoint operator, whilst the functions on $\mathbb{A}$ act multiplicatively by even operators. Second, all commutators $[D,\underline{a}]$, $a\in \mathbf{A}$, are bounded operators. Third, the rate of growth of the eigenvalues of $|D|=(D^*D)^{1/2}$ is such that the sum $\sigma_N(\ |D|^{-d})$ of the first N eigenvalues of $|D|^{-d}$ (arranged in decreasing order and supposed not to include 0)) is O(logN): more precisely, $|D|^{-1}$ belongs to the ideal $\mathcal{L}^{d+}$ of compact operators whose n^{th} eigenvalue is $O(n^{1/d})$ (see Section **[I.2]** for detailed definitions). Finally, the *phase* (=sign) $F=D|D|^{-1}$ of D is such that all $[F,\underline{a}]$, $a\in \mathbf{A}$, belong to the ideal $\mathcal{L}^{d+}$.

[I.2] ***SKETCH OF NON COMMUTATIVE DIFFERENTIAL GEOMETRY***. Our review of the algebraic features of usual differential geometry now gives us the necessary motivation for a natural introduction ((i) through (iv) below) of Alain Connes's "non commutative" substitutes of the items (i) through (iv) above.

(i): As announced at the beginning, our basic object is now a complex unital *-algebra A replacing the algebra $C^\infty(M)$ of classical differential geometry .[3]

---

[3] The philosophy that non-commutative C*-algebras are (duals of) "non- commutative spaces" arose after Gelfand's recognition that unital abelian C*-algebras are algebras of continuous functions on compact spaces, the space being recovered from the algebra as its "*spectrum* " - set of "*characters*"(a character is a homomorphism of the algebra onto the algebra of complex numbers). The study of non-commutative C*-algebras parallels that of spaces: the theory of their Hilbert space representations (resp covariant representations) is *"non-commutative measure theory"* (resp." *non-commutative ergodic*

(ii) THE DIFFERENTIAL ENVELOPE $\Omega(\mathbb{A})$ OF A COMPLEX ALGEBRA $\mathbb{A}$. AND ITS AUGMENTATION $\Omega(\tilde{\mathbb{A}})$. We want to construct a $\mathbb{Z}/2$-graded algebra $\Omega(\mathbb{A})$ generated by the $a\in\mathbb{A}$ and their "differentials" da , so as to have the map $a\to da$ the restriction to A of a differential of $\Omega(\mathbb{A})$ (=graded derivation with vanishing square): to this aim it is natural to construct $\Omega(\mathbb{A})$ via symbols

$$\text{(I,12)}\qquad \begin{cases} a\in\mathbb{A}. \\ da\ ,\ a\in\mathbb{A}. \end{cases}$$

and relations[4]

$$\text{(01,13)}\qquad \begin{cases} \lambda\bullet a \dot{+}\mu\bullet b \dot{-} (\lambda a+\mu b)=0 \\ a\bullet b \dot{-} (ab)=0 \\ \lambda\bullet da \dot{+}\mu\bullet db \dot{-} d(\lambda a+\mu b)=0 \\ da\bullet b+a\bullet db \dot{-} d(ab)=0 \end{cases} ,\quad \begin{cases} a,b\in A \\ \lambda,\mu\in\mathbb{C} \end{cases},$$

(the operations written with a • are the "formal" ones within the free algebra - ordinary notation refers to operations within A).
Clearly, by reordering any "word" with letters (I,12) by means of the last relation (I,13) (so as to have all symbols da standing to the right of the symbols a, the latter then conglomerate by the second relation (I,13)), we see that $\Omega(A)$ is linearly generated by symbols of the type

•

---

theory"). In the topological realm classical topological K-theory very naturally generalises as *C*-K-theory*. Further on this line is our present subject: *"non-commutative differential structures"*.

[4] In other terms $\Omega(A)$ is by definition the quotient of the free algebra over $\mathbb{C}$ generated by the a and da, $a\in A$, through the ideal generated by the expressions on the left side of the relations (01,2). Clearly, the first two relations (01,2) aim at having A a subalgebra of $\Omega(A)$; and the two last ones at making d a graded derivation.

(I,14) $$\begin{cases} a_o da_1 \dots\dots da_n \\ \qquad\qquad , \ a_o, a_1, \dots, a_n \in A, n \in N, \\ da_1 \dots. da_n \end{cases}$$

which, together with a unit $\tilde{1}$ added formally [5] (with the ensuing respective augmentations $\tilde{A}=C\tilde{1}\oplus A$ and $\tilde{\Omega}(A)=C\tilde{1}\oplus\Omega(A)$ of A and $\Omega(A)$), can be more economically written:

(I,15) $$a_o da_1 \dots da_n \ , \begin{cases} a_o \in \tilde{A}=C\tilde{1}\oplus A \\ \qquad\qquad , n \in N \ . \\ a_1, \dots, a_n \in A \end{cases}$$

We are thus led to the following constructive definition of $\tilde{\Omega}(A)$: the latter is built as the vector space

(I,16) $$\tilde{\Omega}(A) = \bigoplus_{n\in N} \tilde{\Omega}(A)^n \quad ,$$

where [6]

(I,17) $$\begin{cases} \tilde{\Omega}(A)^0 = \tilde{A} = C\tilde{1}\oplus A \\ \tilde{\Omega}(A)^n \cong \tilde{A}\otimes A^{\otimes n} = \{a_o da_1 \dots da_n ; \ a_o \in \tilde{A}, a_1, \dots a_n \in A\}, \ n \geq 1 \end{cases} ,$$

and is endowed with an associative bilinear product determined by the rule :

(I,18) $$(a_o da_1 \dots da_n) a_{n+1} = (-1)^n \ a_o a_1 \, da_2 \dots da_{n+1},$$

---

[5] even though A might already possess a unit e (the latter then becomes the generating idempotent of the ideal $0 \oplus A$ in the augmented algebra $\tilde{A}=C\,\tilde{1}\oplus A$).

[6] The free construction via (1,1), (1,2) makes it intuitive that we have a linear isomorphism : $a_o da_1 \dots da_n \leftrightarrow a_o \otimes a_1 \otimes \dots . a_n$, $a_o \in$ \O(A,$\tilde{}$), $a_1, \dots, a_n \in A$ (for proofs see [7] Chapter 1). Note that $\tilde{\Omega}(A)^\circ = \tilde{A}$, $\tilde{\Omega}(A)^n = \Omega(A)^n$.

$$+\sum_{j=1}^{n}(-1)^{n+j}\ a_o da_1 \dots d(a_j a_{j+1}) \dots da_{n+1}$$

$a_o \in \tilde{A}$, $a_1, \dots a_n, a_{n+1} \in A$.

It is intuitive that the above constructive definition of $\tilde{\Omega}(A)$ yields a N-graded complex algebra with N-grading[7]

(I,19) $\partial\omega = n$ for $\omega \in \tilde{\Omega}(A)^n$,

(formal proofs using recursion are easy to construct). It is also intuitive that the constructive definition yields back the augmentation of the algebra defined via the symbols and relations (I,12), (I,13) .

In addition to being a N-graded complex algebra, $\tilde{\Omega}(A)$ *possesses a differential* d *obtained as follows* :

(I,20) $$d\{(\lambda\tilde{1} + a_o)da_1 \dots da_n\} = da_o\, da_1 \dots da_n \quad ; a_o, a_1 \dots a_n \in A,$$

this definition implying the graded derivation property

(I,21) $$\begin{cases} d(\omega_1,\omega_2) = (d\omega_1)\omega_2 + (-1)^{\partial\omega_1}\omega_1 d\omega_2 \\ \omega_1,\omega_2 \in \tilde{\Omega}(A)\ ,\ \omega_1 \text{ of grade } \partial\omega_1 \end{cases},$$

and the fact that

(I,22) $$d^2 = 0.$$

The symbol first line of (I,14) now represents the product of a factor $a_o \in \tilde{A}$ times n factors da_k obtained by applying the differential d to $a_k \in \mathbb{A}$. Note that we then have

(I,23) $$\Omega(A) = A\Omega(A) \oplus d\Omega(A),$$

(with the first (resp. second) direct summand respectively generated by elements of the first (resp. second) line (01,14)).

For a *unital* algebra A (=having a unit 1 - not to be confused with the added unit $\tilde{1}$ above), there is another attribute, the **unital differential envelope** $(\Omega\mathbb{A},\delta)$ **of** $\mathbb{A}$, obtained as the quotient of $\Omega(\mathbb{A})$ by the ideal generated by d1, with δ the operator

---

[7] We endow $\tilde{A}$ with the natural grading $\tilde{A}^\circ = \mathbb{C}\tilde{1} \oplus A^\circ$, $\tilde{A}^1 = 0 \oplus A^1$.

on $\Omega\mathbb{A}$ resulting from d by passage to the quotient (one has $\delta 1=0$). In practical calculations can act as follows: work with $\Omega(\mathbb{A})$ (or rather $\tilde{\Omega}(A)$) instead of $\Omega\mathbb{A}$, and with d instead of δ, replacing d1 by 0 whenever it occurs (this will yield automatically the above-mentioned elements of $\underline{\Omega}\mathbb{A}^1$, replacing $\Sigma_i a_i db_i$ by $\Sigma_i a_i db_i - (\Sigma_i a_i b_i) d1$).

Let us mention that if **A** is a *-algebra (i.e.equipped with an antilinear involution * s.t. $(ab)^*=b^*a^*, a,b\in\mathbb{A}$), the algebra $\Omega(\mathbb{A})$ has a unique *-operation extending that of $\mathbb{A}$ and such that $(da)^*=da^*$, $a\in\mathbb{A}$ (hence turning $a_0 da_1 ... da_n$ into $da_n^* ... da_1^* \cdot a_0^*$), and yielding by restriction a *-operation of $\underline{\Omega}\mathbb{A}$ such that $(\underline{\delta}a)^*=da^*$, $a\in\mathbb{A}$.

Note, in comparison with the classical De Rham complex, that *the graded commutativity (of the wedge product) has to be abandoned in the non-commutative frame* (a fact whose realization was one of the essential points in Alain Connes's initial discovery!).

Before closing this paragraph, I should answer a question which naturally springs to mind. We just defined, for any (*-)algebra **A**, using the above formal construction, a differential algebra $\Omega(\mathbb{A})$ of "**quantum differential forms**". Returning to the classical case $A=C^\infty(M)$, we have the usual (classical) differential forms (elements of $\Omega(M)$), but now also the novel **quantum forms** (elements of $\Omega(\mathbb{A})=\Omega(C^\infty(M))$. What is the relationship between those two different types of objects? In fact, the quantum forms "sit over the classical forms" in the sense that the algebra $\Omega(M)$ is a quotient of the algebra $\Omega(\mathbb{A})$: we have an exact sequence of complex algebras:

$$0 \to K_{cl}(\mathbb{A}) \to \Omega(\mathbb{A}) \xrightarrow{cl} \Omega(M) \to 0 , \tag{I,23}$$

whose canonical map, the **classical projection** cl (with kernel $K_{cl}(\mathbb{A})$), is given by:

$$\begin{cases} cl(a_0 da_1 \ldots da_n)=a_0 da_1\wedge\ldots\wedge da_n \\ \qquad\qquad\qquad , a_0,a_1\ldots,a_n\in\mathbb{A} \\ cl\,(da_1\ldots da_n)=da_1\wedge\ldots\wedge da_n \end{cases} , \tag{I,24}$$

and enjoys the following reality property:

$$cl(\omega^*)=(-1)^{\frac{p(p-1)}{2}}\, cl(\omega) \qquad , \omega\in\Omega(A)^p. \tag{I,25}$$

(iii).FINITE PROJECTIVE MODULES, ABSTRACT CONNECTIONS AND CURVATURE. From what we saw in the previous paragraph (iii) above, it is clear that the non-commutative substitutes for (the sections of) vector bundles should be the **finite**

projective A-modules[8]. As already explained above, the latter are the A-modules E possessing a "coordinatization" - dual set of generators $\{e_i,\varepsilon^i\}_{i=1...n}$ fullfilling condition (I,10).The module E is **finite** (or **of finite type**) in the sense of being finitely generated: in other terms we have a module- homomorphism:

(I,26) $$\gamma: (\xi^i) \rightarrow e_i\, \xi^i$$ (summation over i)

from the free module A^n onto E; E is **projective** in the sense that γ has a *lift* λ: E→ A^n (such that $\gamma\lambda = id_E$ - as follows from (01,10)), namely:

(I,27) $$\lambda : \xi \rightarrow (\xi^i) \quad \text{with } \xi^i = \varepsilon^i \xi.$$

As a result, $p = \lambda\gamma$ is an idempotent endomorphism of the free A-module A^n, i.e. a projection $(p^i{}_k)$ of the algebra $M_n(A)$ of n × n matrices with entries in A (those matrices acting from the left obviously yield the endomorphisms of the right A-module A^n - in fact one has $p^i{}_k = \varepsilon^i e_k$). We see that our "coordinatization" yields the following description of the module E and of its endomorphisms (practical in applications!): one has:

(I,28) $$E = \{(\xi^i) \in A^n;\ p(\xi^i) = (\xi^i)\},$$

whilst :

(I,29) $$End_A(E) = \{(a_i{}^k) \in M_n(A)\ ;\ p(a_i{}^k)p = (a_i{}^k)\}.$$

A **hermitean structure of** E is an A-antilinear bijection *: E→E* s.t. $(\xi^*\eta)^* = \eta^*\xi, \xi,\eta \in E$, thus yielding an A-valued scalar product $<\xi,\eta> = \xi^*\eta$ s.t. $<\xi a,\eta b> = a^*<\xi,\eta>b, \xi, \eta \in E, a,b \in A$. The above "coordinatization" yields the hermitean structure

(I,30) $$<p(\xi^i),p(\eta^i)> = \Sigma_i\, \xi^{i*}\eta^i .$$

The generalized notion of **hermitean connection of** E is easily abstracted from **[I,1] (iii)** above: a **connection** is a ℂ-linear map: $E \rightarrow E \otimes_A \Omega(\mathbb{A})^1$ fullfilling :

(I,31) $$\nabla(\eta a) = (\nabla\eta)a + \eta \otimes da \qquad ,\eta \in E, a \in \mathbb{A},$$

---

[8] We recall that a *right (left)* *A-module* E is an additive group with a biadditive map: E A→E (A E→E) fullfilling $(\xi a)b = \xi(ab)$ *(* $a(b\xi = (ab)\xi)$*)* and $\xi 1 = \xi)$ *(* $1\xi = \xi$*)*, $\xi \in E$, a,b∈ A. If E is a right A-module the *dual* A-module E* (=set of A-valued A-linear forms on E) is a left A-module. In general finitely generated A-modules do not have bases as vector spaces (A not being a field!) ; but finite projective A-modules have "coordinatizations" rendering analogous services.

Introducing the bundle of E-valued exterior forms:

(I,32) $$E_\Omega = E \otimes_{\mathbb{A}} \Omega(\mathbb{A})$$

(with $\mathbb{N}$-grading determined by that of $\Omega(\mathbb{A})$),∇ uniquely extends to a *graded derivation of* E_Ω, i.e.a map: $E_\Omega \to E_\Omega$ fulfilling:

(0I,33) $$\nabla(Xa) = (\nabla X)a + X \otimes da \qquad ,X \in E_\Omega, a \in \mathbb{A}.$$

The **curvature** θ is then defined as :

(I,34) $$\theta = \nabla^2,$$

it is again an endomorphism of E_Ω:

(I,35) $$\theta(Xa) = (\theta X)a, \qquad ,X \in E_\Omega, a \in \mathbb{A}.$$

The **hermitean connexions** (or covariant derivatives) are singled out by the requirement:

(I,36) $$(\nabla\eta, \eta') + (\eta, \nabla\eta') = d(\eta, \eta') \qquad ,\eta, \eta' \in E$$

entailing for the exterior covariant derivative and for the curvature the properties:

(I,37) $$(\nabla X, X') + (X, \nabla X') = d(X, X') \qquad ,X, X' \in E_\Omega,$$

(I,38) $$(\theta X, X') = (X, \theta X') \qquad ,X, X' \in E_\Omega.$$

All the latter formulae are the mere repetition of those encountered above in the classical case.

(iv) K-CYCLES. The non-commutative subtitutes for the elliptic operators are again naturally abstracted from the classical considerations in **[I,1] (iv)** above: for **A** a unital *-algebra, a **K-cycle** (H,D,ε) **of A** (formerly called *unbounded Fredholm module*) is specified by:

— a **Z**/2-graded Hilbert space H (with grading involution ε) carrying a unital *-representation $\mathbb{A} \ni a \to \underline{a} \in B(H)$ of **A** (i.e. $\underline{\alpha a + \beta b} = \alpha\underline{a} + \beta\underline{b}$, $\underline{\alpha(ab)} = (\alpha\underline{a})(\alpha\underline{b})$, $\underline{a^*} = (\underline{a})^*$, $a, b \in \mathbb{A}$, and $\alpha 1 = 1$) by even operators:

(I,39) $$\underline{a}\varepsilon = \varepsilon\underline{a}, \qquad ,a \in \mathbb{A},$$

— an odd self-ajoint operator D=D* on H:

(I,40) $$D\varepsilon = -\varepsilon D,$$

such that $[D, \underline{a}] \in B(H)$ for all $a \in \mathbb{A}$, and D^{-1} (supposed to exist) is compact.

The K-cycle (H,D,ε) is called **d^+-summable** whenever D^{-1} belongs to the ideal $L^+(H)$ of compact operators whose characteristic values yield an at most logarithmically divergent series :

(I,41) $$L^+(H)=\{T\in B(H);\ T \text{ compact}, \sigma_N(T)=O(\mathrm{Log}N)\}$$

($\sigma_N(T)$ is the sum of the N first eigenvalues of $|T|=(T^*t)^{1/2}$ arranged in decreasing order). Positive elements of $L^+(H)$ are traceable for the **Dixmier trace** Tr_ω defined as:

(I,42) $$Tr_\omega(T)=\omega\text{-lim}\ \sigma_N(T)/(\mathrm{Log}N),$$

with ω-lim an appropriate limiting process which picks up the coefficient of LogN for $N\to\infty$: $Tr_\omega(T)$ thus vanishes if the operator T has a trace in the usual sense: unlike the usual trace *the Dixmier trace is hence not faithful* - it is a "renormalising" device!). We conclude this paragraph by describing an important "mutiple integral" attached to the d^+-summable K-cycle (H,D,ε). If we set, for $\omega=a_o da_1 \ldots da_n \in \Omega(A)$:

(I,43) $$\pi_D(\omega)=i^n\, \underline{a}_0[D,\underline{a}_1]\ldots\ldots[D,\underline{a}_n],$$

we get a representation π_D of the algebra $\Omega(A)$ (in fact a *-representation).*Defining:*

(I,44) $$\tau_D(\omega)=Tr_\omega\{\pi_D((\omega)D^{-n}\},$$

then yields a trace of $\Omega(A)$ (due to the fact that the commutator of D^{-n} with $\pi_D((\omega),\omega\in\Omega(A)$, vanishes under the Dixmier trace).

We conclude this lecture by displaying an object of central importance for the model-building scheme of the next lecture.

[I.3]. ***THE TRACE*** $\mathfrak{r}_D$ ***OF*** End E_Ω. We first notice the following easy fact:

Lemma. *Let A be an algebra , with E a projective finite right A--module: given a trace* τ *of A , setting:*

(I,45) $$\mathfrak{r}(\xi\phi)=\tau(\phi\xi) \qquad , \xi\in E,\ \phi\in E^*,$$

yields a trace $\mathfrak{r}$ *of the algebra* $\mathrm{End}_A E$. *In the particular case of* $E=pA^n$,*hence* $\mathrm{End}_A E=pM_n(A)p\in M_n\otimes A$, p *a projection in* $M_n(A)$, *we have:*

(I,45a) $$\mathfrak{r}=Tr_n\otimes\tau.$$

Note that (I,44) defines $\mathfrak{r}$ on all dyads: however the latter linearly generate $\mathrm{End}_A E$,due to the finite projective property of *E*. The fact that $\mathfrak{r}$ is a trace is immediate (it is indeed a mild generalisation of the well known form of the trace on rank-one linear operators - in which case one has $A=\mathbb{C}$, with τ the identity of $\mathbb{C}$).

Applying this lemma to the right $\Omega(A)$-module E_Ω (cf.(I,3)) and the trace τ_D of $\Omega(A)$ (cf.(I,44)), we see that *each d-summable K-cycle* D *of a *-algebra* A *yields a trace* τ_D *of* E_Ω *for each finite projective module* E *of* A.

II. GAUGE-FIELD MODEL-BUILDING VIA NON-COMMUTATIVE DIFFERENTIAL GEOMETRY.

In this lecture I shall attempt to give you an idea of how Alain Connes incorporates the Salam-Weinberg action into his non-commutative geometry: a challenging essay which I hope is germinal; this is the first step of an attemt at recasting quantum field theory in terms of non-commutative geometry. Quite naturally, before attempting to reform space, it is natural to apply the non-commutative apparatus to the classical case of a (euclidean, compact) manifold M[1] (with the algebra $A=C^\infty(M)$ and the Dirac operator as the K-cycle). As a matter of general philosophy, I want to stress that it seems quite reasonable to attempt a renewal of physics at the root of the electroweak model: indeed, the latter is the modern methamorphosis of the theory of photons *which has always led the development of physics* - since the time Maxwell discovered the photon wawe-equation[2] - think of the birth of relativity, the threefold inception of quantum physics (Planck, Heisenberg, De Broglie[3]), the inception of quantum field theory with Dirac's photon quantization, and the line Weyl -Yang-Mills!.

Since it is out of question to describe in one lecture Alain Connes's derivation of the Salam-Weinberg action within his "quantum Yang-Mills scheme" (the calculation is precludingly lengthy!), all I can do is: (i): explain the principles of Connes's general model-building scheme (based on the equipment sketched in the first lecture); (ii): describe as an illustration the derivation of usual Yang-Mills from "quantum Yang-Mills" for a (trivial) U(n)-bundle; (iii): fly off with a skinny sketch of how the Salam-Weinberg action is obtained by applying the general scheme to the "doubled" algebra $C^\infty(M)\oplus C^\infty(M)$, and a $U(1)\times U(2)$ module - with U(2) turned into SU(2) at a later stage.

[II.1] THE GENERAL MODEL-BUILDING SCHEME. The construction uses the following three ingredients:

_(i): a *basic algebra* $\mathbb{A}$ (derivate of $C^\infty(M)$, **M** the usual space-time in compact euclidean guise - compactness eliminates infra-red divergences, and the euclidean frame is required by the needed ellipticity of the Dirac operator)

_(ii): *a 4⁺-summable K-cycle* (H,D,χ) of $\mathbb{A}$ (derivate of the K-cycle of $C^\infty(M)$ given by the Dirac operator acting on $L^2(S_M)$, $S_M$the spin bundle of **M**): as explained in the first lecture, the K-cycle $((H,D,\chi)$ produces (via Dixmier trace) a trace τ_D of $\Omega(\mathbb{A})$.

_(iii): *a finite projective hermitean right module* E *of* $\mathbb{A}$ (describing the inner symmetries of the theory, which one thereby puts in in by hand).

From these data, one proceeds as follows: E provides "quantum potentials" (=compatible connections) represented by elements $\rho\in\Omega=\Omega(\mathbb{A})^1$ (in fact $\Omega\mathbb{A}^1$),

[1] euclidean is reqired to get ellipticity, compact to insure easy spectral properties).

[2] even before, with the futuristic Newton-Huyghens controversy about wawes versus particles!

[3] not to speak about Einstein's contributions (photoelectric effect and spontaneous emission of light).

with corresponding curvature $\theta \in \mathrm{End}_\Omega E_\Omega$. The quantum Yang-Mills action then arises by "*integrating the square of the curvature*" in the following sense: one sets:

(II,1) Quantum Yang-Mills action$=\underline{\tau}_D(\theta^*\theta)$

where $\underline{\tau}_D$ is the trace of $\mathrm{End}_\Omega E_\Omega$ defined at the end of the last lecture (cf.[II.3]); we recall that one obtains $\underline{\tau}_D$ as:

(II,2) $\underline{\tau}_D(\xi\phi)=\tau_D(\phi\xi)$, $\quad$, $\xi \in E$, $\phi \in E^*$,

(concretely for $E=pA^n$:

(II,2a) $\underline{\tau}_D = \mathrm{Tr}_n \otimes \tau$),

from the trace τ_D of $\Omega(A)$ given by:

(II,3) $\tau_D(\omega)=\mathrm{Tr}_\omega\{\tau_D(\omega)D^{-4}\}$, $\quad$ $\omega=a_o da_1 \dots da_n \in \Omega(A)$,

where

(II,4) $\tau_D(a_o da_1 \dots da_n)= (-i)^n \underline{a}_0[D,\underline{a}_1]\dots[D,\underline{a}_n]$, $a_o,a_1 \dots ,a_n \in A$.

[II.2] THE CASE OF U(n) YANG-MILLS GAUGE FIELDS. In the usual (euclidean compact) U(n)-Yang-Mills case, the above items are the following:

__(i): $A=C^\infty(M)$, M a compact, orientable, 4-dimensional, spin^c, riemannian manifold (e.g. S^4).

__(ii): $(H,D,\chi)=(L^2(S_M),D,\chi)$, D the Dirac operator acting on L^2 of the the spin bundle S_M of M.

__(iii): $E=A^n$ (for n=1 one gets pure electrodynamics case with the gauge group U(1) - in that case $E=A$..

We recall that the Dirac operator D gives rise to the 4^+-summable K-cycle (H,D,χ) of A , yielding the above trace τ_D of $\Omega(A)$, with the following analitically important peculiarity: one has

(II,5) $\pi_D(a_o da_1 \dots da_n)= (-i)^n \underline{a}_0[D,\underline{a}_1]\dots[D,\underline{a}_n]$, $a_o,a_1 \dots ,a_n \in A$,

$=$ multiplication by $a_o\gamma^\mu\partial_\mu a_1 \dots \gamma^\mu\partial_\mu a_n$, and *the trace* τ_D *is expressible as* [1]

(II,6) $$\tau_D(\omega)=1/2\pi^2\int tr\{a_o\gamma(\mathbf{d}a_1)\ldots\gamma(\mathbf{d}a_n)\}\,dv,$$

where tr stands for the normalized trace of the Clifford algebra[2], dv for the volume element of M, and we have $\gamma(\mathbf{d}a)=\gamma^\mu\partial_\mu a$, $a\in A$, denoting with a boldface **d** the usual exterior derivative (in order to distinguish it from the differential d of $\Omega(A)$).

The above prescription tells us that the non-commutative Yang-Mills action is $\{Tr_n\otimes\tau_D\}(\theta^*\theta)$. Here the curvature θ is that pertaining to the A-module $E=A^n$ in non-commutative geometry (cf.(I,34)): θ is thus a functional of "quantum potentials" ρ (compatible connections ρ of $E=A^n$), with values in $End_{\Omega(A)}(E_\Omega)=M_n(\Omega(A)=M_n\otimes_C\Omega(A)$, M_n the set of complex n×n matrices - observe that $E_\Omega=E\otimes_A\Omega(A)=\Omega(A)^n$. And$\tau_D=Tr_n\otimes\tau$ (cf. (II,2a)).

Quantum potentials and curvature. We obtain a parametrization of the quantum potentials by noting that the differential d of $\Omega(A)$, applied coordinatewise

(II,7) $$d\eta=(d\eta^i) \quad ,(\eta^i)\in A^n,$$

is a compatible connection of $E=A^n$: indeed d has grade one and obviously fulfills conditions (I,33), (I,36); therefore an arbitrary connection ∇ of E differs from d by a module-endomorphism. Hence ∇ (resp. its canonical extension to $E_\Omega=\Omega(A)^n$ is of the type

(I2,8) $$\nabla X=dX+\rho X \quad ,X\in\Omega(A)^n,$$

for some $\rho=(\rho^i{}_k)\in M_n(\Omega(A)^1)$, which,for a compatible$\nabla$, is such that

(II,9) $$\rho^i{}_k=-\rho^k{}_i{}^* \quad ,i,k=1,\ldots n,$$

Iterating ∇ then yields:

---

[1] This is because of the identity of the Dixmier trace and the ***Wodzicki residue*** for pseudodifferential operators of order - dimM.

[2] denoted ν in [IIIB.4].

(II,10) $\nabla^2 X = d(dX+\rho X) + \rho(dX+\rho X) = d^2 X + (d\rho)X - \rho dX + \rho dX + \rho^2 X \quad , X \in \Omega(A)^n,$

$= (d\rho+\rho^2)X,$

whence, for the curvature:

(II,11) $$\theta = d\rho + \rho^2,$$

i.e., with $\rho = (\rho^i{}_k)$, $\rho^i{}_k = \Sigma_{r=1,\dots,p} a^i{}_{kr} db^i{}_{kr} \in \Omega(A)^1$ (in fact $\in \underline{\Omega}A^1$),i,k=1,...n:

(II,11a) $$\theta^i{}_k = d\rho,i,k=1,\dots n + \rho^i{}_l \rho^l{}_k.$$

Note that the hermiticity of θ (cf.(I,36)

(IV2,12) $$\theta^i{}_k{}^* = \theta^k{}_i \qquad , i,k=1,\dots n \qquad (\Leftrightarrow \theta^* = \theta),$$

proceeds as follows from (II,9) and the postulated commutation $(da)^* = da^*$:[1]

(II,13) $$\theta^i{}_k{}^* = (d\rho^i{}_k + \rho^i{}_l \rho^l{}_k)^* = -d(\rho^i{}_k{}^*) + \rho^l{}_i{}^* \rho^i{}_l{}^* = d\rho^k{}_i + \rho^k{}_l \rho^l{}_i = \theta^k{}_i.$$

Computation of the Yang-Mills action. One has:

(II,14) $$\{Tr_n \otimes \tau_D\}(\theta^*\theta) = \Sigma_{i=1,\dots,n}\, \tau_D\{(\theta^*\theta)^i{}_i\} = \Sigma_{i,k=1,\dots,n} \tau_D\{\theta^{*i}{}_k \theta^k{}_i\}$$

$$= {}^k{}_i \tau_D\{\theta^k{}_i{}^* \theta^k{}_i\}$$

$$= 1/2\pi^{2k}{}_i \int tr\{\pi_D(\theta^k{}_i{}^* \pi_D(\theta^k{}_i)\}\, dv.$$

$$= 1/2\pi^2\, \Sigma_{i,k=1,\dots,n} \int (\pi_D(\theta^k{}_i), \pi_D(\theta^k{}_i))_{Cl_C}\, dv,$$

---

[1] This is the prescription for extending the *-operation of A to a *-operation of Ω(A).

where dv stands for the riemannian volume element of **M**, and $(\, , \,)_{Cl_C}$ for the scalar product of the complexified Clifford algebra $\mathbb{C}l_C(\mathbb{R}^4)$ (we recall that $\mathbb{C}l_C(\mathbb{R}^4)$ *is the orthogonal direct sum of subsets* $\mathbb{C}l_C{}^k$, *images under* γ *of homogeneous antisymmetric* k-*tensors*, k=1,2,3,4, (this property greatly eases our computation). Using the results:

$$(II,15)\qquad \begin{cases} \pi_D(\delta\rho)=1/2\gamma(\mathbf{d}\rho_{cl})+X(\rho)1 \\ 1/2\pi_D(\rho\rho'+\rho'\rho)=(\rho_{cl},\rho'_{cl})1 \\ \pi_D(\rho\rho'-\rho'\rho)=\gamma(\rho_{cl}\otimes\rho'_{cl}-\rho'_{cl}\otimes\rho_{cl}) \end{cases},$$

where **d** denotes an exterior derivative, and we use the notation:

$$(II,16)\qquad X(\rho)=\Sigma_{r=1,\dots,p}(da_r,db_r) \qquad , \rho=\Sigma_{r=1,\dots,p}\, a_r db_r\in\Omega(\mathbb{A}),\ a_r,b_r\in\mathbb{A}, r=1,\dots,p.$$

we then have from (II,13): [1]

$$(II,17)\qquad \pi_D(\theta^i{}_k)=1/2\gamma\{\mathbf{d}(\rho^i{}_k)_{cl}\}+X(\rho^i{}_k)1+1/2((\rho^i{}_l)_{cl},(\rho^l{}_i)_{cl})1$$

$$+1/2\gamma((\rho^i{}_l)_{cl}\otimes(\rho^l{}_i)_{cl}-(\rho^l{}_i)_{cl}\otimes(\rho^i{}_l)_{cl}$$

$$=i/2\ \gamma\{\mathbf{d}A^i{}_k+i/2\ A^i{}_l\wedge A^l{}_i\}+\{X(\rho^i{}_k)-(A^i{}_l,A^l{}_i)\}1,$$

$$=i/2\gamma\{F^i{}_k\}\ +\{X(\rho)^i{}_k-(A^\mu A_\mu)^i{}_k\}1\ ,$$

where we used the notation :

$$(II,18)\qquad A^a{}_b=-i\ (\rho^i{}_k)_{cl}=(A_\mu)^i{}_k dx^\mu\ ,\ a,b=1,\dots,n,$$

($(\rho^i{}_k)_{cl}$ denoting the classical projection of $\rho^i{}_k$ (cf.(I,24)), with

---

[1] with the map γ extended to tensors,so that $\gamma(u)\ \gamma(v)=\gamma(u\otimes v)$.We define $u\wedge v = u\otimes v - v\otimes u$, and denote (.,.) the riemannian scalar product of $T_M{}^*$.

(II,19) $F^i{}_k=(F_\mu)^{ai}{}_k dx^\mu = dA^i{}_k + i/2 A^i{}_l \wedge A^l{}_i$, a,b=1,....,n,

We interpose a comment on these objects: $A=(A^i{}_k)$,resp. $F=(F^i{}_k)$ are a classical one-, resp, two-form on **M** with values in the Lie algebra $\{(m^i{}_k)_{i,k=1,\dots,n}; m^i{}_k \in \mathbb{C}\}$ of the classical gauge group U(n); A is the classical "potential" with curvature $F=\nabla_{cl}A=(F^i{}_k)$, with ∇_{cl} the classical covariant exterior derivative. Note that, owing to (I,24) (II,16), one has

(II,20) $\overline{(\rho^i{}_k)_{cl}} = -\overline{(\rho^k{}_i{}^*)_{cl}} = -(\rho^k{}_i)_{cl}$, i,k=1,...,n,

and

(II,21) $\overline{X(\rho^i{}_k)} = -\overline{X(\rho^k{}_i{}^*)} = X(\rho^k{}_i)$, , i,k=1,...,n,

therefore $F_{\mu\nu}$ and $X(\rho)$ are hermitean matrices with entries in $C^\infty(M)$:

(II,22) $\overline{A_\mu{}^i{}_k} = A_\mu{}^k{}_i$, $\overline{F_{\mu\nu}{}^a{}_b} = F_{\mu\nu}{}^b{}_a$,i,k=1,...,n, ,μ,ν= 1,2,3,4,

(II,23) $\overline{X(\rho)^i{}_k} = X(\rho)^k{}_i$.

We now resume our calculation. The last line (II,15) is an orthogonal sum of an element of $\mathbb{C}l_C{}^2$ and an element of $\mathbb{C}l_C{}^0 = \mathbb{C}\,1$; using the fact

(II,24) $tr\{\gamma(u\otimes v - v\otimes u)^* \gamma(u'\otimes v' - v'\otimes u')\} = 2(u\otimes v - v\otimes u, u'\otimes v' - v'\otimes u')$

we thus have the Clifford scalar product

(II,25) $(\pi_D(\theta^k{}_i), \pi_D(\theta^k{}_i))_{Cl_C} = 1/2\, \overline{F_{\mu\nu}{}^k{}_i} F_{\mu\nu}{}^k{}_i$

$$+\overline{(X(\rho)-A^\mu A_\mu)^k{}_i}(X(\rho)-A^\mu A_\mu)^k{}_i$$

$$=1/2 F^{\mu\nu i}{}_k F_{\mu\nu}{}^k{}_i + (X(\rho)-A^\mu A_\mu)^i{}_k)(X(\rho)-A^\mu A_\mu)^k{}_i$$

$$=1/2F^{\mu\nu i}{}_i+(X(\rho)-A^{\mu}A_{\mu})^i{}_i,$$

(we write $X(\rho^i{}_k)=X(\rho)^i{}_k$) whence

(II,26) $$\{Tr_n\otimes\tau_D\}(\theta^*\theta)=1/2\pi^2\Sigma_{i,k=1,\dots,n}\int(\pi_D(\theta^k{}_i),\pi_D(\theta^k{}_i))_{Cl_C}\,dv,$$

$$=1/2\pi^2\int Tr\{1/2\ F^{\mu\nu}F_{\mu\nu}+(A^{\mu}A_{\mu}+X(\rho))\}\ dv\ ,$$

where Tr_n denotes the usual trace of n×n matrices.

One thus recovers the familiar Yang-mills action plus a second term involving the "field" $X(\rho)$: however this term can be ignored as unphysical, since it vanishes whilst minimizing the action (one can also show that the quantum potential ρ can be modified, without altering its classical projection A, so as to yield $X(\rho)=A^{\mu}A_{\mu}$, this also showing the spurious nature of the last term of (II,24)).

[II.3] SKETCH OF THE SALAM-WEINBERG CASE. The Salam-Weinberg model is obtained by "doubling the space" in a way which we now describe. For this we interpose a description of:

Quantum Yang-Mills of the two-point space: Consider the two-point space $S_2=\{P,P'\}$: the algebra of functions on S_2 is $A_2=\mathbb{C}^2=\{(f,f'):f,f'\in\mathbb{C}\}$. Consider the K-cycle (H_2,D_2,χ_2) of A_2 specified as follows (the"Dirac K-cycle of the two-point space"):

(II,27) $$\begin{cases} H_2=\mathbb{C}^N\oplus\mathbb{C}^N \text{ (as a left } A_2\text{-module)} \\ \chi_2=\begin{pmatrix}1 & 0\\ 0 & -1\end{pmatrix} \\ D_2=\begin{pmatrix}0 & M^*\\ M & 0\end{pmatrix}\end{cases},$$

where M is a N× N matrix. And consider the right A_2-module $E=A_2$ given by A_2 itself. Applying the general scheme **[I.1]** to this simple two-point case case, one easily sees that the connections are of the type $\rho=\lambda e\delta e+\bar{\lambda}(1-e)\delta e,\lambda\in\mathbb{C}$, e the projection in A_2 onto the first component ; and that the Yang-Mills action then equals $(1-|\phi|^2)Tr_N(M^*M)^2$: one recognizes in this expression the characteristic form of a Higgs potential! This suggest that the Higgs boson is connected with a doubling of the space, and makes natural the following:

Recipe for building the Salam-Weinberg model via non-commutative differential geometry. Apply the the general scheme **[I.1]** to the following building blocks:

__(i): the algebra $A=C^\infty(M)\oplus C^\infty(M)=C^\infty(M)\otimes C^2$, **M** a compact, orientable, 4-dimensional, spin^c, riemannian manifold ,e.g. S^4 ("doubling the space":).
__(ii): the K-cycle of A obtained by tensoring the Dirac K-cycle $(H,D,\chi=\gamma^5)$ of $C^\infty(\mathbf{M})$ (cf.[**II.2**]) by the K-cycle (H_2,D_2,χ_2) of the two-point space (here one uses the natural notion of tensor product of K-cycles).
__(iii): the "inner symmetry module" E by which the U(1)×U(2) inner symmetry is put in by hand, namely $E=pA^2$, where:

$$\text{(II,28)} \qquad e=\begin{pmatrix}(1,1) & (0,0)\\ (0,0) & (0,1)\end{pmatrix}\in \mathrm{End}_A E=M_2(A),$$

(One shows that the corresponding structural group is U(1)×U(2), restriction to U(1)×SU(2)-symmetry is subsequently forced by an extra assumption). Sure enough, this leads to the Salam-Weinberg model.

BIBLIOGRAPHY

[1] A.Connes. Essay on physics and non-commutative geometry. *The interface of mathematics and particle physics*. Clarendon press, Oxford (1990).
[2] A.Connes and J.Lott. Particle models and non-commutative geometry. *Preprint*. IHES (1989).
[3] A.Connes. Géometrie non commutative. *Intereditions*. Paris (1990).
[4] A.Connes. The action functional in non-commutative geometry. *Comm.Math. Phys.* 117 673 (1988).
[5] A.Connes. Non commutative differential geometry, I. The Chern character in K-homology, II. De Rham homology and non-commutative algebra, Publ. Math. IHES N°62 (1985) 257-360.
[6] D.Kastler. Cyclic cohomology within the differential envelope. An introduction to Alain Connes's non commutative differential geometry. Travaux en cours, Ed. Scient. Herman, Paris (1988)
[7] D.Kastler and M.Mebkhout. Lectures on non commutative differential geometry. In preparation.

2. Quantum groups and integrable systems

MEASURING COALGEBRAS, QUANTUM GROUP-LIKE OBJECTS, AND NON-COMMUTATIVE GEOMETRY

Marjorie Batchelor
Department of Pure Mathematics & Mathematical Statistics
Cambridge University, 16 Mill Lane, Cambridge CB2 1SB

Classical geometry is commutative because of the way in which multiplication of functions has been defined. The pointwise multiplication of two (real valued) functions of a variable x,

$$fg(x) = f(x)g(x) = g(x)f(x)$$

commutes in part because the range is the commutative ring $\mathbf{R}$, and in part because x is assumed to have a group-like comultiplication

$$x \longrightarrow x \otimes x$$

which is evidently cocommutative. The idea of this paper is that geometry may be generalized to a non-commutative setting by considering generalized 'point sets' which include elements whose comultiplication is neither group-like nor cocommutative.

Such generalized spaces exist; they arise naturally as a 'space of smooth maps' between two ordinary manifolds, and they provide us with Hopf algebras with properties reminiscent of quantum groups. The purpose of this paper is to introduce the concept of universal measuring coalgebra, and to show that quantum group-like objects arise naturally as examples. The paper is organized as follows:

Section 1. Measuring coalgebras; definitions and examples.

Section 2. Universal measuring coalgebras $P(B_1, B_2)$; existence and general properties.

Section 3. Manifolds and jet bundles; the interpretation of $P(B_1, B_2)$ as a 'non-commutative space'.

Section 4. Quantum group-like objects and universal measuring coalgebras.

Section 5. Conclusions.

Section 1. Measuring coalgebras

Given two algebras B_1, B_2 over an algebra A, a measuring coalgebra C is a generalized set of morphisms from B_1 to B_2, in that elements in C can be considered as linear maps from B_1 to B_2 which, while they need not be algebra homomorphisms, do have a valid product rule given by comultiplication in C.

2.1 *Definition.* A pair (C, ψ) is a measuring coalgebra for (B_1, B_2) if C is an A-coalgebra with comulticlication Δ, counit ε and

$$\psi : C \longrightarrow \mathrm{Hom}_A(B_1, B_2)$$

is an A-linear map which satisfies the following properties:

(i) $$\psi(c)(bb') = \sum_{(c)} \psi(c_{(1)})(b)\psi(c_{(2)})(b')$$

where the comultiplication is written $\Delta c = \sum_{(c)} c_{(1)} \otimes c_{(2)}$ (see Sweedler [7]).

(ii) $$\psi(c)(1_{B_1} = \varepsilon(c)1_{B_2}.$$

Notice that (i) is precisely the statement that the coproduct in c gives a valid pjroduct rule in ψC.

The definition is stated in general algebraic terms, but all the examples in this paper have geometric interpretation. Throughout B_1, B_2 will be thought of as function rings, over $\mathbf{R}$ or $\mathbf{C}$ in the classical case, and over a suitable subalgebra in the 'quantum' case. The following four examples of measuring coalgebras demonstrate their geometric interpretation.

2.2 *Example 1.* (Classical.) Let C be a free A module generated by the single generator K. Give C the coalgebra structure given by letting K be group-like. That means

$$\Delta K = K \otimes K, \qquad \varepsilon K = 1.$$

Then a linear map $\psi : C \to \mathrm{Hom}(B_1, B_2)$ measures if and only if ψK is an algebra homomorphism. More generally, if (C, ψ) is any measuring coalgebra, group-like elements c in C are those for which ψc is an algebra homomorphism.

Now suppose $B_i = C^\infty(M_i)$, and recall that smooth maps $\alpha : M_2 \to M_1$ correspond to algebra homomorphisms $\psi_\alpha : C^\infty(M_1) \to C^\infty(M_2)$. Then $(\mathbf{T}^0(M_2, M_1), \psi)$ defined by

$$\begin{aligned}
&\mathbf{T}^0(M_2, M_1) = \bigoplus_{\alpha: M_2 \to M_1} \mathbf{R}\alpha \\
&\Delta\alpha = \alpha \otimes \alpha, \qquad \varepsilon\alpha = 1, \\
&\psi : \mathbf{T}^0(M_2, M_1) \longrightarrow \mathrm{Hom}_{\mathbf{R}}(C^\infty(M_1), C^\infty(M_2)) \\
&\alpha \mapsto \psi_\alpha
\end{aligned}$$

is a measuring coalgebra. This coalgebra recovers the 'point set' for which the comultiplication is group-like.

2.3 *Example 2.* (Classical.) Let C be a free A module on generators K, e, with a coalgebra structure given by

$$\begin{aligned}
&\Delta K = K \otimes K, \qquad \varepsilon K = 1 \\
&\Delta e = e \otimes K + K \otimes e, \qquad \varepsilon e = 0.
\end{aligned}$$

Suppose $\psi : C \to \mathrm{Hom}(B_1, B_2)$ is an A linear map such that ψK is an algebra homomorphism and ψe is a derivation with respect to ψK. that means

† $$\psi e(bb') = \psi e(b)\psi K(b') + \psi K(b)\psi e(b').$$

Then (C, ψ) is a measuring coalgebra. Hypothesis (i) is satisfied by K by Example 1, and † is precisely the statement that e also satisfies hypothesis (i). Again (ii) is easily verified.

Now let $B_i = C^\infty(M_i)$ as before and consider smooth maps $r : M_2 \to TM_1$ which project under $\pi_1 : TM_1 \to M_1$ to a map $\alpha : M_2 \to M_1$

Figure 1.

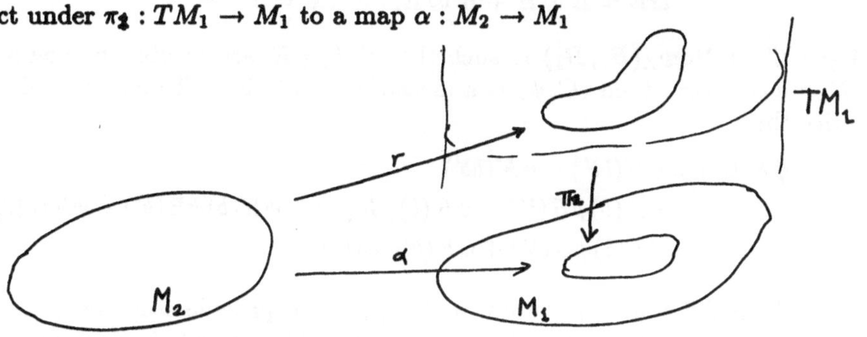

For f in $C^\infty(M_1)$ define $\psi r(f)$ in $C^\infty(M_1)$ via

$$\psi r(f)(m) = r(m)f.$$

The effect of ψr on a product fg in $C^\infty(M_1)$ can be computed:

§
$$\begin{aligned}\psi r(fg)(m) &= r(m)fg \\ &= r(m)fg(\alpha(m)) + f(\alpha(m))r(m)g.\end{aligned}$$

Notice that, for a given α, the set of all smooth maps r covering α form an $\mathbf{R}$-vector space.

Now form a coalgebra $\mathbf{T}^1_\alpha(M_2, M_1)$

$$\begin{aligned}\mathbf{T}^1_\alpha(M_2, M_1) &= \mathbf{R}\alpha \oplus \{r : M_2 \longrightarrow TM_1 : \pi r = \alpha\} \\ \Delta\alpha &= \alpha \otimes \alpha, \qquad \varepsilon\alpha = 1 \\ \Delta r &= r \otimes \alpha + \alpha \otimes r, \qquad \varepsilon r = 0.\end{aligned}$$

Define

$$\psi : \mathbf{T}^1_\alpha(M_2, M_1) \longrightarrow \mathrm{Hom}_{\mathbf{R}}(C^\infty(M_1), C^\infty(M_2))$$
$$\alpha \longrightarrow \psi_\alpha, \qquad r \longrightarrow \psi r.$$

It is not hard to verify that ψ measures, using §. Let

$$\mathbf{T}^1(M_2, M_1) = \bigoplus_{\alpha : M_2 \to M_1} \mathbf{T}^1_\alpha(M_2, M_1).$$

This coalgebra can be given the structure of a measuring coalgebra. It generalizes the point set to include tangents, and provides an example of a 'manifold' whose 'points' are still cocommutative, but are no longer group-like.

1.4 *Example 3.* (Classical, but silly.) Let C be freely generated as an A module by the elements H, K and E, with comultiplication

$$\Delta H = H \otimes H, \qquad \Delta K = K \otimes K, \qquad \varepsilon H = 1 = \varepsilon K$$
$$\Delta E = E \otimes H + K \otimes E, \qquad \varepsilon E = 0.$$

If $\psi : C \to \mathrm{Hom}_A(B_1, B_2)$ is such that ψH, ψK are algebra homomorphisms, and $\psi E = \psi H - \psi K$, then (C, ψ) is a measuring coalgebra. To check condition (i) for E notice that

$$\begin{aligned}\psi E(bb') &= \psi H(bb') - \psi K(bb') \\ &= \psi H(b)\psi H(b') - \psi K(b)\psi K(b') - \psi K(b)\psi H(b') + \psi K(b)\psi H(b') \\ &= \psi E(b)\psi H(b') + \psi K(b)\psi E(b').\end{aligned}$$

Thus difference operators would appear to provide a simple example of a non-cocommutative measuring coalgebra. However this example is not entirely satisfactory since ψ has a kernel, generated by $E + K - H$, and this kernel can be shown to be a coideal. In some sense, the non-commutative aspect of this measuring coalgebra is fictitious, since the quotient coalgebra is cocommutative.

The following example gives an example of a non-cocommutative measuring coalgebra.

1.5 *Example 4.* (Neither classical nor silly.) Let $B_1 = B_2 = \mathbb{C}[q, z] = B$ and let $A = \mathbb{C}[q]$. Let C be the coalgebra of Example 3, generated by H, K and E. Define

$$\begin{aligned}\psi &: C \longrightarrow \mathrm{Hom}_A(B, B) \\ \psi H &= 1 \\ \psi K z^m &= q^m z^m \\ \psi E z^m &= \frac{\psi K z^m - z^m}{(q-1)z} \\ &= [m] z^{m-1}\end{aligned}$$

where

$$[m] = \frac{q^m - 1}{q - 1}.$$

The same calculations as in Example 3 show that (C, ψ) is a measuring coalgebra.

Here ψ is injective since if, for some a, b, c in $\theta[q]$,

$$\psi(aH + bK + cE)z^m = 0$$

for all m, then

$$az^m + bq^m z^m + c[m]z^{m-1} = 0.$$

But this would imply that

$$\begin{aligned}a + bq^m &= 0 \\ c[m] &= 0\end{aligned}$$

for all m, hence ψ is injective.

This provides an archetypal example of a non-cocommutative measuring coalgebra, and all the examples in this paper are of this type. The universal coalgebras described in Section 2 are simply 'maximal' measuring coalgebras, including Examples 1, 2 and 4. From exaples 1 and 2, the geometric nature of this construction begins to be apparent: the whole picture is described briefly in Section 3. If, in this example, q is set equal to 1, the map ψE becomes the ordinary derivative $\frac{d}{dz}$. It is in this sense that the elements like E are deformations of derivations; the algebra generated by such elements are deformations of enveloping algebras. This is discussed in Section 4.

Section 2. Universal measuring coalgebras

Given algebras B_1, B_2 over A, the universal measuring coalgebra is, in the following sense, the maximal measuring coalgebra for B_1, B_2.

2.1 *Definition.* Given algebras B_1, B_2 over A, a universal measuring coalgebra for B_1, B_2 is a measuring coalgebra $(P(B_1B_2), \pi)$ such that for any measuring coalgebra (C, ψ) there exists a unique coalgebra map $j : C \to P(B_1, B_2)$ such that the following diagram commutes

$$(A)\qquad \begin{array}{ccc} P(B_1, B_2) & \xrightarrow{\pi} & \mathrm{Hom}_A(B_1, B_2) \\ j\uparrow & \nearrow_{\psi} & \\ C & & \end{array}$$

2.2 **Theorem**: Universal measuring coalgebras exist and are unique.

Proof. The theorem rests on two facts.

(1) Direct sums of measuring coalgebras are measuring coalgebras.

(2) The category of measuring coalgebras has colimits. This means that if (C_1, ψ_1), (C_2, ψ_2) are measuring coalgebras, and if

$$f, g : C_1 \longrightarrow C_2$$

are such that the following diagram commutes,

$$\begin{array}{ccc} C_1 & \searrow^{\psi_1} & \\ f \text{ or } g \downarrow & & \mathrm{Hom}_A(B_1, B_2), \\ C_2 & \nearrow_{\psi_2} & \end{array}$$

then there is a measuring coalgebra (D, φ) and a map of measuring coalgebras

$$p : C_2 \longrightarrow D$$

such that the following diagram commutes:

$$* \qquad C_1 \underset{g}{\overset{f}{\rightrightarrows}} C_2 \xrightarrow{p} D.$$

Moreover if $q : (C_2\psi_2) \to (D', \varphi')$ is any other map of measuring coalgebras such that $*$ holds (replacing (D, φ) by (D', φ')) then there is a unique map k such that

$$\begin{array}{ccccc} C_1 & \overset{f}{\underset{g}{\rightrightarrows}} & C_2 & \overset{p}{\longrightarrow} & D \\ & & & \searrow^{q} & \downarrow k \\ & & & & D' \end{array}$$

commutes.

Fact 1 is not hard to verify; check that in all cases the required maps can be defined using the defining property of direct sums. For the second deserve that the A module

$$J = \{f(c) - g(c) : c\varepsilon C_1\} \subset C_2$$

is a coideal,and check that C_2/J is itself a measuring coalgebra. Now let $\{(C_\lambda, \psi_\lambda)\}_\Lambda$ be the set of finitely generated (over A) measuring coalgebras containing one representative from each isomorphism class. Form the direct sum

$$(C_2, \psi_2) = \oplus_\Lambda (C_\lambda, \psi_\lambda).$$

Let $\{g_\rho\}_R$ be the set of all maps of measuring coalgebras

$$g_\rho : C_{\rho_1} \longrightarrow C_{\rho_2}$$

between coalgebras in $\{(C_\lambda, \psi_\lambda)\}_\Lambda$. Form

$$(C_1, \psi_1) = \oplus_{\rho \in R}(C_{\rho_1}, \psi_{\rho_1}).$$

Now define two maps, $f, g : C_1 \to C_2$. Define

$$f : C_1 \longrightarrow C_2$$

by the inclusion of C_{ρ_1} in $\oplus_\Lambda C_\lambda$. Define

$$g : C_1 \longrightarrow C_2$$

to be the map determined by the g_ρ composed with the inclusion of C_{ρ_2} in C_2.

The claim is that $P(B_1, B_2)$ is the coequalizer (D, φ) of f and g. Check that (D, φ) has the desired property. Let (C, ψ) be a finitely generated measuring coalgebra and show that there is a unique map

$$j : C \to D$$

making the diagram A commute. Since C is finitely generated, (C, ψ) is isomorphic to some $(C_{\rho_1}, \psi_{\rho_1})$, via an isomorphism

$$t : C \to C_{\rho(1)}.$$

Composing t with the projection $p : C_2 \to D$ provides a map $j : C \to D$. Since j is a map of measuring coalgebras diagram A must commute.

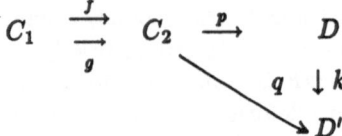

Now suppose $j' : C \to D$ is another map satisfying diagram A. The image of C under j' is a finitely generated measuring coalgebra isomorphic to some $(C_{\rho_2}, \psi_{\rho_2})$. Thus j' provides a map of measuring coalgebras

$$g_\rho : C_{\rho_1} \to C_{\rho_2}.$$

But this means that for c in C $p(g_\rho tc - tc) = 0$ or

$$j'c = pg_\rho tc = ptc = jc$$

as desired.

Now write an arbitrary measuring coalgebra (C, ψ) as the sum of its finitely generated subcoalgebras

$$(C, \psi) = \Sigma(C_\alpha, \psi_\alpha).$$

For each subcoalgebra (C_α, ψ_α) there is a unique map $j_\alpha : C_\alpha \to D$ satisfying A. Since j_α is unique, it must be that $j_\alpha = j_\beta$ on the $C_\alpha \cap C_\beta$. Thus the j_α can be patched together to give a map

$$j : C \to D.$$

Uniqueness again follows by considering the effect of a contending map j' on finitely generated subcoalgebras.

This construction has a number of pleasant functional properties, but one in particular will be needed to construct the desired Hopf algebras in Section 4.

Corollary. Let $B_1 = B_2 = B$. Then $P(B, B)$ is a bialgebra.

Proof. Composition of linear maps provides the map μ in the bottom row of the following diagram. The map $\bar{\pi}$ is defined to be $\mu \circ \pi \otimes \pi$.

$$\begin{array}{ccc}
P(B,B) \otimes P(B,B) & \overset{m}{- - - >} & P(B,B) \\
\pi \otimes \pi \downarrow & \searrow \bar{\pi} & \downarrow \pi \\
\mathrm{Hom}_A(B,B) \otimes_A (B,B) & \underset{\mu}{\longrightarrow} & \mathrm{Hom}_A(B,B)
\end{array}$$

The map m will exist (and be a unique coalgebra map) once $(P(B,B) \otimes P(B,B), \bar{\pi})$ is shown to be a measuring coalgebra. Check this: for p, q in $P(B,B)$, b, b' in B

$$\begin{aligned}
\pi \otimes \pi(p \otimes q)(bb') &= \pi p(\pi q(bb')) \\
&= \pi p \left(\sum_{(q)} \pi q_{(1)}(b) \pi q_{(2)}(b') \right) \\
&= \sum \pi \otimes \pi(p_{(1)} \otimes q_{(1)})(b) \pi \otimes \pi(p_{(2)} \otimes q_{(2)})(b')
\end{aligned}$$

as desired. Property 1 of measuring coalgebras is again easy.

Section 3. Manifolds and jet bundles

Measuring coalgebras generalize the notion of the dual coalgebra B° of an algebra B (definition 3.1 below). In the case $B = C^\infty(M)$, $C^\infty(M)^\circ$ is closely related to the dual of the jet bundle of M. In working with graded manifolds, the dual coalgebra is used to provide a working candidate for a jet bundle [5]. In [2], the cocommutative part of $B(C^\infty(M_1), C^\infty(M_2))$ is used to supply jet-bundle information. My suggestion is that those wishing for a model for non-commutative geometries might try using $B(C^\infty(M_1), C^\infty(M_2))$, without discarding its non-cocommutative parts. This section summarizes the geometric interpretation of $B(C^\infty(M_2), C^\infty(M_2))$. For more details, see [2].

The dual coalgebra is defined as a distinguished subspace of the full linear dual.

3.1 *Definition.* Let A be an algebra over a field F, and define

$$A^\circ = \{\alpha : A \to F : \text{Ker}\alpha \supset \text{I} \\ \text{I is an ideal} \\ A/\text{I is finite dimensional.}\}$$

This construction has many attractive properties, the principal being that if A, B are two algebras, then $(A \otimes B)^\circ = A^\circ \otimes B^\circ$. Functoriality also holds, and thus multiplication in A induces a comultiplication in A°. Its geometric significance is the following.

3.2 Let M be a smooth manifold. Then

$$C^\infty(M)^\circ = \bigoplus_{m \in M} \mathbf{T}_m,$$

$$\mathbf{T}_m = \bigcup_k \mathbf{T}_m^k, \qquad \text{with} \quad \mathbf{T}_m^k \subset \mathbf{T}_m^{k+1}$$

and $\mathbf{T}_m^k$ is dual to the fibre of the kth jet bundle.

Proofs. See [2]. That $C^\infty(M)^\circ$ decomposes (as a coalgebra) into a sum of pointed irreducible subcoalgebras $\mathbf{T}_m$ is a consequence of the fact that $C^\infty(M)^\circ$ is cocommutative. The filtration of $\mathbf{T}_m$ by $\mathbf{T}_m^k$ is the coradical filtration of $\mathbf{T}_m$ (see [7]). If

$$J_m = \{f \in C^\infty(M) : f(m) = 0\}$$

it is easy to verify that an element c in $\mathbf{T}_m^k$ will vanish on J_m^{k+1}. This provides the identification $\mathbf{T}_m^k$ with the dual to the kth jet bundle.

The following theorem describes the sense in which $P(B_1, B_2)$ is a generalization of the dual coalgebra.

3.3 Theorem. For A any algebra over $\mathbf{R}$, $P(A, \mathbf{R}) \cong A^\circ$.

Proof. Consider the map

$$\pi : P(A, \mathbf{R}) \to \text{Hom}_{\mathbf{R}}(A, \mathbf{R}).$$

The claim is that the image of π is contained in A°. For c in $P(A, \mathbf{R})$ let C be the (finite dimensional) subcoalgebra generated by c. Then π induces a map

$$\bar{\pi} : A \to \mathrm{Hom}(C, \mathbf{R}) = C'.$$

Comultiplication in C provides an algebra structure for C', the measuring property states that π is an algebra homomorphism. Thus πc vanishes on $\ker\bar{\pi}$, and πc is in A° as desired.

Now verify that A° itself has the desired universal property of universal measuring coalgebras, hence $A^\circ = P(A, \mathbf{R})$.

Consider $P(C^\infty(M_1), C^\infty(M_2))$. If M_2 is a point, the above theorem says that $P(C^\infty(M_1), C^\infty(M_2)) = C^\infty(M_1)^\circ$. Regarding M_1 as the space of smooth maps from M_2 (a point) into M_1, $P(C^\infty(M_1), C^\infty(M_2))$ recovers the jet-bundle information of the space of smooth maps.

This leads to the hypothesis that $P(C^\infty(M_1), C^\infty(M_2))$ may be used to provide jet-bundle information about the space of smooth maps between two arbitrary smooth manifolds, M_2 and M_1. Since all the statements in 3.2 are coalgebraic, a similar theorem holds for $P(C^\infty(M_1), C^\infty(M_2))$, or at least its cocommutative part.

3.4 **Theorem.** Let $P_c(C^\infty(M_1), C^\infty(M_2))$ denote the sum of all cocommutative subcoalgebras of $P(C^\infty(M_1), C^\infty(M_2))$. Then

$$P_c(C^\infty(M_1), C^\infty(M_2)) = \bigoplus_{\substack{\alpha : M_2 \to M_1 \\ \alpha \text{ smooth}}} \mathbf{T}_\alpha$$

$$\mathbf{T}_\alpha = \bigcup_k^\alpha \mathbf{T}_\alpha^k, \quad \mathbf{T}_\alpha^k \subset \mathbf{T}_\alpha^{k+1}.$$

For a proof and further discussion, see [2]. Notice that from 2.2

$$\mathbf{T}^0(M_2, M_1) = \bigoplus_{\alpha : M_2 \to M_1} \mathbf{T}_\alpha^0$$

and from Example 2.3

$$\mathbf{T}_\alpha^1(M_2, M_1) = \mathbf{T}_\alpha^1.$$

For arbitrary M_2, M_1, $P(C^\infty(M_1), C^\infty(M_2))$ will not be cocommutative (Example 4 of Section 1 can be modified to produce an example of a non-cocommutative measuring coalgebra which includes in $P(C^\infty(M_1), C^\infty(M_2))$. As $C^\infty(M)^\circ$ generalizes the point set M, so may $P(C^\infty(M_1), C^\infty(M_2))$ be interpreted as the generalized space of smooth maps from M_2 to M_1. The application of this technique to quantum group-like objects demonstrates the payoff in accepting this viewpoint.

Section 4. Quantum group-like objects and measuring coalgebras

The purpose of this section is to obtain examples of 'deformations' of universal enveloping algebras in the following sense. So far, the measuring coalgebra $P(B,B)$ pjrovides a bi-algebra with a map $\pi : P(B,B) \rightarrow \mathrm{Hom}(B,B)$ such that the coproduct in $P(B,B)$ defines a good product rule. Sub-coalgebras of $P(B,B)$ generate sub-bialgebras at $P(B,B)$. In this section sub-bialgebras generated by derivations (as in 2.3) and deformed derivations (as in 2.5) are considered. The first construction recovers the universal enveloping algebra; the second recovers Hopf algebras with behaviour similar to quantum groups. The construction of the latter reduces to the former for $q = 1$.

4.1 **Proposition.** Let (C, ψ) be a measuring coalgebra for (B, B). Then jC generates a sub-bialgebra at $P(B,B)$.

Proof. Certainly jC is a subcoalgebra of $P(B,B)$, and will generate a subalgebra Q of $P(B,B)$. Since, for p, p' in jC, $\Delta pp' = \Delta p \Delta p'$, $\Delta pp'$ is in $Q \otimes Q$. Thus Q itself is a subcoalgebra, and hence a bialgebra.

4.2 **Proposition.** Suppose D is a linear space (after $\mathbb{C}$) of derivations (with respect to the identity) in B and give $\mathbb{C}K \oplus D$ the structure of a coalgebra by setting K to be group-like and elements of D to be primitive with respect to K. Then if $\psi C \rightarrow \mathrm{Hom}(B,B)$ sends K to the identity and includes D in $\mathrm{Hom}(B,B)$, (C, ψ) is a measuring coalgebra. Moreover, (C, ψ) includes in $P(B,B)$ and generates the universal enveloping algebra of the Lie algebra of derivations generated by D.

Proof. Since ψ is injective, so is j. Let Q be the subalgebra generated by C, and let L be the subspace generated by commutators of elements in C. The aim is to show that Q is isomorphic to $U(L)$.

Since Q is an associative algebra containing L as a Lie algebra there is a unique algebra homomorphism

$$t : U(L) \longrightarrow Q.$$

Clearly t is surjective (since it maps onto L, hence onto C, which generates Q). Notice that t is a coalgebra map as well. By a theorem about pointed irreducible cocommutative coalgebras ([7], prop.11.02), t will be injective if and only if t restricted to L is injective.

To show that t is injective, show that ψt is injective. Let J be the kernel of ψt restricted to L. Considered as a subspace at Q, J generates an ideal $\langle\langle J \rangle\rangle$ of $P(B,B)$. Check that $\langle\langle J \rangle\rangle$ is also a coideal, and finally that $\psi \langle\langle J \rangle\rangle = 0$. Thus ψ factors through the quotient

$$\begin{array}{ccc} P(B,B) & \longrightarrow & \mathrm{Hom}(B,B) \\ & \searrow \quad \nearrow & \\ & P(B,B)/\langle\langle J \rangle\rangle & \end{array}$$

Thus by uniqueness, $\langle\langle J \rangle\rangle = 0$, and t itself must be injective as well.

Thus the universal measuring coalgebra provides an alternate construction of the universal enveloping algebra, exploiting its coalgebraic properties rather than its algebraic properties. The trade-off is this: instead of specifying the complete Lie algebra and its bracket, one need only specify generators of the Lie algebra, and a representation of those generators as derivations.

The virtue of this plan with respect to quantum groups is that any set of generators, not just derivations, can be used provided these generators are represented as operators on an algebra with a good product rule. In particular, operators with product rule as in 1.5 generate bialgebras which resemble quantum groups.

4.3 Example 1. Let $B = \mathbf{C}[q, g^{-1}, x_1]$ and let $A = \mathbf{C}[q, q^{-1}]$. Consider the coalgebra C generated by $\{E, K, 1, K^{-1}, F\}$:

$$\begin{aligned} \Delta E &= E \otimes 1 + K \otimes F & \varepsilon E &= 0 \\ \Delta F &= F \otimes K^{-1} + 1 \otimes F & \varepsilon F &= 0 \\ \Delta K^j &= K^j \otimes K^j & \varepsilon K^j &= 1, j \varepsilon \{1, -1\}. \end{aligned}$$

Then C is a coalgebra over A. Define $C :\to \mathrm{Hom}_A(B, B)$ by setting

$$\begin{aligned} &\psi 1 = 1, \quad \psi K^j x^i = q^{ji} x^i, j \in \{1, -1\} \\ &\psi E x^i = [i] x^{i-1}, \quad \psi F x^i = [-i] x^{i+1}. \end{aligned}$$

Direct computation shows that (C, ψ) is a measuring coalgebra. Moreover C includes in $P(B, B)$ generating a subcoalgebra Q. By its construction, multiplication in $P(B, B)$ is such that the measuring map into $\mathrm{Hom}_A(B, B)$ is an algebra homomorphism. Thus its kernel is an ideal which contains the elements

$$qKE - EK, \quad q^{-1}KF - FK,$$

$$[E, F] + \left(\frac{K - K^{-1}}{q - 1} \right).$$

These elements generate a coideal, thus by the argument used to show injectivity in 4.2 these elements define relations in Q. Compare these relations with the usual ones for quantum $s\ell(21)$.

4.4 Example 2. Let $B = \mathbf{C}[q, q^{-1}, x, x^{-1}]$. A quantum version of the charge-zero Virasoro algebra can be represented by setting C to be the free $A = \mathbf{C}[q, q^{-1}]$ module generated by $\{1, K, K^{-1}, L_k,\ k \neq 0\}$. Comultiplication is given, as usual, by

$$\begin{aligned} &\Delta 1 = 1 \otimes 1, \quad \Delta K^i = K^i \otimes K^i \quad \varepsilon(1) = \varepsilon(K^i) = 1 \\ &\Delta L_k = L_k \otimes 1 + K \otimes L_k \quad \varepsilon L_k = 0. \end{aligned}$$

The measuring map ψ is defined by setting

$$\begin{aligned} &\psi 1 = 1, \quad \psi K^j x^i \quad q^{ij} x^i, \\ &\psi L_k x^i = [i] x^{i+k}. \end{aligned}$$

As in the previous example, C includes in $P(B, B)$, generates a bialgebra, which can be shown to satisfy the relations

$$KL_k = q^k L_k K, \quad [L_k L_j] = ([j] - [k])L_{j+k}.$$

4.5 **Example 3.** Let G be a Lie group and let $\mathcal{F}$ be the group of smooth maps of S' into G. Let B be $C^\infty(S' \times G)$, and let A be $\mathbf{C}[q, q^{-1}]$ with the action on B given by

$$pf(w, g) = p(w)f(w, g)$$

for p in A, f in $C^\infty(S' \times G)$ where S' is considered as a subset of $\mathbf{C}$.

Given any map σ in $\mathcal{F}$, the map

$$E_\sigma : C^\infty(S' \times G) \to C^\infty(S' \times G)$$

$$E_\sigma f(w, g) = \frac{f(w, \sigma(w)^{-1}g) - f(w, g)}{(w - 1)}$$

turns out to be well defined. Moreover, the set $\{1, \sigma, E_\sigma\}$ spans a coalgebra C_σ given by

$$\begin{aligned} &\Delta 1 = 1 \otimes 1, \quad \Delta\sigma = \sigma \otimes \sigma, \Delta E_\sigma = E_\sigma \otimes 1 + \sigma \otimes E_\sigma \\ &\varepsilon(1) = 1 = \varepsilon(\sigma), \quad \varepsilon E_\sigma = 0, \end{aligned}$$

and the map $\psi : C_\sigma \to \mathrm{Hom}_A(B, B)$ given by

$$\psi 1 = 1 \quad \psi\sigma(f)(w, g) = f(w, \sigma(w)^{-1}g), \quad \psi E_\sigma = E_\sigma$$

measures.

Now observe that the group algebra of $\mathcal{F}$, $A\mathcal{F}$ is a measuring coalgebra for (B, B). Thus

$$\begin{array}{ccc} A\mathcal{F} & \longrightarrow & \mathrm{Hom}_A(B, B) \\ & \nearrow & \\ C_\sigma & & \end{array}$$

Notice that C_σ does not include in $A\mathcal{F}$; however, the sub-bialgebra generated by $\{1, \sigma, (w - 1)E_\sigma\}$ does map into $A\mathcal{F}$. In some sense, the extension of $A\mathcal{F}$ to the sub-bialgebra of $P(B, B)$ generated by elements E_σ has the effect of localizing at $(w - 1)$.

Quantum group-like objects are thus easy to construct, and occur amongst other places, wherever groups and maps from S' into groups are found. Finally the connection between such a quantum group-like object and its classical relations can be described as follows.

Let $A = \mathbf{C}[q, q^{-1}]$. Let J be the kernel of the map $A \to \mathbf{C}$ determined by setting $q = 1$. Notice that if

$$\bar{B} = B/JB$$

there is a map

$$R : \mathrm{Hom}_A(B, B) \longrightarrow \mathrm{Hom}_{\mathbf{C}}(\bar{B}, \bar{B}).$$

Also, if C is an A coalgebra,

$$\bar{C} = C/JC$$

is a coalgebra over $\mathbf{C}$. Finally a measuring map

$$\psi : C \longrightarrow \mathrm{Hom}_A(B, B)$$

gives rise to a measuring map

$$\bar{\psi} : \bar{C} \longrightarrow \mathrm{Hom}_{\mathbf{C}}(\bar{B}, \bar{B}).$$

In some cases the quantum group-like objects obtained as measuring coalgebras are deformations of enveloping algebras in the usual sense but in other cases they do not easily present themselves as such. Using the map R above one can, however, say when a quantum group-like object reduces to an enveloping algebra.

4.6 **Definition.** Let (C, ψ) be a measuring coalgebra (over $A = \mathbf{C}[q, q^{-1}]$) for (B, B), and let Q be the bialgebra generated by C in $P(B, B)$. Say Q reduces to $U(L)$ if $\bar{C}$ is of the form $\bar{C} = \mathbf{C}K \oplus D$ (as in 4.2) and L is the Lie algebra of derivations of $\bar{B}$ generated by D.

In this sense, Q of 4.3 reduces to $U(s\ell(2))$, and Q of 4.4 reduces to the universal enveloping algebra of the charge-0 Virasoro algebra. From 4.5 if C is defined by

$$C = \bigoplus_{\substack{\sigma \\ \sigma(1)=1_G}} C_\sigma,$$

then $\bar{C} = \mathbf{C}1 \oplus g$, where g is the Lie algebra of G. In this last example the bialgebra generated by C reduces to $U(g)$, but is not a deformation of $U(g)$ in the usual sense.

Section 5. Conclusions

In closing there are three points I would like to emphasize.

1. Coalgebraic language offers an attractive alternative to the treatment of non-commutative geometry through purely algebraic (function-ring theoretic) means. Coupled with the theory of the function rings, I believe it can be a very effective tool. This paper demonstrates that non-commutative geometry arises 'naturally' when maps between (commutative) spaces are considered, and that quantum groups can be considered in some cases as examples.

2. The universal measuring coalgebra invites interpretation as an extension or generalization of the universal enveloping algebra to other operator algebras. In this setting, for example, difference operators and differential operators appear on the same footing. Moreover, the construction offers other possibilities than just quantum group-like objects. for example, let $\pi : P \rightarrow B$ be a principal G bundle. The map π provides an algebra homomorphism

$$\pi : C^\infty(B) \longrightarrow C^\infty(P).$$

It would be interesting to consider the structure of $P(C^\infty(P), C^\infty(P))$ with $C^\infty(P)$ considered as an $A = C^\infty(B)$ module.

3. The correspondence between quantum group-like algebras found in measuring coalgebras and the classical quantum groups is not yet complete; I have yet to find the right representation (and presentation) for some groups. However, the variety and generality of this construct is appealing, as is the flexibility of working with universal, presentation-independent constructions.

References

[1] Batchelor, M. Difference operators, measuring coalgebras and quantum group-like structures. To appear in *Adv. in Math.*

[2] Batchelor, M. In search of the graded manifold of maps between graded manifolds. In *Complex Differential Geometry and Supermanifolds in Strings and Fields*, P.J.M. Bongaarts and R. Martini (eds), Springer Lecture Notes in Physics 311 (1988).

[3] Drinfeld, V.G. Quantum groups. In *Proc. ICM, Berkeley*, Amer. Math. Soc., 1988.

[4] Exton, H. *q-Hypergeometric Functions and Applications.* Ellis Horwood, 1983.

[5] Kostant, B. *Graded Manifolds, Graded Lie Groups and Prequantisation.* Springer Lecture Notes in Maths. 570 (1975).

[6] Lusztig. Quantum deformations of certain simple modules over enveloping algebras. *Adv. in Math.*

[7] Sweedler, *Hopf Algebras*, Benjamin, 1969.

Tensor Operator Structures in Quantum Unitary Groups*,**

L. C. BIEDENHARN
Department of Physics, Duke University
Durham, NC 27706 U.S.A.

Abstract

Tensor operators acting on model spaces for the quantum group $SU_q(n)$ are defined ("q-tensor operators") and the fundamental theorem for q-tensor operators (a generalization to non-commutative co-products of the Wigner-Eckart theorem) is proved. Examples from $SU_q(2)$ are discussed.

The symmetry structure characterizing quantum groups[1,2,3] has in the past few years been found to play an important rôle in many distinct fields of physics and mathematics, from knot theory[4] to rational conformal field theory[5,6]. A quantum group is, more precisely, an algebra, a deformation[3] of the universal enveloping algebra of an underlying classical Lie group (here a unitary group).

The great importance of symmetry techniques in quantum physics is associated with a much larger algebraic structure, the algebra of tensor operators (which includes the universal enveloping algebra as a sub-algebra). The problem we wish to pose, and resolve, is the extension to this larger algebraic structure of the symmetry associated with quantum groups.

Contrary to folk wisdom, the Hilbert space structure of quantum physics does not, *a priori*, guarantee the existence of a tensor operator structure (in particular, a

* Supported in part by the Department of Energy and the National Science Foundation.

** Invited paper presented at the XIXth International Conference on Differential Geometric Methods in Theoretical Physics, Rapallo (Genova, Italy) 19-24 June 1990.

version of the Wigner-Eckart theorem) for any given symmetry structure, but depends rather upon the specific way in which the symmetry is realized[7]. In the prototypical symmetry structure in quantum physics—the quantal rotation group $SU(2)$—there are two required properties of the realization[7].

Equivariance: The action of the group generators on a tensor operator (a set of operators) realizes a linear representation defined by transformations on this set.

Derivation: The generators act on products as a derivation: $J_i(ab) = J_i(a)b + aJ_i(b)$.

It is consequence of these two properties that the Wigner-Clebsch-Gordan (WCG) coefficients for $SU(2)$ occur in two logically distinct ways:

(a) as coupling coefficients for the addition of angular momentum carried by kinematically independent constituent systems, and,

(b) as matrix elements (up to a rotationally invariant scale factor) of physical transition operators.

Conversely, if these two properties do not obtain, then this latter result fails[8].

It is not obvious that one can extend these results to quantum groups, particularly when one realizes that the derivative property corresponds to a *commutative co-product*, which is invalid for a general quantum group. Moreover, the equivariance condition is problematic as well. For a Lie group, equivariance is effected by the adjoint action: $ad_x(y) = [x, y]$. This cannot work for a quantum group, since the quantum group irrep corresponding to the adjoint representation is finite dimensional in contrast to the infinite number of linearly independent generators obtained under commutation.

To sort out the problems that occur it is helpful to examine the standard realization of $SU(2)$ and see how this works. In this realization (the Jordan-Schwinger mapping[7]) the generators are:

$$J_+ = a_1\bar{a}_2, \quad J_- = a_2\bar{a}_1, \quad J_z = \tfrac{1}{2}(a_1\bar{a}_1 - a_2\bar{a}_2), \tag{1a}$$

with

$$[\bar{a}_i, a_j] = \delta_{ij}, \tag{1b}$$

and all other commutators vanishing. The set of vectors $\{|jm\rangle\}$, (for $-j \le m \le j$; $j = 0, \frac{1}{2}, \ldots$), carrying all (unitary) irreps in this realization is given by:

$$|jm\rangle = ((j+m)!(j-m)!)^{-1/2} a_1^{j+m} a_2^{j-m} |0\rangle. \tag{2}$$

An elementary spin-$\frac{1}{2}$ tensor operator $O_m^{(\frac{1}{2})}$ can be realized as the pair of operators:

$$O^{(\frac{1}{2})}_{m=\frac{1}{2}} = a_1; \quad O^{(\frac{1}{2})}_{m=-\frac{1}{2}} = a_2. \tag{3}$$

Consider the action of J_+ on the product $a_i|jm\rangle$. We find

$$J_+\left(a_i|jm\rangle\right) \equiv [J_+, a_i]\,|jm\rangle + a_i J_+|jm\rangle. \tag{4}$$

If we identify the action $J_+(a_i)$ to be commutation, as in Eq. (4) above, both the derivation property and the equivariance property are seen to be verified. The fundamental theorem accordingly assures us that matrix elements of $O_m^{(\frac{1}{2})}$ are (to within an invariant normalizing factor) fundamental WCG coefficients.

The characteristic feature of a quantum group, as mentioned already, is the existence of a non-commutative co-product; for example, the co-product for the quantum group $SU_q(2)$ is defined by [9]:

$$\Delta(J_\pm) \equiv q^{-\frac{J_z}{2}} \otimes J_\pm + J_\pm \otimes q^{+\frac{J_z}{2}}, \tag{5a}$$

$$\Delta(J_z) \equiv \mathbf{1} \otimes J_z + J_z \otimes \mathbf{1}. \tag{5b}$$

Let us now re-analyze the standard case using the concept of a co-product. Examining eq. (4) critically shows that there are ***two distinct mappings involved***, which to distinguish properly requires a more precise notation. First we formalize the Hilbert space $\{|jm\rangle\}$ to be a *model space*[10,11], **M**, defined to be the direct sum of vectors carrying (unitary) irreps of the group G, each equivalence class of irreps occurring *once and only once.* The operators on $\{|jm\rangle\}$ are defined to belong to the linear space **T** with the action:

$$\mathbf{T}: \quad \mathbf{M} \to \mathbf{M}.$$

We introduce now the key concept of an *induced action*; this is a mapping:

$$\mathbf{T} \otimes \mathbf{T} \longrightarrow \mathbf{T},$$

which we will denote by $A(B) = C$, that is, A acts on B to give C, with $A, B, C \in \mathbf{T}$. (One must carefully distinguish this action from the natural product of operators on Hilbert space, which we denote as usual by juxtaposition: AB.)

Using these concepts, we can now recognize that eq. (4) consists first of the ***induced co-product*** (Δ):

$$\Delta(J_+)(a_i \otimes |m\rangle) = J_+(a_i) \otimes \mathbf{1}(|m\rangle) + \mathbf{1}(a_i) \otimes J_+(|m\rangle), \tag{6}$$

followed by a mapping c:

$$c: \quad \mathbf{T} \otimes \mathbf{M} \longrightarrow \mathbf{M}, \qquad c: \quad \Delta(\mathbf{T}) \longrightarrow \mathbf{T}. \tag{7a, b}$$

Applying this operation to eq. (6) yields:

$$J_+(a_i|m\rangle) = J_+(a_i)|m\rangle + a_i J_+|m\rangle, \tag{8}$$

which is eq. (4) with the *induced action* $J_+(a_i) \equiv [J_+, a_i]$.

The novelty in this construction is that the co-product $\Delta(\mathbf{T})$ acts on the direct product $\mathbf{T} \otimes \mathbf{M}$ of *different* spaces and, in addition, there is a further operation (the mapping c) which requires that the induced co-product and the induced action be compatible. The compatibility of these two structures is guaranteed for the standard (Lie group) case since the standard matrix action of operators in Hilbert space induces (by commutation) a commutative co-product (a derivation) and the induced commutator action of the generators realizes, by equivariance, an irrep carried by the tensor operators (this compatibility is the content of the tensor operator theorem (generalized Wigner-Eckart theorem)).

The compatibility requirement can be expressed most succintly by using the language of diagrams. Consider the following diagram (where E is a generator):

$$\begin{array}{ccc} \mathbf{T} \otimes \mathbf{M} & \xrightarrow{\Delta(E)} & \mathbf{T} \otimes \mathbf{M} \\ c\big\downarrow & & c\big\downarrow \\ \mathbf{M} & \xrightarrow{E} & \mathbf{M}\,. \end{array} \tag{9}$$

The requirement of compatibility is that the diagram above be commutative. Assuming a given co-product determines a compatible action (or vice-versa). Thus, for example, for a Lie group one uses the diagonal (commutative) co-product and determines the commutator as the compatible induced action.

It should now be clear how to extend this structure to tensor operators for quantum groups (q-tensor operators). Let us formalize these considerations:

DEFINITION: *Let* $\mathbf{T}$ *denote the vector space of operators mapping the model space* $\mathbf{M}$ *of the quantum group* G_q *into itself:* $\mathbf{M} \xrightarrow{\mathbf{T}} \mathbf{M}$. *An irreducible q-tensor operator is a set of operators,* $\{\mathbf{t}_{\Xi,\xi}\} \in \mathbf{T}$ *which carries a finite-dimensional irrep* Ξ, *with vectors* ξ, *of the quantum group* G_q. *That is:*

$$E_\alpha(\mathbf{t}_{\Xi,\xi}) = \sum_{\xi'} \langle \Xi, \xi' | E_\alpha | \Xi, \xi \rangle \mathbf{t}_{\Xi,\xi'}, \tag{10}$$

where E_α is a generator of G_q, $E_\alpha(\mathbf{t}_{\Xi,\xi})$ denotes an action of E_α on **T**, *and* $\langle \cdots \rangle$ *denotes the matrices of the generators for the irrep Ξ. A q-tensor operator is accordingly a linear combination of irreducible q-tensor operators, with coefficients invariant under the q-group action.*

THEOREM. *If $\{\mathbf{t}_{\Xi,\xi}\}$ is a q-tensor operator of the quantum group G_q such that the co-product of G_q is compatible with the action $E_\alpha(\mathbf{t}_{\Xi,\xi})$, that is, diagram (9) is commutative, then the matrix elements of $\{\mathbf{t}_{\Xi,\xi}\}$ in* **M** *are proportional to the q-WCG coefficients of G_q with the constant of proportionality an invariant.*

Conversely, if $\{\mathbf{t}_{\Xi,\xi}\}$ is a q-tensor operator and the matrix elements of $\{\mathbf{t}_{\Xi,\xi}\}$ are proportional to the q-WCG coefficients, then diagram (9) is commutative.

Proof: The proof is given in detail in ref. (12) to which we refer, for brevity.

An intuitive idea of the essential elements in the proof may be obtained in this way. The q-WCG coefficients are the matrix elements of an *invertible* map (denoted W) from **M**⊗**M** into **M**. Expressing this diagrammatically we have the following *commutative* diagram:

$$\begin{array}{ccc} \mathbf{M}\otimes\mathbf{M} & \xrightarrow{W} & \mathbf{M} \\ \Delta(E_i)\big\downarrow & & \big\downarrow E_i \\ \mathbf{M}\otimes\mathbf{M} & \xleftarrow{W^{-1}} & \mathbf{M}. \end{array} \tag{11}$$

Now, because of the way the action has been defined on the q-tensor operator $\{\mathbf{t}_{\Xi,\xi}\}$, for $\mathbf{t}_{a,\alpha}$ acting on the vector $|{}^{b}_{\beta}\rangle$—an element of $\mathbf{T}\otimes\mathbf{M}$—we have a well-defined compatible co-product action $\Delta \circ E_i$, with $\mathbf{T}\otimes\mathbf{M}$ replacing $\mathbf{M}\otimes\mathbf{M}$ in diagram (11) above. It follows from diagram (9) that the mapping c can be identified (to within an invariant factor) with the mapping W, that is, with the q-WCG coefficients. The content of the proof is the detailed working out of these steps in terms of a basis for the operators and state vectors.

COROLLARIES:

(a) *There exists an algebra of q-tensor operators:*

$$\mathbf{T}\otimes\mathbf{T} \xrightarrow{W} \mathbf{T},$$

carrying products of irreducible q-tensor operators into irreducible q-tensor operators.

(b) *In particular, Cor.(a) implies that there exists a product carrying an irreducible q-tensor operator into an invariant having the properties of an inner product. Thus a norm exists and irreducible unit q-tensor operators* ($\hat{\mathbf{T}}$) *are well-defined whose matrix elements, by the fundamental tensor operator theorem, are q-WCG coefficients.*

(c) *Denote the mapping in Cor.(a) by* $\mathbf{T} \circledW \mathbf{T} \in \mathbf{T}$, *and denote the invariant product in Cor. (b) by* $\mathbf{T} \cdot \mathbf{T}$. *Then the (6-j) operators are defined by:* $\hat{\mathbf{T}} \cdot (\hat{\mathbf{T}} \circledW \hat{\mathbf{T}})$. *Similarly (3n-j) operators can be defined.*

Remarks:

(a) As mentioned earlier, the generators of a quantum group are not irreducible q-tensor operators. In the examples below we give explicitly the q-tensor operators carrying the adjoint representation for $SU_q(2)$.

(b) From the definition of a compatible induced action, we see that this action by generators on the adjoint q-tensor operator irrep carries this irrep into itself. In particular, for this adjoint q-tensor operator irrep there is no need to use the tri-linear q-analog of the Serre relations for the induced action.

(c) Since the action of operators on Hilbert space vectors is modelled by matrix action and hence fixed once and for all—it is rather surprising that compatible actions for both commutative and non-commutative co-products can be realized in this same overall structure.

As an example of these concepts let us consider the q-boson realization[13,16] of $SU_q(2)$ and construct explicitly the fundamental (two-dimensional) q-tensor operators. We may realize the quantum group $SU_q(2)$ by the q-boson analog to the Jordan-Schwinger map defined on two *mutually commuting* q-bosons a_1^q and a_2^q:

$$J_+^q = a_1^q \bar{a}_2^q, \quad J_-^q = (J_+^q)^+ = a_2^q \bar{a}_1^q, \quad \text{and} \quad J_z^q = \tfrac{1}{2}(N_1^q - N_2^q). \tag{12}$$

Here we have defined two q-boson *creation operators*, a_i^q, and their Hermitian conjugates, the q-boson *destruction operators* $\bar{a}_i^q$ satisfying the relation:

$$\bar{a}_i^q a_i^q - q^{\mp\frac{1}{2}} a_i^q \bar{a}_i^q = q^{-\frac{1}{2}(\mp N_i^q)}, \quad (\text{with} - \text{for } i = 1, + \text{ for } i = 2), \tag{13}$$

where N_i^q is the (Hermitian) *number operator*, defined to satisfy:

$$[N_i^q, a_j^q] = \delta_{ij} a_i^q, \qquad [N_i^q, \bar{a}_j^q] = -\delta_{ij} \bar{a}_i^q, \text{ with } N_i^q|0\rangle \equiv 0. \tag{14}$$

Defining the q-boson *vacuum ket* by $a_i^q|0\rangle \equiv 0$, we get the n-quanta eigenstates $\{|n\rangle_q\}$ given by:

$$|n\rangle_q = ([n]_q!)^{-1/2}(a^q)^n|0\rangle, \tag{15}$$

where $[n]_q \equiv q^{\frac{n-1}{2}} + q^{\frac{n-3}{2}} + \ldots + q^{-(\frac{n-1}{2})}$ with $[n]_q!$ the corresponding q-factorial.

It follows from eqs. (13-15), that the generators, eq. (12), realize the commutation relations of $SU_q(2)$:

$$[J_z^q, J_\pm^q] = \pm J_\pm^q, \quad [J_+^q, J_-^q] = \frac{q^{J_z^q} - q^{-J_z^q}}{q^{1/2} - q^{-1/2}}, \; q \in I\!R, \tag{16}$$

when acting on the eigenkets $\{|jm\rangle_q\}$. The eigenkets are defined by:

$$|jm\rangle_q = ([j+m]_q![j-m]_q!)^{-1/2}(a_1^q)^{j+m}(a_2^q)^{j-m}|0\rangle_q, \tag{17}$$

and define all finite-dimensional unitary irreps of $SU_q(2)$.

ASSERTION[17]. *Let the operator pair be given by:*

$$\mathbf{t}_{\frac{1}{2},\frac{1}{2}} \equiv a_1^q q^{\frac{-N_2}{4}}, \qquad \mathbf{t}_{\frac{1}{2},-\frac{1}{2}} \equiv a_2^q q^{\frac{N_1}{4}}. \tag{18}$$

Then this pair is a q-tensor operator of $SU_q(2)$.

To verify this assertion, we use the commutativity of diagram (9) which implies:

$$J_\pm \mathbf{t}|jm\rangle = q^{-\frac{J_z}{2}}(\mathbf{t})J_\pm|jm\rangle + J_\pm(\mathbf{t})q^{\frac{J_z}{2}}|j,m\rangle. \tag{19}$$

Equation (19) can be expressed as the operator relation:

$$J_\pm(\mathbf{t}) = J_\pm \mathbf{t} q^{-\frac{J_z}{2}} - q^{-\frac{J_z}{2}}(\mathbf{t})J_\pm q^{-\frac{J_z}{2}}. \tag{20}$$

The action of $q^{-\frac{J_z}{2}}$ on $\mathbf{t}$ is the usual action because $\Delta(J_z)$ is commutative; hence we have $q^{-\frac{J_z}{2}}(\mathbf{t}) = q^{-\frac{J_z}{2}}\mathbf{t}q^{\frac{J_z}{2}}$. Accordingly eq. (20) becomes:

$$J_\pm(\mathbf{t}) = J_\pm \mathbf{t} q^{-\frac{J_z}{2}} - q^{-\frac{J_z}{2}\pm\frac{1}{2}}\mathbf{t}J_\pm. \tag{21}$$

(Note that for $q \to 1$ we cover the commutator action.)

It can be verified now that the induced action, eq. (21), on $\mathbf{t}_{\frac{1}{2},\pm\frac{1}{2}}$ obeys the equivariance condition with $j = \frac{1}{2}$, which proves the assertion.

The matrix elements of $\mathbf{t}_{\frac{1}{2},\pm\frac{1}{2}}$ are easily computed on the eigenkets, eq. (17), which (up to an overall invariant factor) are found to be exactly the spin-$\frac{1}{2}$ q-WCG coefficients[17,18,19], verifying the tensor operator theorem for this example.

Let us consider next irreducible q-tensor operators carrying the adjoint irrep, which corresponds to the generators; we denote these operators by $\mathbf{t}_{1,M}$, $M = \pm 1, 0$. It is convenient to define first an operator realization for the total angular momentum j, namely:

$$J^q \equiv \tfrac{1}{2}(N_1^q + N_2^q), \tag{22a}$$

which obeys:

$$J^q|jm\rangle = j|jm\rangle. \tag{22b}$$

We find:

$$\mathbf{t}_{1,M=1} = -([2]_q)^{\frac{1}{2}} q^{\frac{J_z^q+1}{2}} J_+^q, \tag{23a}$$

$$\mathbf{t}_{1,M=0} = \left(q^{-(\frac{J^q+1-J_z^q}{2})}[J^q + J_z^q]_q - q^{(\frac{J^q+1+J_z^q}{2})}[J^q - J_z^q]_q\right), \tag{23b}$$

$$\mathbf{t}_{1,M=-1} = ([2]_q)^{\frac{1}{2}} q^{\frac{J_z^q-1}{2}} J_-^q. \tag{23c}$$

Using the q-WCG coefficients to define the Casimir invariant:

$$I_2 \equiv \mathbf{t}_1 \cdot \mathbf{t}_1, \tag{24a}$$

we find:

$$I_2 = [2J^q]_q[2J^q + 2]_q \longrightarrow [2j]_q(2j + 2]_q. \tag{24b}$$

Remark:

It is interesting to note that the eigenvalues of the Casimir operator above, involve "integer q-numbers" for both j integer and half-integer, in sharp contrast to Casimir operators obtained by less systematic techniques. Note that this integer property of the q-numbers is preserved under all operations of the group $SU_q(2)$.

Acknowledgements:

This paper is based in large part on ref. (12), which reports on work done in collaboration with Dr. Marco Tarlini. We wish to acknowledge his help with thanks. This paper was written during a stay at T-Division of Los Alamos National Laboratory. We wish to thank Drs. Jim Louck and Max Lohe for helpful discussions and the Director, Dr. Richard Slansky, for the courtesies extended.

References:

1. Sklyanin, E. K., *Funct. Anal. Appl.* **16**, 263 (1982).
2. Kulish, P. P., *J. Sov. Math.* **19**, 1596 (1982); Kulish, P. P., and Reshetikhin, N. Y., *J. Soviet Math.* **23**, 2435 (1983).
3. Drinfeld, V. G., *Quantum Groups*, Proc. of Int. Congr. of Mathematicians. MSRI, Berkeley 798 (1986); Drinfeld, V. G., *Soviet Math. Dokl.* **36**, 212 (1988).
4. Yang, C. N. and Ge, M. L., (Editors), *Braid Group, Knot Theory and Statistical Mechanics*, World Scientific, Singapore, 1989.
5. Faddeev, L., *Integrable Models in* $(1+1)$*-Dimensional Quantum Field Theory*, J-B Zuber and R. Stora, Eds., Les Houches XXXIX, Elsevier Science Publishers B.V., Course 8 (1984).
6. Manin, Yu. I., *Quantum Groups and Non-Commutative Geometry*, Centre de Recherches Mathématiques, University of Montreal, (1988).
7. Biedenharn, L. C. and Louck, J. D., *Angular Momentum in Quantum Physics*, Vol. 8 *Encyclopedia of Mathematics and Its Applications*, Addison-Wesley, Reading, MA (1981).
8. D. Aebersold and L.C. Biedenharn, *Phys. Rev.* **A15**, 441 (1977).
9. Jimbo, M., *Lett. Math. Phys.* **10**, 63 (1985); Jimbo, M., *Commun. Math. Phys.* **102**, 537 (1986).
10. Gel'fand, I. M. and Zelevinsky, A. V., *Societé Math. de France, Astérique, hors serie*, 117 (1985).
11. Flath, D. and Biedenharn, L. C., *Can. J. Math* **37** 710 (1985); Biedenharn, L. C. and Flath, D., *Commun. Math. Phys.* **93**, 143 (1984).
12. L. C. Biedenharn and M. Tarlini, *Lett. Math. Phys.* **20**, 271 (1990).
13. Macfarlane, A. J., *J. Phys. A. Math. Gen.* **22**, 4581 (1989).
14. Biedenharn, L. C., *J. Phys. A. Math. Gen.* **22**, L873 (1989).
15. C.-P. Sun and H.-C. Fu, *J. Phys. A. Math. Gen.* **22**, L983 (1989).
16. Y. J. Ng, *J. Phys. A. Math. Gen.* **23**, 1023, (1990).
17. Biedenharn, L. C., Invited paper at the 1989 Clausthal Summer Workshop on Mathematical Physics (Quantum Groups), to be published in the Proceedings (Springer-Verlag).
18. Kirillov, A. N. and Reshetikhin, M. Yu, "Representations of the Algebra $U_q(sl(2))$, q-Orthogonal Polynomials and Invariants of Links", USSR Academy of Sciences (preprint) 1988).
19. Nomura, M., *J. Math. Phys.* **30**, 2397 (1989).

QUANTUM GROUPS AND QUANTUM COMPLETE INTEGRABILITY: THEORY AND EXPERIMENT

R.K. Bullough, Department of Mathematics, UMIST, P.O. Box 88, Manchester, M60 1QD, UK

and

J. Timonen, Department of Physics, University of Jyväskylä SF-40100, Jyväskylä, Finland

1. INTRODUCTION

This paper is concerned to do three things: first to review once again [1] some of our recent work on the finite temperature quantum statistical mechanics of integrable models in *two* dimensions (1+1 dimensions, one space and one time) and thereby fill in still a little more detail on the Fig. 1 attached; second to show how there may be natural anyon theories [2,3] within this theory; third to illustrate aspects of quantum integrable models by actual experiments involving low-energy quantum electrodynamics. These three things are done in §§ **2**, **3** and **4** of the paper. Since the free anyon gas is almost certainly superfluid, this in §**3** and the §**4** are the two connections with experiments motivating our title. In the EXPERIMENTS box (outgoing box at extreme right in the Fig. 1 of the paper [1]), High-T_c superconductivity appeared already with reference to [4] in particular. The work on anyon theory in the present paper (§3) which has this High-T_c connection [2,3] is new. The work on Rydberg atoms (§4) in the context of quantum integrable models is also new. In this paper we are concerned once again [1,5-7] with *specific* quantum integrable models and specific functional integrals: we are concerned to derive generalities from these (e.g. the Fig. 1 now attached here!) but our view point is more concrete (perhaps) than those advanced in [3,8]. Certainly there is much to do to draw together the various different viewpoints provided in the presentations at this very interesting meeting.

Fundamental to this paper are the quantum groups [1]: this may be a restriction [3]. At the immediate predecessor of this Conference* we showed

---

* 18th Intl. Conf. on Diff. Geom. Methods in Theor. Phys., Lake Tahoe, Calif., July 2-9, 1989.

"SOLITONS"

U, V and Ψ are $N \times N$ Matrices $\left\{ \begin{array}{l} sl(N,C) \\ gl(N,C) \end{array} \right\} = g \ni U, V$

Periodic b.c.'s on
Integrable Lattice Models
in 1+1 dimensions
monodromy matrices
$\psi_{n+1} = L_n \psi_n$
$T = \overline{\prod} L_n$

Lax Pair
$\psi_x = U\psi$
$\psi_t = V\psi$
or
$d\psi = \Omega\psi$

Spectral Problem
$\psi_x = U\psi$
and $U(x,0)$
plus rapidly vanishing b.c.'s at ∞

canonical
transformation

Spectral
Data
$S(0)$

$\psi_t = V\psi$

$\ln a$

compatibility
$d\Omega + \Omega \wedge \Omega = 0$
(Zero Curvature)

Hamiltonian
$H[U]$

Spectral
Data
$S(t)$

SUSY

Integrable Model
(=NEE)
$U_t - V_x + [U,V] = 0$
e.g. $u_t = u_{xxx} - 6uu_x$

Marchenko
Equation

Riemann Problem
$\Phi \overset{\dagger}{1} - \Phi \overset{-}{1} = R(k)\Phi \overset{\dagger}{2}$

Riemann
theta-functions

$U(x,t)$
Solves integrable models
in 1+1 Dimensions

Bose Fermi
Equivalence

$Z \times Z$
Matrices
$\in gl(\infty)$
$\Phi^+ = \Phi^- H$

K-P Equations

$U_{m,t_n} - U_{n,t_m} + [U_m, U_n] = 0$
also Hamiltonians

$gl(\infty)$

SUSY

KSA Theorem

Quantum
Deformations

Saito

Loop Algebras
$sl(N,C) \otimes [\lambda, \lambda^{-1}]$

KSA
theorem

$gl(\infty)$

Saito

?
3+1
Dimensional
Integrable
models

Affine(Kac-Moody)
Lie algebras
Virasoro Algebras

Conformal
Field
Theory

Elliptic
Modular
Functions

Strings

Potts Model
Ising Model

Hard
Hexagon

Theory
of
Partitions
(Wadati)

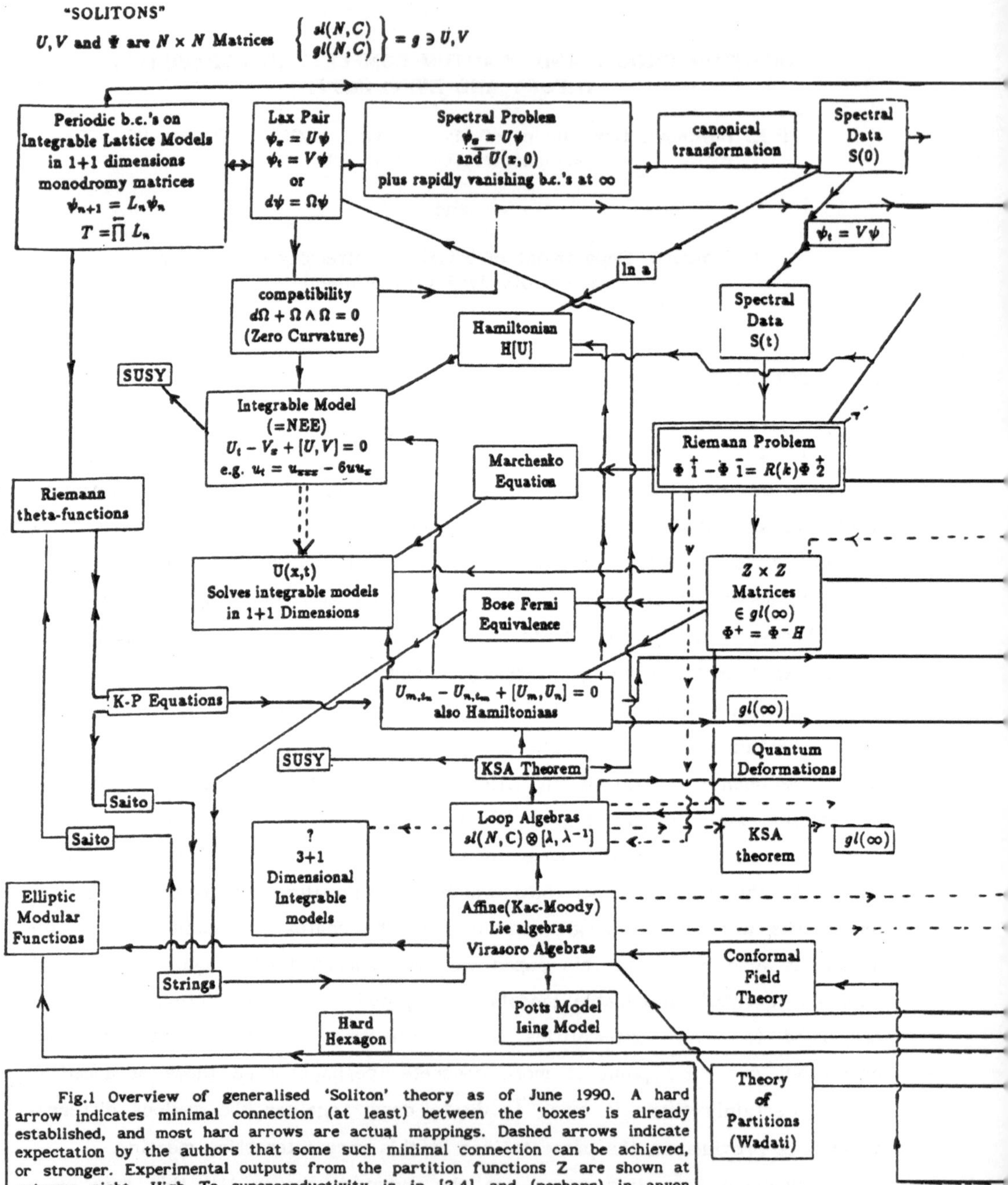

Fig.1 Overview of generalised 'Soliton' theory as of June 1990. A hard
arrow indicates minimal connection (at least) between the 'boxes' is already
established, and most hard arrows are actual mappings. Dashed arrows indicate
expectation by the authors that some such minimal connection can be achieved,
or stronger. Experimental outputs from the partition functions Z are shown at
extreme right. High Tc superconductivity is in [2,4] and (perhaps) in anyon
theory (§3). Rydberg atoms are in §4 (and [29–31]). For CsNiF$_3$ see e.g.
[40]. Free energies $F = -\beta^{-1}\ln Z$ yield specific heats e.g. [41].

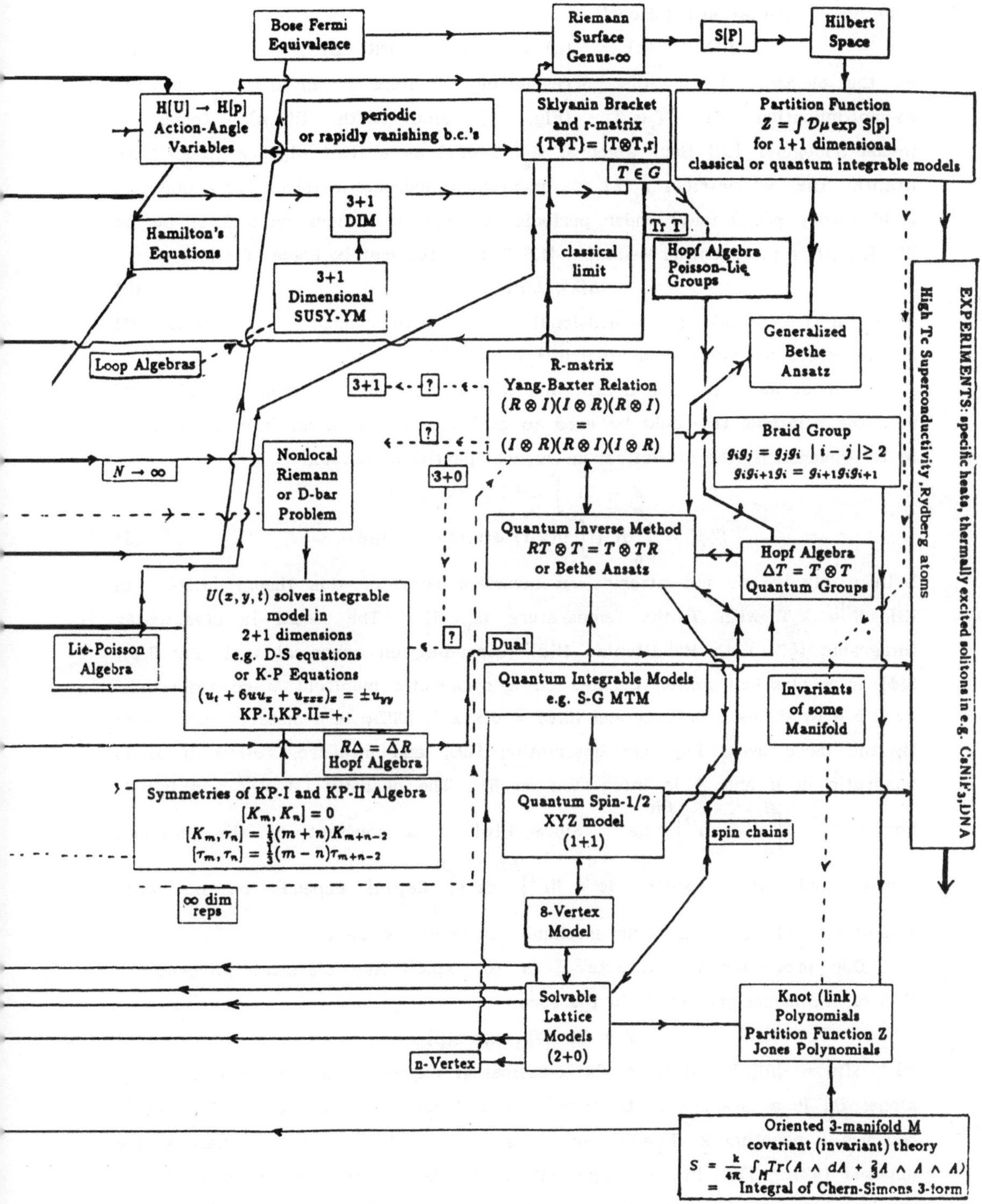
Bose Fermi Equivalence
Riemann Surface Genus-∞
S[P]
Hilbert Space
H[U] → H[p] Action-Angle Variables
periodic or rapidly vanishing b.c.'s
Skiyanin Bracket and r-matrix {T⊗T}= [T⊗T,r]
Partition Function Z = ∫ Dμ exp S[p] for 1+1 dimensional classical or quantum integrable models
T ∈ G
3+1 DIM
Tr T
Hamilton's Equations
classical limit
Hopf Algebra Poisson-Lie Groups
3+1 Dimensional SUSY-YM
EXPERIMENTS: specific heats, thermally excited solitons in e.g. CsNiF3,DNA
High Tc Superconductivity ,Rydberg atoms
Generalized Bethe Ansatz
Loop Algebras
R-matrix Yang-Baxter Relation (R ⊗ I)(I ⊗ R)(R ⊗ I) = (I ⊗ R)(R ⊗ I)(I ⊗ R)
3+1
?
?
Braid Group gigj = gjgi | i - j | ≥ 2 gigi+1gi = gi+1gigi+1
N → ∞
Nonlocal Riemann or D-bar Problem
3+0
Quantum Inverse Method RT ⊗ T = T ⊗ TR or Bethe Ansats
Hopf Algebra ΔT = T ⊗ T Quantum Groups
U(x, y, t) solves integrable model in 2+1 dimensions e.g. D-S equations or K-P Equations (ut + 6uux + uxxx)x = ±uyy KP-I,KP-II=+,-
?
Dual
Lie-Poisson Algebra
Quantum Integrable Models e.g. S-G MTM
Invariants of some Manifold
RΔ = ΔR Hopf Algebra
Symmetries of KP-I and KP-II Algebra [Km, Kn] = 0 [Km, τn] = ⅓(m + n)Km+n-2 [τm, τn] = ⅓(m - n)τm+n-2
Quantum Spin-1/2 XYZ model (1+1)
spin chains
∞ dim reps
8-Vertex Model
Solvable Lattice Models (2+0)
Knot (link) Polynomials Partition Function Z Jones Polynomials
n-Vertex
Oriented 3-manifold M covariant (invariant) theory
S = k/4π ∫M Tr(A ∧ dA + ⅔A ∧ A ∧ A)
= Integral of Chern-Simons 3-form

[1] how the deformed loop algebra $g_q = sl(2,\mathbb{C})_q \otimes [\lambda,\lambda^{-1}]$, $\lambda\in\mathbb{C}$ induced the quantum 'commutation relations'

$$R(\lambda,\mu)\hat{T}_L(\lambda) \otimes \hat{T}_L(\mu) = \hat{T}_L(\mu) \otimes \hat{T}_L(\lambda)R(\lambda,\mu) \tag{1}$$

on the elements $\hat{T}_L$ of the corresponding quantum group G_q: G_q has the co-multiplication $\Delta T = \hat{T}_L \otimes \hat{T}_L$ (Fig. 1) and, with $R(\lambda,\mu)$, forms a (quasi-triangular [9]) Hopf algebra: G_q is dual as a Hopf algebra to g_q with $R(\lambda,\mu)$. The 2x2 matrices $\hat{T}_L(\lambda)$ are quantum monodromy matrices for a quantum field theory $\phi(x,t)$ (say) under periodic boundary conditions on $x\in\mathbb{R}$ of period 2L; $R(\lambda,\mu)$ is the quantum R-matrix [1,5-7,10]. The matrix trace of (1) is

$$[\hat{\Delta}(\lambda),\ \hat{\Delta}(\mu)] = 0 \tag{2}$$

where $\hat{\Delta}(\lambda) \equiv \mathrm{Tr}\hat{T}(\lambda)$: 1+1 dimensional models with commutation relations (1) are therefore 'completely quantum integrable' (§2).

Without much specific detail, and with many gaps, we then showed in [1] how the relations (1) could be used to evaluate, for such an integrable model in 1+1 dimensions, the statistical mechanical partition function

$$Z = \mathrm{Tr} \int \mathcal{D}\Pi\ \mathcal{D}\phi\ \exp S[\phi]$$

$$S[\phi] = \int_0^{\beta} dt\ [i\int_{-L}^{L} \Pi(x,t)\ \phi(x,t)_t dx\ -\ H[\phi]]. \tag{3}$$

Deliberately we do not attempt an invariant form of this theory [3,11]. In (3), $\beta^{-1} = k_B T$ with T the temperature ($k_B = 1$). The model is *classically* integrable (§2) with Hamiltonian $H[\phi]$ (depending on canonical variables Π,ϕ); $S[\phi]$ is a classical action defined on a symplectic manifold M co-ordinatised by $\Pi,\phi \in \mathbb{R}$; M has a bracket and $\{\Pi,\phi\} = \delta(x-x')$; $\mathcal{D}\Pi\mathcal{D}\phi$ is a measure defined on M; the more usual Feynman description [12], and measure, follow if H is quadratic in Π and Z is integrated on Π. The quantity $i \equiv \sqrt{-1}$ and $\hbar = 1$. For $\hbar \neq 1$, $\int_0^{\beta} dt \to \hbar^{-1}\int_0^{\beta\hbar} dt$. Since $H[\phi]$ is a constant of the classical motion, only the quantity $i\Phi \equiv i\hbar^{-1}\int_0^{\beta\hbar} dt\ [\int_{-L}^{L} \Pi\phi_t dx]$ depends on $\hbar$, while $\Phi \in \mathbb{R}$: thus all of the quantum mechanics is in the *phase* Φ.

Our procedure to evaluate Z is to exploit the classical integrability (§2) of the model and put it in the form

$$Z = \mathrm{Tr} \int \mathcal{D}\mu\ \exp S[p] \tag{4}$$

with $S[p] = -\beta H[p]$ and H is the classical H expressed in action variables p alone: $\mathcal{D}\mu$ is a measure to be found. Expression (4) is a classical Z; but it becomes a quantum Z if additional constraints derived from the phase Φ are imposed (§2). This line of argument is sketched in the Fig. 1 where (4) is reached (from various starting points!) in the top right corner. The Fig. 1 attached now updates that in [1] which itself updates that in [5,13]. All of the additions since [5,13] concern the quantum groups or the content of the EXPERIMENTS box (where Rydberg atoms (§4) is now included). The non-abelian, topological only, pure Chern-Simons theory [3,11] was added to Fig. 1 in [1] Reference [11] provides the connections from it marked now on Fig. 1. Further connections with dynamical theories, in 2+1 dimensions in particular, classical or quantum, are not yet explored (by us).

We use the phase Φ as a possible source of anyon theory (§3). Notice that $H[\phi]$ in (3) is parametised by a real time t, $0 \le t < \beta$: the Tr means $S[\phi]$ is periodic in t of period β (as $T \to 0$, for quantum *mechanics*, $\beta \to \infty$). We need to work at *finite density* for a proper thermodynamic limit on (3). Thus, conveniently, we use periodic boundary conditions on x of period 2L and take $L \to \infty$ in finite density limit. The manifold (x,t) therefore forms the space-time torus $-L \le x < L$, $0 \le t < \beta$ with metric diag (1,1). Notice that the functional integral (3) *is* (identically) $\mathrm{Tr}e^{-\beta\hat{H}}$ where $\hat{H}$ is quantum integrable (§2) [6,15,16]. Notice too there is the Wick rotated zero temperature quantum propagator

$$G(\phi, \phi_0; T) = \int \mathcal{D}\Pi\mathcal{D}\phi \exp i\, S[\phi]$$
$$S[\phi] = \int_0^T dt \left[\int_{-L}^{L} \Pi\phi_t dx - H[\phi]\right] \tag{5}$$

with paths $\phi(x,0) = \phi_0(x)$ to $\phi(x,T) = \phi(x)$ [15,16]. Paths $\delta S = 0$ are Hamilton's classical equations of motion for real time t parameterising Π, ϕ. To this extent 'Wick rotation' is formal and the metric on the manifold (x, t) is diag (1, 1): $-L \le x < L$ and $0 \le t \le T$ and a finite density limit is taken in x. When $\hbar \ne 1$ $iS[\phi] \to i\hbar^{-1}S[\phi]$ only and there is no pure phase Φ in (5) describing the quantum mechanics. We derive constraints on Φ in §3 for the 'anyon theory'. The quantum groups impose further constraints which then depend on the constraints on Φ. To evaluate the otherwise classical expression (4) for Z as a quantum Z we need to impose this whole collection of constraints. Surprisingly, perhaps but as already sketched in [1,5,6], these make it all possible to *evaluate* Z as a quantum Z!

2. QUANTUM AND CLASSICAL COMPLETE INTEGRABILITY

A classical Hamiltonian dynamical system with N degrees of freedom is 'completely integrable' if there are N independent constants I_k commuting under the bracket: $\{I_k, I_l\} = 0$; k, l = 1,2,...,N [17]. If the manifold of level lines $\{I_k = \text{const};\ k = 1,\ldots,N\}$ is compact and connected, motions are on N-dimensional tori: N canonical pairs of constant action variables P_i ($0 \le P_i < \infty$), and angle variables Q_i ($0 \le Q_i < 2\pi$), can be found: the Q_i define the tori. H is any function of the I_k. We extend to an integrable classical field theory $\phi(x,t)$, $x \in \mathbb{R}$, by requiring a 'sufficient' infinite number of constants I(k) commuting under the bracket: $\{I(k), I(k')\} = 0$; $k, k' \in \mathbb{R}$. The semiclassical limit of (1) is the Sklyanin bracket ([1], and see top right Fig. 1)

$$\{T(\lambda), \otimes T(\mu)\} = [T(\lambda) \otimes T(\mu), r(\lambda,\mu)] \tag{6}$$

where $r(\lambda,\mu)$ is the 'little r-matrix' [1]: $T(\lambda)$, $\lambda \in \mathbb{C}$ are classical monodromy matrices and $T(\lambda) \in G$, the loop group with undeformed loop algebra $g \otimes [\lambda, \lambda^{-1}]$ [1]. The matrix trace of (6) yields

$$\{\Delta(\lambda), \Delta(\mu)\} = 0 \tag{7}$$

with $\Delta(\lambda) \equiv \mathrm{Tr}\, T(\lambda)$, and there is a large infinity of constants commuting under the bracket: $\Delta(\lambda)$, $\ln \Delta(\lambda)$ and $\ln(f(\lambda)\, \Delta(\lambda))$, $f(\lambda)$ analytic in λ, are generators of completely integrable Hamiltonian systems with classical Hamiltonians H.

Our definition of complete quantum integrability is that there are, for N *quantum* degrees of freedom (§4), N commuting operator constants $\hat{I}_k$: $[\hat{I}_k, \hat{I}_l] = 0$, $k,l = 1,\dots,N$ and $[\dots]$ means commutator. Motions are no longer on well defined tori: the canonical commutation relations $[\hat{P}_i, \hat{Q}_j] = -i\, \delta_{ij}$ ($\hbar = 1$) induce an N-dimensional uncertainty relation on the *states* of the system. These states span an N-fold infinite dimensional Hilbert space, one each for each quantum degree of freedom. There are other more (or much more) complicated classical brackets and quantum commutators e.g. spin systems are more. These entail some extension of the idea of degree of freedom (§4).

We have now reported [6,18] evaluation of quantum or classical Z for at least eleven integrable modes in 1+1 by the methods summarised again now, in this paper. In the paper we confine attention to the particularly simple sinh-Gordon (sinh-G) and repulsive non-linear Schrödinger (repulsive NLS) models. Classically these are

$$\phi_{xx} - \phi_{tt} = m^2 \sin \phi, \qquad (m > 0,\ \phi \in \mathbb{R}) \tag{8}$$

and

$$-i\phi_t = \phi_{xx} - 2c\phi^*\phi^2; \quad c > 0 \quad (\phi \in \mathbb{C}). \tag{9}$$

As quantum theories these are normally ordered. Their natural commutation relations are the bose relations $[\phi, \phi^\dagger] = -i\delta(x-x')$ (sinh-G) and $[\phi, \phi^\dagger] = \delta(x-x')$ (repulsive NLS): in particular quantum repulsive NLS is the bose gas [19]. The quantum group commutation relations (1) are equivalent to these commutation relations for appropriate R-matrices [1,5-7,10]. Moreover eqn. (2) means that these field theories are quantum completely integrable in so far as $\lambda,\mu \in \mathbf{C}$ and there is a large infinity of commuting constants 'sufficient' to ensure quantum integrability by extension from a finite number of quantum degrees of freedom. Ref. [19] evaluated $F = -\beta^{-1}\ln Z$ for the bose gas by the method of quantum Bethe Ansatz: the methods sketched in this paper generalise both the Bethe Ansatz method and the quantum inverse method [1,5-7,10] to functional integral methods (through (3) and (4)) and to the *classical* statistical mechanics of the classical integrable models. Both the BA and the QIM work in terms of fermi-like particles obeying fermi statistics. However the Fig. 1 show how bose-fermi equivalence emerges from a box marked $\mathbf{Z} \times \mathbf{Z}$ matrices $\in gl(\infty)$ (the box below the central double-lined box on the Riemann problem [5]) and how this is imposed (top right) on the functional integrals.

Our understanding of this bose-fermi equivalence is that $gl(\infty)$ has representation in terms of $\mathbf{Z} \times \mathbf{Z}$ matrices [5] (matrices with elements $n,m \in \mathbf{Z}$, all of the integers). But $gl(\infty)$ also has representations in terms of vertex operators on a bosonic space and in terms of products of free fermi operators [20]. Representation theory of $gl(\infty)$ seems incomplete [21]. Still it is natural to look for bose-fermi equivalent descriptions of the quantum integrable models in 1+1. Here we have a gap in our present understanding: the undeformed loop algebras $g \otimes [\lambda, \lambda^{-1}]$ generate [5,22] all of the classical integrable models and $gl(\infty)$ generates [5,20,22] the integrable classical Kadomtsev-Petviashvili (KP) equations in 2+1 (Fig. 1): the deformed loop algebras $g_q \otimes (\lambda,\lambda^{-1})$ have, with $R(\lambda, \mu)$, dual Hopf algebras G_q [1] which are the quantum groups with their co-multiplicaton. Apparently a deformed $gl_q(\infty)$ would yield a $G_q(\infty)$ for the corresponding quantum integrable theory in 2+1, but no such theory in 2+1 dimensions is available yet. Despite [3], no 'good' integrable quantum model is found yet in 2+1. Evidently the structure is there [3,23] e.g. Reshitikhin [23] finds *quantum* symmetries directly comparable (perhaps) with the classical symmetries of integrable models in 1+1. One of us reported classical symmetries of the K-P eqns in [24] and references (this is the significance of the box below the box "U(x,y,t) solves integrable model in 2+1 dimensions" in Fig. 1). Despite this lack of understanding in terms of a $gl_q(\infty)$ for quantum models in 2+1, there *are* bose-fermi equivalent quantum models in 1+1 handed down in some sense from $gl(\infty)$ as are handed down the classical models [5,20,22]. Two of the simplest are the quantum sinh-G and repulsive NLS models, eqns (8), (9)

[1,5-7,25]. For vanishing boundary conditions on ϕ_x or ϕ as $x \to \pm\infty$ these two *classical* models have Hamiltonians H[p], depending on action variables P(k) alone: $H[p] = \int_{-\infty}^{\infty} \omega(k)\, P(k)\, dk$ and $\omega(k) = (m^2 + k^2)^{\frac{1}{2}}$ for sinh-G, $\omega(k) = k^2$ for NLS. Evidently $\{P(k), H[p]\} = 0$ for all $k \in \mathbb{R}$. Proof that this number of constants is sufficient for complete classical integrability relies on the sinh-G and NLS integrable *lattices*, manifestly with enough constants under *periodic* boundary conditions, through which the evaluation of Z in the form (4) is carried out [1,16,18,25]. Our procedure is to establish the connection of the P(k) (and Q(k)) under vanishing b.c.s. to the action-angle variables under periodic b.c.s. [1,5-7] (and §3).

Our point of view for the anyon theory in §3 is to investigate in 1+1 dimensions equivalent descriptions in terms of fermions, bosons *and* anyons. Classification of these theories under the braid groups B_N [2,3,11,26] is not achieved *as we go to press* - despite the connections from eqns (1) for 1+1 dimensional quantum models to the Yang-Baxter relation and then to the braid groups [26] we displayed in the Fig. 1 since [5,13]. Our point of view on anyon theory is new to us, and it is still to be subsumed within representation theory of the B_N. One *firm* result in §3 is however that only bose-fermi equivalence (bose statistics or fermi statistics) is possible for quantum models in 1+1 which have classical H[p]'s $H[p] = \int_{-\infty}^{\infty} \omega(k)\, P(k)\, dk$ on vanishing b.c.s. These models are models with spin-less particles but do not break parity and time reversibility [2,3], so the result is expected. Note again we are in 1+1 not 2+1 dimensions.

3. ANYON AND SEMION THEORY

We first review the bose-fermi equivalence of quantum sinh-G and quantum repulsive NLS [1,5-7,25]. To reach the classical form (4) for Z with quantum constraints we might *remove* the phase Φ in (3) and replace it by constraints. Evidently after discretizing Φ to

$$\Phi = \int_0^{\beta} dt \sum_{n=1}^{N} \Pi_n\, \phi_{n,t} \tag{10}$$

one set of possibilities is

$$\int_0^{\beta} dt\ \Pi_n\, \phi_{n,t} = \oint \Pi_n\, d\phi_n = 2\pi m_n \tag{11}$$

and m_n = integer $\in \mathbb{Z}$. This is Bohr-like semiclassical quantisation

($\hbar = 1 \Rightarrow h = 2\pi$). But it does not damage the quantum group quantisation (1) - see below.

To impose (1) we make a canonical transformation to action-angle variables on the classical action $S[\phi]$: $S[\phi]$ goes to $S[p]$, eqn. (4) if the phase is removed. It turns out that periodic b.c.s. for a finite density thermodynamic limit as period $2L \to \infty$ can still be achieved using action-angle variables under vanishing b.c.s. at $\pm\infty$ for $S[p]$ and pushing the finite density limit problem into the measure $\mathcal{D}\mu$ (vanishing b.c.s. with $\pm L$ *a priori* equal to $\pm\infty$ produces a zero density limit unless the measure $\mathcal{D}\mu$ is chosen appropriately [1,5-7,25]).

For the two models considered the discretized Φ becomes

$$\Phi = \int_0^\beta dt \sum_{n=1}^{N} P_n Q_{n,t} \tag{12}$$

and now we can eliminate Φ by choosing

$$\int_0^\beta dt\ P_n Q_{n,t} = \oint P_n\ dQ_n = 2\pi m_n, \tag{13}$$

m_n a non-negative integer, instead of (10) and (11): P_n, Q_n are discretized action-angle variables for the models, $0 \le P_n < \infty$, $0 \le Q_n < 2\pi$, $\{P_n, Q_m\} = \delta_{n,m}$. Since the P_n are constants, (13) includes the exact Bohr semi-classical quantisations $P_n = m_n = 0, 1$ (fermions) and $P_n = m_n = 0, 1, 2, \ldots$ (bosons).

The connection between action-angle variables under vanishing b.c.s. (zero density limit) and periodic b.c.s. (finite density limit) is that only the labels n which satisfy the conditions [1,5-7,25]

$$\tilde{k}_n = k_n - L^{-1} \sum_{m \neq n}^{N} \Delta(\tilde{k}_n, \tilde{k}_m) P_m, \tag{14}$$

in which $\Delta(k,k')$ is a 2-body S-matrix phase shift, contribute to the measure $\mathcal{D}\mu$ in (4). For fermions $P_m = 0, 1$ and $\Delta = \Delta_f$; $\Delta_f = -2 \tan^{-1} \{m^2\gamma_0'' [k\omega(k') - k'\omega(k)]^{-1}\}$ for sinh-G ($\gamma_0'' \equiv \gamma_0[1 + \frac{\gamma_0}{8\pi}]^{-1}$, with $\gamma_0 > 0$ the bare coupling constant, is the renormalised coupling constant for quantum sinh-G [25]); $\Delta_f = -2 \tan^{-1} \{c(k-k')^{-1}\}$ for repulsive NLS ($c > 0$). For bosons $P_m = 0, 1, 2, \ldots$ and $\Delta = \Delta_b$; $\Delta_b = \Delta_f + 2\pi\theta\ (k' - k)$ where θ is the unit step. The *smooth branch* $-2\pi < \Delta_f < 0$ is taken for the Δ_f's: Δ_b thus has a jump of 2π as k goes through k'. Thus $e^{i\Delta_f} = e^{-i\pi} = -1$ at $k = k'$ and $e^{i\Delta_b} = 0$, consistent with fermi and bose statistics respectively. The *fermi* constraint (14) with $\Delta = \Delta_f$ and $P_n = 0, 1$ is implicit in the BA and QIM methods for repulsive NLS [10, 19], as comparison with the more general expressions (14) shows. The *bose* description was new [25] (but see also Wadati [27]).

One can *demonstrate* bose-fermi equivalence: the collection of constraints (13) with (14) mean that Z can be taken in the form (4) which is discretized to

$$Z = \lim_{L\to\infty} \int \prod_{n=1}^{N} \frac{dP_n dQ_n}{2\pi} \exp [-\beta \sum_n \omega(\tilde{k}_n)P_n]. \tag{15}$$

By *iterating* the constraints (13) with (14) through (15) one can obtain an iterated, strictly asymptotic, form for Z which can be summed (!) for a fermion description to [1,5–8,25]

$$\lim_{L\to\infty} -\beta^{-1} \frac{\ln Z}{L} = \mu\bar{n} - (2\pi\beta)^{-1} \int_{-\infty}^{\infty} \ln (1 + e^{-\beta\tilde{\varepsilon}(k)})dk \tag{16}$$

where the temperature dependent excitation energies $\tilde{\varepsilon}(k)$ are given by

$$\tilde{\varepsilon}(k) = \omega(k) - \mu - (2\pi\beta)^{-1} \int_{-\infty}^{\infty} (d\Delta_f(k, k')/dk) \ln(1 + e^{-\beta\tilde{\varepsilon}(k')})dk' \tag{17}$$

in which μ is a chemical potential and $\bar{n} > 0$ a finite density, $\lim_{L\to\infty} NL^{-1}$, of fermions. There is the bose form obtained by iterating (15) with (14) and *bose* constraints (13). This is

$$\lim_{L\to\infty} \frac{-\beta^{-1} \ln Z}{L} = \mu\bar{n} + (2\pi\beta)^{-1} \int_{-\infty}^{\infty} \ln(1 - e^{-\beta\varepsilon(k)}) dk \tag{18}$$

with

$$\varepsilon(k) = \omega(k) - \mu + (2\pi\beta)^{-1} \int_{-\infty}^{\infty} (d\Delta_b(k, k')/dk) \ln(1 - e^{-\beta\varepsilon(k')})dk'. \tag{19}$$

By defining $\ln(1 + e^{-\beta\tilde{\varepsilon}(k)}) = -\ln(1 - e^{-\beta\varepsilon(k)})$ and using the δ-function in $d\Delta_b/dk$ induced by the step function $\theta(k'- k)$ one shows that (16) with (17) are identically equivalent to (18) with (19).

We turn to possible anyon theories. The quantisation conditions (13) admit any set of non-negative integers m_n. Then since P_n is constant, $P_n = m_n$. On the other hand it seems that *arbitrary* sets of m_n do not necessarily lead to equivalent theories (if not these theories still have fundamental interest). However we can, somewhat differently, introduce homotopy classes characterised by a winding number ν_n winding round the classical tori ν_n times. Evidently $\nu_n = 0$ is the classical statistical mechanics, expression (4), for Z with no quantum constraints; P_n is the classical action variable and the constraints (14) now take $\Delta = \Delta_c$, the *classical* 2-body phase shift [1,25]. This 2-body phase shift Δ_c follows from the quantum boson description with $\gamma_0 \to 0$ (sinh-G) and $c \to 0$ (repulsive NLS) [1,5–7,25].

If $\nu_n = 1$ *for all n*, and $P_n = 0, 1$ or $P_n = 0, 1, 2,\ldots$ for all n, we regain the fermi- and bose- descriptions and results (16)–(19). For a *semion* theory in the sense of [2] in the label n we choose $\nu_n = 2$. Then $P_n = \frac{1}{2}m_n$ with $m_n = 0,1$ is a fermion semion description, while $m_n = 0, 1, 2,\ldots$ is a boson semion description [2]. Indeed since (14) seems to be generic this condition is now

$$\tilde{k}_n = k_n - L^{-1}\sum_{m\neq n} \Delta_s(\tilde{k}_n, \tilde{k}_m)m'_m \tag{20}$$

with $m'_m = 0, 1,$ or $0, 1, 2, \ldots$ for integer labels m, and where $\Delta_s = \frac{1}{2}\Delta_{st}$ in which Δ_{st} is the true semion phase shift. We have not yet calculated any 2-body semion (or other anyon) phase shifts. Still presumably $\Delta_{st} = -\frac{1}{2}\pi$ at $k = k'$ for fermion semions (compare Zee [2]). Or it is $\nu_n^{-1}\Delta_{st} = -\frac{1}{2}\pi$ (see below).

Since $m'_n = 0, 1$ (fermion semions) or $0, 1, 2, \ldots$ (boson semions) Z, expression (15) for the models considered, can be put in the form of (15) identically with a new $\beta \to \beta' = \beta\nu^{-1}$: $\nu_n = \nu = 2$ for *all* labels n by assumption. Evidently (20) for fermion semions means that for fermion semions in particular

$$\lim_{L\to\infty} -\beta^{-1}L^{-1} \ln Z = \mu\bar{n} - (2\pi\beta)^{-1}\int_{-\infty}^{\infty} \ln(1 + e^{-\beta'\tilde{\varepsilon}(k)})\, dk \tag{21}$$

$$\tilde{\varepsilon}(k) = \omega(k) - \mu - (2\pi\beta)^{-1}\int_{-\infty}^{\infty} [d\Delta_{st}(k, k')/dk] \ln(1 + e^{-\beta'\tilde{\varepsilon}(k')})\, dk'. \tag{22}$$

We have used $\beta'^{-1}\Delta_s = \beta^{-1}\Delta_{st}$ for the second expression (22), but $\beta'\tilde{\varepsilon}(k) \equiv \beta\nu^{-1}\tilde{\varepsilon}(k)$ appears elsewhere in (21), (22). There is a similar pair of equations for boson semions. But we need not write this down for we now show that, for the *equivalence* of the semion theory (22) to either a true bose theory or a true fermion theory, $\nu = 1$, and each semion theory we can choose, bose or fermi, is a *true* bose or a *true* fermion theory with its usual statistics.

For the equivalence of the fermion semion theory, (21), (22) for example, it seems that we can only use the jumps in Δ at $k = k'$. For then $d\Delta_{st}(k,k')/dk$ is the same at all points k' except at $k' = k$. By defining

$$\ln(1 + e^{-\beta\nu^{-1}\tilde{\varepsilon}(k)}) = -\ln(1 - e^{-\beta\varepsilon(k)}) \tag{23}$$

we thus arrange the bose form for the integrals in both (22) and (21). The left side of (22) is to become $\varepsilon(k)$. We assume $\Delta_{st} = \pi\lambda$ at $k = k'$. From the proposed jump

$$\Delta_{st}(k,k') = \Delta_b(k,k') + 2\pi\lambda\theta(k'-k) \tag{24}$$

($\lambda \in \mathbb{R}$) on Δ_{st}, we then have on the left side of (22)

$$-\frac{1}{\beta}\ln(e^{-\beta\varepsilon}) = -\frac{1}{\beta}\ln[x^{\nu}(1 + x)^{\lambda}] \tag{25}$$

where $x \equiv e^{-\nu^{-1}\beta\tilde{\varepsilon}}$.

Put $y = e^{-\beta\varepsilon}$. Then we need

$$x^{\nu}(1 + x)^{\lambda} = y \tag{26}$$

while, from (23), $y = 1 - \dfrac{1}{1 + x} = \dfrac{x}{1 + x}$. Thus (26) means $x^{\nu-1} = (1 + x)^{-(\lambda+1)}$, and this is true for all x if, and only if, $\nu = 1$, $\lambda = -1$.

Then $\Delta_s = \Delta_b - 2\pi\theta(k'-k) = \Delta_f$. A similar argument asking for equivalence between the fermion semion and true fermion description produces $\nu = 1$ and $\lambda = 0$. There is only the true fermion description.

We have therefore demonstrated equivalent fermion and boson descriptions for this class of models: there are no equivalent anyon descriptions (by definition here - for semions $\nu_n = \nu = 2$ for all n; and for other anyons $\nu = \nu_n > 2$ for all n in our interpretation). This result hinges on choosing the same homotopy class $\nu_n = \nu$ for every label n. Many other possible choices (different ν_n for different *groups* of labels n) remain open from this particular analysis. These theories are all spin-less and could apparently break parity and time reversal invariance so braid group statistics is possible [2,3]. On the other hand the requirement of an equivalence to bose-fermi descriptions would apparently eliminate this possibility. There is a different but parallel analysis which reaches the same conclusion that only fermi-bose equivalence is possible if $\nu = \nu_n$ for all n [28].

We note other (still unclassified) possibilities remain: one is $\nu_n = \nu_m$ for *groups* m of mode labels n as noted. A more interesting possibility, with more possible physics in it, is to *pair* modes n, n' so that e.g. for $\nu_n = \nu_{n'} = 1$, $m_n = 0, \frac{1}{2}$, $m_n = 0, \frac{1}{2}$ *together* i.e. $P_n = P_{n'} = 0, \frac{1}{2}$ together. A paired semion boson description with $\nu_n = \nu_{n'} = 1$ would be $m_n = 0, \frac{1}{2}, 1, \ldots$ and $m_{n'} = 0, \frac{1}{2}, 1, \ldots$ together. With $\nu_n = \nu_{n'}$ there is also the pairing $m_n = \frac{1}{3}$, $m_{n'} = \frac{2}{3}$, and so on. Indeed with $\nu_n = \nu_{n'} = \nu_{n''}$, etc. we can choose any positive integer l and the set l_1, l_2, l_3, etc. relatively prime to l and set $m_n = l_1 l^{-1}$, $m_{n'} = l_2 l^{-1}$, $m_{n''} = l_3 l^{-1}$, etc. - so roughly there is any anyon description in the sense of [2] (which in effect uses m_n, $m_{n'}$ etc. in the form q_n^{-1}, q_n an odd integer for anyons) and there is apparently much more - still unclassified by us so far; while there is any corresponding fractional statistics. Formal anyon theories can be written down (as the example (21) with (22) indicates - even if the equivalent anyon description in this case does not occur) but we have so far only evaluated the phase shifts Δ_f (fermions) and Δ_b (bosons) explicitly for these models. These follow from the quantum group conditions (1) [1,16,18]. Notice that we have not explicitly shown yet that anyon descriptions like (21) and (22) with a braid group statistics (different homotopy classes ν_n) on groups of modes cannot be equivalent to but distinct from the bose-fermi equivalent descriptions. The repulsive NLS model has a free boson limit $c \to 0$ and a free fermion limit $c \to \infty$ (sinh-G is different as $\gamma_0 \to 0$ and $\gamma_0 \to \infty$ [6,7]). There may be integrable models with equivalent anyon descriptions with free anyon limits: such models would be superfluids in the free anyon limit to the extent demonstrated [2,3]. Such integrable models can apparently be constructed by the methods of this paper even if they are *not* to be equivalent to bose-fermi equivalent theories. We can say nothing about equivalent theory in 2+1 dimensions since quantum integrable models are not available yet.

4. QUANTUM INTEGRABLE MODELS AND EXPERIMENTS

We develop two quantum theories for Rydberg atoms in microwave cavities in this section (see EXPERIMENTS box, Fig. 1): one is the micromaser [29-31], where the cavity has very little damping (is of very high Q in the jargon [29-34]) and concerns one atom and one quantised electromagnetic field cavity mode; this is a quantum integrable system with two *quantum* degrees of freedom for $Q = \infty$. The other theory concerns an arbitrary number N_A of Rydberg atoms in a low-Q (heavily damped) microwave cavity, [32,33,34]. This system is quantum integrable with N_A quantum degrees of freedom for $Q \to \infty$. In the thermal *equilibrium* considered damping plays no role and to that extent the system is quantum integrable. It displays features of bose-fermi equivalence to be seen *in the experimental observations* [32].

Both experimental situations concern Rydberg atoms (atoms in a 'high Rydberg state' with large principal quantum number $n \gtrsim 30$) making microwave transitions to adjacent Rydberg states (with principal quantum number n little changed: the micromaser [29-31] makes e.g. $63P_{3/2} \to 61D_{5/2}$ transitions on ^{85}Rb in spectroscopic notation; the experiments [32] used $30S_{1/2} \to 30P_{1/2}$ on Na). In both cases the cavity is tuned to induce the microwave transition, i.e. cavity mode and atoms are in resonance. A physically good atomic model is then the 2-level atom with states $|g\rangle$, $|e\rangle$ (ground and excited) and energy ω_0 ($\hbar\omega_0$) between them. There is a dipole matrix element p between $|g\rangle \leftrightarrow |e\rangle$. We consider the bose-fermi equivalence of the 2-level atom.

The atom has a spin-$\frac{1}{2}$ description whose algebra is the 2×2 representation of the algebra su(2,**C**): set $S^z = \frac{1}{2}(|\uparrow\rangle\langle\uparrow| - |\downarrow\rangle\langle\downarrow|)$ $S^+ = |\uparrow\rangle\langle\downarrow|$, $S^- = |\downarrow\rangle\langle\uparrow|$ where $|\downarrow\rangle = |g\rangle$, $|\uparrow\rangle = |e\rangle$. Then $[S^\pm, S^z] = \mp S^\pm$, $[S^+,S^-] = 2S^z$, an su(2) Lie algebra. The Hamiltonian is $\omega_0 S^z$ (with eigenvalues $\pm\frac{1}{2}\omega_0$). This is a 2S + 1 = 2 dimensional representation of the algebra in terms of Pauli matrices $S^\mp = \sigma_\mp$, $2S^z = \sigma_3$. Since su(2) is isomorphic to sl(2) we can see from [1] that the 2 dimensional representation of $su_q(2)$ *is* su(2) (this is true only of 2 dimensional representations). In Cartan-Weyl basis

$$[h,e] = 2e,\ [h,f] = -2f,\ [e,f] = h$$

for su(2) goes to

$$[h,e] = 2e,\ [h,f] = -2f,\ [e,f] = (k^2 - k'^2)/(q-q^{-1})$$
$$= (\sinh ht)/(\sinh t) \qquad (27)$$

for $su(2)_q$, where $k^2 = q^h$; $q = e^t \neq 1$ ($t \neq 0$ and nor is $q = -1$). For $t \to 0$ $su_q(2) \to su(2)$. Other relations are [1] $kek^{-1} = qe$, $kfk^{-1} = q^{-1}f$ ($t \neq 0$). The two dimensional representation of su(2), $h = \sigma_3$, $e = \sigma_+$, $f = \sigma_-$ has the corresponding representation of $su_q(2)$ $h = \sigma_3$, $e = \sigma_+$, $f = e_-$ with $k^2 = q^{\sigma_3}$ = diag (q,q^{-1}). Then $k^2 - k^{-2}$ = diag $(q - q^{-1}, q^{-1} - q) = (q - q^{-1})\sigma_3$, so $[e,f] = h$ in 2-dimensional representation. We take this to mean that G_q has the usual 2-dimensional representation of the group SU(2) and that its algebra su(2) acts as canonical commutation relations. We can therefore quantise the 2-level atom with $S^{\pm}$, S^z and its su(2) Lie algebra as explained.

This system has a 2-fermion description; use $a^{\dagger}(a)$ and $b^{\dagger}(b)$ to create (annihilate) a particle in $|\uparrow\rangle$ and $|\downarrow\rangle$ respectively. Take the (usual) 2-fermion algebra $aa^{\dagger} + a^{\dagger}a = 1$, $bb^{\dagger} + b^{\dagger}b = 1$, $(a^{\dagger})^2 = (a)^2 = 0$, $a^{\dagger}b + ba^{\dagger} = 0$, $(b^{\dagger})^2 = b^2 = 0$. Construct $S^+ = a^{\dagger}b$, $S^- = b^{\dagger}a$, $S^z = \frac{1}{2}(a^{\dagger}a - b^{\dagger}b)$. These have the su(2) Lie algebra. The 2-level atom is thus equivalent to *two* fermions.

There is also the two boson description: replace the anticommutators by commutators: $[a,a^{\dagger}] = 1$, $[b,b^{\dagger}] = 1$, etc. Then $S^{+} = a^{\dagger}b$, $S^- = b^{\dagger}a$, $S^z = \frac{1}{2}(a^{\dagger}a - b^{\dagger}b)$ have the su(2) Lie algebra. But these bosons cannot act as normal bosons: for, for one electron involved in the transition, $a^{\dagger}a + b^{\dagger}b = 1$. This means the 2-level atom is *one* fermion or *one* boson (with constraint).

The Primakoff-Holstein transformation for arbitrary S sets $S^- = (2S - \hat{n})^{\frac{1}{2}}a$, $S^+ = a^{\dagger}(2S - \hat{n})^{\frac{1}{2}} = (2S+1 - \hat{n})^{\frac{1}{2}}a^{\dagger}$ in which $\hat{n} \equiv a^{\dagger}a$ and (bosons) $[a,a^+] = 1$: $S^{\pm}, S^z$ now form a $2S + 1$ dimensional representation of the algebra su(2), and for $S = \frac{1}{2}$ it is 2 dimensional. For $S = \frac{1}{2}$ use $S^- = (1 - \hat{n})^{\frac{1}{2}}a$, $S^+ = a^{\dagger}(1 - \hat{n})^{\frac{1}{2}}$ with $\hat{n} = a^{\dagger}a$ and $aa^{\dagger} + a^{\dagger}a = 1$, $a^2 = a^{\dagger 2} = 0$ (fermions). This is a 2 dimensional representation of su(2) and the 2-level atom is equivalently one fermion.

The Hamiltonian $\hat{H}$ for the micromaser is taken to be

$$\hat{H} = \omega_0 S^z + \omega a^{\dagger}a + g(S^+a + a^{\dagger}S^-); \qquad (28)$$

g is a coupling constant; $a(a^{\dagger})$ describe a single cavity mode and $[a,a^{\dagger}] = 1$. Modes are well spaced in the microwave cavity and we can model it by one mode with frequency $\omega \approx \omega_0$. Notice we quantised the mode with the usual $[a,a^{\dagger}] = 1$. We leave as a 'homework problem' quantisation by deformation of the Heisenberg algebra [35]. Only energy conserving terms are in $\hat{H}$ and total

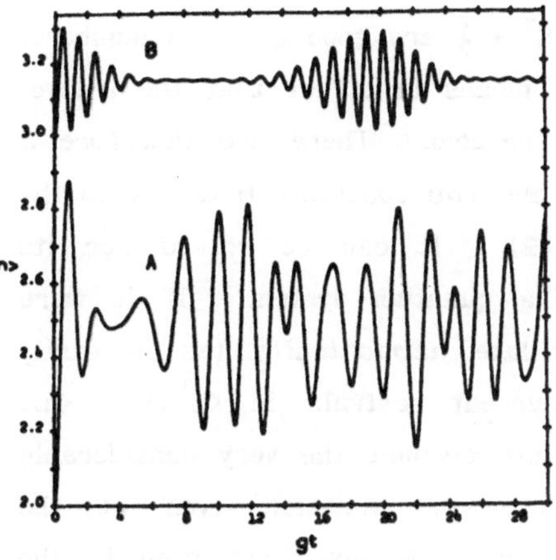

Fig. 2 Recurrent quantum *revivals* in the mean number ⟨n⟩ of photons in the microwave cavity: A on resonance; B off resonance.

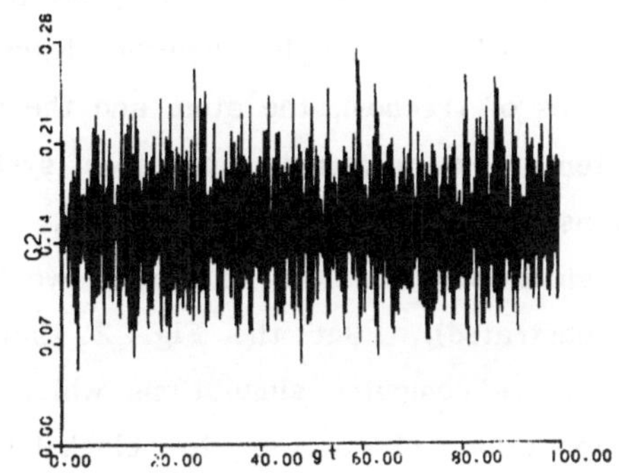

Fig. 3 'Chaotic' dynamics in $g^{(2)}$ (see text) for 5 atoms in the microwave cavity: gt is scaled time.

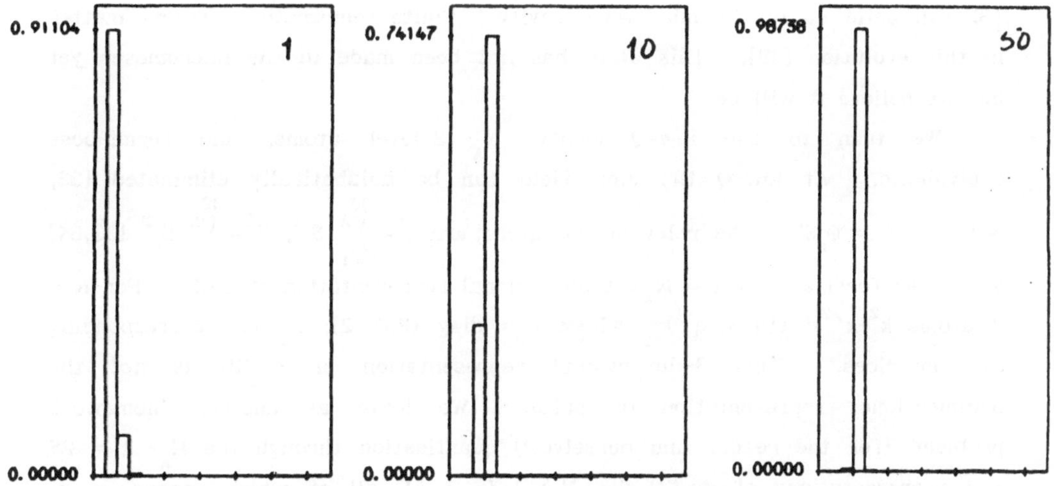

Fig. 4 Build up of the Fock state |3⟩ in the microwave cavity field from initial black-body noise after 1,10,50 inverted atoms have passed through the cavity. The probability $P_n = 0.98738$ for n = 3 in Fig. 4c.

spin $\hat{S}^2$ and the number operator $\hat{N} \equiv a^\dagger a + S^z + \frac{1}{2}$ are good quantum numbers: $[\hat{H}, \hat{S}^2] = 0$, $[\hat{H}, \hat{N}] = 0$. $\hat{S}^2$ good ($S = \frac{1}{2}$) means in effect that the 2-level atom provides a single quantum degree of freedom. There are therefore 2 degrees of freedom, the atom and the mode, and two constants $\hat{H}$ and $\hat{N}$ so the system is a *quantum integrable* system (§2). It can be *solved* for its eigenstates and eigenvalues [34,36]. It *is* a quantum system. If it were classical, $\langle S^z \rangle$ or $\langle \hat{N} \rangle$ or $\langle a^\dagger a \rangle$ would oscillate *harmonically* (as is easily demonstrated). But the Fig. 2 shows recurrent *revivals* in $\langle a^\dagger a \rangle = \langle n \rangle$. These are computer simulations which take into account the very considerable complication of access to black-body microwave radiation even at the temperatures 0.5°K now used [33,34]. Such revivals have been seen in the experiments [30]. The Fig. 3 shows how 5 2-level atoms in the cavity (here actually making 2-photon transitions [37]) produce a 'chaotic' looking dynamics in $g^{(2)} \equiv \langle a^\dagger a^\dagger a a \rangle / (\langle a^\dagger a \rangle^2)$ ($Q = \infty$ for this calculation [37]). The Fig. 4 shows our first report that a pure n-photon state (Fock state) $|n\rangle = |3\rangle$ can build up in the microwave cavity. Finite temperature effects matter in this evolution [38]. This state has *not* been made in the micromaser yet but we believe it will be.

We turn to the low-Q cavity, N_A 2-level atoms, and fermi-bose equivalence. At low-Q the e.m. field can be adiabatically eliminated [33, 34] and $\hat{H} = \omega_0 S^z$. The relevant operators are $S^z = \sum_{i=1}^{N_A} S_i^z$; $S^\pm = \sum_{i=1}^{N_A} S_i^\pm$ [33,34] and these form a $2S + 1 = N_A + 1$ dimensional representation of su(2). For $N_A = 2$ atoms $k^2 - k'^2 = \frac{1}{2}(q + q^{-1})\sigma_3$ where $\sigma_3 = \mathrm{diag}\,(2,0,-2)$: $\sigma_\pm$ are corresponding 3×3 matrices. This 3-dimensional representation of $su_q(2)$ is *not* the 3-dimensional representation of su(2). We leave as another 'homework problem' (for the reader and ourselves!) quantisation through the $N_A + 1 = 2S + 1$ representations of $su_q(2)$ for $N_A > 1$ ($S > \frac{1}{2}$), all of which depend on q, and follow orthodox spin quantum mechanics.

For consistency we return again to the use of the circumflex notation for operators. There are N_A mutually commuting operators $\hat{S}_i^z$, $i = 1,\ldots,N_A$ commuting with $\hat{H} = \omega_0 \hat{S}^z$. Each $\hat{S}_i^z$ is a spin-$\frac{1}{2}$ operator with total spin $S = \frac{1}{2}$ so there are N_A quantum degrees of freedom. The N_A atom system (without damping) is therefore quantum completely integrable. But the $\hat{S}_i^z$ are not observables. Since $[\hat{H}, \hat{\vec{S}}_i^{\,2}] = 0$ one can construct states $|S,m\rangle$ (Dicke states [33,34]): $\hat{\vec{S}}^2\,|S,m\rangle = S(S + 1)|S,m\rangle$, $\hat{S}^z|S,m\rangle = m|S,m\rangle$. The largest S is $S = \frac{1}{2}N_A$ and the states $|\frac{1}{2}N_A,m\rangle$ form a basis for a $2S + 1 = N_A + 1$ representation of SU(2).

For example, if $N_A = 4$ the $2^{N_A} = 16$ states fall into the one set of $N_A + 1 = 5$ states $|\frac{1}{2}N_A, m\rangle$, 3 sets (i.e. $N_A - 1$ sets) of $2S + 1 = 2(\frac{1}{2}N_A - 1) + 1 = 3$ states, and 2 sets (i.e. $\frac{1}{2}N_A(N_A - 1)$ sets) of $2S + 1 = 2(\frac{1}{2}N_A - 2) + 1 = 1$ states (the total number of states is then $5 + 9 + 2 = 16$). These 6 *sets* of states each act like a 'quantum pendulum' (SU(2) is a double cover of SO(3) the group of the classical pendulum). However, only the single set with $S = \frac{1}{2}N_A$ contains the true ground state (all spins down) and the fully excited state (all spins up), both of which are relatively easy to reach in the laboratory. We can therefore focus on N_A Rydberg atoms which enter the microwave cavity in that total ground state (each atom was in the lower $30S_{1/2}$ state in [32]). In [32] some of these atoms make $30S_{1/2} \rightarrow 30P_{1/2}$ transitions to reach thermal equilibrium with black-body radiation in the cavity ($T = 300^{\circ}$K, mean black-body photon number $\bar{n} = 47$, and $T = 900^{\circ}$K, $\bar{n} = 137$ were used [32]).

The partition function Z at equilibrium is

$$Z = \mathrm{Tr}\, e^{-\beta\omega_0 S^z} = \sum_{n=0}^{2S=N_A} e^{-n\beta\omega_0}. \tag{29}$$

The mean number N_+ of atoms excited in equilibrium is

$$N_+ = \sum_{n=1}^{2S=N_A} n e^{-n\beta\omega_0}/Z. \tag{30}$$

Set, conveniently, $X \equiv \bar{n}(1 + \bar{n})^{-1}$ where $\bar{n}$ is the Planck function $\bar{n} = (e^{\beta\omega_0} - 1)^{-1}$ for resonant black-body radiation. Then one finds from (29) that

$$N_+ = X[1-(N_A + 1)X^{N_A} + N_A X^{N_A+1}] / [1 - X - X^{N_A+1} + X^{N_A+2}]. \tag{31}$$

For $N_A = 1$, $N_+ = X(1 + X)^{-1} = \bar{n}(1 + 2\bar{n})^{-1}$, the Einstein formula: the system is acting as one fermion [Check: fermions with 2 states have $N_+ = e^{-\beta\omega_0} / (1 + e^{-\beta\omega_0}) = \bar{n} / (1 + 2\bar{n})$]. However, for $N_A \rightarrow \infty$, $N_+ = X(1-X)^{-1}$ from (31) and $N_+ = \bar{n}$. This is a *bose* occupation number and the system is acting as one giant collective boson (a giant quantum oscillator with eigenenergies ω_0, $2\omega_0$, $3\omega_0$,...). Fig. 5 reproduces the observations in the experiments [32]. There is the complication that, for $N_A \lesssim 2000$, atoms do not reach equilibrium during their transit time of 2.5μ sec across the cavity: collective damping times depend on N_A [32,33]. Thus Fig. 5 does not show the equilibrium curve N_+ against N_A, which is (31), for $N_A \lesssim 2000$. But the two curves, the experimental curve Fig. 5 and the curve of formula (31), prove to be very similar for the parameters of the experiments [33] (see the Fig. 6 of [33]).

The single collective boson emerges from the infinite dimensional representations of su(2) (Another homework problem: find the deformed Heisenberg algebra from $su_q(2)$ this way!). Use $S^- = (2S - \hat{n})^{\frac{1}{2}}a$, $S^+ = a^\dagger(2S - \hat{n})^{\frac{1}{2}}$ with $\hat{n} \equiv a^\dagger a$ and $[a,a^\dagger] = 1$. Now define $\bar{S}^\pm$, $\bar{S}^z$ by $(2S^{\frac{1}{2}})^{-1}S^\pm$, $(2S)^{-1}S^z$ so that $[\bar{S}^+, \bar{S}^-] = 2\bar{S}^z = (\hat{n}S^{-1} - 1) \to -1$ as $S = \frac{1}{2}N_A \to \infty$, while $[\bar{S}^\pm, \bar{S}^z] = \mp (2S)^{-1}\bar{S}^\pm \to 0$ as $S \to \infty$ and $\bar{S}^z$ acts as a c-number. Evidently $S^+ \to a^\dagger$, $S^- \to a$ as $S \to \infty$.

To this extent the system *is* displaying bose-fermi equivalence: it 'acts' like (i.e. can be observed as) one free fermion for $N_A = 1$ and as one collective free boson for $N_A \to \infty$. There is a boson *description* for each value of N_A in terms of the bose operators a, $a^\dagger$ (with $[a,a^\dagger] = 1$). However there is apparently a fermion description only for $N_A = 1$: this is the bose-fermi equivalence of the 2-level atom.

This experimental 'completely integrable quantum system' with N_A degrees of freedom has other connections with the quantum integrable models in 1+1: in particular, and as remarked at the end of §3, for the bose gas when $c \to 0$ in Δ_b in (19) one finds the *free* bose gas with dispersion $\varepsilon(k) = k^2 - \mu$. But from (19), or from (17), one finds free fermions $\tilde{\varepsilon}(k) = k^2 - \mu$ as $c \to \infty$. And consider quantum sine-Gordon (s-G): we (and others) have shown [1,16,18] that the quantum integrable s-G model, bose-fermi equivalent to the quantum massive Thirring model (MTM), has a free energy determined by one quantum kink-antikink pair and $N_{b'} - 1$ quantum breathers. The number $N_{b'} = [8\pi\gamma_0''^{-1}]$ (= integral part) and $\gamma_0'' = \gamma_0(1-\gamma_0/8\pi)^{-1}$ in terms of a bare coupling constant $\gamma_0 > 0$ (compare sinh-G and its renormalised phase Δ_f given below (14)). In semiclassical limit the quantum s-G system reduces to a kink-antikink pair and one collective boson [39] - wholly analogous to that described above!

The quantisation of the quantum s-G derives from (1) and the quantum groups. It would be interesting to see how the simpler quantum models considered in this §4 derive from the deformed universal covering algebra $U(su(2))_q = su(2)_q$ - perhaps through the three 'homework problems' put forward in this section.

REFERENCES

[1] Bullough R.K., Olafsson S., Chen Yu-zhong. In: Differential Geometrical Methods in Theoretical Physics, Chau Ling-Lie and Nahm Werner (eds.) (Plenum, New York 1990) To appear.

[2] Zee A. In: Bedell K.S., Coffey D. Keltzer D.E., Pines D., Schrieffer J.R. (eds.) High Temperature Superconductivity, Proc. Los Alamos

Symposium 1989, (Addison Wesley Publ. Co., Redwood City, Calif., 1990) pp. 248-298.
[3] Froehlich J. This meeting.
[4] Zimanyi G.T., Kivelson S.A., Luther A., Phys. Rev. Letts. 60: 2089 (1988)
[5] Bullough R.K., Olafsson S. In: Solomon Allan I. (ed.) Proc. 17th Intl. Conference on Diff. Geometrical Methods in Theoretical Physics (World Scientific, Singapore, 1989)
[6] Bullough R.K., Chen Yu-zhong, Timonen J. In: Zakharov V.E., Sitenko A.G., Erokhin N.S., Chernousenko V.M. (eds.) Proc. IV Intl. Workshop on Nonlinear and Turbulent Processes in Physics (World Scientific, Singapore, 1990) To appear.
[7] Bullough R.K., Timonen J. In: Bishop A., Campbell D. (eds.) Proc. 10th CNLS Conference, Los Alamos (Physica D, Nonlinear Phenomena, North Holland, 1991) To appear.
[8] Kasteler D. This meeting
[9] Majid S. Quasitriangular Hopf algebras and the Yang-Baxter equations. Int. J. Mod. Phys. A (1989). And this meeting.
[10] For example Kulish P.P., Sklyanin E.K. In: Hietarinta J., Montonen C. (eds.) Proc. of the Tvärminne Symp. Finland, 1981. (Springer-Verlag, Heidelberg, 1982).
[11] Witten E. In: Simon B., Truman A., Davies I.M. (eds.) IXth Intl. Congress on Math. Phys. 17-27 July, 1988, Swansea, Wales (Adam Hilger, Bristol, 1989) pp. 77-116
[12] Feynman R.P., Hibbs A.R. Quantum Mechanics and Path Integrals (McGraw Hill Book Co., New York, 1965)
[13] Bullough R.K., Chen Yu-zhong, Olfasson S., Timonen J. In: Balabane M., Lochak P., Sulem S. (eds.) Integrable Systems and Applications,Lecture Notes in Physics 342 (Springer-Verlag, Heidelberg, 1989) pp. 12-26
[14] Bullough R.K., Timonen J. To be published
[15] Itzykson C., Zuber J.B. Quantum Field Theory (McGraw Hill Book Co., New York, 1980)
[16] Bullough R.K. In: Claro F. (ed.) Nonlinear Phenomena in Physics (Springer-Verlag, Heidelberg, 1985) pp. 70-102. Also see the pp. 103-126 for historical interest.
[17] Arnold V.I. Mathematical Methods of Classical Mechanics (Springer-Verlag, Heidelberg, 1978)
[18] Chen Yu-zhong. Ph.D. Thesis, University of Manchester, July, 1989
[19] Yang C.N., Yang C.P. J. Math. Phys. 10: 1115 (1969)
[20] Jimbo M., Miwa T. In: D'Ariano G.M., Montorsi A., Rasetti M.G. (eds.) Integrable Systems in Statistical Mechanics (World Scientific, Singapore, 1985)
[21] Leites D. Discussions at this meeting.
[22] Olafsson S., Bullough R.K. To be published.
[23] Reshitikhin N., Smirnov F., Hidden Quantum Group Symmetry and Integrable Perturbations of Conformal Field Theories. To be published
[24] Cheng Yi, Li Yi-shen, Bullough R.K. J. Phys. A: Math. Gen 21: L433 (1988)
[25] Bullough R.K., Pilling D.J., Timonen J. J. Phys. A: Math Gen. 19: L955 (1986)
[26] Wadati M., Deguchi T., Akutsu Y., Phys. Reps. 180, Nos. 4 & 5 (1989) pp.247-332
[27] Wadati M. J. Phys. Soc. Japan 54: 3727 (1985)
[28] Bullough R.K., Timonen J. To be published
[29] Meschede D., Walther H., Müller G. Phys. Rev. Lett. 54: 551 (1985)
[30] Rempe G., Walther H., Klein N. Phys. Rev. Lett 58: 353 (1987)
[31] Rempe G., Schmidt-Kaler F., Walther H., Phys. Rev. Lett. 64: 2783 (1990)

[32] Raimond J.M., Goy P., Gross M., Fabre C., Haroche S. Phys. Rev. Lett. 49: 117 (1982)
[33] Bullough R.K. Hyperfine Interactions (J.C. Baltzer, A.G., Basel) 37: 71 (1987)
[34] Bullough R.K. et al. In: Tombesi P., Pike E.R. (eds.) Squeezed and Nonclassical Light (Plenum, New York, 1989) pp. 81-106
[35] Biedenbarn L.C. This meeting
[36] Jaynes E.T., Cummings F.W. Proc. IEEE 51: 89 (1963)
[37] Puri R.R., Bullough R.K., Nayak N. In: Eberly J.H., Mandel L., Wolf E. (eds.) Proc. 6th Rochester Conference on Coherence and Quantum Optics (CQO6) (Plenum, New York, 1990) pp. 943-947
[38] Nayak N., Thompson B.V., Bullough R.K. To be published
[39] Timonen J., Bullough R.K., Pilling D.J. Phys. Rev. B 34: 6525 (1986)
[40] Timonen J., Bullough R.K, Phys. Lett. 82A: 183 (1981)
[41] Timonen J., Stirland M., Pilling D.J., Bullough R.K., Phys. Rev. Lett. 56:2233 (1986)

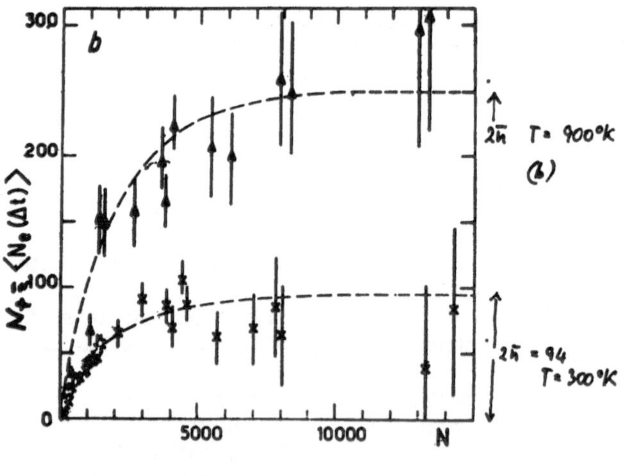

Fig. 5 Observations (from [32]) of N_+, the mean number of excited atoms as a function of atoms $N \equiv N_A$.

SOME IDEAS AND RESULTS ON INTEGRABLE NONLINEAR EVOLUTION SYSTEMS

F.Calogero(*)
Dipartimento di Fisica, Università di Roma "La Sapienza", 00185 Roma, Italy
Istituto Nazionale di Fisica Nucleare, Sezione di Roma

0.Introduction

In this paper we outline some recent ideas and results on integrable nonlinear evolution equations, referring to the literature for full presentations of these findings.

1. Universality, Applicability, Integrability.

A major impediment to the flourishing of nonlinear science, and in particular of the investigation of nonlinear evolution equations, had been the widespread feeling that the interest of such a study was doomed by the following dilemma. One research line might focus on fairly general classes of nonlinear evolution equations, likely to capture many cases of applicative and theoretical importance; but then, it could hardly aim beyond general results, such as proofs of existence and uniqueness; without any hope to get much real understanding of the behaviours described by the actual solutions of the equations in question. Alternatively, the research might focus on special equations which could be, in some sense, solved, so that a substantial understanding of the behaviours described by their solutions could actually be gleaned; but then, it was expected that the equations in question would be just flukes, unlikely to have much relevance, applicatively or otherwise.

Recent findings - which consolidate ideas that had been widely diffused for quite some time - have overcome this perverse dichotomy. The main point has been the identification of certain nonlinear evolution equation which have a universal character, inasmuch as they are obtainable by certain limiting procedures from very large classes of equations. Moreover, these limiting procedures are physically justified in many circumstances; hence, since the "large classes" of course contain (just because they are large) very many equations, including several which are of applicative interest, these universal equations (of which the nonlinear Schroedinger equation is a prototype) are widely applicable. On the other hand, because these limiting procedures have an exact asymptotic character and therefore generally preserve integrability, and because the large classes, just because they are large, are likely to contain at least one integrable equation, it generally turns out that these universal equations are integrable.

(*) On leave while serving as Secretary General, Pugwash Conferences on Science and World Affairs, Geneva London Rome.

Universal nonlinear evolution equations which are both widely applicable and integrable constitute of course a most attractive focus for investigation. The existence of such equations has indeed underlined much of the revival of nonlinear science that has occured over the last quarter century (the other, separate, main motive for this revival has been the emerging understanding of chaotic behaviour, largely caused by the availability of powerful computers). The recent clarification of the close interplay among the universality, wide applicability and integrability of certain nonlinear evolution equations - indeed the fact that these three features can, at least in some cases, now be traced to a common origin, as outlined above - constitutes therefore an important milestone.

The reader who has become sufficiently intrigued by the above discourse to develop a desire for a more precise understanding of these ideas will find a more complete and detailed presentation in [1] and in the literature referred to there.

2. C-integrable Nonlinear PDEs

In the preceding Section the notion of integrable nonlinear evolution equation has been introduced, without any attempt to define precisely its meaning. Indeed, for infinite dimensional systems, such as nonlinear PDEs, the provision of such a precise, and universally accepted, definition, is still an open task. It has, however, turned out to be convenient to introduce the heuristic notions of "S-integrable" and "C-integrable" nonlinear PDEs (see [1] and the literature quoted there).

S-integrable nonlinear PDEs are those solvable by the Spectral transform technique, or some analogous method. Prototypical examples are the Korteweg-de Vries, nonlinear Schroedinger, sine-Gordon, Kadomtzev-Petviashvili, Davey-Stewartson, Benjamin-Ono equations (see, for instance, [2]).

C-integrable nonlinear PDEs are those solvable by an appropriate Change of variables. Prototypical examples are the Burgers and Eckhaus [3] equations.

C-integrable equations are generally easier to solve than S-integrable equations; they may nevertheless exhibit a very interesting phenomenology (see, for instance, [3]), and, just for the reasons outlined in the preceding Section, they may feature a universal character and enjoy wide applicability. Their identification and their study constitute therefore an interesting, if sometimes rather elementary, program. For some recent findings in that direction the interested reader is referred to [2], [3], [4] and [5].

3. Integrable Systems of Coupled Nonlinear ODEs and PDEs, and Solvable Nonlinear Integrodifferential Equations

In this Section we introduce a simple technique which may be used to manufacture integrable systems of coupled nonlinear ODEs and PDEs and solvable nonlinear integrodifferential equations of "Boltzmann type". This presentation is limited to an illustration of the main idea of this approach in a simple context.

The general motivation for this line of research stems from the considerations outlined in the two preceding Sections, which suggest a need for identifying as many integrable or solvable equations as possible, both because these might be themselves of applicative or theoretical interest (even if they have been manufactured by rather simple tricks), or because more interesting, but still solvable, equations might subsequently be derived from them via suitable limiting procedures.

Let the function f(x,t) satisfy the nonlinear "Riccati" equation

$$\dot{f}(x,t) = \alpha_2(x)\, f^2(x,t) + \alpha_1(x)\, f(x,t) + \alpha_0(x). \tag{1}$$

Here, and always below, a superimposed dot denotes (partial) differentiation with respect to the variable t ("time"). Note that the variable x enters in (1) only parametrically.

It is easily seen that the solution of this nonlinear ODE reads

$$f(x,t) = \{f(x,0)+ [2\,\alpha_0(x) + \alpha_1(x)\, f(x,0)]\, T(x,t)\} \,/\, \{1-[2\alpha_2(x)\, f(x,0) + \alpha_1(x)]\, T(x,t)\} \quad , \tag{2a}$$

$$T(x,t) = \{\mathrm{tgh}\,[t\,\Delta(x)/2]\}/\Delta(x), \tag{2b}$$

$$[\Delta(x)]^2 = [\alpha_1(x)]^2 - 4\,\alpha_2(x)\,\alpha_0(x) \quad . \tag{2c}$$

Now introduce an orthonormal complete set of functions in the x variable:

$$\int_{x_0}^{x_1} dx\ \varphi_n(x)\,\psi_m(x) = \delta_{nm} \quad , \tag{3a}$$

$$\sum_n \varphi_n(x)\,\psi_n(y) = \delta(x-y) \ , \ x_0 \le x,y \le x_1 \quad . \tag{3b}$$

Note that we omit here to specify the range of the integration (from x_0 to x_1; these constants need not be finite), as well as the range of the summation.

We now set

$$f(x,t) = \sum_n u_n(t)\,\psi_n(x) \quad , \tag{4a}$$

$$u_n(t) = \int_{x_0}^{x_1} dx\ \varphi_n(x)\, f(x,t) \quad . \tag{4b}$$

Insertion of these expressions in (1) yields, for the quantities $u_n(t)$, the following system of coupled nonlinear ODEs:

$$\dot{u}_n(t) = \sum_{m,\ell} A^{(2)}_{nm\ell}\, u_m(t)\, u_\ell(t) + \sum_m A^{(1)}_{nm}\, u_m(t) + A^{(0)}_n \quad , \tag{5}$$

with

$$A^{(2)}_{nm\ell} = \int_{x_o}^{x_1} dx\, \alpha_2(x)\, \varphi_n(x)\, \psi_m(x)\, \psi_\ell(x) \quad , \tag{6a}$$

$$A^{(1)}_{nm} = \int_{x_o}^{x_1} dx\, \alpha_1(x)\, \varphi_n(x)\, \psi_m(x) \quad , \tag{6b}$$

$$A^{(0)}_{n} = \int_{x_o}^{x_1} dx\, \alpha_0(x)\, \varphi_n(x) \quad . \tag{6c}$$

The system (5) with (6) is clearly solvable. Indeed the solution of the initial value ("Cauchy") problem is given by (4b), with f(x,t) given by (2a,b,c) and f(x,0) in (2a) given (see (4a)) by

$$f(x,0) = \sum_n u_n(0)\, \psi_n(x) \quad . \tag{7}$$

Let us emphasize that the solvability of (5), as demonstrated above, holds for any determination of the coefficients $A^{(2)}_{nm\ell}$, $A^{(1)}_{nm}$ and $A^{(0)}_{n}$, which is consistent with the representation (6) (for any choice of the 3 functions $\alpha_p(x)$, p=2,1,0, and of the orthonormal basis, see (3)). For instance setting $x_0=-\pi$, $x_1=\pi$ and selecting as orthonormal system

$$\psi_n(x) = \exp(i\, n\, x) \ , \tag{8a}$$

$$\varphi_n(x) = (2\pi)^{-1} \exp(-i\, n\, x) = (2\pi)^{-1}\, \psi_{-n}(x) \ , \tag{8b}$$

one obtains for the coefficients $A^{(2)}_{nm\ell}$, $A^{(1)}_{nm}$ and $A^{(0)}_{n}$ the following simple formulae:

$$A^{(2)}_{nm\ell} = a^{(2)}_{n-m-\ell} \quad , \tag{9a}$$

$$A^{(1)}_{nm} = a^{(1)}_{n-m} \quad , \tag{9b}$$

$$A^{(0)}_{n} = a^{(0)}_{n} \quad . \tag{9c}$$

Hence the system (5), with the indices n,m,ℓ running from $-\infty$ to $+\infty$, and the coefficients $A^{(2)}_{nm\ell}$, $A^{(1)}_{nm}$, $A^{(0)}_{n}$ given by these formulae, is easily solvable. Note that the coefficients $a^{(p)}_{n}$, p=2,1,0, that appear in the r.h.s. of (9), can be chosen arbitrarily, being of course related to the 3 functions $\alpha_p(x)$ be the formulae

$$a_n^{(p)} = (2\pi)^{-1} \int_{-\pi}^{\pi} dx \; \alpha_p(x) \exp(-i\,n\,x) \quad , \quad p=2,1,0 \; , \tag{10a}$$

$$\alpha_p(x) = \sum_{n=-\infty}^{+\infty} a_n^{(p)} \exp(in\,x) \quad , \quad p=2,1,0 \; . \tag{10b}$$

Of course, many other (less obviously trivial) examples can be produced, by making a different choice for the orthonormal basis; in particular, the choice of a finite-dimensional basis produces finite-dimensional dynamical systems (i.e., systems of type (5) but with the indices n,m,ℓ spanning only a finite range), while the choice of a basis labeled by a continuous variable yields, instead of the dynamical system (5), an integrodifferential equation "of Boltzmann type". More importantly, the same approach can be used taking as starting point, rather than the first order Riccati equation (1), a second order nonlinear ODE, or an integrable PDE with quadratic nonlinearity (such as the Burgers, Korteweg-de Vries, Kadomtzev-Petviashvili equations), or an integrable integrodifferential equation with quadratic nonlinearity (such as the Benjamin-Ono equation) or, for that matter, solvable equations with higher-than-quadratic nonlinearities. For an exploration of some of these possibilities, and a display of the corresponding findings, the interested reader is referred to [6].

References

[1] F.Calogero: "Why are certain nonlinear PDEs both widely applicable and integrable?", in: What is integrability? (V.E.Zakharov, editor), Springer, 1990, pp.1-62.

[2] F.Calogero and A.Degasperis: Spectral Transform and Solitons, I. North Holland, 1982.

[3] F.Calogero and S.De Lillo: "The Eckhaus PDE $i\psi_t + \psi_{xx} + 2\,(|\psi|^2)_x\;\psi + |\psi|^4\;\psi = 0$". Inverse Problems 3, 633-681 (1987); 4, 571 (1988).

[4] F.Calogero and Ji Xiaoda: "C-integrable nonlinear PDEs. I". J.Math.Phys. (in press).

[5] F.Calogero and Ji Xiaoda: "C-integrable nonlinear PDEs. II". J.Math.Phys. (submitted to).

[6] F.Calogero: "Integrable systems of coupled nonlinear ODEs and PDEs, and solvable integrodifferential equations of Boltzmann type" (in preparation).

An algebraic characterization of complete integrability for Hamiltonian systems

S. De Filippo[a,d], G. Landi[b,d], G. Marmo[c,d], G. Vilasi[a,d]

[a]Dipartimento di Fisica Teorica e s.m.s.a., Università di Salerno, via S. Allende, I-84081 - Baronissi (SA), Italy.
[b]SISSA, Strada Costiera 11, I-34014 Trieste, Italy.
[c]Dipartimento di Scienze Fisiche - Università di Napoli, Mostra d'Oltremare Pad.19, I-80125 Napoli, Italy.
[d]Istituto Nazionale di Fisica Nucleare - Sezione di Napoli, Mostra d'Oltremare Pad.20, I-80125 Napoli, Italy.

Introduction

The main idea behind non commutative differential geometry relies on the existing duality between a manifold M and the ring of smooth functions on M, denoted by $\mathcal{F}(M)$. This can briefly expressed by saying that $M \equiv \text{Hom}\ (\mathcal{F}(M), \mathbf{R})$. To what extent one can reconstruct a manifold M from a ring $\mathcal{F}$ is not our main concern here, the asserted duality should be taken only as a motivation for what we are going to say.

Non commutative differential geometry, in a sense, begins when the starting ring $\mathcal{F}$ is non commutative. In this respect see [Ka, Du].

Here we concentrate our attention on the algebraic description of "classical physics", a kind of "pre-non commutative differential geometry". To write down a description of classical physics we have to define:

1) An "algebraic differential calculus".
2) An "algebraic setting" for Lagrangian and Hamiltonian formalism.
3) An algebraic framework for gauge theories.

This program has been already carried over to a large extent [LaMa1, LaMa2, DLMV]. At the moment, the program of quantizing this approach is under investigation.

Here we shall concentrate our attention on the problem of complete integrability. One of the main advantage of this "algebraic approach" is the possibility of a natural generalization to the super symmetric situation.

The starting point for the "algebraic differential calculus" is a theorem by R. Palais [Pa] stating that the only natural operations on a manifold M are the Lie derivative L_X, the inner derivative i_X, and the exterior derivative d. These operations are connected by the Cartan's identity $L_X = i_X d + d i_X$. Thus, an algebraic differential calculus obtains as soon as these operations are reproduced at the algebraic level. Tangent and cotangent bundles are recovered by further qualifying the ring $\mathcal{F}$ by selecting additional structures: one has to deal with "tangent rings", "symplectic rings" and "Poisson rings". As one could anticipate from the operations involved, our "algebraic differential calculus" will reproduce much of what is known as De Rham cohomology, Chevalley cohomology and related topics.

Algebraic differential calculus

We start with a Lie algebra g on a field $\mathbf{K}$ of characteristic zero. The exterior algebra $\Lambda(g) = \oplus_j \Lambda^j(g)$ can be thought of as a differential complex over g by introducing a homology operator ∂ [Ko]. We have

$$\cdots \rightarrow \Lambda^p(g) \xrightarrow{\partial} \Lambda^{p-1}(g) \xrightarrow{\partial} \cdots \Lambda^1(g) \xrightarrow{\partial} 0 \,,$$

with $\Lambda^1(g) = g$, and ∂ defined by

$$\partial(X_1 \wedge X_2 \wedge \cdots \wedge X_k) =: \sum_{i<j} (-1)^{i+j+1} [X_i, X_j] \wedge X_1 \wedge \overset{i}{\check{\cdots}} \overset{j}{\check{\cdots}} \wedge X_k \,, \quad X_i \in g \,. \tag{1}$$

the Jacobi identity implies that $\partial \circ \partial = 0$ so that we can define homology spaces as usual by $H_k(g) = \dfrac{\text{Ker } (\partial : \Lambda^k \rightarrow \Lambda^{k+1})}{\text{Im } (\partial : \Lambda^{k-1} \rightarrow \Lambda^k)}$. We notice that, as ∂ is not a derivation for the wedge product, $H_*(g) = \oplus_k H_k(g)$ is not an algebra but only a vector space. The fact that ∂ is not a derivation allows to define a bracket

$$[\ ,\ ] : \Lambda^q \times \Lambda^p \rightarrow \Lambda^{p+q-1} \,,$$

$$[G, H] =: (-1)^{q+1} \{\partial(G \wedge H) - \partial G \wedge H - (-1)^q G \wedge \partial H\} \,, \quad G \in \Lambda^q \,,\ H \in \Lambda^p \,, \tag{2}$$

which makes $\Lambda(g)$ a graded Lie algebra.
The bracket (2) is the Schouten-Nijenhuis bracket. It is the only bracket on $\Lambda(g)$ which is a derivation with respect to the wedge product and reduces to the Lie bracket on vectors. This also implies that it cannot be deformed.
We can also define the operation

$$\varepsilon_{(\cdot)} : g \times \Lambda^p \rightarrow \Lambda^{p+1} \,,$$
$$\varepsilon_X(X_1 \wedge \cdots X_p) =: X \wedge X_1 \wedge \cdots \wedge X_p \,, \quad X, X_i \in g \,, \tag{3}$$

with properties

$$\varepsilon_X \varepsilon_Y = \varepsilon_{X \wedge Y} = -\varepsilon_{Y \wedge X} = -\varepsilon_Y \varepsilon_X \,. \tag{4}$$

For any $X \in g$ it is natural to define the operator

$$L_X =: \varepsilon_X \partial + \partial \varepsilon_X : \Lambda^p \rightarrow \Lambda^p \,. \tag{5}$$

It turns out that $L_{(\cdot)}$ is the extension to $\Lambda(g)$ of the adjoint representation, more specifically,

$$\text{ad}_{(\cdot)} : g \rightarrow \text{Der} g \,,$$
$$\text{ad}_X Y = [X, Y] \,. \tag{6}$$

Given any endomorphism

$$S : g \rightarrow g \,, \tag{7}$$

we can extend it in a natural way to all of $\Lambda(g)$

$$\delta_S : \Lambda^p \to \Lambda^p ,$$

$$\delta_S(X_1 \wedge X_2 \wedge \cdots \wedge X_p) =: \sum_j X_1 \wedge \cdots \wedge S(X_j) \wedge \cdots \wedge X_p . \tag{8}$$

One can also consider the operator

$$\partial_S := \delta_S \circ \partial - \partial \circ \delta_S : \Lambda^p(g) \to \Lambda^{p-1}(g) . \tag{9}$$

In general $\partial_S \circ \partial_S \neq 0$, and $\partial_S \circ \partial_S(X \wedge Y \wedge Z) =: \mathbf{M}_S(X, Y, Z) \in g$ defines a **K**-linear map associated with S. ∂_S is a boundary operator, i.e. $\partial_S \circ \partial_S = 0$ iff $\mathbf{M}_S = 0$. Again, ∂_S is not a derivation and we can define a new bracket

$$[\ ,\ ]_S : \Lambda^q \times \Lambda^p \to \Lambda^{p+q-1} ,$$

$$\begin{aligned} [G, H]_S \ &=: \ (-1)^{q+1}\{\partial_S(G \wedge H) - \partial_S G \wedge H - (-1)^q G \wedge \partial_S H\} , \\ & \qquad G \in \Lambda^q , \ H \in \Lambda^p , \end{aligned} \tag{10}$$

On elements of g it gives

$$[X, Y]_S = S[X, Y] - [SX, Y] - [X, SY] . \tag{11}$$

This process can be iterated.

It is possible to "dualize" these operations to any "space" which carries an action (or a representation) of g. To this aim, let us consider now a complex of vector spaces (Ω, d), $\Omega = \oplus_j \Omega^j$, $d\Omega^p \subset \Omega^{p+1}$, $d \circ d = 0$. We say that (Ω, d) is a complex over the Lie algebra g, if with any $X \in g$ there is associated a linear map

$$i_X : \Omega \to \Omega , \quad i_X \Omega^p \subset \Omega^{p-1} , \quad i_X \Omega^0 = 0 , \tag{12}$$

with properties

$$i_X i_Y + i_Y i_X = 0 , \quad [i_X d + d i_X, i_Y] = i_{[X,Y]} . \tag{13}$$

The operator $L_X := i_X d + d i_X$, which carries Ω^p into itself, will be called Lie derivative. The map

$$L_{(\cdot)} : g \to \mathrm{Der}\Omega^p \subset \mathrm{Lin}\ (\Omega^p, \Omega^p) \tag{14}$$

is a representation of g in Ω^p since $L_{[X,Y]} = [L_X, L_Y]$.

It is easy to exhibit complexes over a Lie algebra g. Let V be a vector space carrying a representation of g ,$g \ni X \to X\cdot \in \mathrm{End}\ V$ and $\Omega^p, (g, V)$ the space of V-valued alternating maps (of degree p) on g. Then, $\Omega(g, V) =: \oplus_p \Omega^p(g, V)$, $\Omega^0(g, V) \equiv V$, can be made into a complex over g by defining the following operators

$$\begin{aligned} d : \Omega^p(g, V) \ &\to \ \Omega^{p+1}(g, V) , \\ (d\alpha)(X_1 \wedge \cdots \wedge X_{p+1}) \ &=: \ \sum_i (-1)^{i+1} X_i \cdot \alpha(X_1 \wedge \cdots \overset{i}{\checkmark} \cdots \wedge X_p) + \\ & \qquad -\alpha(\partial(X_1 \wedge \cdots \wedge X_{p+1})) ; \end{aligned} \tag{15}$$

$$\begin{aligned} & i_{(\cdot)} : g \times \Omega^p(g, V) \to \Omega^{p-1}(g, V) , \\ & (i_X \alpha)(X_1 \wedge \cdots \wedge X_{p-1}) =: \alpha(\varepsilon_X(X_1 \wedge \cdots \wedge X_{p-1})) ; \end{aligned} \tag{16}$$

$$\begin{aligned} L_{(\cdot)} : g \times \Omega^p(g,V) &\rightarrow \Omega^p(g,V)\,, \\ (L_X\alpha)(X_1 \wedge \cdots \wedge X_p) &=: X \cdot \alpha(X_1 \wedge \cdots \wedge X_p) + \\ &\quad - \sum_i \alpha(X_1 \wedge \cdots \wedge [X, X_i] \wedge \cdots \wedge X_p)\,. \end{aligned} \tag{17}$$

That $d^2 = 0$ follows from the fact that the action of g on V is a representation; i.e. from the Jacobi identity. Again $L_{(\cdot)} = i_{(\cdot)}d + di_{(\cdot)}$ which gives $[L_{(\cdot)}, d] = 0$.

A particular case is $V = \mathbf{K}$ with the trivial representation of g on $\mathbf{K}$. Now $\Omega(g, \mathbf{K})$ can be identified with the exterior algebra $\Lambda(g^*)$ over $g^* =: \mathrm{Lin}_{\mathbf{K}}(g, \mathbf{K})$. Since g acts trivially on $\mathbf{K}$, the first term in the definition of d and $L_{(\cdot)}$ in (15) and (17) are now absent. If $<,>$ denotes the natural pairing between g and g^*, it can be extended to a pairing between $\Lambda(g)$ and $\Lambda(g^*)$

$$< X_1 \wedge \cdots X_p \,,\; \alpha^1 \wedge \cdots \wedge \alpha^p >=: \det \| < X_i, \alpha^j > \| \,.$$

With this pairing the operators d, $i_{(\cdot)}$ and $L_{(\cdot)}$ in ((15)-(17)) are the adjoint operators of $-\partial$, $\varepsilon_{(\cdot)}$ and $L_{(\cdot)}$ defined in (1), (3) and (5) respectively.

Remark 1. In many cases of interest in physics, V can be a Lie algebra, a ring of functions, a ring of operators, an exterior algebra, a complex over g. If V is the ring $\mathcal{F}(M)$ of functions over a manifold M and g is the Lie algebra of all derivations of $\mathcal{F}(M)$, the latter can be identified with vector fields on M and definitions (15)-(17) give the usual exterior calculus over M. An example of non-commutative ring are the bounded operators on a Hilbert space. A more simple example is provided by the sections of the endomorphism-bundle associated with any vector bundle over M. We mention that if the ring $\mathcal{F}$ is commutative, derivations are a left $\mathcal{F}$-module otherwise they are a left module only for the centre of $\mathcal{F}$.

Remark 2. In the sequel we shall concentrate our attention to a ring $\mathcal{F}$ with unital element and we shall take $g = Der\mathcal{F}$ and $g^* = Lin_{\mathcal{F}}(g, \mathcal{F})$.

In this context we shall use "a generating set for $\mathcal{F}$" to mean that every derivation of $\mathcal{F}$ is uniquelly determined by the its action on the "generating set".

Derivations as algebraic dynamical systems

Any derivation X on $\mathcal{F}$ can be used to define formally an "algebraic dynamical system" by setting

$$\dot{f} = L_X f\,, \qquad \forall f \in \mathcal{F}\,. \tag{18}$$

Without additional structures on $\mathcal{F}$ a formal integration of this equation is provided by the formal series

$$\Phi_t^X(f) = f + \sum_{k=1}^{\infty} \frac{t^k}{k!}(L_X)^k f\,, \quad t \in \mathbf{R}\,, \; f \in \mathcal{F}\,. \tag{19}$$

When a particular generating set for $\mathcal{F}$ can be found with the property that for any element of such a set $(L_X)^k f = 0$, for some integer k indipendent of f, the series is no more formal and

$$\Phi^X : \mathbf{R} \times \mathcal{F} \rightarrow \mathcal{F} \tag{20}$$

is a 1-parameter group of automorphisms for $\mathcal{F}$.

For symplicity we say that $\mathcal{F}_X \subset \mathcal{F}$ is an *integrating set* for the derivation X if it generates $\mathcal{F}$ and $L_X L_X f = 0$ for all $f \in \mathcal{F}_X$. In this case the 1-parameter group of authomorphisms is given by

$$\Phi_t^X(f) = f + tL_X f\,, \quad t \in \mathbf{R}\,, \; f \in \mathcal{F}\,. \tag{21}$$

To deal with *Hamiltonian systems* we need the notion of *Poisson brackets* on the ring $\mathcal{F}$.

Poisson rings

A ring $\mathcal{F}$ with a Lie algebra bracket

$$\{\,,\,\} : \mathcal{F} \times \mathcal{F} \longrightarrow \mathcal{F} \tag{22}$$

over $\mathbf{K}$, is said to be a *Poisson ring* if in addition

$$\{f_1 f_2, g\} = f_1\{f_2, g\} + \{f_1, g\} f_2\,. \tag{23}$$

In this way the map

$$\mathcal{F} \longrightarrow Der\mathcal{F}\,, \quad f \mapsto ad_f\,, \quad ad_f \cdot g =: \{f, g\}\,, \tag{24}$$

is a Lie algebra homomorphism. To symplify notation we set

$$ad_f = X_f \in Der\mathcal{F}\,. \tag{25}$$

A Poisson structure could also be given with a map

$$B : (Der\mathcal{F})^* \to Der\mathcal{F} \tag{26}$$

which is "skew-symmetric", in the sense that

$$i_{B\alpha}\beta = -i_{B\beta}\alpha\,, \quad \forall\, \alpha, \beta \in (Der\mathcal{F})^*\,. \tag{27}$$

Then the Poisson brackets on $\mathcal{F}$ are given by

$$\{f_i, f_j\}_B =: i_{Bdf_i} df_j = -\{f_j, f_i\}_B\,. \tag{28}$$

The commutant of f, $C_f \subset \mathcal{F}$, is a Poisson subring of $\mathcal{F}$. It is the set of *constants of the motion* for X_f .

Given a Poisson subring $\mathcal{A}$ we define its *polar* or *reciprocal* set $\mathcal{A}'$ to be the set of all $f \in \mathcal{F}$ such that $\{f, \mathcal{A}\} = 0$.

We say that a Poisson subring $\mathcal{A}$ is *regular* if the following two subalgebras of derivations

$$\mathcal{X}_{\mathcal{A}'} =: \{X_f, f \in \mathcal{A}'\}\,, \tag{29}$$

$$\mathcal{N}_{\mathcal{A}} =: \{Y \in Der\mathcal{F} : L_Y \mathcal{A} = 0\} \tag{30}$$

coincide as $\mathcal{F}$-modules.

Mimicking the standard definitions [LM] we say that

a. $\mathcal{A}$ is isotropic if $\mathcal{A} \subset \mathcal{A}'$,

b. $\mathcal{A}$ is coisotropic if $\mathcal{A}' \subset \mathcal{A}$.

Definition 1. We say that an element $f \in \mathcal{F}$ defines a completely integrable system X_f if its set of constants of the motion $\mathcal{C}_f$ containes a regular Poisson ring $\mathcal{A}_f$ which is isotropic and coisotropic.

On a Poisson ring we can define a *Darboux set* to be a generating set for $\mathcal{F}$, denoted by $\mathcal{F}_1 \oplus \mathcal{F}_2$, with properties

1. $\mathcal{F}_1$ is regular isotropic and coisotropic,

2. $\mathcal{F}_2$ is regular isotropic and coisotropic,

3. for all $f \in \mathcal{F}_1$ there exists $g_f \in \mathcal{F}_2$ such that $\{f, g_f\} \in \mathbf{K}$.

Remark. A Poisson ring need not have Darboux generating sets.

Definition 2. A Darboux set for $\mathcal{F}$ defines *action-angle* functions for X_f if it is an integrating set for X_f.

The integration of dynamical systems on a smooth manifold M lacks general global teorems already in low dimensions. We should therefore not expect to provide them in our algebraic setting. It is however possible to connect complete integrability with bi-Hamiltonian description and (1-1)-tensor fields.

Bi-Hamiltonian systems

A given ring $\mathcal{F}$ may have two different structure of Poisson ring, say B_1 and B_2. We say that they are compatible if

$$\lambda\{\ ,\ \}_{B_1} + \mu\{\ ,\ \}_{B_2} : \mathcal{F} \times \mathcal{F} \longrightarrow \mathcal{F} \tag{31}$$

defines a Poisson ring structure on $\mathcal{F}$ for any $\lambda, \mu \in \mathbf{K}$.

A given derivation X is said to have two Hamiltonian description if there exist two elements of $\mathcal{F}$, f_1 and f_2 such that

$$X \cdot g = \{f_1, g\}_{B_1} = \{f_2, g\}_{B_2}\ , \quad \forall g \in \mathcal{F}\ . \tag{32}$$

With these two brackets we can construct $\mathcal{C}_X^1 = \{f \in \mathcal{F} : \{f, f_1\} = 0\}$, and $\mathcal{C}_X^2 = \{f \in \mathcal{F} : \{f, f_2\} = 0\}$. In general $\mathcal{C}_X^1$ (resp. $\mathcal{C}_X^2$) is not a Poisson subring with respect to B_2 (resp. B_1) so that we can "construct" new constants of the motion for X by using $\mathcal{C}_X^1$ and $\mathcal{C}_X^2$ with both brackets.

Given a pair (B_1, B_2) such that one of them, say B_1, is invertible, we can define an endomorphism

$$R = B_1^{-1} \circ B_2 : \quad \begin{array}{ccc} (Der\mathcal{F})^* & \xrightarrow{R} & (Der\mathcal{F})^* \\ B_2 \searrow & & \nearrow B_1^{-1} \\ & Der\mathcal{F} & \end{array} . \tag{33}$$

and extend it to an operator δ_R on $\Lambda(Der\mathcal{F}, \mathcal{F})$ as in (8).
By using the operator d of the complex $(\Lambda(Der\mathcal{F}, \mathcal{F}), d)$, we can also define the operator

$$\begin{aligned} &d_R : \Lambda^p(Der\mathcal{F}, \mathcal{F}) \to \Lambda^{p+1}(Der\mathcal{F}, \mathcal{F}) , \\ &d_R =: \delta_R d - d\delta_R , \end{aligned} \tag{34}$$

along with

$$\partial_R =: \delta_R d - d\delta_R , \tag{35}$$

by using R on derivations, $\alpha(RX) = (R\alpha)(X)$.
From the compatibility of the two Poisson structures B_1 and B_2 it follows that the Nijenhuis tensor of R vanishes so that $(d_R)^2 = 0$ and $(\partial_R)^2 = 0$.

Proposition 1. Let $\alpha \in (Der\mathcal{F})^*$ be such that $d\alpha = 0$ and $dR\alpha = 0$. Then $dR^k\alpha = 0 , \ \forall\, k \in N$.

Proof. Using the closure of α and $R\alpha$ and $\mathbf{N}_R = 0$ one easily proves that $dR^2\alpha = 0$, and so on.

Let us assume that $H^1(d) = 0$ on the complex associated with the 1-forms $\{df_0, Rdf_0, \ldots, R^k df_0, \ldots\}$. Then, from Prop.1, it follows the following proposition

Proposition 2. If $f_0, f_1 \in \Omega^0$ are such that

$$Rdf_0 = df_1 , \tag{36}$$

then, there exist $f_k \in \Omega^0$ such that

$$R^k df_0 = df_k \ , \quad k \in N . \tag{37}$$

Proposition 3. The set $\{f_0, \ldots, f_k, \ldots\}$ is an isotropic Poisson ring with respect to any structure $\lambda B_1 + \mu B_2$,

$$\{f_j, f_k\}_{\lambda B_1 + \mu B_2} = 0 \ , \quad j, k = 0, 1, \ldots . \tag{38}$$

Proof. The relations (38) follows from the following, easily verifiable relations

$$\begin{aligned} &\{f_j, f_k\}_{B_1} = \{f_{j-1}, f_k\}_{B_2} \\ &\{f_j, f_k\}_{B_2} = \{f_j, f_{k+1}\}_{B_1} \ , \quad j > k . \end{aligned} \tag{39}$$

As B_1 and B_1^{-1} both exist, we can define a bivector field $\widetilde{B}_1$ on $(Der\mathcal{F})^*$ and a 2-form ω_{B_1} on $Der\mathcal{F}$. They are given by

$$\widetilde{B}_1(\alpha,\beta) =: i_{B\alpha}\beta \ , \qquad \forall\, \alpha,\beta \in (Der\mathcal{F})^* \ , \tag{40}$$
$$\omega_{B_1}(X,Y) =: i_X(B_1^{-1}(Y)) \ , \qquad \forall\, X,Y \in Der\mathcal{F} \ . \tag{41}$$

In particular,

$$\widetilde{B}_1(df_1, df_2) = \{f_1, f_2\}_{B_1} \ , \tag{42}$$
$$\omega_{B_1}(B_1 df_1, B_1 df_2) = \{f_1, f_2\}_{B_1} \ . \tag{43}$$

Condition $(\partial_{B_1})^2 = 0$ implies that $d\omega_{B_1} = 0$ and $\partial(\widetilde{B}_1 \wedge \widetilde{B}_1) = 0$. Moreover one can easily verify that

$$\omega_{B_1}(\widetilde{R}^k X_0, \widetilde{R}^{k-1} X_0) \;=\; 0 \ , \tag{44}$$
$$\omega_{B_1}(\widetilde{R}^k X_0, \widetilde{R}^l X_0) \;=\; \omega_{B_1}(\widetilde{R}^{k-1} X_0 \ , \ \widetilde{R}^{l+1} X_0) \ , \quad k > l \ , \ \forall\, X_0 \in Der\mathcal{F} \ . \tag{45}$$

Here $\widetilde{R} =: B_2 \circ B_1^{-1} : Der\mathcal{F} \to Der\mathcal{F}$ is the "dual" of R.

Finally, one can prove the proposition

Proposition 4. Given $X_f \cdot = \{f, \cdot\}$, and a (1,1) tensor R such that $(d_R)^2 = 0$, we have

1. $L_{X_f} R = 0 \Rightarrow \mathcal{A}_{X_f} = \{X_f, \widetilde{R} X_f, \ldots, \widetilde{R}^k X_f, \ldots\}$ is an abelian Lie subalgebra,
2. $\partial_R(B_1 \wedge B_1) = 0 \ \Rightarrow \ \delta_R B_1$ is a Poisson structure compatible with B_1 which provides an alternative Hamiltonian description for X.

When B_1 is invertible, $\mathcal{A}_{X_f}$ is associated with a Poisson subring with respect to B_1, assuming $H^1 = 0$, that is isotropic. Thus it remains to check that it is regular, isotropic and coisotropic to conclude that X_f is completely integrable. Obviously for the most general situation this check can be highly non trivial.

Remark. By using a symplectic structure one can define a symplectic ring that turns out to be a special kind of Poisson ring. In this case one can define isotropic Lie subalgebras in $Der\mathcal{F}$ and Lagrangian ones in a way that is easy to figure out. In this case the *polar* or *reciprocal* subalgebra is always such that together (algebra and polar) generate the full algebra of derivations. This is quit different from the situation with functions.

To show how to make sense of all these definitions we are going to discuss an example.

Example: Periodic Toda Lattice.

We consider a ring $\mathcal{F}$ with the following properties:

1. $\mathcal{F}$ is generated by
$h_1, h_2, \ldots, h_N; f_1, f_2, \ldots, f_N$
with elements f_i's admitting an inverse that we denote by f_i^{-1} or $\frac{1}{f_i}$ and $f_{N+1} = f_1$.
2. Derivations Der$\mathcal{F}$ are generated over $\mathcal{F}$ by
$\{X_i, Y_i,\ i = 1, \ldots, N\}$
defined by

$$\begin{aligned} L_{X_i} f_j &= \delta_{ij} f_i \quad \text{(no sum)}\ , \\ L_{Y_i} f_j &= 0\ , \\ L_{X_i} h_j &= 0\ , \\ L_{Y_i} h_j &= \delta_{ij}\ . \end{aligned} \tag{46}$$

We define a pair of invertible Poisson operators B_1, and B_2 by giving the associated 2-forms on Der$\mathcal{F}$

$$\omega_1 : -\sum_i f_i^{-1} df_i \wedge dh_i, \tag{47}$$

$$\omega_2 : \sum_i (-\lambda_i df_i \wedge df_{i+1}^{-1} + h_i f_i df_i^{-1} \wedge dh_i) - \sum_{i<j} dh_i \wedge dh_j\ . \tag{48}$$

The 'dynamics' of the Toda Lattice is described by the following derivation

$$\Gamma = \sum_i (-h_i X_i) + \sum_i \frac{\lambda_i f_i}{f_{i+1}} (Y_i - Y_{i+1})\ . \tag{49}$$

Here the λ_i *'s* are constant parameters.
The derivation (49) is an Hamiltonian derivation with respect to both B_1 and B_2 with Hamiltonian functions

$$H_1 = \sum_i \frac{1}{2} h_i^2 + \sum_i \lambda_i f_i f_{i+1}^{-1}\ , \tag{50}$$

$$H_2 = \sum_i \frac{1}{3} (h_i)^3 + \sum_i \lambda_i (h_i + h_{i+1}) f_i f_{i+1}^{-1}\ . \tag{51}$$

One can show that B_1 and B_2 are compatible.
The operator $R = B_2^{-1} \circ B_1 : (\text{Der}\mathcal{F})^* \to (\text{Der}\mathcal{F})^*$ is given by

$$\begin{aligned} R(df_i) &= h_i df_i - f_i \sum_j \varepsilon_{ij} dh_j\ , \\ R(dh_i) &= \lambda_i f_i df_{i+1}^{-1} - \lambda_{i-1} f_i^{-1} df_{i-1} + h_i dh_i\ , \quad i = 1, \ldots, N\ . \end{aligned} \tag{52}$$

The simbol ε_{ij} is antisymmetric for any pair of indices ij, with $\varepsilon_{ij} = +1$ if $i > j$.
The dual operator $\tilde{R} = B_1 \circ B_2^{-1} : \text{Der}\mathcal{F} \to \text{Der}\mathcal{F}$ is given by

$$\begin{aligned} \tilde{R}(X_i) &= \lambda_i f_i f_{i+1}^{-1} Y_{i+1} - \lambda_{i-1} f_{i-1} f_i^{-1} Y_{i-1} + h_i X_i\ , \\ \tilde{R}(Y_i) &= h_i Y_i + \sum_j \varepsilon_{ij} X_j\ , \quad i = 1, \ldots, N\ . \end{aligned} \tag{53}$$

The algebra $\mathcal{A}_\Gamma$ is an abelian Lagrangian subalgebra. The first few elements of it are

$$
\begin{aligned}
A_1 &= \Gamma = \sum_i(-h_iX_i) + \sum_i \lambda_i f_i f_{i+1}^{-1}(Y_i - Y_{i+1}) \,, \\
A_2 &= \tilde{R}(\Gamma) \\
&= \sum_i(-h_i^2X_i) \\
&\quad + \sum_i \lambda_i f_i f_{i+1}^{-1}\{\, (h_i + h_{i+1})(Y_i - Y_{i+1}) - (X_i + X_{i+1}) \,\} \,, \\
A_3 &= \tilde{R}^2(\Gamma) \\
&= \sum_i(-h_i^3X_i) \\
&\quad - \sum_i \lambda_i f_i f_{i+1}^{-1}(2h_iX_i + 2h_{i+1}X_{i+1} + h_iX_{i+1} + h_{i+1}X_i) \\
&\quad + \sum_i \lambda_i f_i f_{i+1}^{-1}(h_i^2 + h_ih_{i+1} + h_{i+1}^2)(Y_i - Y_{i+1}) \\
&\quad + \sum_i \lambda_i^2 (f_i f_{i+1}^{-1})^2 (Y_i + Y_{i+1}) \\
&\quad + \sum_i \lambda_{i-1}\lambda_i f_{i-1} f_{i+1}^{-1}(Y_{i-1} - Y_{i+1}) \,, \\
&\ldots\ldots\ldots
\end{aligned}
\tag{54}
$$

The previous derivations are associated with the following functions in pairwise involution

$$
\begin{aligned}
F_1 &= H_1 = \sum_i \frac{1}{2}h_i^2 + \sum_i \lambda_i f_i f_{i+1}^{-1} \,, \\
F_2 &= H_2 = \sum_i \frac{1}{3}(h_i)^3 + \sum_i \lambda_i (h_i + h_{i+1}) f_i f_{i+1}^{-1} \,, \\
F_3 &= \sum_i \frac{1}{4}h_i^4 + \sum_i \lambda_i f_i f_{i+1}^{-1}(h_i^2 + h_ih_{i+1} + h_{i+1}^2) \\
&\qquad + \sum_i \lambda_i^2 (f_i f_{i+1}^{-1})^2 + \lambda_{i-1}\lambda_i f_{i-1} f_{i+1}^{-1} \,. \\
&\ldots\ldots\ldots
\end{aligned}
\tag{55}
$$

By explicit calculations it is not hard to show that they are indipendent and constitute a maximal set.

In this way we have associated a Poisson complex with our completely integrable derivation.

REFERENCES

DLMV S. De Filippo, G. Landi, G. Marmo, G. Vilasi, Tensor Field Defining a Tangent Bundle Structure, *Ann. Inst. H. Poinc.* **53** (1989) 205.
An Algebraic Description of the electron-monopole dynamics, *Phys. Lett.* **220B** (1989) 576-580.

Du M. Dubois-Violette, Noncommutative differential geometry, quantum mechanics and gauge theory, in these Proceedings.

Ka D. Kastler, Introduction to non-commmutative geometry and Yang-Mills model-building, in these Proceedings.

Ko J.L. Koszul, Homologie et cohomologie des algèbres de Lie, *Bull. Soc. Math. Fr.* **78** (1953) 65-130.

LaMa1 G. Landi, G. Marmo, Algebraic Differential Calculus for Gauge Theories, in: Integrability and Quantization, Proceedings of the GIFT XXth International Seminar on Theoretical Physics, edited by J.F. Cariñena, M. F. Rañada, L.A. Ibort, to be published.
Lie Algebra Extensions and Abelian Monopoles, *Phys. Lett.* **195B** (1987) 429-434.
Extensions of Lie Superalgebras and Supersymmetric Abelian Gauge Fields, *Phys. Lett.* **193B** (1987) 61-66.
Graded Chern-Simons terms, *Phys. Lett.* **192B** (1987) 81-88.
Algebraic Reduction of the 't Hooft-Polyakov Monopole to the Dirac Monopole, *Phys. Lett.* **201B** (1988) 101-104.
Einstein Algebras and the Algebraic Kaluza-Klein Monopole, *Phys. Lett.* **210B** (1988) 68-72.
Algebraic Instantons, *Phys. Lett.* **215B** (1988) 338.

LaMa2 G. Landi, G. Marmo, Algebraic Lagrangian Formalism, Preprint SISSA, Trieste, SISSA 90/90/FM.

LM P. Libermann, C.-M. Marle, Symplectic Geometry and Analytical Mechanics (Dordrecht, Reidel 1987).

Pa R. Palais, Natural Operations on Differential Forms, *Trans. AM. Math. Soc.* **92** (1959) 125-141.

INTEGRABLE LATTICE MODELS AND THEIR SCALING LIMITS : QFT AND CFT

H. J. de Vega
LPTHE, Université Paris VI, Tour 16, 1er. etage, 4, Place Jussieu,
F-75230,PARIS Cedex 05, FRANCE.

Integrable massive QFT and conformal invariant models follow from lattice integrable models in suitable scaling limits. There are Yang-Baxter algebras (YBA) associated with all these two-dimensional models. These YBA allow one to construct the exact solution (spectrum, S-matrix, form-factors,...) for this class of theories. Braid groups and quantum groups are derived as limiting cases of YBA when θ (spectral parameter) goes to $\pm\infty$.

Integrable lattice models (on vertex or faces), integrable (massive) quantum field theories and conformal (massless) theories are the subject of this short review[1]. The interrelation between these three subjects is described by the triangle depicted in fig. 1. The two upper sides of the triangle constitue the "royal way" to construct massless and massive QFT : starting from a critical lattice model in a finite volume, the continuum limit can be rigorously constructed letting the lattice spacing a to zero and the volume V to infinity. Moreover the exact solution of the continuum theory follows as the $a = 0$ limit of the lattice model solution.

The light-cone approach[2,3,4] is probably the more powerful one to derive massive QFT as scaling limits of integrable lattice models. No parameters are loosed in the scaling limit within the light-cone approach.

It must be stressed that both the massless limits (Conformal Field Theories, CFT) and the massive limits are **universal**. That is, all models in the same universality class (integrable and **non**-integrable) posses the same scaling behavior. The CFT describe the behavior **at** criticality : $a << x << \xi$, where ξ stands for the correlation length. Massive QFT describes the physics in the larger domain : $a << x \approx \xi$. This additional information around $x \approx \xi$ is absent in CFT.

The lower line in the triangle [fig.1] stands for the connection between CFT and massive QFT through the addition of relevant -bations[5,6].

In the center of the triangle in fig.1 are the Yang-Baxter algebras. These mathematical structures underly the solvability of all integrable models (classical or quantum, on the lattice and on the continuum).

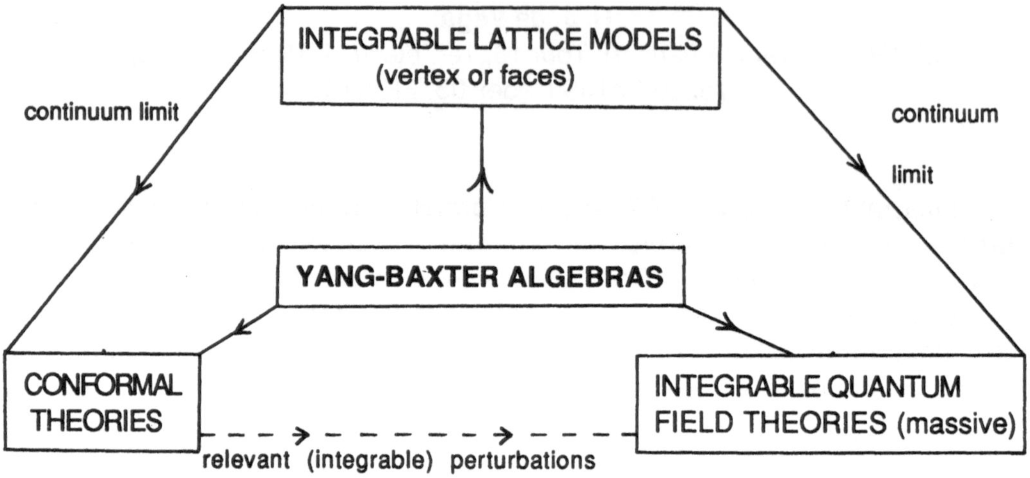

Fig. 1

Let us start by defining a Yang-Baxter algebra (YBA). We consider a set of lines of different types . A vector space $\mathcal{V}_I$ ($I \in I$) is associated with each type of line. These lines may intersect and a YB operator is associated with each intersection:

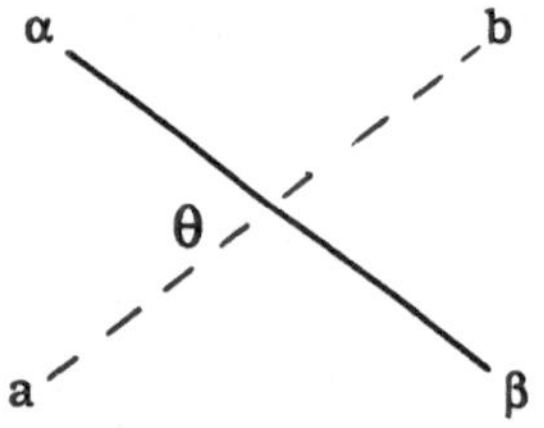

$$= \quad [\, t_{ab}^{(\mathcal{A},\mathcal{V})}(\theta)\,]_{\alpha\beta} \tag{1}$$

Here the lines ——— and ~~~~~~~~~~ are associated with the vector spaces $\mathcal{A}$ and $\mathcal{V}$ respectively. $t(\theta)$ is an operator acting on this couple of spaces , that is on $\mathcal{A} \otimes \mathcal{V}$. The indices a, b ($1 \le a, b \le \dim\mathcal{A}$) and α, β ($1 \le \alpha, \beta \le \dim\mathcal{V}$) label the basis vectors (states) in $\mathcal{A}$ and $\mathcal{V}$ respectively. The complex variable θ is called the spectral parameter and describes geometrically the angle between the lines.

The operators $t(\theta)$ fulfil a YBA when the following relation holds :

$$t^{(K,I)}(\theta-\theta')\; t^{(K,J)}(\theta)\; t^{(I,J)}(\theta') = t^{(I,J)}(\theta')\; t^{(K,J)}(\theta)\; t^{(K,I)}(\theta-\theta') \tag{2}$$

for any choice of vector spaces $\mathcal{V}_I$, $\mathcal{V}_J$ and $\mathcal{V}_K$ (I, J and K ε I) and any value of θ , $\theta' \varepsilon$ **C.** Graphically eq.(2) reads :

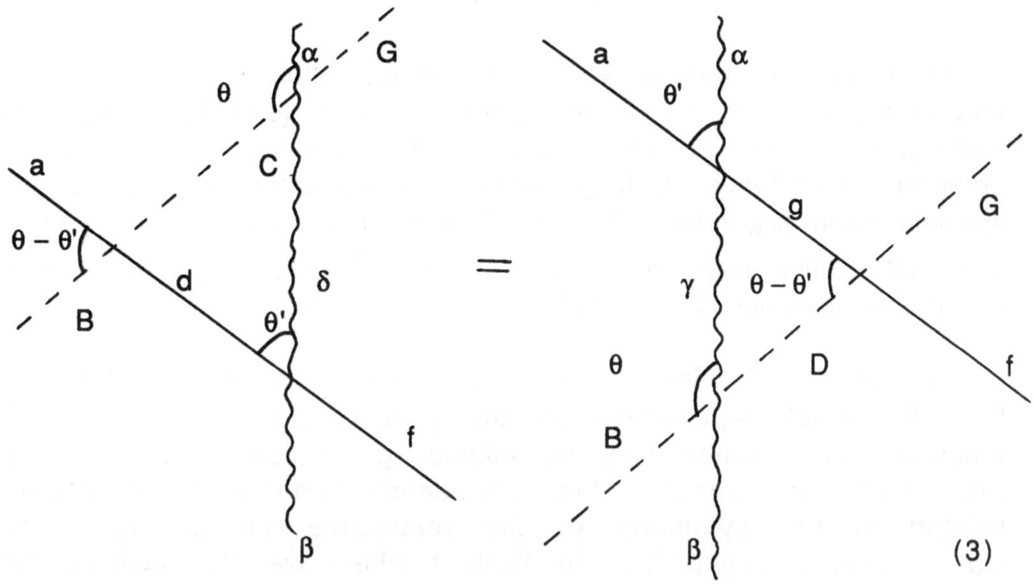

(3)

Here we use the convention that one must sum over all states in internal lines. That is, lines linking two intersections. The graphical meaning of the YBA is clear : one can parallely displace any line through the intersection of two others keeping the value of the expression invariant. This invariance holds provided all angles θ between the lines are kept fixed in the process. Eqs.(2) or (3) are usually called the Yang-Baxter equations (YBE). They have different physical meanings in different contexts (see below and ref.[1]).

The YBA (2) takes a particularly simple form when two vector spaces are identical. That is $\mathcal{V}_K = \mathcal{V}_I = \mathcal{A}$ and $\mathcal{V}_J = \mathcal{V}$ in eq.(2). One finds

$$R(\theta - \theta')\,[\,t(\theta) \otimes t(\theta')\,] = [\,t(\theta') \otimes t(\theta)\,]\, R(\theta - \theta') \qquad (4)$$

where $R = t^{(\mathcal{A},\mathcal{A})}$ and $t = t^{(\mathcal{A},\mathcal{V})}$ and we have used the tensor product notation $(A\otimes B)_{ab,cd} \doteq A_{ac}\, B_{bd}$. In eq.(4) there is an operatorial product of the T's in the space $\mathcal{V}$. The numbers $R^{ab}{}_{cd}(\theta)$ play the role of "structure constants" and the $T_{ab}(\Theta)$ that of generators of the YB algebra. It should be noticed that the YB algebras are not in general Lie algebras but Hopf algebras (see ref.[1]). In summary, a YB algebra, is a set of operators $T^{(I,J)}(\theta)$ acting on a couple of vector spaces $(\mathcal{V}^I, \mathcal{V}^J)$. $\mathcal{V}^I$ $(\mathcal{V}^J)$ stands for the auxiliary space (quantum space = $\mathcal{V}$) for this case. This set of operators is such that relation (2) holds for any choice (I,J,K) of vector spaces.

In summary, the crux of the integrability (and consequent solvability) lies in the Yang-Baxter equation [eq.(2)]. Hence the systematic search of YB solutions and their classification is of primordial importance. A large number of solutions in known today, specially when $\mathcal{V}_K = \mathcal{V}_I = \mathcal{V}_J = \mathcal{V}$. That is, R-matrices. Starting from a given YB solution acting on some $\mathcal{V}$, one can produce solutions acting on spaces of higher dimensionality by the fusion method[7].

The known YB solutions can be conveniently classified according to their functional dependence on the spectral parameter θ. This functional dependence may be elliptic (genus one), hyperbolic or trigonometric or rational. This dependence happens to be directly related to the symmetry of the respective YB solution. This classification is summarized in Table I where we also indicate the critical (gapless) or non-critical character of the associated lattice statistical model. The parameters labelling the different solutions are also given there.

Besides the spectral parameter θ always present, we have the anisotropy parameter γ in elliptic and hyperbolic/trigonometric solutions. When the elliptic modulus k, characteristic of elliptic solutions tends to zero or to one, the elliptic YB solutions become trigonometric or hyperbolic, respectively.

In the $\theta \to 0$, $\gamma \to 0$, limit with fixed ratio θ/γ, the hyperbolic /trigonometric YB solutions become rational solutions. Rational YB solutions are invariant under the full non-abelian group (SU(n), O(n), etc.) underlying them. Since only the invariance under the Cartan subalgebra holds for the trigonometric/hyperbolic YB solutions, it is reasonable to call γ the anisotropy parameter. These YB solutions contain quantum groups in their asymptotic limits $\theta \to \pm\infty$ (for the hyperbolic case) and $\theta \to \pm i\infty$ (in the trigonometric case) [see below eqs.(21-22)].

That is, quantum groups are the underlying θ-independent structure to the trigonometric/hyperbolic YB solutions. Ordinary Lie algebras underly rational YB solutions. The deformation parameter q in the quantum group literature (and in the q-functions[8] and q-strings[9-10]) is connected with γ by $\exp(i\gamma)$ or $\exp(-\gamma)$ in the trigonometric and hyperbolic cases respectively.

Furthermore, it can be shown that the trigonometric/hyperbolic YB solutions are invariant under the full quantum group using the corresponding coproduct definition[1].

$R_{\mathcal{A}\mathcal{V}}$ in table I stands for $\mathcal{A} \otimes \mathcal{V}$, the representation space where $T^{(\mathcal{A},\mathcal{V})}(\theta)$ acts. As the reader has noticed, no underlying θ-independent structure for elliptic YB solutions is indicated. This is still an open problem. A partial answer is the Sklyanin algebra which does not have any known coproduct.

YB algebras can be formulated in face language[1,11]. It is possible to associate to each YB vertex solution a face YB solution and viceversa. However, many models can be formulated in a simpler and more natural way using face language. This is specially true for the reduced models that appear when γ/π is a rational number. The protypes are the Andrews-Baxter-Forrester RSOS models obtained from the eight vertex model. A very large set of RSOS models is known at present[12]. As it is the case for vertex models, face models are also classified by Lie algebras, more precisely by affine Lie algebras. The local states of the faces being dominant integral weights.

θ-dependence	**Solutions labels**	**Invariance Group $\mathcal{G}$**	**Gap $\approx \xi^{-1}$**	**θ-independent content**
Elliptic	$k, \gamma, \theta, \mathcal{G}, R_{\mathcal{A}\mathcal{V}}, \ldots$	discrete Z_n	$\neq 0$	?
Hyperbolic - - - - - - - - - - - - Trigonometric	$\gamma, \theta, \mathcal{G}, R_{\mathcal{A}\mathcal{V}}, \ldots$	continuous abelian Cartan algebra of $\mathcal{G}$ Quantum Group $\mathcal{G}_\gamma$	$\neq 0$ - - - - $= 0$	Quantum Groups (from $\theta = \pm\infty$)
Rational	$\theta, \mathcal{G}, R_{\mathcal{A}\mathcal{V}}, \ldots$	$\mathcal{G}$	$= 0$	Lie Groups (from $\theta = \pm\infty$)

Table I

In table I the dots stand for the solutions obtained by transformations like the duality $q \to -q^{-1}$ performed on some matrix elements and according to precise rules. In this way solutions of ref.[13] for $\mathcal{G} = \mathcal{A}_n$ follows from the basic $\mathcal{A}_n$ solution found in ref.[14]. This duality generalizes for all Lie algebras[155]. It must be noticed that probably there are still more YB solutions to be found.

In the case of the YB solutions of ref.[16] where the spectral parameter belongs to a higher genus curve, the R-matrix depends on θ **and** θ' and not just on $\theta - \theta'$ as in the cases here considered [see eq.(4)]. Anyway, since these solutions can be related to the six-vertex model[17], we do not need to consider them as a separate class.

To find and classify YB solutions is indeed a fundamental problem, but it is only the very first step from a physical point of view. YB solutions usually have several physical meanings : statistical lattice models, two dimensional S-matrices, generating functionals of one dimensional quantum spin hamiltonians, etc. Moreover, the scaling limit of the (gapless) statistical models yield CFT and QFT. In order to extract all this physics one must solve the models defined by the YB solutions. The more powerful method is the Bethe Ansatz (BA). Actually, it is not an ansatz, but the explicit construction of exact eigenvectors and eigenvalues of the transfer matrix $\tau(\theta)$. There are other related ways to find the transfer matrix eigenvalues, like the analytic Bethe Ansatz[18] and the inverse method[19]. These methods are simpler than the BA and allowed to find eigenvalues for models where the full BA is not yet known. However, they are less powerful methods since they do not provide eigenvectors.

YB solutions have different physical interpretations in different contexts. The first one being statistical weights in two-dimensional classical statistical mechanics on the lattice. That is $[\, t_{ab}{}^{(\mathcal{A},\nu)}(\theta)\,]_{\alpha\,\beta}$ defines the Boltzmann weight for the configuration (1). Then the operator

$$T_{ab}{}^{[N]}(\theta) = \sum_{a_1,\dots,a_{N-1}}^{\dim\mathcal{A}} t_{aa_1}(\theta) \otimes t_{a_1a_2}(\theta) \otimes \dots \otimes t_{a_{N-1}b}(\theta) \qquad (5)$$

is associated to a N-points horizontal line of the lattice. It is easy to prove that $T_{ab}{}^{[N]}(\theta)$ obeys a YB algebra provided the local $t_{ab}{}^{(\mathcal{A},\nu)}(\theta)$ do that [eq.(4)]. For $N = 2$ this is the coproduct property of YB algebras. It should be recalled that the coproduct in quantum groups can be just considered a consequence of this YB coproduct.

The notion of coproduct has a very simple physical meaning. Suppose a physical system formed by two subsystems of the same kind (two atoms, two molecules, etc.). Each subsystem having its own physical magnitudes obeying some algebra like angular momentum, Poincaré or Yang-Baxter. Then, the coproduct is the way magnitudes of both systems must be combined in order that the quantities associated to the whole sistem fullfil the same algebra as those associated with the subsystems do.

In all known cases (besides Yang-Baxter and quantum groups) the coproduct is just a trivial sum:

$$\vec{P} = \vec{P}_1 + \vec{P}_2 \qquad \vec{L} = \vec{L}_1 + \vec{L}_2$$

$$M_{\mu\nu} = M^{(1)}{}_{\mu\nu} + M^{(2)}{}_{\mu\nu} \tag{6}$$

where the indices (1) and (2) refer to the subsystems 1 and 2 respectively and $\vec{p}$, $\vec{L}$ and $M_{\mu\nu}$ stand for the momentum, angular momentum and Lorentz generators, respectively. In the Yang-Baxter algebra case we have a non-commutative coproduct [eq.(5) for N = 2]

$$T_{ab}(\theta) = \sum_{c=1}^{\dim \mathcal{A}} t_{ac}(\theta) \otimes t_{cb}(\theta) \tag{7}$$

Coming back to the operator $T_{ab}{}^{[N]}(\theta)$, it plays a central rôle in lattice models and in their BA solution[1]. Its trace on $\mathcal{A}$, $\tau[\theta] = \sum_a T_{aa}(\theta)$ fulfils

$$[\,\tau[\theta]\,,\tau[\theta']\,] = 0 \tag{8}$$

$\tau[\theta]$ is called the transfer matrix and can be directly related to the partition function through

$$Z = \mathrm{Tr}[\,\tau[\theta]^M\,] \tag{9}$$

Here Tr stands for trace over the space $\mathcal{V}^N$. Eq.(8) tells us that the operators $\tau(\theta)$ may have the same eigenvectors for values of θ. The BA allows to prove that and to explicitly construct them[1]. The free energy per site follows from eq.(9) as

$$f = -\lim_{(N,M)\to\infty} (1/NM) \operatorname{Log} Z = -\lim_{N\to\infty} (1/N)\log \Lambda_{max}(\theta) \tag{10}$$

where $\Lambda_{max}(\theta)$ stands for the eigenvalue of $\tau(\theta)$ with maximun modulus. In this example we see how physical magnitudes directly connect with YB operators. Let us now describe how the continuous limit may be performed using the light-cone approach.

The light-cone approach is the more general and precise way to construct integrable QFT and conformal invariant theories. One starts from integrable lattice models such as vertex models on a diagonal lattice. This diagonal lattice is a discretization of Minkowski space-time in light-cone coordinates $X_\pm = X \pm T$. The matrix elements

$$[T_{ab}(\theta)]_{\alpha\beta} \qquad (1 \le \alpha, \beta \le \dim \mathcal{V},\ 1 \le a, b \le \dim \mathcal{A}) \qquad (11)$$

are now interpreted as quantum mechanical transition amplitudes for bare particles propagating to the right or to the left by the bonds at the speed of light. In the simplest case $\dim \mathcal{V} = \dim \mathcal{A} = 2$, and we interpret these particles as bare fermions without internal degrees of freedom. The allowed microscopical amplitudes, assuming a U(1) charge conservation, correspond to the six-vertex model weights in the statistical mechanical language. We can organize the microscopic amplitudes at a site into a unitary bare scattering matrix:

$R_{ab}{}^{cd}(\theta) =$

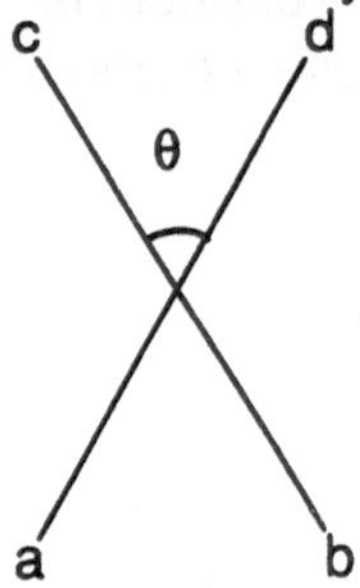

$$= \begin{pmatrix} 1 & 0 & 0 & 0 \\ 0 & c & b & 0 \\ 0 & b & c & 0 \\ 0 & 0 & 0 & \omega \end{pmatrix} \qquad (12)$$

where $b = \sin(i\theta)/\sin(i\theta + \gamma)$ and $c = \sin(i\theta)/\sin(i\theta + \gamma)$.

We can now form the operators describing the evolution by one lattice step in the diagonal directions

$$U_R = \overset{1\ \ 2}{\underset{2N\ \ 1}{X}}\ \overset{3\ \ 4}{\underset{2\ \ 3}{X}}\ \overset{5\ \ 6}{\underset{4\ \ 5}{X}}\ \overset{7\ \ 8}{\underset{6\ \ 7}{X}} \cdots\cdots\cdots \overset{2N-1\ \ 2N}{\underset{2N-2\ \ 2N-1}{X}} \qquad (13)$$

$$U_L = \overset{1\ \ 2}{\underset{2\ \ 3}{X}}\ \overset{3\ \ 4}{\underset{4\ \ 5}{X}}\ \overset{5\ \ 6}{\underset{6\ \ 7}{X}}\ \overset{7\ \ 8}{\underset{8\ \ 9}{X}} \cdots\cdots\cdots \overset{2N-1\ \ 2N}{\underset{2N\ \ 1}{X}} \qquad (14)$$

where the numbers 1, 2, 3,...,2N label the sites.

In this way the massive Thirring model is constructed both at the bare and at the renormalized levels[3]. The appropiate scaling limits are defined as $a \to 0$, $\theta \to \infty$:

$$\text{Bare} : \sin\gamma \exp[-\theta]/a = \text{fixed.}$$
$$\text{Renormalized} : \sin\gamma \exp[-K\theta]/a = \text{fixed} \qquad (15)$$

Here a is the lattice spacing , $K = \pi/\gamma$ and γ is the anisotropy parameter. The light-cone transfer matrices rigorously define the Hamiltonian and momentum as

$$H \pm P = (2i/a) \operatorname{Log} U_{R_L}(\theta) \qquad (16)$$

Using U_R and U_L, the Heisenberg equations of motion are derived in the lattice and their scaling limit obtained[3]. They lead to the MTM bare equations, providing a precise identification of the fields, the coupling constant and the mass[3] .The exact spectrum of U_R and U_L follows from the usual row-to-row transfer matrices and provides the particle spectrum of the theory in the renormalized scaling limit (eq.(15)).

Let us now study the ultrarelativistic limit of hyperbolic-trigonometric YB algebras and its connection with braids, knots and quantum groups. It is useful to introduce the operators

$$X_i(\theta) = \underset{1}{1} \otimes \ldots\ldots\ldots \otimes \underset{(i,i+1)}{R(\theta)} \otimes \ldots\ldots \otimes \underset{n}{1} \qquad (17)$$

They fulfil the relations

$$[\, X_i(\theta)\, ,\, X_j(\theta')\,] = 0 \text{ if } |i-j| \geq 2, \;\forall\, \theta, \theta'$$

$$X_i(\theta)\, X_{i+1}(\theta+\theta')\, X_i(\theta') = X_{i+1}(\theta')\, X_i(\theta+\theta')\, X_{i+1}(\theta) \qquad (18)$$

The last two being a consequence of the YB equation (2) . There are two different limits for hyperbolic (trigonometric) YB algebras : $\theta \rightarrow \pm\infty$ ($\pm i \infty$) . In an appropiate gauge we find

$$b_i = \lim_{\theta\rightarrow\infty} X_i(\theta), \quad b_i^{-1} = \lim_{\theta\rightarrow -\infty} X_i(\theta) \qquad (19)$$

and analogous relations for trigonometric YBA. The operators b_i fulfil:

$$b_i\, b_j = b_j\, b_i \quad \text{when} \quad |i-j| \geq 2\,,$$

$$b_i\, b_{i+1}\, b_i = b_{i+1}\, b_i\, b_{i+1} \qquad (20)$$

This is precisely a braid group. In other words, the ultrarelativistic limit of the YB algebras provide braid group representations. This fact has been recently exploited successfully to compute link and knot invariants[20].

Let us study the $\theta = \pm\infty$ limit of the six-vertex YB generators. The six-vertex model being the simplest non-trivial trigonometric/hyperbolic YB solution. We find[1,21]

$$T_{11}(\theta) = (y_{\pm})^N \exp(\pm\gamma S_z)\; [\, 1 + O(y_{\pm}^{-2})\,]$$

$\theta \rightarrow \pm\infty$

$$T_{22}(\theta) = (y_\pm)^N \exp(+\gamma S_z) [1 + O(y_\pm^{-2})]$$
$\theta \rightarrow \pm\infty$ (21)

$$T_{12}(\theta) = (y_\pm)^{N-1} \operatorname{sh}\gamma \, J_-(+\gamma) [1+O(y_\pm^{-2})]$$
$\theta \rightarrow \pm\infty$

$$T_{21}(\theta) = (y_\pm)^{N-1} \operatorname{sh}\gamma \, J_+(\pm\gamma) [1 + O(y_\pm^{-2})]$$
$\theta \rightarrow \pm\infty$

where $y_\pm = \pm \exp[\pm(\theta + \gamma/2)]$, $S_z = (1/2) \sum_{1 \leq a \leq N} (\sigma_a)_z$ and

$$J_\pm(\gamma) = \sum_{k=1}^{N} \prod_{j=1}^{k} \exp\{\gamma(\sigma_j)_z/2\} (\sigma_\pm)_k \prod_{l=k+1}^{N} \exp\{-\gamma(\sigma_l)_z/2\}$$

The ultrarelativistic limit of the YB algebra relations (4) with the six-vertex R-matrix yields the algebra of the operators $J_\pm(\gamma)$ and S_z. We find [1]

$$[J_+(\gamma), J_-(\gamma)] = \operatorname{sh}(2\gamma S_z) / \operatorname{sh}\gamma$$

$$[S_z, J_\pm] = \pm J_\pm \qquad (22)$$

Analogous relations hold with $\gamma \rightarrow i\gamma$ in the trigonometric regime. Eqs.(22) define the so-called $SU_\gamma(2)$ quantum group.

To conclude, I want to stress that YB algebras are more general and powerful tools than the quantum groups. Moreover, the elliptic YB algebras provide additional structures .

Concerning the physical implications, the work on critical properties of integrable theories shows that for any conformal field theory there exists at least one lattice integrable theory having this conformal model as critical (continuous) limit. This means that in each universality class (set of all lattice models with a given critical behavior, that is yielding the same CFT at criticality) there is at least one integrable representative. This shows that integrable models are not just mathematical curiosities, but they are deeply linked with physics.

Therefore, as remarked at the begining, both the CFT and the massive QFT derived as scaling limits of lattice integrable models contain **universal information** that applies to **all** models in the universality class (almost all of them being non-integrable).

The conformal properties of integrable lattice models are efficienttly investigated through the calculation of their finite-size corrections. This can be done numerically or analitically. There is a vast literature on this subject[22]. The basic methods to solve the Bethe Ansatz equations for finite size can be found in refs.[1] and [22]. In the last reference under [22] the problem of finite size corrections when the ground state is formed by complex roots of the BA is solved at dominant order. This applies to integrable models with spin higher than 1/2, for example.

The attainment of integrable QFT by adding relevant perturbations to CFT is a recently developed subject[5,6]. As a relevant example we have the identification of the scaling theory for the Ising model in an external magnetic field (which is a nonintegrable model) with the E_8 affine Toda field theory (ATFT) at the special value $i\sqrt{4\pi}$ of the coupling constant[5,6]. This unexpected connection allows to obtain physical information (like the long range behavior of correlations) for the Ising model in a field from the E_8 ATFT. Preliminary results are reported in ref.[24].

Elliptic YB algebras provide more general integrable statistical models which are in general non-critical. It can be shown that their scaling limits are the same as those of their trigonometric limits[23]. In other words the elliptic modulus is an irrelevant parameter. It is interesting to notice that an analogous situation happens for knot and link invariants. Elliptic YB solutions do not provide new invariant polynomials besides those obtained from hyperbolic YB solutions. There must be somehow a deep connection between universality in statistical mechanics and invariant objects in knot theory.

1 See for example for more detailed accounts
H. J. de Vega, a)Int. J. Mod. Phys. **4**, 2371 (1989),
b) Adv. Stud. in Pure Math. **19**, 567 (1989) and
c) Int. J. Mod. Phys. **B4**, 735 (1990).

2 T. T. Truong and M. D. Schotte, Nucl. Phys. **B220**, 77 (1983) and **B230**, 1 (1984).
T. T. Truong in Springer Lecture Notes in Physics, vol. **226** , N. Sanchez, Editor.

3 C. Destri and H. J. de Vega, Nucl. Phys. **B290**, 363 (1987), Phys. Lett. **B201**, 261(1988) and J. Phys. **A 22**, 1329 (1989).

4 H. J. de Vega and E. Lopes, J. Phys. **A** (to appear)

5 A.B. Zamolodchikov, Int. J. Mod. Phys. **A3**, 746 (1988) , **A4**, 4235 (1989) and Adv. Stud. in Pure Math. **19**, 64 (1989).

6 H. W. Braden, E. Corrigan, P.E. Dorey and R. Sasaki, Nucl. Phys.**B** (1990).
P. Christe and G. Mussardo, Nucl. Phys. **B330**, 465 (1990).
C. Destri and H. J. de Vega, Phys. Lett. **233 B**, 236 (1989).
H.J. de Vega and V.A. Fateev, LPTHE preprint 90-36.
V.A. Fateev and A.B. Zamolodchikov, Int. J. Mod. Phys. **A5**,1025 (1990).
V.A. Fateev, Landau Institute preprint, 1990.
N. Y. Reshetikhin and F. A. Smirnov, Comm. Math. Phys. **131**, 157 (1990).
F. A. Smirnov, Nucl. Phys. **B337**, 156 (1990).
G. Sotkov and C.J. Chu, Phys. Lett. **229B**, 391 (1989).

7 P.P. Kulish and E.K. Sklyanin, in Springer Lect. in Phys. **151**, (1981)

8 E. Heine, J. Math. **34**, 285 (1847).
See for example H. Exton, q-Hypergeometric functions and applications, New York, 1983 and references therein.

9 D. D. Coon, Phys. Lett. **29B**, 669 (1969).
M. Baker and D. D. Coon, Phys. Rev. **D2**, 2349 (1970).

10 H. J. de Vega and N. Sanchez, **216B**, 97 (1989).

11 H. J. de Vega, Int. J. Mod. Phys. **A5**, 1611 (1990).

12 See for instance, E. Date, M. Jimbo, T. Miwa and M. Okado, RIMS-590, 1987.

13 C. L. Schultz, Phys. Rev. Lett. **46**, 629 (1981).

14 O. Babelon, H. J. de Vega and C.M. Viallet, Nucl. Phys. **B190**, 542 (1981).

15 M. L. Ge, Wu Y. S. and K. Xue, preprint ITP-SB-02,
M. Couture, Y. Cheng, M. L. Ge and K. Xue, preprint ITP-SB-05,
Y. Cheng, M. L. Ge and K. Xue, preprint ITP-SB-38,
M. L. Ge and K. Xue, preprint ITP-SB-20 (Stony Brook).

16 H. Au-Yang, B.M.McCoy, J.H.H. Perk, S. Tang and M.L. Yan, Phys.Lett. **123A**, 219 (1987).
B.M.McCoy, J.H.H. Perk, S. Tang and C.H. Sah, Phys.Lett. **125A**, 9 (1987).
R.J. Baxter, J.H.H. Perk and H. Au-Yang, Phys.Lett. **128A**, 138 (1987).

17 V.V. Bazhanov and Yu. G. Stroganov, ANU Canberra preprint, 1989.

18 N. Yu. Reshetikhin, L.M.P., **7**, 205 (1983).

19 R. J. Baxter, Exactly solvable models in Statistical Mechanics, Academic Press (1982).

20 Y. Akutsu, T. Deguchi and M. Wadati, Phys. Rep. **180**, 247 (1989).

21 H. J. de Vega, Nucl. Phys. **B290**, 363 (1984).

22 H. J. de Vega and F. Woynarovich, Nucl. Phys. **B251**, 439 (1985).
H. P. Eckle and F. Woynarovich, J. Phys. **A20**, L97 (1987) and L443 (1987).
C. J. Hamer et al., J. Phys. **A20**, 5677 (1987).
H.J. de Vega and M. Karowski, Nucl. Phys. **B285**, 619 (1987).
M. Karowski, Nucl. Phys. **B300**, 473 (1988).
H. J. de Vega, J. Phys. **A20**, 6023 (1987) and J. Phys. **A21**, L1089 (1988).
L. V. Avdeev and B.D. Dörfel, Theor. Math. Phys. **71**, 272 (1987).
F. A. Alcaraz et al. Phys. Rev. Lett. **58**, 771 (1987) and Ann. Phys. **182**, 280 (1988).
F. A. Alcaraz and M.J. Martins, J. Phys. **A 21**, 4397 (1988) and **A 22**, 1829 (1989).
F. Woynarovich, Phys. Rev. Lett. **59**, 259 (1987).
H. J. de Vega and F. Woynarovich, J. Phys. **A 23** 1613 (1990).

23 C. Destri and H. J. de Vega, Mod. Phys. Lett. **A 27**, 2595 (1989).

24 C. Destri and H. J. de Vega, LPTHE preprint, 90-08.

Quantum groups, Riemann surfaces and conformal field theory

C. Gómez†‡¶ and G. Sierra§

† Départment de Physique Théorique, Université de Genève, CH-1211 Genève 4, Suisse

§ Instituto de Estructura de la Materia, CSIC, Serrano 119, Madrid, España

Abstract: An explicit representation is constructed, starting with the operator algebra corresponding to the Coulomb gas representation of conformal field theories. By "quantizing" the uniformization theory of Riemann surfaces, the geometric interpretation of such a representation is obtained.

---

‡ Permanent address: Departamento de Física Teórica, Universidad de Salamanca, Salamanca, España.

¶ Partially supported by the Swiss National Science Foundation.

1. Introduction: Quantum Groups and Ice–Type Models

Historically, quantum groups appear as the most natural mathematical framework to use in the construction of solutions to the quantum Yang–Baxter equation [1, 2, 3]. More precisely quantum groups provide a way to construct generalized ice–type models [4].

To be more concrete let us consider the simple example of Baxter's six– vertex model. The lattice variables in this case can take only two values ± 1 with the Boltzmann weights satisfying the Yang–Baxter equation:

$$\begin{aligned} &\sum_{\mu'\nu'\gamma} W(\mu\alpha\mu'\alpha|u)W(\nu\gamma\nu'\beta|u+v)W(\nu'\mu'\mu''\nu''|v) \\ &= \sum_{\mu'\nu'\gamma} W(\nu\mu\mu'\nu'|v)W(\mu'\alpha\mu''\gamma|u+v)W(\nu'\gamma\nu''\beta|u) \\ &\qquad\qquad\qquad \mu,\nu,\ldots = \pm 1 \end{aligned} \tag{1}$$

An important type of solutions to (1) are the trigonometric ones:

$$W(\mu\nu\rho\delta|u) = \delta_{ac} + \delta_{bd}\left[\frac{\sin u}{\sin(\lambda - u)}\right] e^{-i\lambda(a+c-2b)/2} \tag{2}$$

where we have used the "IRF–variables" a,b,..living on the semiinfinite Coxeter graph of type A:

$$0 \quad 1 \quad 2 \quad 3 \quad \to \infty \tag{3}$$

and the vertex variables $\mu, \nu, \ldots$ are defined according to the vertex–IRF Kadanoff–Wegner map [5]:

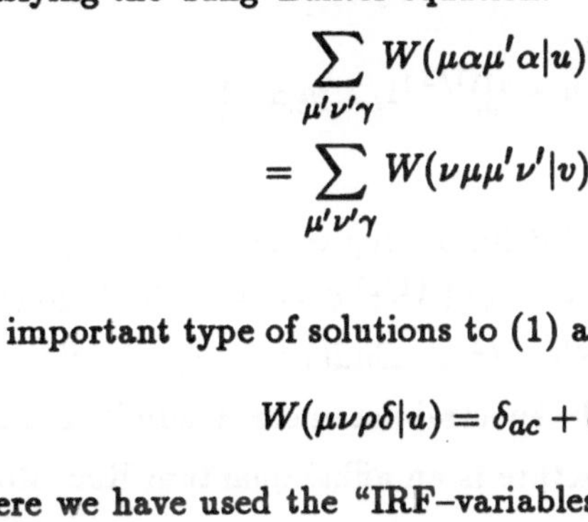

$$\begin{aligned} \mu &= d - a \\ \delta &= b - a \\ \rho &= c - b \\ \nu &= c - d \end{aligned} \tag{4}$$

In the limit $u \to \infty$ this trigonometric solution defines the R–matrix of $SU(2)_q$:

$$R = \begin{pmatrix} q & & & \\ & 1 & q-q^{-1} & \\ & & 1 & \\ & & & q \end{pmatrix} \tag{5}$$

where q depends on the free parameter λ appearing in (2). The quantum group $SU(2)_q$ is defined by:

$$\begin{aligned} [X^{\pm}, H] &= \mp 2X^{\pm} \\ [X^{+}, X^{-}] &= [H] \end{aligned} \tag{6}$$

with the q–number $[a]$ given by $(q^a - q^{-a})/(q - q^{-1})$. The comultiplication compatible, as an homomorphism, with the conmutation relations(6) is:

$$\begin{aligned} \Delta(X^{\pm}) &= X^{\pm} \otimes q^{H/2} + q^{-H/2} \otimes X^{\pm} \\ \Delta(H) &= H \otimes 1 + 1 \otimes H \end{aligned} \tag{7}$$

and Drinfeld's universal R–matrix is defined as the element in $\otimes^2 U_q(SU(2))$ such that:

$$\sigma \circ \Delta(a) = R\Delta(a)R^{-1} \qquad a \in U_q(SU(2)) \tag{8}$$

with σ the permutation map $\sigma(a \otimes b) = b \otimes a$.

The quantum R–matrix (5) we have obtained from the trigonometric solution of the six–vertex model is simply $\rho^{1/2} \otimes \rho^{1/2}[\sigma R]$ where $\rho^{1/2}$ denotes the two dimensional representation $V^{(1/2)}$ of $SU(2)_q$:

$$\begin{aligned} X^{\pm}\,|1/2, m\rangle &= ([1/2 \mp m][1/2 \pm m + 1])^{1/2}\,|1/2, m \pm 1\rangle \\ H\,|1/2, m\rangle &= 2m\,|1/2, m\rangle \end{aligned} \tag{9}$$

This example shows in what way an ice–type model defines a representation of a quantum group. The quantum states are now elements in $V^{(1/2)} \otimes V^{(1/2)} \otimes \cdots$ and the Boltzmann weights are defined by the quantum R–matrix $R^{1/2,1/2} \in End(V^{(1/2)} \otimes V^{(1/2)})$. The trigonometric solution to (1) with finite value for the spectral parameter u admits a similar representation but now the relevant algebraic structure is an affine quantum Kac–Moody algebra [6].

2.Contour Representations of Quantum Groups: Conformal Field Theories

In the previous section we have considered the realization of quantum groups as integrable vertex models. In this section we will briefly summarize another representation of quantum groups, the one defined by conformal field theories [7]. We will start with a set of local operators $V_a(z)$, $J_i(z)$ satisfying the following exchange algebra:

$$\begin{aligned} J_i(z)J_j(\zeta) &= e^{i\pi\Omega_{ij}}J_j(\zeta)J_i(z) \\ J_i(z)V_a(\zeta) &= e^{i\pi\Omega_{ia}}V_a(\zeta)J_i(z) \\ V_a(z)V_b(\zeta) &= e^{i\pi\Omega_{ab}}V_b(\zeta)V_a(z) \end{aligned} \tag{10}$$

where we will require $\Omega_{ij} = \Omega_{ji}$ and $\Omega_{ab} = \Omega_{ba}$. A simple example of the exchange algebra (10) is given by the Coulomb gas version of $c < 1$ conformal field theories. In this case we have:

$$\begin{aligned} V_a(z) &=: e^{ia\phi(z)} : \\ J_{\pm}(z) &=: e^{i\alpha_{\pm}\phi(z)} : \qquad i = \pm \end{aligned} \tag{11}$$

where :: denotes the usual normal ordering and $\alpha_{\pm}$ are given in terms of c by:

$$\begin{aligned} c &= 1 - 24\alpha_o^2 \\ \alpha_+ + \alpha_- &= 2\alpha_o \\ \alpha_+\alpha_- &= -1 \end{aligned} \tag{12}$$

In this example we obtain for the "monodromy matrices" the following values:

$$\begin{array}{lll} \Omega_{++} = 2\alpha_+^2 & \Omega_{+-} = -2 & \Omega_{ab} = 2ab \\ \Omega_{+a} = 2\alpha_+ a & \Omega_{-a} = 2\alpha_- a & \Omega_{--} = 2\alpha_-^2 \end{array} \tag{13}$$

In terms of the local operators $V_a(z)$, $J_i(z)$ we define a new set of non local operators, that we will call "screened vertices".

Definition 1.

$$\begin{aligned} e^a_{r_i}(z) &= \int_C dt_1 \cdots \int_C dt_{r_i} J_i(t_1) \cdots J_i(t_{r_i})\, V_a(z) \\ e^a_{r_i,r_j}(z) &= \int_C dt_1 \cdots \int_C dt_{r_i} \int_C dw_1 \cdots \int_C dw_{r_j} J_i(t_1) \cdots J_i(t_{r_i}) J_j(w_1) \cdots J_j(w_{r_j})\, V_a(z) \end{aligned} \tag{14}$$

with C the contour of figure 1.

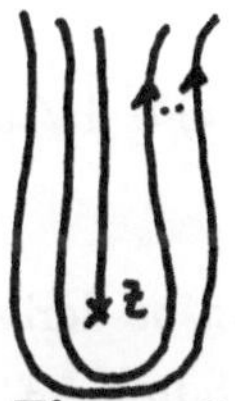

Figure 1.

It is now easy, using the relations (10) to prove that the dimension n^a_i of the vector spaces $\mathcal{V}^a_i$ generated by the $e^a_{r_i}(z)$ are determined by the monodromy matrices as follows:

$$2\Omega_{ia} + \Omega_{ii}(n^a_i - 1) = 0 \mathrm{mod} 2 \tag{15}$$

For rational theories:

$$a \in \left\{ \frac{1-p}{2}\alpha_+ + \frac{1-q}{2}\alpha_- \equiv \alpha_{p,q}\ ; \quad p, q \in \mathbf{Z} \right\} \tag{16}$$

we obtain:

$$n^a_+ = p \qquad n^a_- = q \tag{17}$$

if $p, q > 1$. For a non rational theory *i.e.*, $c \notin \mathbf{Q}$, some of the spaces $\mathcal{V}^a_i$ have infinite dimension. As an example consider $a = 2\alpha_\circ - \alpha_{p,q}$ with $2\alpha_\circ$ non rational. After these technical preliminaries we can prove the following theorem.

Theorem 1. For $c < 1$ rational theories the spaces $\mathcal{V}^a_i$ with $i = \alpha_\pm$ and $a \in \{\frac{1-n}{2}\alpha_\pm,\ n \in \mathbf{N}\}$ are irreducible representations of spin $j = \frac{n-1}{2}$ of $SU(2)_q$ with $q = e^{2\pi i \alpha_\pm^2}$.

The complete proof of this theorem can be found in reference [8], here we simply present the main steps of the construction:

i) Definition of the quantum group generators (Chevalley basis):

$$\begin{aligned}
F_i e^a_{r_i}(z) &= \int_C dt_{r_i+1} J_i(t_{r_i+1}) e^a_{r_i}(z) = e^a_{r_i+1}(z) \\
K^2_i e^a_{r_i}(z) &= e^{i\pi(r_i\Omega_{ii}+\Omega_{ia})} e^a_{r_i}(z) \\
E_i e^a_{r_i}(z) &= [[r_i]]_i \frac{1 - e^{2\pi i\Omega_{ia}} q_i^{r_i-1}}{1-q_i^{-1}} e^a_{r_i-1}(z) \\
i = \pm \quad q_\pm &= e^{2\pi i\alpha^2_\pm}, \quad [[r]]_i = \frac{1-q_i^r}{1-q_i}
\end{aligned} \tag{18}$$

which in words can be described as contour creation, anhilation operators (E,F) and monodromy counting operator (K). The definition (18) of the contour anhilation operator E is implicit in the conformal transformation law of the screened vertex operators $e^a_{r_i}(z)$:

$$\delta_\xi e^a_{r_i}(z) = e^a_{r_i}(\delta_\xi V_a(z)) - (1-q_i^{-1})\xi(\infty)J_+(\infty)E_i e^a_{r_i}(z) \tag{19}$$

where

$$\delta_\xi V_a(z) = (\xi(z)\partial_z + \Delta_a \xi'(z))V_a(z) \tag{20}$$

with Δ_a the conformal weight of $V_a(z)$.

ii) Commutation Relations. They can be derived using equations (18) and (19). The final result coincides with the defining relations of $SU(2)_q$ in the E, F, K basis:

$$E_i F_i - q_i F_i E_i = \frac{1-K_i^4}{1-q_i^{-1}} \tag{21}$$

iii) Comultiplication rules. They can be easily obtained from (18), using contour deformation techniques.

As a simple generalization of theorem 1 it is not hard to prove the following result [8]:

Theorem 2. For $c < 1$ rational theories the spaces $\mathcal{V}^a_{i,j}$ generated by the screened vertex $e^a_{r_i,r_j}(z)$ define irreps of the following Hopf algebra Q:

$$\begin{aligned}
E_\pm F_\pm - q_\pm F_\pm E_\pm &= \frac{1-K_\pm^4}{1-q_\pm^{-1}} \\
E_\pm F_\mp &= F_\mp E_\pm \\
E_+ E_- &= E_- E_+ \\
F_+ F_- &= F_- F_+ \\
K_\pm F_\mp &= -F_\mp K_\pm \\
K_\pm F_\pm &= q_\pm^{\frac{1}{2}} F_\pm K_\pm \\
K_+ K_- &= K_- K_+ \\
K_\pm E_\pm &= q_\pm^{-\frac{1}{2}} E_\pm K_\pm \\
K_\pm E_\mp &= -E_\mp K_\pm
\end{aligned} \tag{22}$$

The Hopf algebra Q embodies the quantum group symmetry of minimal models. The explicit connection with conformal field theory can be established through the following main theorem.

Theorem 3. For $c < 1$ rational theories the space $Inv(\mathcal{V}^{a_1} \otimes \cdots \otimes \mathcal{V}^{a_n})$ of quantum group invariant elements in $\mathcal{V}^{a_1} \otimes \cdots \otimes \mathcal{V}^{a_n}$ with respect to the Hopf algebra Q, coincide with the space of conformal blocks with n external legs saturated by the primary fields $V_{a_i}(z)$.

To prove this theorem we must simply remember that for $c < 1$ rational theories the conformal blocks are defined by solving the decoupling of null vector equations. These equations are hypergeometric and admit an integral representation. To map any of their solutions to the space $Inv(\mathcal{V}^{a_1} \otimes \cdots \otimes \mathcal{V}^{a_n})$ we only need to open the contours appearing in the corresponding integral representation. The coefficients we obtain by the opening contour manipulations are in fact the q–Clebsch–Gordan coefficients (see reference [8] for details). Generalizations of the previous results to WZWN–models can be found in ref. [9].

3. Riemann Surfaces and Quantum Groups:Quantum Uniformization

The theorem 3 of the previous section reflects a deep connection between the quantum group symmetry and the integrability properties of a rational conformal field theory. In this section we will briefly summarize some recent progress in the direction of understanding this interplay from a more geometrical point of view [10].

Let us start by reviewing some well–known classical facts about uniformization of Riemann surfaces [11]. Let us consider a Riemann surface Σ with genus equal to zero and n punctures $(z_1, z_2, \ldots. z_n)$. We fix the puncture z_n at infinity. Associated with this Riemann surface we define the uniformization equation:

$$\frac{d^2y}{dz^2} + \frac{1}{2}Q_\Sigma(z)y = 0 \qquad z \in \Sigma \tag{23}$$

with $Q_\Sigma(z)$ defined as the Schwartzian derivative of the uniformization map π^{-1} which embeds the Riemann surface Σ into the upper half plane H:

$$Q_\Sigma(z) = S(\pi^{-1}) \tag{24}$$

$$\pi_{-1} : \Sigma \to H \tag{25}$$

A classical result due to Poincaré establishes that equation (23) is fuchsian with $Q_\Sigma(z)$ of the following form:

$$Q_\Sigma(z) = \sum_{i=1}^{n-1} \left(\frac{1}{2} \frac{1}{(z - z_i)^2} + \frac{c_i}{z - z_i} \right) \tag{26}$$

The auxiliary uniformization parameters c_i, which depend on the particular Riemann surface we are considering, satisfy the conditions

$$\sum_{i=1}^{n-1} c_i = 0 \quad ; \quad \sum_{i=1}^{n-1} c_i z_i = 1 - \frac{n}{2} \tag{27}$$

Moreover, the uniformization map π^{-1} can be represented as the quotient of two linearly independent solutions of equation (23):

$$\pi^{-1}(z) = y_2/y_1(z) \, . \tag{28}$$

To fix ideas, let us consider the trivial example of the Riemann sphere with two punctures, one located at zero and the other one at infinity. The uniformization map π^{-1} in this case is simply $\pi^{-1}(z) = \log z$ and the uniformization equation reads:

$$y'' + \frac{1/4}{z^2} y = 0 \tag{29}$$

whose solutions $y_1 = \sqrt{z}$, $y_2 = (\log z)\sqrt{z}$ are in agreement with equation (28). A different characterization of a Riemann surface can be obtained using Jenkins–Strebel differentials. This description of a Riemann surface is more appropiate from the point of view of string field theory and it will be of some relevance for the following discussion. Denoting by $\sqrt{\phi}$ the Jenkins–Strebel differential associated with Σ, equation (28) becomes:

$$\frac{\mathrm{d}\left(\pi^{-1}(z)\right)}{\mathrm{d}z} = \sqrt{\phi(z)} \tag{30}$$

which for the Riemann sphere with two punctures described by (29) implies $\phi(z) = 1/z^2$. A Jenkins-Strebel differential is nothing but a one–differential of the form $\sqrt{\phi(z)}dz$ where $\phi(z)$ is a quadratic differential on the Riemann surface.

Our aim now will be to prove that the screened vertex operators we have used in the previous section to define the "contour–representation" of a quantum group can be interpreted as the quantum ancestors of the solutions to the uniformization equation. More precisely, we will show that the solutions y_1, y_2 to the uniformization equation (29) are the classical limit of the screened vertex operators $V_{2,1}(z)$, $\int_C J_+(t)V_{2,1}(z)dt$ in terms of which we have defined the fundamental 1/2 representation of $SU(2)_q$. To motivate this result, we will first make some preliminary comments concerning the conformal properties of the two solutions $z^{\frac{1}{2}}$ and $(\log z)z^{\frac{1}{2}}$ to equation (29). The first thing to be noticed is that $z^{\frac{1}{2}}$ has the conformal properties we should expect for the classical limit (defined as $c \to \infty$) of the $V_{2,1}(z)$ primary field. In fact, from the BPZ–Kac formula for the conformal weights we obtain

$$\lim_{c\to\infty} \Delta_{2,1} = \lim_{c\to\infty} \frac{1}{16}\left[5 - c + \sqrt{(c-1)(c-25)}\right] = -1/2 \tag{31}$$

which coincides with the conformal weight of the solution $y_1 = z^{\frac{1}{2}}$. The second solution $y_2 = (\log z)z^{\frac{1}{2}}$ does not correspond to a conformal field. Moreover, its transformation law

$$y_2(\lambda z) = \sqrt{\lambda z}\log z + \log\lambda\sqrt{\lambda}\sqrt{z} \tag{32}$$

shares a striking simliarity with the conformal transformation law (19) for screened vertex operators. Recall that the screened vertex operators are not conformal fields, and that it is this fact which provided the mechanism to define the contour annihilation operator E and to prove the commutation relations (22). The previous two comments can serve to illustrate why solutions to uniformization equations can be interpreted as the classical limit of screened vertex operators.

We will first obtain a quantum field theoretical representation to the solutions of uniformization equations. This problem is in a certain sense analogous to the theory of holonomic quantum fields worked out by the Kyoto school. To get this representation, we will consider for the Riemann sphere with n punctures ($n > 2$) the following Fuchsian equation:

$$\frac{d^2y}{dz^2} + Q_\phi(z)y(z) = 0 \tag{33}$$

where $Q_\phi(z)$ is defined using equations (34) and (30), with ϕ the quadratic differential $\phi = \prod_{i=1}^{n-2}(z - z_i)^2$. The main difference between equation (23) and (33) is that in the second case we concentrate the curvature on the punctures. This second representation is more appropriate for obtaining a quantum field theoretical representation of the solutions as the classical limit of vertex correlators. Notice that for $n = 2$ the two representations coincide. Our main result can be summarized in the following way:

Proposition 1. Solutions to the uniformization equations (33) can be obtained as the classical limit of the "Feigin–Fuks" correlators

$$\int_C dt\, \langle V_{a_\infty}(\infty)J_+(t)V_{2,1}(z)V_{1,0}(z_1)\cdots V_{1,0}(z_{n-1})\rangle$$

where we have inserted a $V_{1,0}(z_i)$ operator at each puncture, the $V_{2,1}(z)$ operator for the z–dependence, and where the "charge" a_∞ is fixed by the condition

$$(n-1)\alpha_{1,0} + \alpha_{2,1} + a_\infty = \alpha_- \tag{34}$$

Each solution will be associated with a particular contour C.

For people familiar with the Coulomb gas representation of conformal blocks, the previous proposition might seem in contradiction with our comments on the conformal properties of the solutions to equation (29). To clarify this point, we will describe in some detail the two–puncture case. Following proposition 1 we consider the correlator

$$\int_c dt \, \langle V_{a_\infty}(\infty) J_+(t) V_{2,1}(z) V_{0,1}(0) \rangle \tag{35}$$

with the contour C still to be specified. The value of (35) can be obtained using the standard operator product expansion for V and J. The result is

$$z^{\frac{1}{2}} \int_C dt \, t^{-1} (z-t)^{-\alpha_+^2} \tag{36}$$

If now we choose C as a closed contour around zero and take the classical limit, we get

$$y_1(z) = z^{\frac{1}{2}} = \lim_{c\to\infty} z^{\frac{1}{2}} \oint_{C_0} dt \, t^{-1} (z-t)^{-\alpha_+^2} \tag{37}$$

The difficult part now is to choose C in (36) in order to obtain the non–conformal solution $z^{\frac{1}{2}} \log z$. We first introduce a reference point z_0, which will act as a regulator for (36), as follows:

$$z^{\frac{1}{2}} \int_z^{z_0} dt \, t^{-1} (z-t)^{-\alpha_+^2} \tag{38}$$

Now to get the solution $z^{\frac{1}{2}} \log(z/z_0)$ (notice that in the previous case we have implicitly set $z_0 = 1$) we must take first the classical limit of (38) ($c \to \infty$):

$$y_2(z) = z^{\frac{1}{2}} \log(z/z_0) = \lim_{c\to\infty} z^{\frac{1}{2}} \int_z^{z_0} dt \, t^{-}1 (z-t)^{-\alpha_+^2} \tag{39}$$

To make explicit now the connection with the screened vertex operators, we can rewrite the results (37), (39) in a more appealing way as follows:

$$\begin{aligned} y_1(z) &= \lim_{c\to\infty} \langle V_{2,1}(z) \rangle_{\Sigma_{(0,2)}} \\ y_2(z) &= \lim_{c\to\infty} \left\langle \int_z^{z_0} J(t) V_{2,1}(z) \right\rangle_{\Sigma_{(0,2)}} \end{aligned} \tag{40}$$

where the "uniformization expectation value" $< >_{\Sigma_{(0,2)}}$ is defined by inserting at the punctures the $V_{1,0}$ operators. The connection with the screened vertex operators is already clear from (40). It would be very interesting to generalize the preious results to generic Riemann surfaces with non–vanishing genus and to find its physical interpretation. Let us indicate a possible geometrical origin of the screening current. In fact, from equations (28), (30) and (40) we see that the classical limit of the screening operator $J_+(z)$ can be interpreted as the Jenkins–Strebel differential $\sqrt{\phi}$, with ϕ a quadratic differential

$$\lim_{c\to\infty} J_+(z) = \sqrt{\phi(z)} \tag{41}$$

To conclude this section, we will make a few comments on how the previous results can be used in the long-standing problem of quantization of the Liouville theory. The connection between quantum groups and Liouville is not new. It first appeared in the papers on quantum Liouville by Gervais and Neveu [12] and has been worked out recently by Gervais [13] and Smirnov and Takhtadjan [14]. Their approach is based on the representation of Liouville solutions

$$\exp -\phi(z,\bar{z}) = \frac{1}{2} \sum_{m=-\frac{1}{2}}^{+\frac{1}{2}} \psi_m^{\frac{1}{2}}(z)\psi_m^{\frac{1}{2}}(\bar{z}) \tag{42}$$

where the fields $\psi_m^{\frac{1}{2}}(z)$ are the solutions to the equation

$$-\psi'' + T(z)\psi = 0 \tag{43}$$

with $T(z)$ the holomorphic component of the Liouville energy–momentum tensor. In references [12] and [13], a quantization procedure for the field ϕ is provided which elevates the classical fields $\psi(z)$ appearing in (42) to quantum operators satisfying exchange relations similar to the ones obtained for soliton operators in factorized S–matrix models.

In this section we have shown how to find quantum group representations, in particular the contour representation, starting with the uniformization equations (33) that play the same role and have the same meaning as equation (43). A technical advantage of our approach is that it does not require, in order to make contact with quantum group objects, any *ad hoc* transformations on the fields ψ in (42) (for details, see [10]).

4. Concluding Remarks

In this lecture, we have constructed an explicit representation of quantum groups starting with the operator algebra corresponding to the Coulomb gas representation of conformal field theories. The geometrical interpretation of our representation was obtained by "quantizing" the uniformization theory of Riemann surfaces. Quantum groups appear then as the appropriate tool to quantize geometry and, consequently, to quantize Liouville theory. This last point was already clear for the cases $c < 1$ and $c > 25$ *i.e.,* no tachyonic states in the spectrum or, more precisely, rational conformal field theories. In fact this comes from the well–known results connecting uniformization of Riemann surfaces and the Liouville representation of the curvature. Many questions remain open, as the one already mentioned on the extension of our results on quantum uniformization to surfaces with non–vanishing genus and of course the analysis of the tachyonic states.

Acknowledgements. One of us, C.G. would like to thank the organizers of the XIXth International Conference on Differential Geometric Methods in Theoretical Physics for their kind invitation to participate in such an interesting meeting.

References

[1] P. Kulish and E. K. Sklyanin, J. Soviet Math. 19 (1982) 1596, Lecture Notes in Physics 151 (1982) 61; P. Kulish and N. Yu. Reshetikhin, Uspekhi Math. Nauk. 40 (1985) 214; E. K. Sklyanin,Funct. Anal. i Appl. 16 (1982) 27; 17 (1983) 273

[2] M. Jimbo, Lett. Math. Phys. 10 (1985) 63.

[3] V.G. Drinfeld, Soviet Math. Doklady 32 (1985) 254 and ICM Proceedings, Berkeley 798, 1986

[4] R.J. Baxter, *Exacty Solved Models in Statistical Mechanics*, 1982

[5] L. Kadanoff and J. Wegner, Phys. Rev. B4 (1971) 3983; Wu F.Y., Phys. Rev. B4 (1971) 2312

[6] See *e.g.*, M. Jimbo in Advanced Series in Math. Physics, Vol. 9 (1989) World Scientific, Singapore

[7] L. Alvarez Gaume, C. Gomez and G. Sierra, Phys. Lett. 220B (1989) 142, Nucl. Phys. B319 (1989)155; G. Moore and N. Reshetikhin, Nucl. Phys. B328 (1989) 557; E. Witten, Nucl. Phys. B330 (1990) 285; I. Todorov, Clausthal Meeting on Quantum Groups; P. Furlan, V. Petkova and I. Ganchev, Trieste preprint (1989)

[8] C. Gomez and G. Sierra, Phys. Lett. 240B (1990) 149 and Genève preprint UGVA–DPT–1990/04/669, to appear in Nucl. Phys. B

[9] C. Ramírez, H. Ruegg and M. Ruiz–Altaba, Genève preprint UGVA–DPT–1990/06/675, to appear in Phys. Lett. B

[10] C. Gomez and G. Sierra, Genève preprint UGVA–DPT–1990/08/694

[11] See *e.g.*, P. Zograf and L. Takhtadjan, Funk. Anal. i Priloz. 19,3 (1985)

[12] J.L. Gervais and A. Neveu, Nucl. Phys. B224 (1983) 329; B238 (1984) 125

[13] J.L. Gervais, Paris preprint LPTENS 90/13

[14] F. Smirnov and L. Takhtadjan, Steklov Institute (Leningrad) preprint (1990)

Some Physical Applications of Category Theory

Shahn Majid

Department of Applied Mathematics & Theoretical Physics, University of Cambridge, Cambridge CB3 9EW, U.K.

Abstract: We explain the physical meaning of some recent results in category theory: Associated to any topological quantum field theory (in the sense of a functor) is a quasi-quantum group of internal symmetries. Associated to any algebraic quantum field theory (where there is no functor) is a braided group. We also mention some joint work relating Chern-Simons theory to quantum mechanics in a bounded domain.

1 Introduction

Category theory is probably anathema to most physicists, partly because it is often presented very abstractly and partly because it seems so general as to be useless for concrete physical situations. We shall see in this paper that this is wrong. Firstly, in this section, we shall see that every theoretical physicist is a native category theorist at an intuitive level. Then in the later sections I shall give several applications of this category theory that make contact with real physical situations. A couple of these are developments[10][11] of the connection with category theory in [18][17][1][3], while others are more novel and based on [8, Sec. 7][9][15][12][13]. These applications aim to make clear that category theory is not just a language but has non-trivial theorems with non-trivial physical consequences.

To understand the point of category theory it is necessary to appreciate that there are many problems in physics which have been around for some time but for which the right concepts just do not seem to be available. The laws of nature that theoretical physicists seek just might not be expressible in known mathematics. In other words, a good theorist has to be ready to create entirely new concepts, new mathematics, in order to formulate some physical ideas in terms that are concrete enough to have testable predictions. In doing this he or she has to have a good intuition of what notions or concepts are "natural", i.e. are well-behaved as we change the situation in which the concept is to be applied. This is also what pure mathematicians do. So a good *theorist* needs to be a good pure mathematician as regards the creative aspect of pure mathematics: This idea of creating "natural" or well-behaved concepts is really at the heart of pure mathematics, much more so than issues of rigour although intimately and inextricably connected with them.

A category $\mathcal{C}$ is just a collection of objects $X, Y, Z, \cdots$ and a specification of what are the "maps" or morphisms, $\phi : X \to Y$ between any two objects X, Y. The set of morphisms between X and Y is denoted $Mor(X, Y)$. There is also specified a way to compose morphisms.

Category theory consists of ideas and results that hold at such a level of generality. Clearly anything that holds at such generality is likely either to be very powerful or else very tautological (or perhaps both). At the very least, I want to explain that category theory is useful to organize ideas, i.e. to formulate physical principles. We shall see as an example that Einstein's principle of equivalence can be formulated in category-theoretic terms as a *naturality* condition. Einstein was lucky in that Riemannian geometry was already invented: if we have some similar principle for which the concepts do not exist, category theory can help.

The first idea that we need is an obvious one, that of a map between categories. This is called a *functor*. A functor $F : \mathcal{C} \to \mathcal{V}$ is just a map of the structure of $\mathcal{C}$ to the structure of $\mathcal{V}$, i.e. it sends objects to objects and morphisms to morphisms, in a way that respects compositions, $F(\phi \circ \psi) = F(\phi) \circ F(\psi)$. It is useful to consider also *contravariant functors*. These send $\phi : X \to Y$ instead to $F(\phi) \in Mor(F(Y), F(X))$ obeying $F(\phi \circ \psi) = F(\psi) \circ F(\phi)$.

We can now introduce our first novel and powerful idea of category theory, that of a natural transformation between functors[7]. If $F_1, F_2 : \mathcal{C} \to \mathcal{V}$ are two functors we say that $h : F_1 \to F_2$ is a natural transformation, $h \in Nat(F_1, F_2)$, if h is a collection $\{h_X\}$ for each X in $\mathcal{C}$ of morphisms $h_X : F_1(X) \to F_2(X)$ in $\mathcal{V}$. These should fit together in a way that respects morphisms in $\mathcal{C}$,

$$h_Y \circ F_1(\phi) = F_2(\phi) \circ h_X, \qquad \text{for all } \phi : X \to Y \tag{1}$$

So a natural transformation $h : F_1 \to F_2$ is basically a function on $\mathcal{C}$ with values in morphisms in $\mathcal{V}$ that is "coherent" or "well-behaved" or "natural" as we change the point in $\mathcal{C}$ in the sense of (1). A useful way of thinking about this is in Section 1.2.

What about some non-trivial theorems? Here is one. An easy example of a functor $\mathcal{C} \to Sets$ (the category of sets) is $Mor(X,)$ defined by an object X in $\mathcal{C}$,

$$Y \mapsto Mor(X, Y), \qquad Mor(X, \phi) = \phi \circ \tag{2}$$

for all objects Y and morphisms ϕ in $\mathcal{C}$. We now come to our first non-trivial tool in category theory: *many functors $F : \mathcal{C} \to Sets$ can be represented in the form $Mor(X, \ )$ for some X.* (This tends to be most useful when $\mathcal{C}$ has direct sums.)

Acknowledgments Work supported by SERC Research Fellowship.

1.1 Einstein's Principle of Equivalence

Here are some geometrical examples of functors. Let $\mathcal{C}$ = Manifolds be the category of smooth manifolds. Morphisms are smooth maps between them. Let $\mathcal{V}$ = Vector − Bundles be the category of vector bundles. Morphisms (bundle maps) are pairs consisting of maps of the base space and compatible linear maps of the fibres. An example of a functor is T : Manifolds → Vector − Bundles that sends X to TX its tangent bundle. A morphism $\phi : X \to Y$ between manifolds maps to $T(\phi) = (\phi, \phi_*) : TX \to TY$. Here $\phi_* : T_xX \to T_{\phi(x)}Y$ is the differential of ϕ defined by $\phi_*(v) = \frac{d}{dt}_{|0}\phi(exp_x(tv))$ where $exp_x(tv)$ is any smooth curve through x with tangent at x given by $v \in T_xX$. Likewise, $T^* : X \mapsto T^*X$, $T^*(\phi) = (\phi, \phi^*)$ is a contravariant functor. Here $\phi^* = (\phi_*)^*$ (the adjoint or transpose).

Consider now some contravariant functors Ω^n : Manifolds $\to$ Vector $-$ Spaces that send X to $\Omega^n(X)$ (the n-forms on X) and $\phi : X \to Y$ to the linear maps $\Omega^n(\phi) = \phi^*$. Here ϕ^* denotes the pullback of ϕ. For example when $n = 1$, $\Omega^1(X)$ denotes sections of the cotangent bundle. If $\omega \in \Omega^1(Y)$ is such a covector field, then $\phi^*(\omega) \in \Omega^1(X)$ is defined by $\phi^*(\omega)(x) = \phi^*(\omega(\phi(x)))$ where ϕ^* on the right is the adjoint of $\phi_* : T_xX \to T_{\phi(x)}Y$ as above.

An example of a natural transformation is the familiar exterior derivative $d \in Nat(\Omega^n, \Omega^{n+1})$. This means the entire family $d_X : \Omega^n(X) \to \Omega^{n+1}(X)$: covariant derivatives make sense on any manifold X. Moreover, these are indeed natural (functorial) for (the contravariant version of) equation (1) becomes

$$d_X \circ \phi^* = \phi^* \circ d_Y, \qquad \text{for all } \phi : X \to Y. \tag{3}$$

This is a key property of exterior derivatives (exterior derivatives commute with pullbacks)[6, Vol. I, Chap. 1.1]. For example, if ϕ is a diffeomorphism (i.e. a co-ordinate transformation) then this just implies that d is co-ordinate invariant. Thus (3) is an ultrastrong form of general co-ordinate invariance.

For another example, let $\mathcal{C}$ = Manifolds again and consider the contravariant functor S^2 : Manifolds $\to$ Vector $-$ Spaces that assigns to X the symmetric 2-covector fields $S^2(X)$ (like $\Omega^2(X)$ above but symmetric. Again, $S^2(\phi) = \phi^*$). Likewise, let M^2 be the contravariant functor sending X to the invertible subset of $S^2(X)$ (i.e. the space of possible metrics). An example of a natural transformation is Ricci $\in Nat(M^2, S^2)$. For this means in fact a family of operators $\text{Ricci}_X : M^2(X) \to S^2(X)$ (i.e. we can compute the Ricci curvature in any manifold) such that

$$\text{Ricci}_X \circ \phi^* = \phi^* \circ \text{Ricci}_Y, \qquad \text{for all } \phi : X \to Y. \tag{4}$$

If we limit our attention to ϕ diffeomorphism, this says that the expression for Ricci is general co-ordinate invariant, $\text{Ricci}(\phi^*(g)) = \phi^*(\text{Ricci}(g))$ for any metric g. (4) is slightly stronger. The subject of Riemannian geometry consists of expressions like Ricci that are well-behaved under such general co-ordinate transformations. Einstein's principle of equivalence is the statement that the equations of physics for a field g (the metric) must be general co-ordinate invariant. For an example, $\text{Ricci(g)} = 0$ is allowed. We see that naturality of Ricci as in (4) corresponds to:

Ultrastrong Equivalence Principle: *expressions must make sense on all manifolds and commute not only with diffeomorphisms ϕ but with all morphisms $\phi : X \to Y$ between manifolds.*

This ultrastrong form is actually used in physics. For example in quantum gravity we want an action that makes sense in all manifolds since we shall "integrate" over them. More about this is Section 1.2. Even in ordinary scalar boson quantum field theory we use the fact that expressions make sense on any manifold, in particular on any dimension, to dimensionally regulate. If mapping to manifolds of different dimension, say, was not well-behaved as in (4) we would not have any natural way to make formal sense of $4-\epsilon$ dimensions.

We see that naturality in the sense of category theory (1) really coincides with "physically natural" in the form of a slightly stronger version of Einstein's equivalence principle. Although this is my own point of view, it is probably something along the lines that the inventors of category theory had in mind in choosing the terminology. So this observation is probably well-known in some context. Category theory then, is a way to make precise our ideas of "physicality" or "naturality" and prove consequences of such restrictions. It is genuinely useful. For example, one can hope to ask what Lagrangians or resulting stress-energy tensors are natural in a precise category-theoretic sense.

A word of warning is in order. Category theory is a tricky subject to do rigourously in full generality. A standard way out is to work with small categories, but this is not really a very physical thing to do so we wont do this, and therefore we'll continue formally. The problem is that to specify a category, say Vec of vector spaces, we have to decide which spaces are the *same* and which are distinct but isomorphic. This is true even at the level of sets and is a problem of set theory. Ultimately, this is a question that physicists are going to have to face up to (i.e. specify it in a physical way). That would involve thinking physically about logic (rather than logically about physics) and is beyond our present scope.

1.2 Physical Picture of Category Theory

The above seems to be all about very general abstract ideas but observe now that categories are in fact only slightly more general than geometry and algebra already familiar to physicists. Here then is a geometric way of thinking about categories, functors and natural transformations. Think of $\mathcal{C}$ as a "space" with objects $X, Y, \cdots$ "points". Think of a morphism $\phi : X \to Y$ as a "path" from the "point" X to "point" Y.

Think of a functor $F : \mathcal{C} \to \mathcal{V}$ as defining a "fibre bundle" over $\mathcal{C}$ and a "gauge field" (connection) on it as follows. For "fibre" above the "point" X take the set $Mor(F(X), F(X))$. Define the gauge field by its parallel transport: the "parallel transport" along "path" $\phi : X \to Y$ is $F(\phi) \circ (\;) \circ F(\phi)^{-1}$ (assuming formally that $F(\phi)$ is invertible). In other words our fibre bundle is like a frame bundle with sections transforming in the adjoint action.

A natural transformation $h \in Nat(F, F)$ is just a "flat section" (i.e. covariantly constant section) of this bundle – a field on $\mathcal{C}$ transforming in the adjoint representation that is covariantly constant. This is just the naturality condition (1). Note that we didn't write (1) this way to avoid assuming $F(\phi)$ is invertible. More generally, two functors $F_1, F_2 : \mathcal{C} \to \mathcal{V}$ define a bundle with fibre above X given by $Mor(F_1(X), F_2(X))$ and connection with parallel transport along ϕ given by $F_2(\phi) \circ (\;) \circ F_1(\phi)^{-1}$. Again $Nat(F_1, F_2)$ is just "flat sections" of this bundle.

This is a geometrical way of thinking about categories. For the examples of Section 1.1 where $\mathcal{C}$ = Manifolds (a "point" in $\mathcal{C}$ is a manifold) we have for example that double differentiation defines functors S^2 and M^2, resulting in a "fibre bundle" and connection. The principle of equivalence then says that a candidate expression for Einstein's equation (such as Ricci) should be a flat section. Note that mathematicians can often compute the dimension of the space of flat sections of a bundle: there are various theorems. If these were generalized to the present setting they might classify all the possibilities. Moreover, in quantum gravity we need to construct measures of integration. If, according to a fundamental principle of quantum theory, we have to integrate over all manifolds, we would have to build geometric structures, solve heat equations etc on bundles over $\mathcal{C}$ = Manifolds. So this physical picture of categories is the level of generality needed for quantum gravity. I intend to pursue this elsewhere. Slightly more conventionally, we can take also $\mathcal{C}$ = Riemann (the category of manifolds equipped with metrics).

Not only is there geometry associated to a category, there is also algebra. Thus given a functor $F : \mathcal{C} \to \mathcal{V}$, the set $Nat(F, F)$ becomes as we saw above, sections, i.e. "matrix-valued functions" on $\mathcal{C}$. This is an algebra with pointwise operations. Namely, if $h, g \in Nat(F, F)$ define $hg \in Nat(F, F)$ by $(hg)_X = h_X \circ g_X$. So, abstract ideas of naturality can be turned into algebra.

And vice-versa: For the converse direction, given an algebra A, the category $\mathcal{C} = Rep(A)$ of representations also has a functor to $\mathcal{V} = Vec$, the forgetful functor. It assigns to a representation its underlying vector space. So representation theory/naturality takes us back and forth between abstract ideas and concrete algebras. It would be very interesting to find examples of natural transformation which as an algebra were physical, associated to a quantum system and as a category were geometrical like the above examples. This is realized to some extent in the examples that follow and is part of the view[16] that quantum mechanics and geometry are representations of each other.

2 Internal Symmetries of Quantum Field Theories

In the following we shall work with categories $\mathcal{C}$ that have a tensor product $\otimes$. They are called monoidal categories. This will give the algebras associated by functors a coproduct Δ making them into quantum groups of one generalised sort or another. The section is an introduction to [9][15][10][11]. These ideas have their origin in Tannaka-Krein reconstruction theorems familiar for groups and for Hopf algebras[19] and quantum groups[8, Sec. 7]. Other relevant preprints (independent of the work cited above) are [20].

We shall assume that the product $\otimes$ has a unit object $\underline{1}$ and is associative and commutative up to isomorphisms, $\Phi_{X,Y,Z} : X \otimes (Y \otimes Z) \to (X \otimes Y) \otimes Z$ and $\Psi_{X,Y} : X \otimes Y \to Y \otimes X$. Here $\otimes, \otimes^{\mathrm{op}} : \mathcal{C} \times \mathcal{C} \to \mathcal{C}$ and $\otimes(\ \otimes\ ), (\ \otimes\ )\otimes : \mathcal{C} \times \mathcal{C} \times \mathcal{C} \to \mathcal{C}$ are functors and Ψ, Φ respectively are natural transformations, i.e. "flat" sections as we change X, Y, Z. They obey obvious pentagon and two hexagon consistency conditions. We do not assume in general that $\Psi^2 = 1$. Such categories are called ***quasitensor***[8, Sec. 7] or ***braided monoidal***[5].

2.1 Symmetries of TQFT are Quasi-quantum Groups

Topological quantum field theories (TQFTs) are a class of quantum field theories closely associated with categories of manifolds. We include here conformal field theories, expressed by Segal in [18] (more or less) in the form of a category $\mathcal{C}$ and a functor $F : \mathcal{C} \to Vec$ (the "modular functor"). Here $\mathcal{C}$ is a category whose objects are oriented d-manifolds and whose morphisms are equivalence classes of cobording $d+1$-manifolds (for CFTs, $d = 2$). The $\otimes$ is given by disjoint union. The functor must associate to each d-manifold X a vector space $F(X)$ and to each $d+1$-manifold ϕ (up to equivalence) with boundaries X and (with reverse orientation) Y, a linear map $F(\phi)$. These matrices $F(\phi)$ are the $d+1$-manifold invariants characteristic of TQFTs. F is required to satisfy

$$c_{X,Y} : F(X)\otimes F(Y) \cong F(X\otimes Y), \qquad \text{for all } X, Y \in \mathcal{C} \tag{5}$$

where the isomorphisms $c_{X,Y}$ are functorial (i.e. "flat sections"). Here $c \in Nat(F^2, F\circ\otimes)$ where $F^2(X,Y) = F(X)\otimes F(Y)$ and $F\circ\otimes$ are functors $\mathcal{C}\times\mathcal{C} \to \mathcal{C}$.

Many authors have looked for quantum groups associated to conformal field theories: for the Wess-Zumino-Witten model the quantum groups $U_q(g)$ appear. We show now that, generalizing quantum groups slightly, this is a general feature of TQFTs. The generalization, that of a quasi-quantum group, was introduced by Drinfeld also in connection with Wess-Zumino-Witten models. A quasi-quantum group $(H, \Delta, \epsilon, \mathcal{R}, \phi)$[2] is like a quantum group $(H, \Delta, \epsilon, \mathcal{R})$ but includes now a special invertible element $\phi \in H\otimes H\otimes H$. H is an associative algebra but now the coproduct $\Delta : H \to H\otimes H$ is not strictly coassociative but obeys

$$(1\otimes\Delta)\Delta = \phi((\Delta\otimes 1)\Delta\ )\phi^{-1}. \tag{6}$$

Here $\mathcal{R} \in H\otimes H$ obeys $\Delta^{\mathrm{op}} = \mathcal{R}(\Delta\ )\mathcal{R}^{-1}$ as usual and $\phi, \mathcal{R}$ obey[2]

$$(1\otimes 1\otimes\Delta)(\phi)(\Delta\otimes 1\otimes 1)(\phi) = (1\otimes\phi)(1\otimes\Delta\otimes 1)(\phi)(\phi\otimes 1) \tag{7}$$

$$(\Delta\otimes 1)(\mathcal{R}) = \phi_{312}\mathcal{R}_{13}(\phi_{132})^{-1}\mathcal{R}_{23}\phi,\ (1\otimes\Delta)(\mathcal{R}) = (\phi_{231})^{-1}\mathcal{R}_{13}\phi_{213}\mathcal{R}_{12}\phi^{-1}. \tag{8}$$

Theorem 2.1 *[10] Associated to every TQFT with category and functor data $(\mathcal{C}, F, c)$ is a certain quasi-quantum group $H = \mathrm{Aut}(\mathcal{C}, F, c)$. It can be defined as a quasi-quantum group of "flat sections" on the category.*

Proof. [10] Briefly, let $H = Nat(F, F)$. It has the algebra structure explained at the end of Section 1.2. Because of the $\otimes$, it also has a coproduct, $(\Delta h)_{X,Y} = c^{-1}_{X,Y} \circ h_{X\otimes Y} \circ c_{X,Y}$. Here the result of the coproduct Δh must lie in $H\otimes H$, i.e. a function of two variables. Define $\phi \in H\otimes H\otimes H$ (i.e. a function of three variables) and $\mathcal{R} \in H\otimes H$ (i.e. a function of two variables) by

$$\phi_{X,Y,Z} = (c^{-1}_{X,Y}\otimes 1) \circ c^{-1}_{X\otimes Y,Z} \circ F(\Phi_{X,Y,Z}) \circ c_{X,Y\otimes Z} \circ (1\otimes c_{Y,Z}) \tag{9}$$

$$\mathcal{R}_{X,Y} = \Psi^{Vec^{-1}}_{F(X),F(Y)} \circ c^{-1}_{Y,X} \circ F(\Psi_{X,Y}) \circ c_{X,Y}. \tag{10}$$

Theorem 2.2 *$H = \mathrm{Aut}(\mathcal{C}, F, c)$ acts on each of the vector spaces $F(X)$ where X is an object in $\mathcal{C}$: There is a functor $\mathcal{C} \to Rep(\mathrm{Aut}(\mathcal{C}, F, c))$.*

Indeed, the action $\triangleright$ of "flat section" $h \in \mathrm{Aut}(\mathcal{C}, F, c) = Nat(F, F)$ on $v \in F(X)$ is $h \triangleright v = h_X(v)$. In a TQFT this means that $\mathrm{Aut}(\mathcal{C}, F, c)$ acts on each of the vector spaces (or Hilbert spaces) associated by the functor F to d-manifolds. For this reason the automorphism quasi-quantum group $\mathrm{Aut}(\mathcal{C}, F, c)$ can be termed the ***internal symmetry quasi-quantum group*** of the TQFT.

It is instructive to consider briefly the artificial situation in which $\mathcal{C} = Rep(H)$, the representations of a quasi-quantum group H already given. For F we take the forgetful functor that assigns to a representation its underlying vector space and we take $c_{X,Y} = 1$. Then

$$\mathrm{Aut}(Rep(H), \mathrm{Forgetful}, 1) \cong H. \tag{11}$$

This is nothing other than a Fourier convolution theorem for quasi-quantum groups cf.[9]. Recall that Fourier's theorem on a group is an isomorphism from the group convolution algebra H to a family of functions on $Rep(H)$ (with pointwise multiplication). For a general TQFT then our category-theoretic construction succeeds in extracting from the category that part that can be identified with the superselection structure of a quasi-quantum group and reconstructs it as $\mathrm{Aut}(\mathcal{C}, F, c)$.

Given any invertible element $f \in H \otimes H$ one can twist H by f to obtain a new quasi-quantum group $\widetilde{H}$[2],

$$\widetilde{\Delta} h = f(\Delta h) f^{-1}, \quad \widetilde{\phi} = f_{23}(1 \otimes \Delta)(f)\phi(\Delta \otimes 1)(f^{-1})(f_{12})^{-1}, \quad \widetilde{\mathcal{R}} = f_{21}\mathcal{R} f^{-1}. \tag{12}$$

This twisting or "gauge transformation" does not change the representation theory and any quantization of the universal enveloping algebra $U(g)$ of a Lie algebra g must be isomorphic to some twisting of the quantum group $U_q(g)$[2].

Theorem 2.3 *[10] Suppose we are given invertible $f \in \mathrm{Aut}(\mathcal{C}, F, c) \otimes \mathrm{Aut}(\mathcal{C}, F, c)$. View it as a function of two variables $f_{X,Y}$. Then $\mathrm{Aut}(\mathcal{C}, F, c \circ f^{-1}) = \widetilde{\mathrm{Aut}(\mathcal{C}, F, c)}$ as twisted by f.*

This explains Drinfeld's observation that many properties of quasi-quantum groups depend only on the equivalence class up to twisting: They are the properties depending only on the category of representations $\mathcal{C}$ and not on F, c.

2.2 Symmetries of AQFT are Braided Groups

Category theory not only helps understand existing structures as in the last section, but helps to define new structures. It is known that in any two or three dimensional algebraic quantum field theory (AQFT) there arises a certain rather abstract braided category $\mathcal{C}$ of endomorphisms[3]. Now there is no functor F to Vec: our abstract category-theoretic approach now becomes indispensable.

In dimension four this category $\mathcal{C}$ has $\Psi^2 = 1$. In this case it is known that $\mathcal{C}$ can be identified with $Rep(G)$ for a compact group G of internal symmetries [1].

In the present two or three dimensional case we showed in [11] that the relevant symmetry is not that of an ordinary group but that of a *braided group* H in $\mathcal{C}$. This is an object H in $\mathcal{C}$ (we assume that $\mathcal{C}$ has been formally extended to include direct sums) with maps $\cdot : H\otimes H \to H$ and $\Delta : H \to H\otimes H$ making it into a cocommutative Hopf algebra in $\mathcal{C}$. The notion of cocommutativity in $\mathcal{C}$ needed here is subtle and given in [11]. The braiding Ψ shows up in the axiom of a Hopf algebra that says that Δ is an algebra homomorphism. This is because it determines the algebra structure on $H\otimes H$ as

$$(a\otimes b)\cdot(c\otimes d) = \sum (a\Psi(b\otimes c)^{(1)}\otimes\Psi(b\otimes c)^{(2)}d). \tag{13}$$

Our notation is to write $\Psi(b\otimes c) = \sum \Psi(b\otimes c)^{(1)}\otimes\Psi(b\otimes c)^{(2)}$. For an ordinary Hopf algebra of course $\mathcal{C} = Vec$ and $\Psi^{Vec}(b\otimes c) = c\otimes b$ (the permutation or twist map).

Example 2.4 Let $\mathcal{C} = SuperVec$, the category of super vector spaces, then a braided group in $\mathcal{C}$ is more or less just the same thing as (the convolution algebra of a) supergroup or (universal enveloping algebra of a) super-Lie algebra. In this example the braiding Ψ is $\Psi_{X,Y}(x\otimes y) = (-1)^{|x||y|}y\otimes x$ for x of degree $|x|$ etc.

Theorem 2.5 *[11] Given a braided monoidal category $\mathcal{C}$ (such as arising in AQFTs) and some representability hypotheses, there is an associated braided group* Aut $(\mathcal{C})$, *which we call the braided group of internal symmetries of $\mathcal{C}$.*

Proof. [11] The first step is to generalize Theorem 2.1 to the case where we are given a functor $F : \mathcal{C} \to \mathcal{V}$. $\mathcal{V}$ here is another braided category. To do this, consider the contravariant functor $\widetilde{F} : \mathcal{V} \to Sets$ that assigns to $V \in \mathcal{V}$ the set $Nat({}_VF, F)$. Here ${}_VF : \mathcal{C} \to \mathcal{V}$ send X to $V\otimes F(X)$. Now as explained in Section 1, equation (2), such functors tend to be representable, i.e. there is an object H in $\mathcal{V}$ such that $Nat({}_VF, F)\cong Mor(V, H)$ for all V (naturally in V). Of course, H will be a (quasi)-quantum group *in the category* $\mathcal{V}$ rather than in Vec[9, Sec. 4]. The second step is to apply this here by setting[9, Sec. 4] $\mathcal{V} = \mathcal{C}$ and $F = \mathrm{id}$ the identity (and $c = 1$) giving H the structure of a Hopf algebra in $\mathcal{C}$, denoted Aut $(\mathcal{C})$. The third step is to show that Aut $(\mathcal{C})$ is cocommutative. This can be seen already in (10) where now Ψ^{Vec-1} is replaced by Ψ^{-1} and $F = \mathrm{id}$ so that $\mathcal{R}$ is trivial. We really need the category theory because in general it is not so easy to explicitly construct the representing object $H = \mathrm{Aut}(\mathcal{C})$ as it was in Section 2.1. Nevertheless, it acts on each of the objects X of $\mathcal{C}$ by maps $\alpha_X : \mathrm{Aut}(\mathcal{C})\otimes X \to X$[9, Sec. 4]. This justifies the term "internal symmetry group".

TQFTs should be examples of AQFTs and in that case we expect a (quasi)-quantum group as in Section 2.1, rather than a braided group. In fact every quantum group can be transmuted into a braided group. The latter concept appears however, to be more general. The quasi case also follows by the same reasoning.

Theorem 2.6 *[11] Every quantum group $(H, \Delta, \epsilon, S, \mathcal{R})$ gives rise to a braided group $(\underline{H}, \underline{\Delta}, \underline{\epsilon})$. We call this process transmutation because it transforms an ordinary algebra ("bosonic object") into a group in a braided category ("braidonic object").*

Proof. [11] We apply Theorem 2.5 in the case $\mathcal{C} = Rep(H)$, defining $\underline{H} = \mathrm{Aut}\,(Rep(H))$. This is therefore a braided group in the category $Rep(H)$. Explicitly, as a vector space we take $\underline{H} = H$. The adjoint action of H,

$$h \triangleright b = \sum h_{(1)} b S h_{(2)}, \qquad h \in H,\ b \in \underline{H} \tag{14}$$

makes $\underline{H}$ an object in $Rep(H)$. Here $\Delta h = \sum h_{(1)} \otimes h_{(2)}$ is the coproduct in H. For $\underline{H}$ we take the algebra the same as H and with $\mathcal{R} = \sum \mathcal{R}^{(1)} \otimes \mathcal{R}^{(2)}$,

$$\underline{\epsilon} = \epsilon, \qquad \underline{\Delta} b = \sum b_{(1)} S\mathcal{R}^{(2)} \otimes \mathcal{R}^{(1)} \triangleright b_{(2)}. \tag{15}$$

3 Chern-Simons Theory and Quantum Crystals

In this section I mention a third physical application of all this category theory. Every finite group G has a natural order $|G|$, the number of elements in G. What is the natural notion of "quantum order" $|H|$ of a quantum group H? Natural means with reasonable properties and can be expressed through category theory.

Let $\mathcal{C} = Rep(H)$ be the category of representations of the quantum group H. As seen above this is a braided monoidal category. In addition, because of the antipode S, every representation X has a conjugate X^*. Now in any such category with conjugates there is a notion of "endomorphisms" $\underline{\mathrm{Hom}}(X, Y) = Y \otimes X^*$. In particular

$$\underline{\mathrm{Hom}}(X, X) = X \otimes X^* \xrightarrow{\Psi} X^* \otimes X \xrightarrow{\mathrm{eval}} \underline{1}. \tag{16}$$

This $\underline{\mathrm{Hom}}(X, Y)$ is a representing object for the functor $Z \mapsto Mor(Z \otimes X, Y)$. So $Mor(X, X) \cong Mor(\underline{1} \otimes X, X) \cong Mor(\underline{1}, \underline{\mathrm{Hom}}(X, X))$. So the functor $Mor(\underline{1}, \ )$ applied to (16) gives a map $Tr : Mor(X, X) \to Mor(\underline{1}, \underline{1})$. Category theorists now define $rank(X) = Tr(\mathrm{id}_X)$. In our case $\underline{1} = \mathbb{C}$ and $Mor(\underline{1}, \underline{1}) = \mathbb{C}$ with the result that $rank(X)$ is a complex number.

$rank(X)$ is the category-theoretic dimension of X. It need not be an integer. It was computed for the present category $\mathcal{C}$ in [8, Sec. 7]. It was found to be closely related to the physicists q-dimension of X in the case when $H = U_q(su(2))$ ($rank(X)$ of course makes sense in generality even when there is no q). One finds that $rank(X) = \mathrm{Tr}_X \underline{u}$. Here $\underline{u} = \sum (S\mathcal{R}^{(2)}) \mathcal{R}^{(1)}$ (notation of the last section) as an operator on X, and the trace is over X.

Given this natural definition of category-theoretic dimension of any object X in $Rep(H)$ it is natural to define $|H| = rank(H)$, i.e. the category-theoretic dimension of H acting on itself by multiplication ("Fock space" representation).

Theorem 3.1 *[12][13]*

$$|U_q(g)| = \frac{\sum_{\Lambda \in \mathrm{Weights}} q^{-(\Lambda, \Lambda)}}{\prod_{\alpha > 0}(1 - q^{-2(\rho, \alpha)})} = \frac{Z(\text{particle in a crystal})}{\prod_{\alpha > 0}(1 - q^{-2(\rho, \alpha)})}.$$

The proof[12] depends on the theory of W-harmonic polynomials where W is the Weyl group of simple Lie algebra g. The inner product here is determined by the Killing form. The α are taken over the positive roots and $\rho = \frac{1}{2}\sum_{\alpha>0}\alpha$. As $q \to 1$, $U_q(g)$ becomes roughly speaking the group algebra on G, the Lie group of g. The quantum order then becomes the infinite number of points in G. This explains the denominator on the right. We see that $q \neq 1$ regularizes this ∞[14].

The numerator is the partition function for a particle living in a crystal consisting of the coroot lattice. The eigenvalues of the Laplacian here are well-known to be proportional to (Λ, Λ). The interpretation works with suitable q. For example, for $su(3)$ the coroot lattice is a rhombic lattice of side $a\sqrt{2}$ (say) and

$$q = e^{\frac{4\pi^2}{3}\frac{\hbar^2}{MkTa^2}} \tag{17}$$

where M is the mass, T the temperature and k Boltzmann's constant. The partition function is also basically proportional (with some zero modes added) to that for a quantum particle confined to an alcove of the Lie algebra. For example, for $g = su(3)$ the alcove is an equilateral triangle, here of side a.

The quantum order has number-theoretic properties as we shall see below (just as the order of a group does). So the theorem explains the origin of number theory in the partition functions of these simple quantum systems. It is hoped to extend this to understand the origin of number theory in more complex models.

The quantum order on the left in the theorem is Tr $\underline{u}$ in a certain representation. This is basically the vacuum expectation value of the Wilson loop of the unknot of framing one in a Chern-Simons theory with group G (in this representation). This vev is thus connected with number theory and with the quantum crystal above[13].

Rather than go into details of the number-theoretic properties of the quantum order, we demonstrate them with an example. $|U_{q^{-\frac{3}{2}}}(su(3))| = \frac{1+6r(q)}{(1-q^3)^2(1-q^6)}$ where, $r(q) = \cdots q^{675} + 2q^{673} + 2q^{669} + 2q^{661} + 2q^{657} + 2q^{652} + 4q^{651} + 2q^{643} + 6q^{637} + 2q^{633} + 2q^{631} + 2q^{628} + q^{625} + 2q^{624} + 2q^{619} + 2q^{613} + 2q^{607} + 2q^{604} + 2q^{603} + 2q^{601} + 2q^{597} + 2q^{592} + 4q^{589} + 3q^{588} + 2q^{579} + 2q^{577} + q^{576} + 2q^{571} + 2q^{567} + 4q^{559} + 2q^{556} + 4q^{553} + 2q^{549} + 2q^{547} + 2q^{543} + 2q^{541} + 4q^{532} + q^{529} + 2q^{525} + 2q^{523} + 2q^{516} + 2q^{513} + 4q^{511} + 2q^{508} + 3q^{507} + 2q^{499} + 2q^{496} + 2q^{489} + 2q^{487} + q^{484} + 4q^{481} + 2q^{475} + 2q^{471} + 4q^{469} + 2q^{468} + 2q^{463} + 2q^{457} + 2q^{453} + 2q^{448} + 2q^{444} + 3q^{441} + 2q^{439} + 2q^{436} + 2q^{433} + q^{432} + 4q^{427} + 2q^{421} + 2q^{417} + 2q^{412} + 2q^{409} + 4q^{403} + q^{400} + 4q^{399} + 2q^{397} + 2q^{388} + 2q^{387} + 2q^{381} + 2q^{379} + 2q^{373} + 2q^{372} + 2q^{367} + 4q^{364} + q^{363} + 3q^{361} + 2q^{351} + 2q^{349} + 4q^{343} + 2q^{337} + 2q^{336} + 2q^{333} + 2q^{331} + 2q^{327} + 2q^{325} + q^{324} + 2q^{316} + 2q^{313} + 2q^{309} + 2q^{307} + 2q^{304} + 4q^{301} + q^{300} + 2q^{292} + 2q^{291} + q^{289} + 2q^{283} + 2q^{279} + 2q^{277} + 4q^{273} + 2q^{271} + 2q^{268} + 4q^{259} + q^{256} + 2q^{252} + 4q^{247} + 2q^{244} + q^{243} + 2q^{241} + 2q^{237} + 2q^{229} + 2q^{228} + q^{225} + 2q^{223} + 2q^{219} + 4q^{217} + 2q^{211} + 2q^{208} + 2q^{201} + 2q^{199} + 3q^{196} + 2q^{193} + q^{192} + 2q^{189} + 2q^{183} + 2q^{181} + 2q^{175} + 2q^{172} + 2q^{171} + 3q^{169} + 2q^{163} + 2q^{157} + 2q^{156} + 2q^{151} + 2q^{148} + 3q^{147} + q^{144} + 2q^{139} + 4q^{133} + 2q^{129} + 2q^{127} + 2q^{124} + q^{121} + 2q^{117} + 2q^{112} + 2q^{111} + 2q^{109} + q^{108} + 2q^{103} + q^{100} + 2q^{97} + 2q^{93} + 4q^{91} + 2q^{84} + q^{81} + 2q^{79} + 2q^{76} + q^{75} + 2q^{73} + 2q^{67} + q^{64} + 2q^{63} + 2q^{61} + 2q^{57} + 2q^{52} + 3q^{49} + q^{48} + 2q^{43} + 2q^{39} + 2q^{37} + q^{36} + 2q^{31} + 2q^{28} + q^{27} + q^{25} + 2q^{21} + 2q^{19} + q^{16} + 2q^{13} + q^{12} + q^{9} + 2q^{7} + q^{4} + q^{3} + q.$

The reader can easily see that if m, n are coprime then the coefficient of q^{mn} is the product of the coefficient of q^m times the coefficient of q^n. This holds to all orders and is connected with the theory of modular functions [12].

4 Representations of CFT and Self-duality

In this section I want to conclude with some further work in category theory in [15] that does not yet have physical applications but has the potential for them. Fix a braided monoidal category $\mathcal{V}$. We saw in Section 2.2 that $F : \mathcal{C} \to \mathcal{V}$ where F is a monoidal functor and $\mathcal{C}$ is a monoidal category, should be thought of as a Hopf algebra H in the category $\mathcal{V}$. A (right) representation of $F : \mathcal{C} \to \mathcal{V}$ is[15] an object V in $\mathcal{V}$ and a natural transformation $\lambda_V \in Nat({}_VF, F_V)$ such that the $\lambda_{V,X} : V \otimes F(X) \to F(X) \otimes V$ obey $\lambda_{V,Y} \circ \lambda_{V,X} = c_{X,Y}^{-1} \circ \lambda_{V,X\otimes Y} \circ c_{X,Y}$ and $\lambda_{V,\underline{1}} = \mathrm{id}$. For simplicity we require the $\lambda_{V,X}$ all to be isomorphisms.

Theorem 4.1 *[15] The right representations of $F : \mathcal{C} \to \mathcal{V}$ form a dual category $\mathcal{C}^\circ$. $Mor((V, \lambda_V), (W, \lambda_W))$ consists of morphisms $\phi : V \to W$ in $\mathcal{V}$ such that*

$$(1 \otimes \phi) \circ \lambda_{V,X} = \lambda_{W,X} \circ (\phi \otimes 1).$$

There is also a dual F° provided by the forgetful functor. Further basic facts, such as a Pontryagin-type theorem for $\mathcal{C}^\circ$ also hold.

Now, such data $F : \mathcal{C} \to Vec$ or more generally $F : \mathcal{C} \to \mathcal{V}$ arise in general TQFTs and AQFTs. Thus to each one of these there is a dual $F^\circ : \mathcal{C}^\circ \to \mathcal{V}$. I do not know if the dual system itself arises as the category of a dual TQFT or AQFT: I want to pose this as an interesting problem for further work. Indeed, it would be interesting to find some self-dual physical examples and also to relate this notion of duality and self-duality to that in the work of Goddard[4].

There do exist self-dual examples, namely ones found by the author in [16]. These consist of a quantum algebra of observables H of a particle moving on certain homogeneous spacetimes. H takes the bicrossproduct form $H = \mathcal{M}(G_1)^\beta \bowtie_\alpha L^\infty(G_2)$ which is a Hopf (von Neumann) algebra of self-dual type. The associated category of representations is thus of self-dual type,

$$Rep(\mathcal{M}(G_1)^\beta \bowtie_\alpha L^\infty(G_2))^\circ = Rep(\mathcal{M}(G_2)^\alpha \bowtie_\beta L^\infty(G_1)). \tag{18}$$

Here α is an action of momentum group G_1 on position group G_2 and β is a matching "back-reaction" of G_2 on G_1.

We would like examples in which $\mathcal{C}$ is not just the representations of an ordinary Hopf algebra and correspondingly in which the physics is not just that of quantum mechanics and gravity of a single particle but that of a quantum field theory. According to the programme of [16] such self-dual models could provide more realistic models relevant to physics at the Planck scale.

References

1. S. Doplicher, J.E. Roberts: Comm. Math. Phys. **131** 51 (1990)
2. V.G. Drinfeld: Quasi-Hopf algebras and Knizhnik-Zamolodchikov equations, Acad. Sci. Ukr. preprint (1989)
3. K. Fredenhagen, K.H. Rehren, B. Schroer: Comm. Math. Phys. **125** 201 (1989); R. Longo: Comm. Math. Phys. **126** 217 (1989)
4. P. Goddard: Meromorphic conformal field theory, in V.G. Kac ed., *Infinite Dimensional Lie Algebras and Lie Groups*, World Sci. (1989)
5. A. Joyal, R. Street: Macquarie Univ. Math. Rep. 86008 (1986)
6. S. Kobayashi, K. Nomizu: *Foundations of Differential Geometry*, Wiley (1969)
7. S. Maclane: *Categories for the Working Mathematician*. Springer (1974)
8. S. Majid: Int. J. Modern Physics A **5** 1–91 (1990)
9. S. Majid: Reconstruction theorems and rational conformal field theories, preprint (1989)
10. S. Majid: Quasi-quantum groups as internal symmetries of topological quantum field theories, preprint (1990)
11. S. Majid: Braided groups and algebraic quantum field theories, preprint (1990)
12. S. Majid, Ya. S. Soibelman: Rank of quantized universal enveloping algebras and modular functions, preprint (1990)
13. S. Majid, Ya. S. Soibelman: Chern-Simons theory, modular functions and quantum mechanics in an alcove, to appear Int. J. Mod. Phys. A (1990)
14. S. Majid: On q-regularization, to appear Int. J. Mod. Phys. A (1990)
15. S. Majid: Representations, duals and quantum doubles of monoidal categories, to appear Supl. Rend. Circ. Mat. Palermo (1989)
16. S. Majid: *Non-commutative-geometric Groups by a Bicrossproduct Construction*, PhD thesis, Harvard (1988), J. Class. Quant. Grav. **5** 1587 (1988), J. Algebra **130** 17–64 (1990), Principle of representation-theoretic self-duality, preprint (1988)
17. G. Moore, N. Seiberg: Comm. Math. Phys. **123** 177 (1989)
18. G. Segal: The definition of conformal field theory, preprint (1988);
M.F. Atiyah, N. Hitchin, R. Lawrence, and G. Segal: Oxford seminar on Jones-Witten theory, notes (1988)
19. K.-H. Ulbrich: On Hopf algebras and rigid monoidal categories, to appear Isr. J. Math (1990)
20. D.N. Yetter: Quantum groups and representations of monoidal categories, preprint (1989);
A. Rozenberg: Hopf algebras and Lie algebras in categories with multiplication, preprint, in Russian (1978)

From Poisson groupoids to quantum groupoids and back

Meinhard E. Mayer

Department of Physics, University of California, Irvine CA 92717

Abstract: This talk reviews the basic definitions of Lie and Poisson groupoids and then proposes Lie Hopf Algebroids as a possible definition for "Quantum Groupoids" — objects which generalize quantum groups on the one hand, and have Poisson groupoids as their classical limits, on the other.

Introduction

At the last two of these Conferences, at Lakes Como and Tahoe [1, 2], I have tried to make the case for the use of Lie groupoids and algebroids in the formulation of gauge theories. I also pointed out their possible use string theory and other nonlinear field theories, such as quantum field theory in a curved background. Since then there has been some activity in this area; I have become acquainted with the work by Alan Weinstein and collaborators [3–5] on Poisson groupoids and their connection to the deformation theory approach to quantization (see, e.g., [6, 7]).

Like many others, I have long been uncomfortable with the traditional way to quantum theory: take a classical model (Lagrangian, Hamiltonian, classical observables, with their symmetries) and then "quantize" it, e.g., by replacing the Poisson brackets by commutators, or, in more modern language, deforming the Poisson manifold into a quantum theory. I felt more comfortable with the other extreme, which for lack of a better word, I will call the "algebraic" approach: the quantum theory is formulated directly in terms of a net of C^*-algebras, acted upon by groups of automorphisms, and subject to the Haag-Kastler (HK) axioms, or a generalization of these, as in the Doplicher-Haag-Roberts (DHR) theory of s uperselection sectors and "internal" symmetries. It is still somewhat of a mystery of how the space-time manifold (and even more so, the gauge bundle), emerges in the algebraic approach; one hoped that the manifolds would emerge as part of the dual. Thanks to Connes' Noncommutative Differential Geometry (NCDG) this now seems closer to realization, but we still need the classical geometric theories as a guide to select among all possible models.

I would like to express the hope that the road to "quantization" should in fact be traversed backward: quantum theories being more fundamental than their classical limits, we should perhaps worry about "dequantizing" quantum theories, rather than the other way around. The main difficulty with this approach is that we usually don't know the quantum theory, but have a pretty good feeling of what its classical limit is supposed to be. So, as is usual in science, the "right path" will contain a certain amount of backtracking and progress by the strategy of "wishful thinking" [13]: we assume we know how to define quantum objects by listing what properties they should have (properties derived from their classical counterparts, wherever they exist), then the classical limit is taken, leading to new classical objects which are extended to quantum objects; the inspiration for a "second guess," etc. The hope is that such a procedure will eventually converge, and a mathematically consistent theory allowing us to go both ways will emerge from these trials and errors. I am presenting a rough sketch of the outcome of one of these trials, attempting to gain some insight into the correct formulation of quantum gauge theories (and by this I mean not only traditional gauge theory, but also possible generalizations, such as supersymmetric gauge theory, Yang-Mills theories in gravitational backgrounds, and the treatment of coupled Einstein-Yang-Mills theories, and maybe various higher-dimensional generalizations, such as strings, membranes, "p-branes," etc.

Another source of inspiration for this research was the rising interest in "quantum groups," which were briefly mentioned in my Como talk, and to which a large number of talks was devoted at Tahoe, as well as here, in Rapallo.

Since crash courses seem to be the order of the day, I will start out in Section 1 with less than a crash course on the theory of Lie-Poisson groupoids (LPG), to establish terminology and notation (space limitations force me just to list definitions, without examples). In Section 2 is devoted to the "wishful thinking" approach to finding an appropriate quantum analog, by using a tentative definition of a quantum groupoid as a "Lie Hopf algebroid" ("Lie-Hopfoid"). Section 3 then discusses some possible physical interpretations, in particular, hints of how to deal with connections, curvature and how the BRS method could be reformulated in this context.

The bibliography at the end is incomplete and I apologize to any authors whose work I may have overlooked. Please let me know (by e-mail[1], or ordinary mail) of any blatant omissions. Space-time constraints have forced me to omit many details which will appear elsewhere.

Acknowledgements. The author would like to express his gratitude to the organizers for their generous hospitality in Rapallo, to Daniel Kastler for critical discussions, and to Alan Weinstein for preprints and a valuable suggestion related to this work.

---

[1] *Bitnet: MMAYER@UCIVMSA; Internet: hardy@golem.ps.uci.edu*

1. A Review of the Basic Definitions

This section briefly recalls some basic definitions relating to Lie groupoids and Poisson groupoids, mainly to establish terminology and notations. We follow mainly the terminology of [9], [3], [5], to which we refer the reader for details.

1.1. Lie Groupoids and Algebroids

Definition 1.1.1 A *groupoid* $\Omega = (\Gamma \rightrightarrows B)$ consists of two sets, the *groupoid* Γ and the *base* B, together with two maps $s, t : \Gamma \rightrightarrows B$ (the *source* and *target* projections), and a map $\varepsilon : B \to \Gamma$ (the *object inclusion map*). Γ can be thought of as a collection of transformations ("arrows") ν, η, ξ, of the set B of *objects* forming the base of the groupoid, with a partially defined composition (multiplication) $(\nu, \xi) \mapsto \nu\xi$. The elements of B will be denoted by $x, y, z \in B$, and their images in Γ by the *object inclusion map* ε are denoted by $\tilde{x}, \tilde{y}, \tilde{z} \in \Gamma$, respectively. The map ε identifies elements x of the base B with arrows in Γ with source and target x, called *units* in (the set of units is often denoted by Γ^0). If γ is an arrow from x to y then $x := s(\gamma)$ is called its *source* and $y := t(\gamma)$ is called its target (or source and target *projection*, respectively. The reversed arrow is called the *inverse* of γ and is denoted by γ^{-1}. The composition of arrows $\xi\eta$ is *defined only* if the target of η is the source of $\xi : t(\eta) = s(\xi)$ (mappings compose from right to left!). Composition is *associative* iff both terms $\xi(\eta\nu)$ and $(\xi\eta)\nu$ are defined.

Obviously, for any $\gamma \in \Gamma$: $s(\gamma^{-1}) = t(\gamma)$, $t(\gamma^{-1}) = s(\gamma)$, $\gamma^{-1}\gamma = \widetilde{s(\gamma)}$, $\gamma\gamma^{-1} = \widetilde{t(\gamma)}$. This set of equations is the reason why the arrows $\tilde{x}$ corresponding to the objects x are called *units* or *identities*. It is easy to see that they actually behave like units, i.e., if $s(\xi) = x$, $t(\xi) = y$, then if $\eta \in \Gamma$ is such that $s(\eta) = y$ and $\xi\eta = \eta$, then $\eta = \tilde{x}$, etc. The set of *composable pairs* of arrows is denoted by Γ^2, or by $\Gamma * \Gamma$. If $\xi \in \Gamma$ we denote the set of its possible sources by $d(\xi) = \xi^{-1}\xi$ and call this set the *domain of* ξ and the set of its possible targets by $r(\xi) = \xi\xi^{-1}$ and call it the *range of* ξ. For $x, y \in B$ we will denote the set of arrows which have x as a source by $\Gamma_x = s^{-1}(x)$, and the set of arrows which have y as target by $\Gamma^y = t^{-1}(y)$. The set of arrows which have x as a source and y as target is denoted by $\Gamma_x^y = \Gamma_x \cap \Gamma^y$. In the language of fibrations, Γ_x is the s-fiber (*source-fiber*) over x, and Γ^y is the t-fiber (*target-fiber*) over y. For two subsets A and C of B one uses the similar notations: $\Gamma_A = s^{-1}(A)$, $\Gamma^C = t^{-1}(C)$ and $\Gamma_A^C = t^{-1}(C) \cap s^{-1}(A)$. The restriction of the groupoid Γ to a subset $A \subset B$ is $\Gamma\,|_A = \Gamma_A^A$. If A consists of a single element $A = \{x\}$ then $\Gamma_x^x = G(x)$ is the *isotropy group* (or *vertex group*) of Γ at $x \in B$.

It is often convenient to identify the objects with the units or identities, i.e., omit the tilde and write, when defined: $x\gamma = \gamma$ and/or $\gamma x = \gamma$. The definition of a *subgroupoid* is the obvious one. The set of units $\Gamma^0 = \tilde{B}$ is sometimes called the *base subgroupoid*.

I like to think of a groupoid as "paths" (more precisely, as *homotopy classes or reparametrization classes of paths*: the inverse path is defined, two paths can be composed if the end of one is the beginning of the next, associativity holds only if both compositions are defined, and left and right identities are defined as the addition of a zero path at the left or right end. There is one important example I like to mention here, for obvious gauge-theoretic reasons:

Example *Groupoid associated to a principal bundle (Gauge Groupoid).* Let $\xi = (P, G, M, \pi)$ be a principal bundle and consider the right action of G on $P \times P$: $(p_2, p_1) \cdot g \mapsto (p_2 \cdot g, p_1 \cdot g)$. Let us denote the orbit of the pair (p_2, p_1) by $\langle p_2, p_1 \rangle$ and the orbit space by $\Pi = (P \times P)/G$. The manifold Π can be turned into a groupoid over the base M by the following definitions: the *source and target projections* are, respectively: $s(\langle p_2, p_1 \rangle) = \pi(p_1)$, $t\langle p_2, p_1 \rangle = \pi(p_2)$; the *object inclusion map* is $x \mapsto \tilde{x} = \langle p, p \rangle$, with $p \in \pi^{-1}(x)$ any element of the fiber over x; the groupoid multiplication, when defined, is given by the following composition of orbits: $\langle p_3, p_2' \rangle \langle p_2, p_1 \rangle = \langle p_3, p_1 \cdot g_{p_2 p_2'} \rangle)$, where $g_{p_2 p_2'} \in G$ is the group element which takes p_2 into p_2' in the same fiber, i.e., $p_2' = p_2 \cdot g_{p_2 p_2'}$. In other words, the groupoid elements are composed in such a way that the "path" in the bundle space is shifted vertically to make "ends meet." This description is equivalent to the use of the *division map* $\delta : P \times_\pi P \to G : (p \cdot g, p) \mapsto g$ which maps the fibered product of P with itself, i.e., pairs of points in the same fiber, into the group elements which map the second point onto the first. Then the condition that the source of the second arrow be equal to the target of the first: $s\langle p_3, p_2' \rangle = t\langle p_2, p_1 \rangle$ guarantees that p_2' and p_2 are on the same fiber, i.e., belong to the fibered product $P \times_\pi P$. By means of a "gauge transformation" one can always choose representatives on the orbits so that $p_2' = p_2$,i.e., $g_{p_2 p_2'} = e \in G$ so that the groupoid multiplication becomes simply "cancellation of the middle": $\langle p_3, p_2 \rangle \langle p_2, p_1 \rangle = \langle p_3, p_1 \rangle$,i.e., the division map becomes the identity $e \in G$. The *inverse* of the orbit $\langle p_1, p_2 \rangle$ is the orbit of the transposed pair $\langle p_2, p_1 \rangle$.

The association of groupoids to principal bundles, and the fact that the Atiyah sequence is naturally defined in Lie groupoids, suggest the use of groupoids in gauge theory, which was discussed in Refs. [1,2].

The terminology and definitions that follow are from [9], to which the reader is referred for full proofs of many of the asserted facts.

Definition 1.1.2 A *differentiable groupoid* (DG) (Γ, B) is a groupoid where both Γ and B are differential manifolds, such that the source and target projections are *surjective submersions* (— *surmersions*), and the object inclusion map $\varepsilon : B \ni x \mapsto \tilde{x} \in \Gamma$, and the partial multiplication $\Gamma * \Gamma \to \Gamma$ are *smooth* maps. A morphism of DG-s is a *smooth morphism*, i.e., one where the pair of maps is smooth.

The tangent bundle to $\Gamma * \Gamma$ is $T\Gamma * T\Gamma = \{Y \oplus X \in T(\Gamma \times \Gamma) | T(s)(Y) = T(t)(X)\}$. This implies that the tangent to the partially defined multiplication follows a Leibniz rule and from this it can be shown [9] that inversion is automatically smooth. Furthermore, ε is an immersion, and therefore a homeomorphism onto $\tilde{B}$, which is therefore a closed embedded submanifold of Γ. The source and target fibers are also submanifold.

Definition 1.1.3 A *Lie groupoid* (LG) is a *locally trivial* differentiable groupoid.

Here *locally trivial* means practically the same thing as in the fiber bundle context: A topological groupoid Γ, B is locally trivial if it is *transitive* and there exists a covering $\bigcup_i U_i = B$ of the base by open sets such that each restriction $\Gamma|_{U_i}$ is isomorphic to a trivial groupoid $U_i \times G \times U_i$ (with s, t respectively the projections on the third and first factor, G a Lie group $\varepsilon : B \ni x \mapsto \tilde{x} = (x, e, x)$, the partial multiplication $(z, g_2, w)(y, g_1 x) = (z, g_2 g_1, x)$ iff $w = y$, and the inverse element is $(y, g, x)^{-1} = (x, g^{-1}, y)$). Just as in the case of principal bundles, local triviality is easiest to describe in terms of *local sections*:

Definition 1.1.4 Let (Γ, B) be a DG. It is *locally trivial* if there exists a point $b \in B$ an open cover $\bigcup_i U_i = B$ and smooth *sections* $\sigma_i : U_i \to \Gamma_b$ such that $t_b \circ \sigma_i = Id_{U_i}$.

It is simple to show that the differentiable subgroupoids $\Gamma_{U_i}^{U_i}$ are isomorphic to the trivial groupoids $U_i \times \Gamma_b^b \times U_i$ under the mapping $\Sigma : (y, \gamma, x) \mapsto \sigma_i(x)\gamma\sigma_i(y)^{-1}$ The collection of trivializing sections $\{\sigma_i\}$ is called a *section-atlas* of the groupoid.

In [9], Chapter II one can find various equivalent criteria for local triviality of topological groupoids. In Chapter III the reader can find a proof that any transitive differentiable groupoid is locally trivial, and hence a Lie groupoid.

The concept of *Lie algebroid* was originally introduced by Pradines in [11] in an attempt to generalize the notion of Lie algebra. In effect, a Lie algebroid is a *Lie-algebra bundle (LAB)* with some additional structure. Lie-algebra bundles made their appearance (implicitly) in Atiyah's definition of a connection, see [10]. Details can be found in Section 3.2 in [9].

Definition 1.1.6 A *Lie-algebra bundle (LAB)* is a vector bundle $\mathcal{L} = (L, \pi, B)$ with a *field of Lie brackets* $[\phi, \varphi]$ defined for the vector space of smooth sections, $\phi, \varphi \in \mathrm{Sec}L$, such that each $[\ ,\ ]_x : L_x \times L_x \to L_x$ is a Lie algebra bracket (i.e., bilinear, antisymmetric and satisfies the Jacobi identity). In addition, there exists a Lie algebra $\mathfrak{g}$ and $\mathcal{L}$ admits an *atlas* $\{\psi_i : U_i \times \mathfrak{g} \to L|_{U_i}\}$ where $\psi_{i,x}$ is a Lie algebra isomorphism at each $x \in B$.

Lie-algebra bundles are a special case of Lie algebroids, namely they are totally transitive algebroids. In order to define Lie algebroids in general Pradines introduced a vector bundle map ("flèche") from the algebroid to the tangent bundle of the base manifold, which Mackenzie has renamed the *anchor* of the algebroid, and which turned out to lead to the following useful definition:

Definition 1.1.5 A *Lie algebroid* over a smooth manifold B is a vector bundle A, p, B, together with a Lie bracket $[\ ,\ ] : \mathrm{Sec}A \times \mathrm{Sec}A \to \mathrm{Sec}A$ (bilinear, antisymmetric and Jacobi), and vector bundle map (linear on each fiber and depending smoothly on the base-point) $q : A \to TB$ called the *anchor* (French: flèche) of A with the following properties: $q([X,Y]) = [q(X), q(Y)]$, $X, Y \in \mathrm{Sec}A$, $[X, fY] = f[X,Y] + q(X)(f)Y$, $X, Y \in \mathrm{Sec}A, f \in C^\infty(B)$. A Lie algebroid A is *transitive* if q is a submersion, *regular* if q is of locally constant rank, and *totally intransitive* if $q = 0$. The latter is obviously the case for a *LAB*.

The *anchor* contains the basic information about the Lie algebroid: It measures the relation between the bracket structure on Sec$A$ and the ordinary Lie or Poisson bracket structure on SecTB. If A is transitive, it will turn out that the right inverse of q defines a connection which is identical to the Atiyah definition introduced in 1957. If A is regular the image of q foliate the base manifold into leaves where A is transitive.

Given a Lie groupoid $\Gamma \rightrightarrows B$, we want the Lie algebroid structure introduced below to be its Lie algebroid, just as a Lie algebra $\mathfrak{g}$ is the algebra of right (or left) invariant vector fields at the identity of a Lie group. However the right translations on the groupoid Γ are diffeomorphisms of the source-fibers only, vector fields will be right invariant only if they are tangent to the source fibers (and left-invariant for the target fibers). This accounts for the complications in the following definition, and ultimately for the recourse to double groupoids. A right invariant vector field is determined by its values on the units $\tilde{x}, \forall x \in B$. We thus introduce the following

Definition 1.1.7 *The Lie algebroid of a Lie groupoid is* $A\Gamma = \bigcup_{x\in B} T(\Gamma_x)_{\tilde{x}}$ with the vector bundle structure induced by the source map s, inherited from the tangent bundle $T\Gamma$. The Lie bracket is defined on the module of sections Sec$A\Gamma$, through the correspondence between sections of $A\Gamma$ and right invariant vector fields on Γ. The bracket is not bilinear with respect to the module structure on Sec$A\Gamma$, but satisfies the identity $[X, fY] = f[X,Y] + q(X)(f)Y$, $X, Y \in \mathrm{Sec}A\Gamma, f \in C^\infty(B)$ of definition 1.1.6, where the *anchor* q maps each section $X \in \mathrm{Sec}A\Gamma$ to the target projection $t * X$ of the appropriate vector field. The vector bundle $A\Gamma \to B$ is the inverse image of $T^s\Gamma \to B$ (the source-tangent bundle) through the object identification $\varepsilon : B \hookrightarrow \Gamma$. The bracket in $A\Gamma$ is inherited from the bracket of right invariant source-vector fields: $[X,Y]_{A\Gamma} = [X,Y]_{RI}$, where the right-hand side is the usual Lie bracket of right-invariant vector fields.

It is legitimate to identify $A\Gamma$ with the restriction of $T^s\Gamma$ to B and the fibers $A\Gamma|_x$ with the tangent spaces $T(\Gamma_x)_{\tilde{x}}$. For more details on Lie algebroids the reader should consult [9].

We now list the basic facts about Poisson groupoids, following mainly the definitions of [3–6].

1.2 Poisson groupoids

Essentially, a Poisson groupoid is a Lie groupoid which is also a Poisson manifold, such that the Poisson structure is compatible with the Lie groupoid structure. We first recall the definition (originally due to Lichnerowicz) of a *Poisson structure* on a manifold [3]:

Definition 1.2.1 A *Poisson structure* **PS** on a commutative algebra A is a Lie algebra structure (with curly brackets used for the Lie bracket) $\{\ ,\ \}$, such that for $\forall h \in A$ the linear operator $X_h : f \mapsto \{f, h\}$ is a derivation of the multiplication in A.[2] A **PS** *on a vector space* V is simply an antisymmetric bilinear form on the dual space V^*. A **PS** *on a smooth vector bundle* E is a smooth field of Poisson structures on each of the fibers E_x. A **PS** on a manifold P is a **PS** on the commutative algebra of smooth functions $C^\infty(P)$.

Definition 1.2.2 A *Poisson manifold or vector bundle* is a manifold or vector bundle equipped with a *Poisson structure.*

Example The tangent bundle TP of a Poisson manifold P is a Poisson bundle with the bracket defined by

$$\pi\{df, dg\} = \{f, g\}$$

where π is the projection in the tangent bundle; conversely, a **PS** on TP will induce a Poisson bracket on P iff the Schouten-Nijenhuis bracket vanishes: $[\pi, \pi] = 0$

Definition 1.2.3. A *Poisson groupoid* $\Pi = (\Pi \rightrightarrows B, s, t)$ is a *Lie groupoid* Π over the *base* B, with a *multiplicative Poisson structure*, such that the graph $\{(z, x, y) | z = xy\}$ of

[2] This means that $X_h(fg) = (X_h f)g + f(X_h g)$, i.e., the Poisson bracket acts as a derivation not only with respect to the Lie algebra structure it induces (Jacobi identity) but also with respect to the commutative multiplication. Think of the example of functions on phase space.

the multiplication (composition) map is a *Poisson map* from $\Pi \times \Pi \to \Pi$. Here the *graph* of a map $f : A \to B$ is defined, following Weinstein, as the set of pairs $\{(f(x), x) | x \in A\}$ (reversed order!), and a map between two Poisson manifolds A, B is a *Poisson map* essentially, if it preserves the Poisson structure. (For complete definitions, see [3–7].)

Definition 1.2.4. (a) A *coisotrope* in a Poisson algebra is a subset which is an ideal for the multiplicative structure, and a subalgebra for the Lie algebra structure. (b) A *coisotropic* subspace W of a Poisson Vector space V is a subspace whose annihilator $W^{\perp} \subset V*$ is *isotropic*, i.e., the symplectic product of any two elements in $W^{\perp}$ vanishes. (c) Similarly, a subbundle F of a Poisson bundle E is coisotropic if $F^{\perp}$ is isotropic in E^*.

Using Weinstein's definition of a graph of $f : P_1 \to P_2$ as the set of pairs $\{(f(y), y) | y \in P_2\}$, he defines a *Poisson relation* $R : P_2 \to P_1$ as a coisotropic submanifold of the product $P_1 \times P_2^-$. Weinstein proves that the graph of a mapping is coisotropic iff the mapping is Poisson. For details and examples of Poisson relations, see [1–4]. Weinstein's definition of a Lie Groupoid differs slightly from the one given above, which is due to Mackenzie; namely he requires only that the source and target maps should be differentiable submersions and multiplication, inversion, and object inclusion maps should be smooth. The distinction is technical and will not bother us here.

For each element $\xi \in \Pi$ the left translation $l_\xi : \eta \mapsto \xi\eta$ is a diffeomorphism from $s^{-1}(t(\xi))$ to $s_{-1}(s(\xi))$, i.e., the source fibers are interchanged. Similarly, a vector field X on Π is left-invariant if $X(\Pi) \subset Ker(Ts)$ and if for reach pair of composable elements $(\xi, \eta) \in |\Pi^2$, $X(\xi\eta) = Tl_\xi(X(\eta))$. The normal bundle $N(B, \Pi)$ becomes a Lie algebroid.

A Lie groupoid with multiplicative Poisson structure is a Poisson Group in the sense of Drinfel'd, [12].

Poisson groupoids have a rich structure, including a complete reduction theory, momentum maps, cohomology theory, etc., which lack of space does not permit us to discuss, and for which we refer to the papers by Weinstein and collaborators.

2. In Search of Quantum Groupoids

This section discusses one possible definition of quantum groupoid without really settling the issue. At the time of writing the most promising definition is that of quantum groupoid as a Lie Hopf algebroid. After experimenting with several possible definitions, Alan Weinstein[3] called my attention to the book [14], where I found a definition of *Hopf algebroids* due to Haynes Miller and Douglas Ravenel, who used it in the context of generalized homology theories. I summarize the basic definitions in S ec. 2.1. This definition seems to lead to the most obvious generalization of quantum groups to quantum groupoids: from Lie Hopf algebras

[3] Just before sending the manuscript off to the Editors (14 September 1990) I received a preprint from A. Weinstein: "Noncommutative Geometry and Geometric Quantization," PAM-507, UCB, August 1990, which deals with the geometric quantization of symplectic groupoids. He arrives at a concept of *symplectic double groupoid* as an object leading to Drinfel'd algebras. I have not had the time to explore the relation of this article to the ideas proposed here. I am grateful to Alan Weinstein for the suggestion, and for sending me his preprints.

to Hopf Lie algebroids, for which I will use the nickname "hopfoids." In Sec. 2.2 I try to merge the concept of Hopf algebroid with Mackenzie's definition of a Lie algebroid coming up with a tentative definition of *Hopf Lie algebroids* (HLA or "hopfoids"— a term suggested by Alan Weinstein) which should be related to Poisson groupoids more or less as the Drinfel'd quantum groups (Lie Hopf algebras). In the following subsection(s) the suitability of these objects for physics is discussed briefly in Sec. 3.

2.1 Hopf algebroids, comodules and cotensor products

The generalization from Hopf algebras to Hopf algebroids is analogous to that of the generalization of groups to groupoids: just as in the latter case the requirement that the product be defined for any pair is relaxed, one obtains a Hopf algebroid from a Hopf algebra by relax ing the requirement that the coproduct should be everywhere defined. For anyone who likes category theory jargon, a *groupoid is a small category where every morphism is invertible, and a Hopf algebroid is a cogroupoid object in the categ ory of (commutative) algebras over a commutative ground ring*. If we want these objects to really qualify as *quantum groupoids* we will have to add three features: *noncommutativity* of the algebra, a *topology* to accommodate operator algebras, and a *Lie bracket structure* in the fibers. We arrive at the tentative definition in several steps, first recalling (almost verbatim — I parenthesized the word *commutative* so that it can be left out when needed) the definition of Hopf algebroid from [14, Appendix A.1].

Definition 2.1. A (*commutative*) **Hopf algebroid** over a commutative ring K is a pair $(A \rightrightarrows \Gamma)$ of (*commutative*) K-algebras, with a pair of maps τ, σ (target and source), a *coproduct* $\Delta : \Gamma \to \Gamma \otimes_A \Gamma$, a *counit* $\varepsilon : \Gamma \to A$, and an *inverse, or conjugation* $c : \Gamma \to \Gamma$, such that for any other (*commutative*) R-algebra B, the sets $\mathrm{Hom}(A, B)$ and $\mathrm{Hom}(\Gamma, B)$ are respectively the *objects and morphisms* of a groupoid.

The *Hopf algebroid space* Γ is a left A-module via the target map (which is a left unit) τ, and a right A-module via the source map σ which acts as a right unit. $\otimes_A$ is a tensor product of bimodules, and Δ, ε are A-bimodule maps. By dualization from the properties of source, target, and object inclusion map of a groupoid, one can derive the following relations among the structure maps of the Hopf algebroid:

i. $\varepsilon\tau = \varepsilon\sigma = \mathrm{Id}_A$ (the source and target of the identity are the object on which it acts – or an "arrow that bites its tail").
ii. $(\Gamma \otimes \varepsilon)\Delta = (\varepsilon \otimes \Gamma)\Delta = \mathrm{Id}_\Gamma$.
iii. $(\Gamma \otimes \Delta)\Delta = (\Delta \otimes \Gamma)\Delta$ (associativity).
iv. $c\tau = \sigma$ and $c\sigma = \tau$ (inversion interchanges source and target).
v. $cc = \mathrm{Id}_\Gamma$ (idempotency of inversion).

vi. There exist maps $c\cdot\Gamma : \Gamma\otimes_K\Gamma \to \Gamma : \gamma_1\otimes\gamma_2 \mapsto c(\gamma_1)\gamma_2$ and $\Gamma\cdot c : \Gamma\otimes_K\Gamma \to \Gamma : \gamma_1\otimes\gamma_2 \mapsto \gamma_1 c(\gamma_2)$ such that the composition of a morphism with its inverse yields the identity.

If the algebras are graded, commutatitvity should be interpreted as graded commutativity ($xy = (-1)^{|x||y|}yx$, where $|\ |$ denotes the grade (degree) of the quantity it surrounds).

Ravenel also defines the notions of *left and right Γ-comodules* as well as *comodule algebra and cotensor product,* which I summarize in the following:

Definition 2.2. A *left Γ-comodule* M is a left A-module with a left A-linear map $\psi : M \to \Gamma\otimes_A M$ which has a counit and is coassociative, i.e, $(\varepsilon\otimes M)\psi = M$ and $(\Delta\otimes M)\psi = (\Gamma\otimes\psi)\psi$. Similarly, one defines a *right Γ-comodule*. An element $m \in M$ is primitive if $\psi(m) = 1\otimes m$.

Definition 2.3. A *comodule algebra* M is a comodule which is also a (commutative) algebra such that the linear map ψ is an algebra morphism. If M, N ar left Γ-comodules their *comodule tensor product* is $M\otimes_A N$ with the structure map being the composite

$$M\otimes N \xrightarrow{\psi_M\otimes\psi_N} \Gamma\otimes M\otimes N \xrightarrow{\iota} \Gamma\otimes\Gamma\otimes M\otimes N \xrightarrow{\mu} \Gamma\otimes M\otimes N,$$

where the second map ι interchanges the second and third factors and the third map μ is multiplication in Γ. The tensor products are over A and use only the left A-module structure. A *differential comodule* C^* is a cochain complex where each C^s is a comodule and the coboundary is a comodule morphism.

Definition 2.4. The *cotensor product over Γ* of the right Γ-comodule M with the left Γ-comodule N is the K-module defined by the exact sequence:

$$0 \longrightarrow M\square_\Gamma N \longrightarrow M\otimes_A N \xrightarrow{\psi\otimes N - M\otimes\psi} M\otimes_A\Gamma\otimes_A N,$$

where ψ are the comodule structure maps for M and N.

The cotensor product $M\square_\Gamma N = N\square_\Gamma M$ is *neither a comodule nor an A-module* but just a K-module.

Definition 2.5. A *Hopf-algebroid morphism* ("hopfoid map") $f : (A,\Gamma) \to (B,\Xi)$ is a pair of K-algebra maps: $f_1 : A \to B$ and $f_2 : \Gamma \to \Xi$ such that: $f_1\varepsilon = \varepsilon f_2$, $f_2\tau = \tau f_1$, $f_2\sigma = \sigma f_1$, $f_2 c = cf$, and $\Delta f_2 = (f_2\otimes f_2)\Delta$.

2.2 Hopf Lie Algebroids ("hopfoids")

We now go beyond the known definitions and try to combine Mackenzie's (or Weinstein's) definition of Lie algebroid with that of Hopf algebroid introduced above. The resulting object is a *Hopf Lie algebroid* (I chose this terminology, so that the abbreviation HLA will not lead to confusion with LHA – used for Hopf Lie algebras = Drinfel'd quantum groups). I will also use the nickname "hopfoid" for these objects,reserving the name "quantum groupoid" for the object which will ultimately prove to meet all the specifications, i.e., the object whose classical limit is indeed a Poisson groupoid. More on the possible physical uses in Sec. 3. We thus arrive at the following tentative definition, the validity of which remains to be tested:

Definition 2.6. A *Lie Hopf algebroid* is a Hopf algebroid $\mathcal{L} \rightrightarrows B$ where $\mathcal{L}$ is equipped with a Lie bracket $[\ ,\ ]$ in each fiber of the fibration induced by the source (or target) map, Definition 1.1.6, such that the following properties hold: i. The Hopf algebroid and Lie algebroid structures are compatible, i.e., the tangent map to the coproduct, counit, transposition, carry over to the bracket operation. ii. The anchor $q : A\Gamma \to TB$ induces a comodule morphism. iii. A connection can be defined on $\mathcal{L}$ as a splitting of the appropriate Atiyah sequence and is compatible with the Hopf algebroid structure on $\mathcal{L}$.

Examples of Lie Hopf algebroids and possible extensions modifications of this definition form the object of a article in preparation.

3. Are Lie Hopf Algebroids Quantum Groupoids?

In this section we indulge in a little wishful thinking, trying to connect the concept of Lie Hopf Algebroid tentatively defined in the previous section with our wishlist for a quantum object, having the right differential-geometric properties. First let me restate the wishlist in a somewhat different form.

Since the penultimate goal is to deal with a differential geometric object having the characteristics of both a gauge bundle and of a field-theory algebra, we would like it to be embedded in an infinite-dimensional algebra $\mathcal{A}$ which in some classical limit becomes the Poisson groupoid of a generalized gauge theory. The algebra $\mathcal{A}$ must contain "its own symmetries,"i.e., it should be possible to extract the holonomy groupoid, superselection structure, etc., from it by an appropriate generalization of the Doplicher-Haag-Roberts construction. This is where the importance of the Hopf algebra structure on $\mathcal{A}$ comes in: it is becoming now accepted that superselection and symmetry are hidden in the Hopf-algebraic aspects of the field-theory algebras. Some related ideas, albeit in a different form, have been expressed in other talks at this conference, notably, by Fröhlich, Kastler, Dubois-Violette, Majid, and Woronowicz (see also the survey [15]). I hope to come back to this relation in a separate publication, and maybe in a talk at the next Conference.

It should be noted that some of what I am saying goes back to vague ideas I expressed 20 years ago, which can be found in my 1971 Schladming lecture and talk I gave at the 1973 Conference in Bonn [16]. At that time I proposed (somewhat prematurely) to use some form of Hopf-von Neumann algebra in place of "broken" groups to describe the symmetries of elementary particles.

Here is a partial list of features that I would like to see incorporated in the putative quantum groupoids, in order to make them useful for quantum field theory:

i. The "hopfoids" should be embedded in some kind of C^*- or W^*-algebra structure, from which they inherit their topology and possibly local net.
ii. Ideally, the quantum groupoid should yield the space-time manifold structure as some kind of dual object, together with a holonomy structure which could open up a way of treating quantum fields in a gravitational background or even (since we are engaged in wishful thinking) treating gravitational holonomy, as an element of our quantum groupoid.
iii. The superselection structure of quantum field theory should be built-in into the quantum groupoid definition. In particular, non-gauge symmetries should arise this way.
iv. There should be a "deformation parameter" (playing the role of the Planck constant) which, when taken to zero should produce the classical Poisson groupoid we are aiming for (in particular, Yang-Mills or super Yang-Mills bundles, etc.).
v. Find additional restrictions which leading to a complete characterization of quantum groupoids.
vi. Find the relation of the description of symmetries within this framework with BRS-cohomology, as it appears in quantum (and classical) gauge theories.

I realize that this is a rather ambitious program and hope that the younger generation of mathematical physicists, which is so well represented here, will find solutions to them. A preliminary discussion of the last topic is in preparation.

Traditionally, symmetry groups acted on the C^*-algebra $\mathcal{A}$ (of observables, or fields) by a crossed product (or Fell-bundle, see, e.g., [15, 17]) of the automorphisms and the algebra. I would like to replace this by a "bundle of algebras" with the automorphisms being replaced by something akin to a holonomy groupoid. Then one could search for a Hopf algebroid structure associated with this algebra and finally try to extract some physical significance from it.

The least one should hope to achieve in a first attempt is to recover the Doplicher-Haag-Roberts superselection structure. The intertwiners which appear in the discussion of superselection structure have groupoid (or more precisely, algebroid) properties.

References

1. M. E. Mayer: "Groupoids and Lie bigebras in gauge and string theories," in K. Bleuler and M. Werner, editors, *Proceedings of the Conference on Differential-Geometric Methods in Physics, Como, August 1987*, Reidel, Dordrecht, 1988.
2. M. E. Mayer: "Groupoids versus principal bundles in gauge theories," In L.-L. Chau and W. Nahm, editors, *Proceedings of the Conference on Differential-Geometric Methods in Physics, Tahoe City, July 1989*, Plenum Press, New York, 1990.
3. A. Coste, P. Dazord, and A. Weinstein: "Groupoïdes symplectiques," Publ. Dép. Math. Univ. de Lyon, **2/A**, 1–62 (1987).
4. A. Weinstein: J. Math. Soc. Japan, **40**,705–727 (1988).
5. K. Mikami and A. Weinstein : Publ. RIMS, Kyoto Univ., **24**, 121–140 (1988).
6. A. Weinstein: "Affine Poisson structures." *Center for Pure and Applied Math, UC Berkeley*, Preprint PAM-489:1–27, February 1990; Cal-Tech Talk, February, 1990.
7. A. Lichnerowicz: "Quantum mechanics and deformations of geometrical dynamics," in A. O. Barut, editor, *Quantum Theory, Groups, Fields, and Particles 3-82*, pages 3–82, Reidel, Dordrecht, 1983.
8. A. Lichnerowicz: "Applications of the deformation of algebraic structures to geometry and mathematical physics," in M. Hazewinkel and M. Gerstenhaber, editors, *Deformation Theory of Algebras and Structures and Applications*, pages 855–896, Kluwer, Dordrecht, 1989.
9. K. Mackenzie: "Lie Groupoids and Lie Algebroids in Differential Geometry." Vol. 124 of *London Math. Soc. Lecture Note Series*, Cambridge University Press, Cambridge, 1987; See the review by Kumpera, Bull. AMS 1988.
10. M. F. Atiyah: *Trans. Amer. Math. Soc.*, **85**, 181–207 (1957). H. Nickerson: *Trans. Amer. Math. Soc.*, **99**, 509–539 (1961).
11. J. Pradines, *C.R. Acad. Sc., Paris, Ser. A* **264**, 245–248 (1967).
12. V. G. Drinfel'd: *Sov. Math. Dokl.*, **27**, 68–71 (1983). Also: International Congress of Mathematicians, Proceedings Berkeley, 1986.
13. H. Abelson and G. J. Sussman: "The Structure and Interpretation of Computer Programs," MIT Press/McGraw-Hill, Cambridge MA, 1985, p.75.
14. D. C. Ravenel: "Complex Cobordism and Stable Homotopy Groups of Spheres," Academic Press, 1986, Appendix 1.1.
15. D. Kastler, M. Mekhbout, and K. H. Rehren: "Introduction to the Algebraic Theory of Superselection Sectors," Luminy Preprint, 1990.
16. M. E. Mayer: "Automorphisms of C*-Algebras, Fell Bundles, W*-Bigebras, and the Description of Internal Symmetries in Algebraic Quantum Theory," *Acta Phys. Austriaca* Suppl. **VIII**, 177–226 (1971). "The Uses of Group–Theoretical Duality Theorems in Quantum Theory," in *Proceedings of the Conference on Differential-Geometric Methods in Physics*, K. Bleuler and A. Reetz, Eds., Bonn 1973, pp. 254–275.
17. M. E. Mayer: "Differentiable Cross Sections in Banach-*-Algebraic Bundles," in *Cargèse Lectures in Physics, Vol. 4*, D. Kastler, Ed. Gordon and Breach, New York 1970.

Quantization on Kähler Manifolds

John Rawnsley

Mathematics Institute, University of Warwick, Coventry CV4 7AL,
United Kingdom

In this talk I want to describe and contrast three methods of quantization

1. Geometric Quantization á la Kostant-Souriau [5,8];
2. Berezin's covariant symbols [2];
3. Deformation Quantization [1].

This is joint work [3,4] with Michel Cahen and Simone Gutt. See also the work of Moreno [6].

1. Geometric Quantization

Let (M, ω) be a symplectic manifold then we say (M, ω) is *quantizable* if there is a complex line bundle L over M with Hermitian structure and having a metric connection ∇ such that, if the curvature 2-form ρ is given by

$$\rho(X,Y) = [\nabla_X, \nabla_Y] - \nabla_{[X,Y]},$$

then the curvature and symplectic structure are related by

$$\omega = i\hbar\rho.$$

It is well-known that such a line bundle with connection exists if and only if the de Rham class $[\omega/h]$ is integral (equivalently, ω/h has integer periods). We now assume that (M, ω) is quantizable, and choose and fix a line bundle with connection (L, ∇). Then we set

$$Q(\phi) = i\hbar\nabla_{X_\phi} + \phi$$

for each smooth function $\phi \in C^\infty(M)$, where X_ϕ is the Hamitonian vector field on M associated to ϕ by

$$i(X_\phi)\omega = d\phi.$$

The differential operators $Q(\phi)$ acting on the vector space of smooth sections of L then have the commutation relations

$$[Q(\phi), Q(\psi)] = i\hbar\, Q(\{\phi, \psi\})$$

where $\{\phi, \psi\}$ denotes the *Poisson bracket* on $C^\infty(M)$ given by

$$\{\phi, \psi\} = X_\phi(\psi).$$

The process of associating the operator $Q(\phi)$ (quantum observable) to the function ϕ (classical observable) is called ***prequantization***. It can be made even more plausible as a candidate for a general quantization rule by observing that we can get a Hilbert space structure on the space of sections by defining

$$\|s\|^2 = \int_M |s|^2 \, vol$$

where *vol* denotes the Liouville volume on M. Each operator $Q(\phi)$ is then formally self-adjoint when ϕ is real-valued.

If we take the most basic example namely the phase space of a particle in one dimension $\mathbb{R}^2$ with its standard symplectic structure $dp \wedge dq$, then

$$\{\phi, \psi\} = \frac{\partial \phi}{\partial q}\frac{\partial \psi}{\partial p} - \frac{\partial \phi}{\partial p}\frac{\partial \psi}{\partial q}$$

and L is the trivial line bundle $M \times \mathbb{C}$. The space of sections can thus be identified with the complex functions on M, and a choice of connection would be

$$\nabla_X s = X(s) + \frac{1}{i\hbar} p X(q) s.$$

If we now quantize position and momentum we obtain

$$Q(q) = i\hbar \frac{\partial}{\partial p} + q, \qquad Q(p) = -i\hbar \frac{\partial}{\partial q}.$$

These operators act on the Hilbert space $L^2(\mathbb{R}^2)$. This clearly differs from the Schrödinger quantization scheme where the Hilbert space would be $L^2(\mathbb{R})$. Note, however, that if we apply the above operators to functions of q alone, then we do get the usual operators of Schrödinger quantization. So one of the basic problems of prequantization is that taking all the sections gives a space which is too big. We need to cut the space of sections down to a space with half as many variables.

The method of doing this cutting down in a coordinate invariant way is accomplished in Geometric Quantization by means of a ***polarization***. This consists of a subbundle F of the complexified tangent bundle of M of half the dimension of M which is isotropic for the symplectic form ω and closed under Lie brackets. Further technical conditions are generally needed to take care of various pathologies which can arise, but we shall ignore these problems in this exposition. In the cases we shall be interested in later there are no difficulties. We now define the space of polarized sections of L to be those sections s which are covariant constant in the directions of F.

For example, we can take F to be spanned by $\partial/\partial p$ in the example above. It is then an easy exercise to see that covariant constancy as a section is the same as being independent of p, and so a function of q alone. Thus Schrödinger quantization is recovered as a special case of prequantization combined with the use of a polarization.

Well, not quite; for not all the operators $Q(\phi)$ have the property that they preserve the subspace of polarized sections. In fact, if we denote by $\mathcal{Q}_F$ those functions ϕ whose Hamiltonian flow preserves F, then it is easy to see that $Q(\phi)$ will preserve the polarized sections. We call $\mathcal{Q}_F$ the space of *quantizable functions* (for the polarization F). If we now work out which functions are quantizable for the momentum polarization of $\mathbb{R}^2$ above then we find they are just the functions linear in the momentum. Since most Hamiltonians we want to quantize are quadratic in the momentum, this illustrates one of the basic difficulties of the geometric approach. There are several ways around this, but generally at the expense of introducing some method dependency. See, for example, [5] where a scheme is described which quantizes all polynomials in momentum, although it no longer preserves Poisson brackets beyond first order.

We might ask if there are other polarizations we can choose which would work with a given Hamiltonian. For example, can we obtain the Fock space quantization this way? The answer is yes, because we have allowed ourselves to have complex polarizations. A Kähler polarization of a symplectic manifold (M,ω) is a polarization F which is the space of (0,1) tangents of a complex structure J on M such that ω is the Kähler form of a Kähler metric g on M. g, J and ω are related by

$$g(X,Y) = \omega(X,JY).$$

If $L \to M$ is a quantization of (M,ω) then L has a unique holomorphic structure such that a (local) section s of L is holomorphic if and only if s is a polarized polarized section. So the space $\mathcal{H}$ of polarized sections of L is then the space of global holomorphic sections. As above we take the L^2 norm on $\mathcal{H}$. If we apply this to $\mathbb{R}^2 \cong \mathbb{C}$, then we do get the Bargmann-Segal-Fock space of holomorphic functions on $\mathbb{C}$ which are square-integrable with respect to a gaussian measure. The gaussian factor comes from the Hermitian structure in the line bundle. For this choice of polarization the harmonic oscillator Hamiltonian is in $\mathcal{Q}_F$.

In this Kähler case the Poisson bracket Lie algebra $\mathcal{Q}_F$ is finite dimensional, an extension by the constants of the Lie algebra of the group of holomorphic isometries of the Kähler manifold. In particular it is independent of which line bundle we take.

2. Berezin's Method

In this approach to quantization we are assume we are in the Kähler manifold situation (M,ω,J) as above, and that it is quantizable with a line bundle L and connection ∇. We give L the holomorphic structure just described. Then pointwise evaluation of sections is continuous in the Hilbert space topology, so can be represented by inner product with a vector in the Hilbert space. The only technical point is that evaluation of a section s at x gives a point in L_x which is an abstract one-dimensional vector space. Choosing a basis u for L_x then converts the value of

s into a complex number and so we can apply the classical Riesz Theorem. Hence for each non-zero vector u in L_x we have a vector e_u in $\mathcal{H}$ such that

$$s(x) = \langle s, e_u \rangle\, u, \qquad \forall s \in \mathcal{H}.$$

We call e_u a *coherent state.* Obviously we have the homogeneity condition

$$e_{cu} = \bar{c}^{-1}\, e_u, \qquad c \in \mathbb{C}^*.$$

See [3] for more details of this geometrical version of Berezin's theory.

It follows easily from the definition that the only vector orthogonal to all of the coherent states is the zero vector, so finite linear combinations of coherent states are dense in $\mathcal{H}$. This makes the coherent states extremely useful in calculations. In particular any linear operator A on $\mathcal{H}$ is determined by what it does Ae_u to the coherent states. Since the coherent states are dense it follows that A will be determined by its matrix entries $\langle Ae_u, e_v \rangle$ between pairs of coherent states. Now we observe that this last function is holomorphic in v and antiholomorphic in u, so is the analytic continuation off the diagonal of the real analytic function $\langle Ae_u, e_u \rangle$ on L, and so A is determined by $\langle Ae_u, e_u \rangle$.

The homogeneity property means that if $A \in \mathrm{End}(\mathcal{H})$ then

$$\frac{\langle Ae_u, e_u \rangle}{\langle e_u, e_u \rangle}$$

depends only on the base point x of the fibre L_x in which u lives. It therefore defines a function $\widehat{A}$ on M which we call the covariant symbol of the operator A. $\widehat{A}$ is clearly a real analytic function on M which determines the original operator A uniquely.

We denote by $\mathcal{E}(L)$ the set of covariant symbols of operators on $\mathcal{H}$. We have just seen that the map $A \mapsto \widehat{A}$ is a bijection of $\mathrm{End}(\mathcal{H})$ with $\mathcal{E}(L)$. It has the two additional properties

$$\widehat{I} = 1, \qquad \overline{\widehat{A}(x)} = \widehat{A^*}(x).$$

The inverse map from functions to operators is ***Berezin Quantization.***

Note that the space of quantizable functions for Berezin Quantization depends on the quantization line bundle L, unlike geometric quantization, and so grows polynomially with the Chern class of L (The Riemann-Roch formula gives the precise relation).

To compare Berezin Quantization with Geometric Quantization we introduce the function

$$\begin{aligned}\epsilon(x) &= |u|^2 \|e_u\|^2, \qquad u \in L_x \setminus \{0\}\\ &= \sum_{i=1}^{\infty} |s_i|^2\end{aligned}$$

where s_i is an orthonormal basis for $\mathcal{H}$.

The second definition shows that ϵ is invariant under automorphisms of the quantization since these act unitarily in $\mathcal{H}$. If the group of automorphisms of the quantization acts transitively on M then we call the quantization ***homogeneous.*** In this case ϵ will have to be constant. Then the first definition allows us to deduce the coherent states simply by analytic continuation of the Hermitian structure in L. An easy calculation shows that the trace of an operator A can be computed in terms of its covariant symbol and ϵ as

$$\mathrm{Tr}(A) = \int_M \widehat{A}\,\epsilon\, vol.$$

Applying this to the identity operator we deduce that when $\mathcal{H}$ is finite-dimensional $\dim \mathcal{H} = \int_M \epsilon\, vol$.

ϵ keeps turning up throughout the theory of coherent states. For example it is shown in [7] that the map $\phi: M \to \mathbb{P}(\mathcal{H})$ which sends a point x of M to the ray spanned by e_u is symplectic (for the Kähler form of the Fubini-Study metric on $\mathbb{P}(\mathcal{H})$) if and only if $\partial\bar{\partial}\log\epsilon = 0$. Here is another result involving ϵ:

Theorem 1. [3] *For ϕ in $\mathcal{Q}_F$ the covariant symbol of $Q(\phi)$ is related to ϕ by*

$$\widehat{Q(\phi)} = i\hbar X_\phi^{(1,0)}(\log\epsilon) + \phi$$

where $X_\phi^{(1,0)}$ denotes the holomorphic component of X_ϕ.

An immediate consequence of this theorem is that Berezin Quantization and Geometric Quantization are, in general, different. However when ϵ is constant then $\mathcal{Q}_F \subset \mathcal{E}(L)$ and the two methods agree on $\mathcal{Q}_F$.

3. Deformation Quantization

The point of view of this theory is that Quantum Mechanics is fundamental, and we obtain Classical Mechanics in the limit as $\hbar$ tends to zero. In order to make a comparison, we have to assume that the two theories have a common set of observables, which for non-zero $\hbar$ are an algebra of operators, and for $\hbar = 0$ an algebra of functions. In other words quantum theory is expressed in terms of a deformation of the algebra of functions on a symplectic manifold. This means we have, for any two functions ϕ, ψ, a family of products $\phi *_\hbar \psi$ which reduces to the product $\phi\psi$ for $\hbar = 0$. In fact this is generally too much to ask, and the general theory deals only with a formal version of this. We assume we have a formal power series

$$\phi *_\hbar \psi = \sum_k (\hbar)^k\, C_k(\phi, \psi)$$

with $C_0(\phi,\psi) = \phi\psi$. This is tied to the classical mechanics by asking that the commutation relations hold at least to first order. That is we require

$$C_1(\phi,\psi) - C_1(\psi,\phi) = \{\phi,\psi\}.$$

Such a deformation is called a (formal) $*$-product.

The best known example of a formal $*$-product is given by the Moyal brackets on $C^\infty(\mathbb{R}^2)$. See [1].

I want to conclude my lecture by showing how the Berezin Quantization can be made to yield a formal $*$-product that is rather special for the case where M is a compact Hermitian symmetric space. This is based on the observation that the algebra structure of End(L) can be transferred to $\mathcal{E}(L)$ by using the bijection given by the covariant symbol map. We define

$$\widehat{A} * \widehat{B} = \widehat{AB}.$$

In fact there is an explicit formula for this multiplication given by Berezin [2]. We introduce the 2-point function

$$\psi(x,y) = \frac{|\langle e_v, e_u\rangle|^2}{\|e_u\|^2\|e_v\|^2}, \qquad u \in L_x, v \in L_y,$$

and for each symbol $f(x)$ we denote by $f(x,\overline{y})$ the analytic continuation of $f(x)$ off the diagonal, so

$$f(x,\overline{y}) = \frac{\langle Ae_v, e_u\rangle}{\langle e_v, e_u\rangle}.$$

Then

$$(f * g)(x) = \int_M f(x,\overline{y})g(y,\overline{x})\,\psi(x,y)\epsilon(y)vol_y.$$

The idea is to examine how this product depends on the quantization L. In particular we can introduce a paramter into the theory by considering the powers of L. L^k will be a quantization for $k\omega$. So varying k is like varying $\hbar$. Roughly $\hbar \sim 1/k$.

Denote by $\epsilon^{(k)}$ the ϵ function for L^k. Then we have

Theorem 2. [3] *If all $\epsilon^{(k)}$ are constant then $\mathcal{E}(L^k) \subset \mathcal{E}(L^{k+1})$ for all k and $\mathcal{E}(L) = \cup_{k=1}^\infty \mathcal{E}(L^k)$ is a dense subalgebra of $C^\infty(M)$.*

This theorem means that if we take two symbols $f, g \in \mathcal{E}(L^l)$, a fixed symbol space, then for every $k \geq l$ $f *_k g$ will be well-defined. It is easy to see that under the constancy assumption on the $\epsilon^{(k)}$ that the 2-point function for L^k is $\psi(x,y)^k$ so that

$$(f *_k g)(x) = \int_M f(x,\overline{y})g(y,\overline{x})\,\psi(x,y)^k\epsilon^{(k)}(y)k^n vol_y$$

where $\dim M = 2n$.

Theorem 3. [4] *$f *_k g$ admits an asymptotic expansion in $1/k$*

$$f *_k g \sim \sum_r k^{-r}C_r(f,g)$$

where the C_r are bi-differential operators and

$$C_0(f,g) = fg, \qquad C_1(f,g) - C_1(g,f) = \{f,g\}.$$

If we set

$$f \tilde{*}_k g = \sum_{r \geq 0} k^{-r} C_r(f,g)$$

as a formal power series, then the fact that the original products were all associative suggests that this formal product will be also, and hence will define a $*$-product. This is true, but we have only managed to prove it in the compact homogeneous cases where M is compact coadjoint orbit with its natural invariant Kähler structure:

Theorem 4. [4] *If M is an integral coadjoint orbit of a compact Lie group then $\tilde{*}_k$ is a formal $*$-product on $C^\infty(M)$. If M is Hermitian symmetric then the series converges for large k to the function $f *_k g$ for $f, g \in \mathcal{E}(L)$ which is a rational function of k.*

We conjecture that the series converges for all the compact coadjoint orbits, but our proof uses some of the fine structure of Hermitian symmetric spaces so will not extend. On the other hand, direct calculations with some non-symmetric cases have only yielded convergent series so far.

Note that the theorem yields infinitely many distinct new examples of convergent $*$-products, and that there is an explicit integral formula for the product since on $\mathcal{E}(L)$, $f \tilde{*}_k g = f *_k g$.

References

1 F. Bayen et al.: Ann. Phys. **111** 61 (1978)
2 F. Berezin: Math. USSR Izvestija **8** 1109 (1974)
5 B. Kostant: In Géométrie symplectique et physique mathématique. CNRS **237** 187 (1975)
6 C.Moreno: Lett. Math. Phys. **11** 361 (1986)
7 J. Rawnsley: Quart. J. Math. Oxford (2) **28** 403 (1977)
8 J.-M. Souriau: Structure des systèmes dynamiques. Dunod, Paris 1970.
3 M.Cahen, S. Gutt and J. Rawnsley: J. Geom. Phys. (To appear)
4 M.Cahen, S. Gutt and J. Rawnsley: Trans. Amer. Math. Soc. (To appear)

A NEW CLASS OF INFINITE-DIMENSIONAL LIE ALGEBRAS (CONTINUUM LIE ALGEBRAS) AND ASSOCIATED NONLINEAR SYSTEMS

M.V.Saveliev
Institute for High Energy Physics, Protvino, 142284 Moscow region, USSR

A.M.Vershik
Leningrad State University, Leningrad, USSR.

ABSTRACT

We give a review of our recent results concerning a new class of infinite-dimensional Lie algebras - the generalizations of Z-graded contragredient Lie algebras with a, generally speaking, infinite-dimensional Cartan subalgebra and a contiguous set of roots (a manifold or a more general space, for example with a measure). We call such algebras "continuum Lie algebras". Special examples of these algebras are the Kac-Moody algebras, the Poisson bracket algebras, algebras of vector fields on a manifold, current algebras, various versions of $gl(\infty)$, algebras of diffeomorphisms of a manifold, more general cross product Lie algebras (including, in particular, various multiindex generalizations of the Virasoro algebra), and other Lie algebras with differential or integro-differential Cartan operator (a continuous extension of the generalized Cartan matrix).

Then the nonlinear dynamical systems associated with the continuum Lie algebras via a zero curvature type representation are considered. We pay particular attention to the special example – the continuous analogues of the generalized (finite nonperiodic) Toda lattices.

INTRODUCTION

In this review we present an axiomatic formulation of continuum generalizations of usual ("discrete case") Z - graded contragredient Lie algebras with a, generally speaking, infinite-dimensional Cartan subalgebra and a contiguous set of roots. After the discovery of the continuum Lie algebras in papers [1,2] containing a preliminary theorem on some continuous limits of semisimple Lie algebras, the list of special examples of the continuum Lie algebras becomes more and more long [3,4]. The same is true for nonlinear dynamical systems generated by these algebras in the framework of a continuous extension of the group-algebraic approach (see e.g.[5]) for one- and two-dimensional integrable systems of PDEs associated with the discrete case. Such an extension allows to investigate not only the corresponding differential systems in more than two dimensions, e.g. the simplest continuous analogue of the Toda lattice – "heavenly equation" (14), but also to consider integro-differential equations, for example like (13). Here, of course, many notions related with the integrability problem acquire a different meaning. For example, the crucial point of the integrability of the systems in the discrete case is that they admit a Lie-Bäcklund algebra of finite growth. For the continuum Lie algebras we were forced to introduce a different definition – the algebras of temperate growth, since the corresponding subspaces of the Z-graded continuum Lie algebras are functional spaces.

Further generalization, as usual, stimulates deeper understanding of the original structure as, for example, the well-known transition from finite-dimensional simple Lie algebras to Kac-Moody algebras. In their turn, the algebras we consider, include those with the generalized Cartan matrices in a broader and rather unusual context. Moreover, there are many types of algebras which, being at the first glance completely different, represent the isomorphic Z-graded continuum Lie algebras, for example, $A_\infty(su(\infty)) \simeq S_0 Diff\, T^2$ (infinitesimal area - preserving diffeomorphisms of the torus T^2) $\simeq$ Poisson bracket algebra on $T^2 \simeq W_\infty$. This fact has far going consequences for physical phenomena in various branches of theoretical physics which are described by the same equation (14), however possess the symmetries which correspond to the different manifestations of the same algebra.

To this end it is remarkable that equation (14) arises independently in general relativity (self-dual Einstein spaces with one rotational Killing vector [6]); in the theory of Hamiltonian systems [7]; for wave phenomena in a shallow water; in long radio-relay lines; for isentropic motion of fluid in a tube; in the diffusion processes in semiconductor, in polymers; in isothermal filtration of a fluid; etc. (see e.g.[8]).

Finally, we hope that our results concerning a new class of infinite-dimensional continuum Lie algebras and associated nonlinear systems will be useful in pure mathematics and mathematical physics (theory of Lie algebras, of Lie algebra-valued distributions, dynamical systems,nonlinear wave and evolution equations, etc.) and in physical applications, for example, in particle physics (extended objects like strings and membranes in gauge theories and in statistical physics; extended conformal symmetries and higher spin fields; classical and quantum gravity, etc.).

AXIOMATIC FORMULATION OF CONTINUUM (CONTRAGREDIENT) Z-GRADED LIE ALGEBRAS.

Let E be an associative commutative algebra (possibly, without unity) over the field R or C; K is a linear operator, $K : E \to E$. Define "a local Lie algebra" $\hat{\mathcal{J}} \equiv \mathcal{J}_{-1} \oplus \mathcal{J}_0 \oplus \mathcal{J}_{+1}$ as follows. Each of $\mathcal{J}_i$, $i = 0, \pm 1$, as a vector space is isomorphic to E, in order words the elements of $\mathcal{J}_i$ are parametrized by the vectors $\varphi \in E$ so that $\mathcal{J}_i = \{X_i(\varphi), \varphi \in E, i = 0, \pm 1\}$. Besides, these elements satisfy the defining relations

$$\begin{gathered}[X_0(\varphi), X_0(\psi)] = 0, \qquad [X_0(\varphi), X_{\pm 1}(\psi)] = \pm X_{\pm 1}(\psi K \varphi), \\ [X_{+1}(\varphi), X_{-1}(\psi)] = X_0(\varphi \cdot \psi),\end{gathered} \tag{1}$$

for all $\varphi, \psi \in E$. Note that the Jacobi identity for $\hat{\mathcal{J}}$ is satisfied automatically.

Definition 1. Let $\mathcal{J}'(E; K)$ be a Lie algebra freely generated by a local part $\hat{\mathcal{J}}$ and J be the largest homogeneous ideal having a trivial intersection with $\mathcal{J}_0$. Then $\mathcal{J}(E; K) = \mathcal{J}'(E; K)/J$ is called *a continuum contragredient Lie algebra* with the local part $\hat{\mathcal{J}}$ and defining relations (1); K is called *a Cartan operator*.

Statement. Lie algebra $\mathcal{J}(E; K)$ is Z-graded, $\mathcal{J} = \bigoplus_{n \in Z} \mathcal{J}_n, \quad [\mathcal{J}_m, \mathcal{J}_n] \subset \mathcal{J}_{m+n}$.

Here $\mathcal{J}_n = [\mathcal{J}_{n-1}, \mathcal{J}_1]$ for $n > 0$, and $\mathcal{J}_n = [\mathcal{J}_{n+1}, J_{-1}]$ for $n < 0$.

Definition 2. The Lie algebra $\mathcal{J}(E;K)$ is called *the algebra of temperate growth* if for each $\mathcal{J}_n$ there exists a finite-demensional subspace $L_n \subset \mathcal{J}_{+1}$, $dim\, L_n < \infty$, such that $\mathcal{J}_n = [\mathcal{J}_{n-1}, L_n]$.

More restrictive is the notion of *polynomial growth* in terms of the Gel'fand–Kirillov dimension, however, implied in a functional sense in the spirit of Kolmogorov's ε-entropy. Finally, we speak about a *constant growth* if $\mathcal{J}_n \simeq \mathcal{J}_1 \simeq E$.

Reduction to the discrete case (The Kac-Moody algebras)

This case, in our approach, corresponds to the finite-dimensional algebra $E = C^r$ with coordinate multiplication in some basis. Here the Cartan operator K coincides with the generalized $r \times r$ Cartan matrix k, the local Lie algebra $\widehat{\mathcal{J}}$ is a linear hull of $3r$ elements: generators h_i of the Cartan subalgebra and Chevalley generators $X_{\pm i}$, $1 \leq i \leq n$, with the defining relations [9]

$$[h_i, h_j] = 0, \quad [h_i, X_{\pm j}] = \pm k_{ji} X_{\pm j}, \quad [X_{+i}, X_{-j}] = \delta_{ij} h_i,$$

to which (1) reduce. The consideration of the quotient algebra $\mathcal{J}(E,K) = \mathcal{J}'(E,K)/J$ in Def.1 is equivalent to imposing the Serre conditions $(ad\, X_{\pm i})^{1-k_{ji}} X_{\pm j} = 0, i \neq j$. Remind also that the finite-dimensional and affine Lie algebras are distinguished among all the contragredient Lie algebras by the fact that they have a finite growth, i.e. $dim\, \mathcal{J}_n$ grow no faster than some polynomial in n. (Just this condition provides integrability of all known integrable one- and two-dimensional nonlinear PDEs.) Naturally, every graded Lie algebra in the discrete case (e.g. affine algebras) has a temperate growth.

Generalizations

The nearest generalization $\mathcal{J}(E;\, K, S)$ of the continuum Lie algebras $\mathcal{J}(E;\, K)$ is obtained if we consider one more linear operator $S : E \to E$, which enters in the last relation in (1), i.e. $[X_{+1}(\varphi), X_{-1}(\psi)] = X_0(S(\varphi\psi))$. However, if the operator S is invertible then the substitution $X_0(S\varphi) \to X_0(\varphi)$ reduces the defining relations for $\mathcal{J}(E;\, K, S)$ to original (standard) form (1) with $K \to KS$, i.e. for the algebra $\mathcal{J}(E;\, KS)$.

If operator S has a kernel and $Ker\, S \subset Ker\, K$, then we have the central extension of $\mathcal{J}(E;\, K)$.

Other generalizations of the algebras $\mathcal{J}(E;\, K)$ or $\mathcal{J}(E;\, K, S)$ are related with three linear operators $K_\pm, S$, i.e. $[X_0(\varphi), X_{\pm 1}(\psi)] = X_{\pm 1}(\psi K_\pm \varphi)$; with bilinear mappings $K, S : E \times E \to E$ (here the Jacobi identity becomes already nontrivial and leads to the relations for K and S); with a consideration of an associative *noncommutative* algebra E, etc. Finally, we can consider, for example, a continuum noncontragredient Lie algebra which is generated by a *modified local algebra* $\widehat{\mathcal{J}}_{m_0} = \bigoplus_{|i| \leq m_0} \mathcal{J}_i$ with $m_0 > 1$. (Remind that in the discrete case the simplest example of noncontragredient (but still of finite growth) algebra is the Virasoro algebra.)

PRINCIPAL EXAMPLES

Here we list several principal examples of the continuum Lie algebras of temperate or even constant growth.

1. Poisson bracket algebra.

Let E be the algebra of trigonometric polynomials on a circle, $K = S = -i\,\partial/\partial z$. Then the continuum Lie algebra $\mathcal{J}(E;\ -i\,\partial/\partial z,\ -i\,\partial/\partial z)$ is isomorphic to the algebra of functions on T^2 with the standard Poisson bracket.

Note that here $\mathcal{J}_n \simeq E$ and $[X_n(\varphi), X_m(\psi)] = i\,X_{n+m}(m\varphi'\psi - n\varphi\psi')$.

The algebra $\mathcal{J}(E;\ -i\,\partial/\partial z,\ -i\,\partial/\partial z)$ and its standard form $\mathcal{J}(E;\ -\partial^2/\partial z^2)$ possess the following sets of roots: $n\,\partial/\partial z\,\delta(z-\tilde{z})$ and $n\,\partial^2/\partial z^2\,\delta(z-\tilde{z})$, respectively.

2. The simplest continuous limit of A_r : $\mathcal{J}(E;\ \partial^2/\partial z^2)$.

3. Vector fields on a manifold $\mathcal{M}$.

Let E be the algebra $C^\infty(\mathcal{M})$ and $K = S = V$ is a vector field on $\mathcal{M}$. Then the Z-graded algebra $\mathcal{J}(E;\ V, V)$ is defined by the monomial brackets

$$[X_n(\varphi), X_m(\psi)] = X_{n+m}(m\varphi V\psi - n\psi V\varphi).$$

4. Current algebra on a manyfold $\mathcal{M}$.

Let E be the space of vector functions on $\mathcal{M}$, $K = k \otimes I$ with k being the Cartan matrix of a simple Lie algebra $\mathcal{J}^f(k)$. Then we obtain the Lie algebra of currents taking values in $\mathcal{J}^f$, i.e. $\mathcal{J}(E;\ k\otimes I) \simeq C^\infty(\mathcal{M};\ \mathcal{J}^f)$.

5. Algebras of diffeomorphisms.

Let $\mathcal{M}$ be a manifold of C-class, T be its C - diffeomorphism; $E = C^\infty(\mathcal{M})$, $T\varphi(z) = \varphi(Tz)$. Let $K = I - T$, $S = I - T^{-1}$. Then the elements $X_n(\varphi)$ of $\mathcal{J}(E;\ I - T, I - T^{-1}) \simeq \mathcal{J}(E;\ 2I - T - T^{-1})$ satisfy the commutation relations $[X_n(\varphi), X_m(\psi)] = X_{n+m}(\varphi T^n\psi - \psi T^m\varphi)$.

Theorem. (On some continuum limits of semisimple Lie algebras)[2].

Let $\mathcal{M}$ be a compact manifold and $\mathcal{J}$ be a simple contragredient graded Lie algebra of a constant growth having $\mathcal{J}_0 \simeq E = C^\infty(\mathcal{M})$ as the Cartan subalgebra. Then, if the Cartan operator has one of the following two forms:

i) **$K\varphi(z) = \varphi(Tz) - 2\varphi(z) + \varphi(T^{-1}z)$, where T is a diffeomorphism of $\mathcal{M}$,**

ii) **$K\varphi(z) = V^2\varphi(z)$, where V is a vector field on $\mathcal{M}$,**

the algebra $\mathcal{J}(E;K)$ is a continuum limit of a discrete case.

6. Cross product Lie algebras

Definition 3. Let E be an associative commutative algebra and G be a group of its automorphisms with the generators T. Then the algebra $\mathcal{J}(E; I-T, I-T^{-1}) \simeq \mathcal{J}(E; 2I-T-T^{-1})$ which consists of finite sums of the form $\sum_n \varphi_n \otimes W^n$ with the bracket

$$[\varphi \otimes W^n, \psi \otimes W^m] = (\varphi T^n \psi - \psi T^m \varphi) \otimes W^{m+n}, \qquad \varphi, \psi \in E, \tag{2}$$

is called *the cross product Lie algebra.*

Let us give several concrete examples of this quite general class of the continuum Lie algebras.

i) Kac-Moody algebras as cross products. For this case $E = C^r$ with a coordinate multiplication, T is any cyclic permutation of the coordinates. Then $\mathcal{J}(E; 2I-T-T^{-1})$ is exactly the centreless Kac-Moody algebra $A_r^{(1)}$.

ii) Lie algebras associated with circle rotation. Let E be a space of trigonometric polynomials, $T \equiv T_{2\lambda}$ is the operator of rotations, $Te^{2\pi inz} = e^{2\pi in(z+2\lambda)}$, where λ is irrational. Then $\mathcal{J}(E; I-T, I-T^{-1})$ is a continuum contragredient Z-graded Lie algebra with $K_\lambda = -i\xi e^{\lambda\partial/\partial z} sh\lambda\partial/\partial z$, $S_\lambda = K_{-\lambda}$, which is isomorphic (up to a factorization over the constants) to $\mathcal{J}(E; \xi^2 sh^2\lambda\partial/\partial z)$. Then, if we choose the basis of $\mathcal{J}_n$, $\mathcal{J} = \bigoplus_{n\in Z} \mathcal{J}_n$, as

$$Y_{\bar{n}} = (\xi/2i)\, exp[in_2(z+n_1\lambda)] \otimes T^{n_1},$$

the commutation relations for the algebra in question take the form [10]

$$[Y_{\bar{m}}, Y_{\bar{n}}] = \xi\, sin[\lambda(\bar{m} \times \bar{n})] Y_{\bar{m}+\bar{n}}. \tag{3}$$

Here $\bar{m} = (m_1, m_2)$ and $\bar{n} = (n_1, n_2)$ are two-dimensional integer vectors, $\bar{m} \times \bar{n} \equiv m_1 n_2 - m_2 n_1$, ξ is some constant; the central terms are omitted. Note that the elements X_ε, $\varepsilon = 0, \pm 1$, entering defining relations (1) are expressed via $Y_{\bar{n}}$ as $X_\varepsilon(e^{inz}) = (2/i\xi) Y_{\varepsilon,n}$, while the roots are identified with $\xi\, n\, s\, h(\lambda\,\partial/\partial z)\delta(z-\tilde{z})$.

Bracket (2) with a cocycle for the algebra in question is

$$[\varphi \otimes T_\lambda^n, \psi \otimes T_\lambda^m] = (\varphi T^n \psi - \psi T^m \varphi) \otimes T_\lambda^{m+n} + \delta_{m+n,0}\, n \int_0^{2\pi} dz\, \varphi(z)\, \psi(z+n\lambda). \tag{4}$$

A direct check allows one to get convinced that this is a nontrivial two-cocycle which defines the central extension of the algebra. However, it is not clear, whether a similar two-cocycle exists for every transformation T, or not.

iii) Limiting case associated with $S_0\, Diff\, T^2$. Taking in (3) $\xi^{-1} = \lambda$, in the limit $\lambda \to 0$ one comes [10] to the algebra $S_0\, Diff T^2$,

$$[Y^0_{\bar{m}}, Y^0_{\bar{n}}] = (\bar{m} \times \bar{n}) Y^0_{\bar{m}+\bar{n}}, \tag{5}$$

i.e. the centreless algebra of the infinitesimal area-preserving diffeomorphisms of the torus T^2, which is isomorphic (as Z-graded algebra) to examples 1 and 2 given above.

In terms of the Poisson bracket algebra on T^2 (see example 1) the corresponding cocycle has the form

$$[\varphi,\psi] = (\partial\varphi/\partial s\,\partial\psi/\partial z - \partial\varphi/\partial z\,\partial\psi/\partial s) + \int ds \int dz\,(\xi_1\,\partial\varphi/\partial s + \xi_2\,\partial\varphi/\partial z)\psi,$$

where $\xi_{1,2}$ are some parameters. Note that there are also other cocycles.

iv). Vector fields on a manifold (see example 3); etc.

7. Linear superposition.

Another interesting example of Lie algebras of diffeomorphism groups of two-dimensional manifolds is a symbiosis of the algebra $S_0\,Diff\,T^2$, see (5), and the algebra of [11] with the relations

$$[Y^1_{\bar m}, Y^1_{\bar n}] = \bar c(\bar m - \bar n)Y^1_{\bar m+\bar n}. \tag{6}$$

Here, for simplicity, we do not consider their central extensions; $\bar c = (c_1, c_2)$ is some constant 2-vector. This new algebra is described by the relations [4]

$$[Y^s_{\bar m}, Y^s_{\bar n}] = [c_0\,\bar m \times \bar n + \bar c(\bar m - \bar n)]\,Y^s_{\bar m+\bar n} \tag{7_1}$$

or, in a continuum form,

$$\begin{aligned}[X_n(\varphi), X_m(\psi)] = -X_{m+n}\Big(& ic_0(n\varphi\psi' - m\varphi'\psi) + \\ &+ c_1(m-n)\varphi\cdot\psi + ic_2(\varphi'\psi - \varphi\psi')\Big)\end{aligned} \tag{7_2}$$

with $Y^s_{\bar m} \equiv X_{m_1}(e^{i m_2 z})$.

8. Lie algebra with the Cartan-Hilbert operator.

Let E be a space of functions φ on C^1 which satisfy the Hölder condition and are expanded into a sum of holomorphic and antiholomorphic parts, $\varphi = \varphi_+ + \varphi_-$, in some domain D. Define the multiplication in E as $\varphi_\circ\psi = \varphi_+\psi_+ + \varphi_-\psi_- \equiv 1/2(H\varphi\cdot\psi + H\psi\cdot\varphi)$ where H is the Hilbert transform, i.e. $H\varphi(z) = P.V. \int dz'\,\varphi(z')/(z'-z)$. Then it is possible to prove that the algebra $\mathcal{J}(E; I \pm iH)$ with the pointwise product $\varphi\cdot\psi$ in E is isomorphic to the algebra $sl(2; E)$ with the product $\varphi_\circ\psi$ in E.

NONLINEAR DYNAMICAL SYSTEMS ASSOCIATED WITH THE CONTINUUM Z-GRADED LIE ALGEBRAS

In accordance with an algebraic approach [5] so as to construct integrable (in the sense of the Cauchy or Goursát problem) nonlinear partial differential equation system in two dimensions, one should consider the connections $A_\pm(u_+, u_-)$ of the zero curvature type representation

$$[\partial/\partial u_+ + A_+,\ \partial/\partial u_- + A_-] \equiv \partial A_-/\partial u_+ - \partial A_+/\partial u_- + [A_+, A_-] = 0, \tag{8}$$

i.e. the components of the Cartan-Maurer 1-form, taking values in the subspaces $\mathcal{J}_{\pm m}$, $0 \leq m \leq m_\pm$, of a Z-graded Lie algebra $\mathcal{J} = \bigoplus_{m \in Z} \mathcal{J}_m$ of finite growth. Here $m_\pm \geq 1$ are some integers,i.e.

$$A_\pm = \sum_{0 \leq m \leq m_\pm} \sum_{\alpha=1}^{dim \mathcal{J}_{\pm m}} X^\alpha_{\pm m} \overset{\pm}{\varphi}{}^\alpha_{\pm m}(u_+, u_-) \tag{9}$$

with X^α_m being the basis elements of $\mathcal{J}_m$, $\overset{\pm}{\varphi}{}^\alpha_{\pm m}$ are the functions, for which we obtain the nonlinear equations in question. Clearly, for one-dimensional case we deal with the Lax-type representation,

$$\partial \mathcal{L}/\partial \tau \equiv \dot{\mathcal{L}} = [\mathcal{L}, \mathcal{M}] \tag{10}$$

where the pair $\mathcal{L} = A_+ + A_-$ and $\mathcal{M} = A_+ - A_-$ depends only on one variable $\tau = (u_+ - u_-)/2$.

The nonlinear systems associated with the continuum Z-graded Lie algebras $\mathcal{J}(E;K)$ are obtained if we replace the summands in (9) by the elements $X^\alpha_{\pm m}(\overset{\pm}{\varphi}{}^\alpha_{\pm m})$ of the subspaces $\mathcal{J}_m$ of $\mathcal{J}(E;\, K)$, which is quite similar to this approach. Remind that here, by definition, $dim \mathcal{J}_{0,\pm 1} = 1$ in the functional sense.

Let us illustrate this general construction by the example when $A_\pm$ take values in the local part $\widehat{\mathcal{J}} = \mathcal{J}_{-1} \oplus \mathcal{J}_0 \oplus \mathcal{J}_{+1}$ of $\mathcal{J}(E;\, K)$, i.e. $m_\pm = 1$, so

$$A_\pm = X_0(f^0_\pm) + X_{\pm 1}(f^1_\pm). \tag{11}$$

Then, substituting (11) into representation (8) and using relations (1) we come to the following system of equations

$\partial f^0_- / \partial u_+ - \partial f^0_+ / \partial u_- + f^1_+ f^1_- = 0, \qquad K f^0_\pm = \pm \partial ln\, f^1_\mp / \partial u_\pm,$

which results in the second order nonlinear equation for the gauge invariant function $\rho \equiv ln\, f^1_+ f^1_-$,

$$\Delta \rho \equiv \partial^2 \rho / \partial u_+ \partial u_- = K\, e^\rho. \tag{12}$$

This is a continuous analogue of the usual ("discrete") Toda lattices [5], i.e. $\Delta \rho_i = \sum_j k_{ij} \exp \rho_j$, associated with a simple Lie algebra $\mathcal{J}$ supplied with the principal gradation.

If K is an invertible and symmetrizable (i.e. there exists such a function $v(z)$ that $v(z)K(z,z') = v(z')K(z',z)$) operator, $K\varphi(z) = \int dz' K(z,z')\varphi(z')$, then equation (12) is a variational Lagrange-Euler equation with the Lagrangian density

$$L = \int dz\, v(z) \Big[\frac{1}{2} \frac{\partial x(u_\pm, z)}{\partial u_+} \int dz'\, K(z,z') \frac{\partial x(u_\pm, z')}{\partial u_-} + e^{\rho(u_\pm, z)} \Big]$$

where $x = k^{-1} \rho$.

If the function ρ depends only on one variable, say τ, then the corresponding equation $\partial^2 \rho / \partial \tau^2 = K e^\rho$ for invertible operator K is obtained with the Lax-pair

$$\mathcal{L} = X_0(K^{-1} \partial \rho / \partial \tau) + X_{+1}(1) + X_{-1}(e^\rho),$$

$$\mathcal{M} = X_0(K^{-1} \partial \rho / \partial \tau) + X_{+1}(1) - X_{-1}(e^\rho).$$

In this case the involutive integrals of motion I_k, $\partial I_k/\partial\tau = 0$, have the form $I_k = \mathrm{Sp}\,\mathcal{L}^k$.

Let us briefly discuss some particular cases of (12).

1) $\underline{K=0}$. Then (12) reduces to the Laplace equation $\Delta\rho_0 = 0$.

2) $\underline{K=2I}$. Then (12) reduces to the Liouville equation $\Delta\rho_L = 2e^{\rho_L}$.

3) $\underline{K=I\pm iH}$. Then (12) takes the form

$$\Delta\rho_H(u_\pm, z) = \int dz'\, e^{\rho_H(u_\pm,z')}(z'-z\pm io)^{-1} \tag{13}$$

which is also integrable [1] due to the corresponding algebra is isomorphic (see Example 8 given above) to $sl(2;E)$ with the product $\varphi\circ\psi$ in E.

4) $\underline{K=\partial^2/\partial z^2}$. Here we come to the so-called heavenly equation discussed in the Introduction

$$\Delta\rho_h = \partial^2 e^{\rho_h}/\partial z^2 \tag{14}$$

which can be rewritten in two other equivalent forms
$\Delta x = e^{\partial^2 x/\partial z^2}$, $\Delta y = e^{\partial y/\partial z}\,\partial^2 y/\partial z^2$, $\rho_h = \frac{\partial^2 x}{\partial z^2} = \frac{\partial y}{\partial z}$.
This system possesses an improved energy-momentum tensor
$W_2^{\pm\pm} = c\int dz\,[\partial^2 x/\partial u_\pm^2 - \frac{1}{2}\frac{\partial x}{\partial u_\pm}\frac{\partial^2}{\partial z^2}\frac{\partial x}{\partial u_\pm}]$
with a vanishing trace, $W_2^{+-} = 0$, on shell.

The Riemannian metric corresponding to such a system is given by the formula (see [6])

$$\begin{aligned} ds^2 =& \partial\rho_h/\partial x_3\left[(dx_1^2 + dx_2^2)e^{\rho_h} + dx_3^2\right] + \\ &+ (\partial\rho_h/\partial x_3)^{-1}\left[\pm[(-\partial\rho_h/\partial x_2)dx_1 + (\partial\rho_h/\partial x_1)dx_2] + dx_4\right]^2 \end{aligned}$$

where $x_1 \equiv u_+ + u_-$, $x_2 \equiv -i(u_+ - u_-) \equiv -2i\tau$, $x_3 \equiv iz$.

Note that equation (12) for an invertible operator K admits a formal solution of the Goursát problem in infinite series which depends on two arbitrary functions of two variables, $\rho_0^+(u_+, z)$ and $\rho_0^-(u_-, z)$, see [1]. For the case with $K = \partial^2/\partial z^2$ the series is absolutely convergent due to that just this case is the continuous limit of the algebra A_r. In other words, the solutions for the algebras A_r are the partial sums of their continuous limit. The corresponding general solution to equation (14) in a quite vivid form was given in [4]. The simplest special solution of (14),
$e^{\rho_h(u_\pm,z)} = (d_0 + d_1 z + z^2)\,e^{\rho_L(u_\pm)}$,
describes, in particular, the Eguchi-Hanson gravitational instanton
$e^{\rho_h} = \frac{1}{2}(x_3^2 - a^2)(1 + x_1^2 + x_2^2)^{-1}$ with $x_3^2 \geq a^2$.

Clearly, the analogous special solution

$$e^{\rho(u_\pm,z)} = (\sum_{n=0}^{N-1} d_n\, z^n + \frac{2}{N!}z^N)e^{\rho_L(u_\pm)}, \tag{15}$$

can be written for equation (12) with $K = \partial^N/\partial z^N$.

5) $\underline{K=\partial/\partial z}$. Among the equations of type (12) it is useful to single out also the equation $\Delta\rho = \partial e^\rho/\partial z$ which is the heat-transfer equation (in the absence of a source of a heat) with a dissipative factor inversely proportional to the temperature $e^\rho \geq 0$

(see e.g.[8]). The corresponding continuum Lie algebra $\mathcal{J}(E; \partial/\partial z)$ is not an algebra of constant growth [2] unlike the algebra with the Cartan operator $\partial^2/\partial z^2$. Besides the evident solution of type (15) with $N = 1$, the equation in question possesses an automodel solution. This solution for the function ρ depending on z and only on one more variable, say $x_1 \equiv u_+ + u_-$, is expressed in the form clarifying those from [12], $e^{-\rho} = z(\mu + \frac{1}{2}x_1^2 z^{-2})$, $\mu = \text{const}$.
6) The last example which is the continuous analogue of the (proper) Bäcklund transformation for equation (14), i.e. the continuous limit of the difference KdV equation, is

$$\partial\rho/\partial\tau = \partial e^{\rho}/\partial z. \tag{16}$$

This equation has the following Lax-pair:

$$\mathcal{L} = X_{+1}(e^{\rho}) + X_{-1}(\varphi),\ \mathcal{M} = X_{+2}(\psi),$$

which leads, up to trivial transformation, to $\psi = e^{2\rho}$ and (16), whose differentiation with respect to τ gives one-dimensional (in $u_{\pm}$) version of equation (14). Note that equation (16) describes also a special case of the transfer equation.

REFERENCES

1. M.V.Saveliev: preprint IHEP 88-39, Serpukhov(1988); Comm.Math.Phys. 121, 283 (1989).
2. M.V.Saveliev and A.M.Vershik: Comm.Math.Phys. 126, 367 (1989).
3. M.V.Saveliev and A.M.Vershik: Phys.Lett. 143*A*,121 (1990).
4. R.M.Kashaev, M.V.Saveliev, S.A.Savelieva and A.M.Vershik: preprint IHEP 90-01, Serpukhov (1990), preprint PAM-497, Berkeley (1990), to be published in the volume "Ideas and methods in mathematics and physics" - in Memory of R.Høegh-Krohn (Cambridge Univ.Press)
5. A.N.Leznov and M.V.Saveliev: Group methods for integration of nonlinear dynamical systems. Moscow, Nauka, 1985; Acta Appl. Math. 16, 1(1989).
6. C.Boyer and D.Finley: J.Math.Phys. 23, 1126 (1982).
7. M.I.Golenisheva - Kutuzova and A.G.Reyman: Zap. Nauch. Semin. LOMI 169,44(1988); Bogoyavlensky, O.I.: Izv.Acad. Nauk SSSR, ser. mat. 52, 712(1988).
8. V.A.Galaktionov et al: in Itogi Nauki i Techniki, Sovremennye problemy matem. Noveishie dostizenia, 28, 95(1987); V.A.Baikov, R.K.Gazizov and N.Ch.Ibragimov: ibid., 34, 85(1989).
9. V.G.Kac: Infinite-dimensional Lie algebras. Boston, Birkhäuser, 1983.
10. D.B.Fairlie, P.Fletcher and C.K.Zachos: preprint DTP - 89/37; Phys.Lett. 224*B*, 101(1989).
11. E.Ramos, C.H.Sah and R.E.Shrock: preprint ITP-SB-89-16.
12. Ya.B.Zeldovich and A.S.Kompaneets: in a volume in honour of A.F.Ioffe, Moscow, 1950.

3. Conformal field theory and related topics

Exchange Algebra in the Conformal Affine sl_2 Toda Field Theory

O.Babelon
Laboratoire de Physique Théorique et Hautes Energies Université Pierre et Marie Curie, Tour 16 1er étage, 4 place Jussieu, 75252 Paris CEDEX 05-France

L.Bonora
International School for Advanced Studies (SISSA/ISAS)
Strada Costiera 11, 34014 Trieste, Italy, and INFN, Sezione di Trieste.

Abstract. We analyse the sl_2 affine Toda field theory introduced elsewhere. In particular we study the chiral splitting of the theory and the relevant Drinfeld-Sokolov equations. We exhibit the chiral exchange algebra and the conformal properties of objects involved.

1 Introduction

It is becoming clear that a close relationship exists between integrability and conformal invariance. In Sine-Gordon, the prototype example of integrable field theory, the main tool which was used to solve the theory is the Yang-Baxter equation, and so only the quantum group structure of the model was exploited. On the other hand, in conformal field theory, only the conformal algebra was first considered, and it took some time to understand how quantum groups intervene. The complete structure is powerfully expressed in the exchange algebra. It remains to understand what role the conformal algebra plays in integrable field theory, i.e. in the Sine-Gordon theory.
To analyse the role of the conformal algebra in integrable models, we introduced in ref.[1] a hierarchy of three models: the simple Liouville model (or simple sl_2 Toda field theory) defined by the equation

$$\partial_+\partial_-\varphi = m^2 e^{2\varphi} \tag{1}$$

the standard affine Sinh-Gordon model

$$\partial_+\partial_-\varphi = m^2\left(e^{2\varphi} - e^{-2\varphi}\right) \tag{2}$$

and the conformal affine Liouville model (conformal affine sl_2 Toda field theory) defined by three equations of motion

$$\partial_+\partial_-\varphi = m^2\left(e^{2\varphi} - e^{2(\eta-\varphi)}\right) \tag{3}$$

$$\partial_+\partial_-\eta = 0 \tag{4}$$

$$\partial_+\partial_-\xi = m^2 e^{2(\eta-\varphi)} \tag{5}$$

Here light-cone coordinates $z_\pm = x \pm t$, $\quad x \in S^1, t \in R$, are being used.

The first and last models are conformal invariant, the second is not. So, the Sinh-Gordon model may be approached from conformal field theory in two different ways. One is through a reduction procedure, setting $\eta = 0$ in eq.(3) we recover the Sinh-Gordon equation. The other consists in regarding the $e^{-2\varphi}$ in the RHS of eq.(2) as a perturbation added to the RHS of eq.(1).
In other words, we may envisage the hierarchy of three equations as follows: eq.(1) is regarded as the initial conformal invariant (unperturbed) theory, which we perturb by adding the second term in the RHS of eq.(2) breaking in such a way conformal invariance; we then recover conformal invariance by adding suitable fields η and ξ. The third theory contains therefore, under suitable limiting conditions, both the initial unperturbed theory and its integrable perturbation. We do not need to stress the relevance of all this in connection with the situation encountered when we perturb rational conformal field theories, where certain perturbations seem, at the lowest perturbative order, to be well mimicked by integrable models.
Finally one should add that the scheme proposed by means of the above hierarchy can be generalized to Toda field theories based on any Lie algebra.
The purpose of this paper is to further carry on the analysis started in ref.[1].
Let us summarize first the results of ref.[1]. In that paper we described a few elementary properties of the conformal affine Liouville model. In particular we showed that it can be written in terms of a Lax pair. In the framework of the $\widehat{sl}_2$ Kac-Moody algebra, let us define (see Appendix A, for the notation)

$$\mathcal{E}_\pm = E_\pm + \lambda^\pm E_\mp$$

the Cartan subalgebra-valued field

$$\Phi = \tfrac{1}{2}\varphi H + \eta d + \tfrac{1}{2}\xi c$$

and the connection

$$A_{z_\pm} = \pm\partial_{z_\pm}\Phi + e^{\pm ad\Phi}\mathcal{E}_\pm \tag{6}$$

We write the linear system

$$(\partial_{z_\pm} + A_{z_\pm})T(z_+, z_-) = 0 \tag{7}$$

The flatness condition for the connection (6) gives eqs.(3,4, 5).
In ref.[1] we gave a general recipe to find the classical solutions of this model. We also introduced chiral Cartan subalgebra valued *free* fields

$$K_\pm \equiv K_\pm(z_\pm) = \theta_\pm H \mp 2\eta_\pm d + \zeta_\pm c$$

and proved that the improved energy momentum tensor can be expressed in terms of these fields

$$T_{++} = -\tfrac{1}{2}\Big((\theta'_+)^2 + \theta''_+ - 2\eta'_+\zeta'_+ + \tfrac{B}{2}\eta''_+ + 2\zeta''_+\Big) \tag{8}$$
$$T_{--} = -\tfrac{1}{2}\Big((\theta'_-)^2 + \theta''_- - 2\eta'_+\zeta'_- + \tfrac{B}{2}\eta''_- + 2\zeta''_-\Big) \tag{9}$$

where B is an arbitrary constant. We remark that the difference in normalization of eqs.(8,9) as well as other equations of this paper with respect to [1] is due to the choice of eqs.(12) as basic Poisson brackets for this paper.
In this paper we are going to construct the periodic and local solutions to the equtions of motion (section 2), by implementing the chiral splitting typical of any conformal field theory in 2 dimensions. In section 3 we study the conformal properties of the chiral objects involved in the construction of the solution. Appendix A is devoted to notation and properties concerning $\widehat{sl}_2$ and the relevant classical r-matrix and Appendix B to the technical problem of diagonalizing the monodromy matrix S (see below).

2 The chiral exchange algebra

The purpose of this section is to construct a periodic and local solution of the equations of motion (3,5), expressed in terms of chiral and antichiral free fields.

2.1 Introduction

Let us define

$$A_+ = P - \mathcal{E}_+, \qquad A_- = -\bar{P} + \mathcal{E}_-$$

where

$$\begin{aligned} P &= p_\theta H + 2p_\eta d + p_\zeta c \equiv \partial_+ K_+ \\ \bar{P} &= \bar{p}_\theta H + 2\bar{p}_\eta d + \bar{p}_\zeta c \equiv \partial_- K_- \end{aligned}$$

We proved in [1] that there exists an object $Q_+(Q_-)$ which is an upper(lower)-triangular element of the group whose Lie algebra is $\widehat{sl}_2$ which can be represented by an infinite upper(lower) triangular matrix, is a function only of $z_+(z_-)$ and satisfies the *Drinfeld-Sokolov equations*

$$\partial_+ Q_+ = (P - \mathcal{E}_+)Q_+ \tag{10}$$

$$\partial_- Q_- = Q_-(-\bar{P} + \mathcal{E}_-) \tag{11}$$

These two equations are the basis of our construction. We write the solutions in the following form

$$\begin{aligned} Q_+(x) &= \overleftarrow{P} e^{\int_0^x A_+(y)dy}, \quad Q_+(0) = 1 \\ Q_-(x) &= \overrightarrow{P} e^{\int_0^x A_-(y)dy}, \quad Q_-(0) = 1 \end{aligned}$$

and for later use we introduce also the left and right monodromy matrices

$$S = Q_+(2\pi), \qquad \bar{S} = Q_-(2\pi)$$

We assume the following Poisson brackets

$$\begin{aligned} \{P(x) \overset{\otimes}{,} P(y)\} &= \tfrac{1}{2}\mathcal{H}_\gamma(\partial_x - \partial_y)\delta(x-y) \\ \{\bar{P}(x) \overset{\otimes}{,} \bar{P}(y)\} &= -\tfrac{1}{2}\mathcal{H}_\gamma(\partial_x - \partial_y)\delta(x-y) \end{aligned} \tag{12}$$

where

$$\mathcal{H}_\gamma = \gamma(H \otimes H + 2d \otimes c + 2c \otimes d) = \gamma\,\mathcal{H}$$

and we have used the definition

$$\delta(x) = \tfrac{1}{2\pi}\textstyle\sum_n e^{inx}$$

$P(x)$ and $\bar{P}(x)$ are assumed to be periodic, so we can Fourier expand them

$$\begin{aligned} P(x) &= \sum_n P_n e^{inx} \\ \bar{P}(x) &= \sum_n \bar{P}_n e^{inx} \end{aligned}$$

As a consequence we have the Poisson brackets

$$\begin{aligned} \{P_n \overset{\otimes}{,} P_m\} &= \tfrac{1}{2\pi} in\delta_{n,-m}\mathcal{H}_\gamma \\ \{\bar{P}_n \overset{\otimes}{,} \bar{P}_m\} &= -\tfrac{1}{2\pi} in\delta_{n,-m}\mathcal{H}_\gamma \end{aligned} \tag{13}$$

In the following an important role will be played by P_0 and $\bar{P}_0$. Eventually we will set

$$P_0 = \bar{P}_0 \tag{14}$$

but for the time being we keep them distinct. We will also need k_+ and k_-, defined as follows

$$\begin{aligned} k_+ &= \sum_{n\neq 0} \frac{iP_n}{n} \\ k_- &= \sum_{n\neq 0} \frac{i\bar{P}_n}{n} \end{aligned}$$

By definition we set

$$\begin{aligned} K_+(x) &= \int_0^x dy P(y) = k_+ + P_0 x + \sum_{n\neq 0} \frac{P_n}{in} e^{inx} \\ K_-(x) &= \int_0^x dy \bar{P}(y) = k_- + \bar{P}_0 x + \sum_{n\neq 0} \frac{\bar{P}_n}{in} e^{inx} \end{aligned}$$

The left and right degrees of freedom are independent, so

$$\{P_n \overset{\otimes}{,} \bar{P}_m\} = 0 \quad \forall n, m \tag{15}$$

2.2 Construction of a periodic local solution

In this subsection we explicitly construct a local periodic solution of (3,5). Here we give an outline of the construction, since many details are the same as in ref.[2]. Some important relations which are needed for the proofs are given in Appendix A.

For any highest weight vector $|\Lambda_r>$ of $\widehat{sl}_2$ we define

$$\begin{aligned}\sigma^{(r)}(x) &= <\Lambda_r|Q_+(x) \\ \bar{\sigma}^{(r)}(x) &= Q_-(x)|\Lambda_r>\end{aligned} \tag{16}$$

These satisfy the exchange algebra

$$\{\sigma(x) \overset{\otimes}{,} \sigma(y)\} = -\tfrac{1}{2}\sigma(x)\otimes\sigma(y)\Big(\theta(x-y)r^+ + \theta(y-x)r^-\Big) \tag{17}$$

$$\{\bar{\sigma}(x) \overset{\otimes}{,} \bar{\sigma}(y)\} = -\tfrac{1}{2}\Big(r^-\theta(x-y) + r^+\theta(y-x)\Big)\bar{\sigma}(x)\otimes\bar{\sigma}(y) \tag{18}$$

A solution to eqs.(3,5) is given (see [1]) by

$$e^{-\xi} = \sigma^{(0)}M\bar{\sigma}^{(0)}, \qquad e^{-\varphi-\xi} = \sigma^{(1)}M\bar{\sigma}^{(1)} \tag{19}$$

where M is a constant matrix to be determined. Since

$$\begin{aligned}\sigma^{(r)}(x+2\pi) &= \sigma^{(r)}(x)S \\ \bar{\sigma}^{(r)}(x+2\pi) &= \bar{S}\,\bar{\sigma}^{(r)}(x)\end{aligned}$$

to get a periodic solution, we must have

$$S\ M\ \overline{S} = M \tag{20}$$

From Appendix B we learn that we can set

$$\begin{aligned}S &= g\kappa g^{-1}, \qquad \kappa = e^{2\pi P_0} \\ \bar{S} &= \bar{g}^{-1}\bar{\kappa}\bar{g}, \qquad \bar{\kappa} = e^{-2\pi\bar{P}_0}\end{aligned} \tag{21}$$

Therefore we see that condition eq(20) will be satisfied if

$$\begin{aligned}M &= gD\overline{g} \\ \kappa\bar{\kappa} &= 1\end{aligned}$$

where $D \in \exp(\mathbf{H})$, $\mathbf{H}$ being the Cartan subalgebra. The second condition simply means that $P_0 = \overline{P}_0$ and will be imposed at the end. The diagonal matrix D has to be chosen in such a way tha the fields ξ and φ are local. The solution is

$$\begin{aligned}D &= \rho\overline{\rho} \\ \rho &= \theta e^{-k_+} \\ \overline{\rho} &= \overline{\theta}e^{k_-}\end{aligned}$$

and θ is conjugate to P_0

$$\begin{aligned}\{\theta \overset{\otimes}{,} P_0\} &= \tfrac{1}{2\pi}\theta\otimes 1\cdot\mathcal{H}_\gamma \\ \{\overline{\theta} \overset{\otimes}{,} \overline{P}_0\} &= \tfrac{1}{2\pi}\overline{\theta}\otimes 1\cdot\mathcal{H}_\gamma\end{aligned} \tag{22}$$

Therefore we can write

$$e^{-\xi} = \psi^{(0)}\,\overline{\psi}^{(0)} \qquad e^{-\varphi-\xi} = \psi^{(1)}\,\overline{\psi}^{(1)} \tag{23}$$

where we define the new objects (Block wave basis)

$$\psi^{(r)}(x) = \sigma^{(r)}(x) g\rho, \qquad \bar{\psi}^{(r)}(x) = \bar{\rho}\bar{g}\bar{\sigma}^{(r)}(x) \tag{24}$$

The ψ and $\bar{\psi}$ have diagonal monodromy

$$\psi^{(r)}(x+2\pi) = \psi^{(r)}(x)\ \kappa, \qquad \bar{\psi}^{(r)}(x+2\pi) = \bar{\kappa}\ \bar{\psi}^{(r)}(x) \tag{25}$$

and obey the exchange algebra

$$\begin{aligned} \{\psi(x) \overset{\otimes}{,} \psi(y)\} &= -\tfrac{1}{4}\psi(x)\otimes\psi(y)\Big\{\epsilon(x-y)(r^+ - r^-) - \\ &\quad -coth(\pi ad_1 P_0)(r^+ - \mathcal{H}_\gamma) - coth(\pi ad_2 P_0)(r^- + \mathcal{H}_\gamma)\Big\} \end{aligned} \tag{26}$$

$$\begin{aligned} \{\overline{\psi}(x) \overset{\otimes}{,} \overline{\psi}(y)\} &= \tfrac{1}{4}\Big\{\epsilon(x-y)(r^+ - r^-) + coth(\pi ad_1 \overline{P}_0)(r^- + \mathcal{H}_\gamma) \\ &\quad + coth(\pi ad_2 \overline{P}_0)(r^+ - \mathcal{H}_\gamma)\Big\}\overline{\psi}(x)\otimes\overline{\psi}(y) \end{aligned} \tag{27}$$

and since the two zero modes are considered as independent, we have

$$\{\psi(x) \overset{\otimes}{,} \overline{\psi}(y)\} = 0 \tag{28}$$

Eq.(23) represents a general solution of eqs.(3,5) which is both periodic and local and implements the chirality splitting.

2.3 Verifying the properties of the solution

The solution (23) has all the required properties.
1 - *It is a solution*, as was proved in [2].
2 - *It is periodic.* To see this one should simply remember (25) and impose (14).
3 - *It is local.* Using eqs.(26,27,28) and eventually identifying P_0 with $\bar{P}_0$, one can verify that

$$\begin{aligned} \{\xi(x), \xi(y)\} &= 0 \\ \{\varphi(x), \varphi(y)\} &= 0 \\ \{\xi(x), \varphi(y)\} &= 0 \end{aligned}$$

Here one uses the fact that

$$\rho\otimes\rho\cdot r^\pm = r^\pm\cdot\rho\otimes\rho$$

which follows from the first three eqs.(38), and the property

$$coth\pi ad_1 P_0(r^+ + r^-) + coth\pi ad_2 P_0(r^+ + r^-) = 0$$

This in turn comes from

$$ad_1 P_0(r^+ + r^-) = -ad_2 P_0(r^+ + r^-)$$

which follows again from eq(38).

3 Conformal properties of σ and ψ

The purpose of this section is to study the conformal properties of the objects introduced before. In particular we want to single out those objects that behave tensorially under a conformal transformation. In this section we work in the sector + only and the corresponding label will be generally understood. In this section we also set, for simplicity, $\gamma = 1$. Let us recall the expression of the energy momentum tensor from eq.(8) and rename it $\mathcal{U}$

$$\mathcal{U} \equiv T_{++} = -\tfrac{1}{2}\left(p_\theta^2 + p_\theta' + 2p_\eta p_\zeta - \tfrac{B}{2}p_\eta' + 2p_\zeta'\right)$$

It is convenient to use the scalar product $(\cdot,\cdot)$ of $\widehat{sl}_2$ (see Appendix B), to express $\mathcal{U}$ in compact form. We recall in particular

$$(H,H) = 2, \quad (d,c) = 1 = (c,d)$$

We introduce also the element Ξ belonging to the Cartan subalgebra, defined by

$$\Xi = H + 4d - \tfrac{B}{2}c \tag{29}$$

It satisfies in particular

$$[\Xi, \mathcal{E}_+] = 2\mathcal{E}_+$$

Using this element we can write

$$\mathcal{U} = -\tfrac{1}{4}\left((P,P) + (\Xi, P')\right) \tag{30}$$

Let us set

$$\mathcal{U}(f) = \int_0^{2\pi} dx f(x)\mathcal{U}(x), \qquad P(f) = \int_0^{2\pi} dx f(x) P(x)$$

for any periodic function f. We find

$$\{\mathcal{U}(f), P(x)\} = \partial\big(f(x)P(x)\big) - \tfrac{1}{2}\Xi\partial^2 f(x) \tag{31}$$

and

$$\{\mathcal{U}(f),\mathcal{U}(g)\} = -\int_0^{2\pi} dx f\left(\partial_x\mathcal{U} + \mathcal{U}\partial_x + \tfrac{1}{2}(\Xi,\Xi)\partial_x^3\right)g \tag{32}$$

Using eq.(31) and with the same methods as in section 2, one finds

$$\{\mathcal{U}(f), Q(x)\} =$$
$$= f(x)\partial Q(x) - \tfrac{1}{2}\partial f(x)\Xi Q(x) - f(0)Q(x)\big(P(0) - \mathcal{E}_+\big) + \tfrac{1}{2}\partial f(0) Q(x)\Xi$$

As a consequence

$$\begin{aligned}\{\mathcal{U}(f), \sigma^{(r)}(x)\} \;=\;& f(x)\partial\sigma^{(r)}(x) - \\ & -\tfrac{1}{2}\partial f(x)\big(\lambda^{(r)}(H) - \tfrac{B}{2}\lambda^{(r)}(c) + 4\lambda^{(r)}(d)\sigma^{(r)}(x)\big) \\ & - f(0)\sigma^{(r)}(x)\big(P(0) - \mathcal{E}_+\big) + \tfrac{1}{2}\partial f(0)\sigma^{(r)}(x)\Xi\end{aligned}$$

Thus σ does not have good tensorial properties (which would be represented by the first two terms in the RHS). But if we define

$$W(x) = Q(x)g\rho$$

we obtain

$$\{\mathcal{U}(f), W(x)\} = f(x)\partial W(x) - \partial f(x)\Xi W(x) \tag{33}$$

It is clear that all the "rows" of $W(x)$ have tensorial character. In particular for

$$\psi^{(r)}(x) = <\Lambda_r|W(x) \tag{34}$$

we have

$$\{\mathcal{U}(f), \psi^{(r)}(x)\} = f(x)\partial\psi^{(r)}(x) - \tfrac{1}{2}\lambda^{(r)}(\Xi)\partial f(x)\psi^{(r)}(x) \tag{35}$$

Therefore

$$h^{(r)} = -\tfrac{1}{2}\lambda^{(r)}(\Xi)$$

is the conformal weight of $\psi^{(r)}(x)$. Therefore the basis with good tensorial properties is given by the ψ's.

4 Appendices

4.1 Appendix A

This Appendix is devoted to the notation concerning $\widehat{sl}_2$ and to the definition of the r matrices we need in the text.
Let us introduce the generators E_+, E_- and H of sl_2, which satisfy the algebra

$$[H, E_\pm] = \pm 2E_\pm \qquad [E_+, E_-] = H$$

The $\widehat{sl}_2$ algebra has generators c, d and $X^k = \lambda^k \otimes X$ where $X = E_+, E_-, H$, which satisfy the algebra

$$\begin{aligned} [H^k, E^l_\pm] &= \pm 2E^{k+l}_\pm \qquad & [H^k, H^l] &= 2k\delta_{k,-l}c \\ [E^k_+, E^l_-] &= H^{k+l} + k\delta_{k,-l}c \qquad & [d, X^k] &= kX^k \end{aligned}$$

The two fundamental highest weight vectors $|\Lambda_0>$ and $|\Lambda_1>$ of $\widehat{sl}_2$ are defined by:

$$\begin{aligned} &H|\Lambda_0>=0, \qquad c|\Lambda_0>=|\Lambda_0>, \qquad d|\Lambda_0>=0 \\ &H|\Lambda_1>=|\Lambda_1>, \qquad c|\Lambda_1>=|\Lambda_1>, \qquad d|\Lambda_1>=0 \end{aligned}$$

Let us normalize the usual inner product $(\cdot,\cdot)$ as follows

$$(c, d) = 1, \qquad (H^n, H^{-n}) = 2, \qquad (E^n_+, E^{-n}_-) = 1, \forall n \tag{36}$$

Next we define the r matrices

$$r^+ = \frac{\gamma}{2}\Big(\mathcal{H} + 4\sum_{\alpha\ positive}\frac{E_\alpha\otimes E_{-\alpha}}{(E_\alpha,E_{-\alpha})}\Big) \qquad (37)$$

$$r^- = -\frac{\gamma}{2}\Big(\mathcal{H} + 4\sum_{\alpha\ positive}\frac{E_{-\alpha}\otimes E_{\alpha}}{(E_\alpha,E_{-\alpha})}\Big)$$

Here α stands for a generic root. So the couples $(E_\alpha, E_{-\alpha})$ with α positive, are (E_-^n, E_+^{-n}), $n \geq 1$, (H^n, H^{-n}), $n \geq 1$ and (E_+^n, E_-^{-n}), $n \geq 0$). We recall that r^+ and r^- differ by a multiple of the Casimir of $\widehat{sl}_2$. We will need in particular the following commutation relations (from now on in this Appendix r will stand both for r^+ and for r^-)

$$\begin{aligned}
[r, H^0\otimes 1 + 1\otimes H^0] &= 0 \qquad (38)\\
[r, d\otimes 1 + 1\otimes d] &= 0\\
[r, c\otimes 1 + 1\otimes c] &= 0\\
[r, E_+^0\otimes 1 + 1\otimes E_+^0] &= \gamma(H^0\otimes E_+^0 - E_+^0\otimes H^0)\\
[r, E_-^0\otimes 1 + 1\otimes E_-^0] &= \gamma(H^0\otimes E_-^0 - E_-^0\otimes H^0)\\
[r, E_-^1\otimes 1 + 1\otimes E_-^1] &= \gamma(E_-^1\otimes H^0 - H^0\otimes E_-^1 + c\otimes E_-^1 - E_-^1\otimes c)\\
[r, E_+^{-1}\otimes 1 + 1\otimes E_+^{-1}] &= \gamma(E_+^{-1}\otimes H^0 - H^0\otimes E_+^{-1} + c\otimes E_+^{-1} - E_+^{-1}\otimes c)
\end{aligned}$$

Using this it is easy to prove

$$[r, 1\otimes\mathcal{E} + \mathcal{E}\otimes 1] = [\mathcal{H}_\gamma, 1\otimes\mathcal{E} + \mathcal{E}\otimes 1]$$

and

$$ad_1 P_0 \cdot r = -ad_2 P_0 \cdot r$$

These relations allow us to extend the proofs of ref.[2] to the affine case.

4.2 Appendix B. Diagonalizing S

In this appendix we deal with the diagonalization of the monodromy matrix S. Diagonalizing $\bar{S}$ is completely analogous and so we limit ourselves to the + sector. Actually in this appendix we solve a more general problem, since diagonalizing S means (see [2]) finding the (unique) periodic strictly upper triangular gauge transformation $\mathbf{g}(x)$ such that

$$P = {}^{\mathbf{g}}A_+ \equiv \partial\mathbf{g}(x)\mathbf{g}^{-1}(x) + \mathbf{g}(x)A\mathbf{g}^{-1}(x)$$

In terms of $\mathbf{g}$ we have, see [2]

$$Q_+(x) = \mathbf{g}^{-1}(x)e^{K_+(x)}\mathbf{g}(0)$$

and

$$S = \mathbf{g}^{-1}(0)\kappa\mathbf{g}(0)$$

So, the g we need in subsection 2.2 is exactly $\mathbf{g}^{-1}(0)$. In order to construct $\mathbf{g}$ we make the ansatz

$$\mathbf{g} = g_0 g_- g_+ \tag{39}$$

where

$$\begin{aligned} g_0 &= e^{X_0 H} \\ g_- &= e^{X_- E_-} \\ g_+ &= e^{X_+ E_+} \end{aligned}$$

Here X_+, X_-, X_0 are functions of x and λ, periodic in x. In particular, as for the λ-dependence we have

$$\begin{aligned} X_+(x,\lambda) &= \sum_{n=0} X_+^{(n)}(x)\lambda^n \\ X_-(x,\lambda) &= \sum_{n=1} X_-^{(n)}(x)\lambda^n \\ X_0(x,\lambda) &= \sum_{n=1} X_0^{(n)}(x)\lambda^n \end{aligned} \tag{40}$$

We have to determine the periodic functions $X_+^{(n)}(x)$, $X_-^{(n)}(x)$ and $X_0^{(n)}(x)$.
Let us start with g_+:

$$\begin{aligned} {}^{g_+}A_+ &= \partial g_+ g_+^{-1} + g_+ A g_+^{-1} = \\ &= P + L_+ E_+ - \lambda E_- - \lambda X_+ H \end{aligned}$$

where

$$L_+ = \partial X_+ - 2X_+ p_\theta - 1 - 2p_\eta \lambda \tfrac{\partial}{\partial\lambda} X_+ + \lambda X_+^2$$

Next it is g_-'s turn

$$\begin{aligned} {}^{g_- g_+}A_+ &= \partial g_- g_-^{-1} + g_-({}^{g_+}A) g_-^{-1} = \\ &= P + L_- E_- + L_+ E_+ - (\lambda X_+ + X_- L_+) H \end{aligned}$$

where

$$L_- = \partial X_- - \lambda + 2p_\theta X_- - 2p_\eta \lambda \tfrac{\partial}{\partial\lambda} X_- - 2\lambda X_+ X_- - L_+ X_-^2$$

Finally

$$\begin{aligned} {}^{g_0 g_- g_+}A_+ &= \partial g_0 g_0^{-1} + g_0({}^{g_- g_+}A) g_0^{-1} = \\ &= P + L_0 + e^{2X_0} L_+ E_+ + e^{-2X_0} L_- E_- \end{aligned}$$

where

$$L_0 = \partial X_0 - \lambda X_+ - X_- L_+ - 2p_\eta \lambda \tfrac{\partial}{\partial\lambda} X_0$$

So that we will have

$$ {}^{g_0 g_- g_+}A = P $$

provided that

$$
\begin{aligned}
L_+ = 0 \; &: \; \partial X_+ - 2X_+ p_\theta - 1 - 2p_\eta \lambda \tfrac{\partial}{\partial\lambda} X_+ + \lambda X_+^2 = 0 && (41)\\
L_- = 0 \; &: \; \partial X_- + 2p_\theta X_- - \lambda - 2p_\eta \lambda \tfrac{\partial}{\partial\lambda} X_- - 2\lambda X_+ X_- = 0 && (42)\\
L_0 = 0 \; &: \; \partial X_0 - \lambda X_+ - 2p_\eta \lambda \tfrac{\partial}{\partial\lambda} X_0 = 0 && (43)
\end{aligned}
$$

To study the solutions of these equations we introduce the following notation. First define the roots of $\widehat{sl}_2$

$$
\begin{aligned}
\alpha_n^{\pm} &= n\delta + \alpha^{\pm}\\
\alpha_n^{0} &= n\delta
\end{aligned}
$$

where $\delta(d) = 1$ and otherwise $= 0$, and $\alpha^{\pm}(H) = \pm 2$ and zero otherwise. So, in particular

$$
\begin{aligned}
\alpha_n^{\pm}(P) &= 2np_\eta \pm 2p_\theta\\
\alpha_n^{0}(P) &= 2np_\eta
\end{aligned}
$$

Let us also define the kernels

$$ K_{\pm,0}^{(n)}(x,y) = \theta(x-y) - \frac{1}{1 - e^{-2\pi\alpha_n^{\pm,0}(P_0)}} $$

and the quantities

$$
\begin{aligned}
Y_+^{(n)}(x) &= Y^{(0)}(x) Z^n(x)\\
Y_-^{(n)}(x) &= \frac{Z^n(x)}{Y^{(0)}(x)}
\end{aligned}
$$

where

$$
\begin{aligned}
Y^{(0)}(x) &= e^{2\int_0^x dy\, p_\theta(y)}\\
Z(x) &= e^{2\int_0^x dy\, p_\eta(y)}
\end{aligned}
$$

Now, eq(41) implies

$$
\begin{aligned}
&\partial X_+^{(0)} - \alpha_0^+(P) X_+^{(0)} - 1 = 0\\
&\partial X_+^{(n)} - \alpha_n^+(P) X_+^{(n)} + \sum_{l=0}^{n-1} X_+^{(n-l-1)} X_+^{(l)} = 0, \qquad n > 0
\end{aligned}
$$

The solutions are respectively

$$
\begin{aligned}
X_+^{(0)}(x) &= Y_+^{(0)}(x) \int_0^{2\pi} dy \frac{K_+^{(0)}(x,y)}{Y_+^{(0)}(y)} && (44)\\
X_+^{(n)}(x) &= -Y_+^{(n)}(x) \int_0^{2\pi} dy \frac{K_+^{(n)}(x,y)}{Y_+^{(n)}(y)} \sum_{l=0}^{n-1} X_+^{(n-l-1)}(y) X_+^{(l)}(y), \qquad n > 0
\end{aligned}
$$

Similarly for eq(42) we have

$$
\begin{aligned}
\partial X_-^{(1)} - \alpha_1^-(P) X_-^{(1)} &= 1 \\
\partial X_-^{(n)} - \alpha_n^-(P) X_-^{(n)} &= 2 \sum_{l=0}^{n-2} X_+^{(l)} X_-^{(n-l-1)}
\end{aligned}
$$

The solutions are

$$
\begin{aligned}
X_-^{(1)} &= Y_-^{(1)}(x) \int_0^{2\pi} dy \frac{K_-^{(1)}(x,y)}{Y_-^{(1)}(y)} \qquad (45) \\
X_-^{(n)} &= 2Y_-^{(n)}(x) \int_0^{2\pi} dy \frac{K_-^{(n)}(x,y)}{Y_-^{(n)}(y)} \sum_{l=0}^{n-2} X_+^{(l)}(y) X_-^{(n-l-1)}(y), \quad n > 1
\end{aligned}
$$

Finally for the case eq(43), we find

$$
\partial X_0^{(n)} - \alpha_n^0(P) X_0^{(n)} = X_+^{(n-1)}, \quad n \geq 1
$$

The corresponding solutions are

$$
X_0^{(n)}(x) = Z^n(x) \int_0^{2\pi} dy \frac{K_0^{(n)}(x,y)}{Z^n(y)} X_+^{(n-1)}(y), \quad n \geq 1 \qquad (46)
$$

References

[1] O.Babelon and L.Bonora, "Conformal Affine sl_2 Toda Field Theory" Phys.Lett. B244(1990) 220, and references therein.

[2] O.Babelon, L.Bonora and F.Toppan, "Exchange algebra and the Drinfeld-Sokolov theorem" I.S.A.S. preprint 65/90/EP.

SOME PROPERTIES OF P-LINES

Krystyna M. Bugajska
Department of Mathematics, York University
Toronto, ONT. M3J 1P3 Canada

I. WHAT IS *P*-LINE.

The main difference between the commonly accepted Polyakov string theory and our approach is that we are not moving from a Lorentzian to an Euclidean structure of the world sheet without lack of concern but, just opposite, we are giving the primary importance to the Lorentzian signature. This approach is based, similarly like some others, on the Teichmüller space formalism.

Let us assume that we have some one dimensional object in $1+D$ dimensional Minkowski space time $\mathbb{R}^{1,D}$. Let us assume that as this object propagates in $\mathbb{R}^{1,D}$ it sweeps out a Lorentzian world sheet Σ^L. Moreover, let us assume that this world sheet forms a connected, orientable manifold Σ of genus $g\geq 2$. Now we can ask which Riemann surfaces (i.e. which complex structures) can be related to Σ^L, or equivalently we can ask how we can describe Σ^L in an appropriate Teichmüller space $\mathcal{T}_{g,n}$.

The set of all conformal structures on $\mathbb{R}^2$ can be given by elements of $GL_+(,\mathbb{R})/GL(1,\mathbb{C}) \cong \Delta = \left\{ z \in \mathbb{C};\ |z| < 1 \right\}$. Now if we fix some Lorentzian structure $\mathbb{R}^{1,1}$ on $\mathbb{R}^2$ then it determines a 1-parameter subset of Δ of those Euclidean structures which are defined by Lorentzian equivalent basis of $\mathbb{R}^{1,1}$. This family of conformally unequivalent $\mathbb{R}^{2,0}$ structures corresponds to the set of the future oriented time-like unit vectors of $\mathbb{R}^{1,1}$ and can be also parametrized by parameter α of the $SO(1,1)$ group.

Let X: $\Sigma^L \to \mathbb{R}^{1,D}$ denote an immersion of our world-sheet into Minkowski space-time. Any fixed time-like vector $e_0 \in \mathbb{R}^{1,D}$ (which we can identify with "a physical observer") defines a time-like vector field $\bar{v}$ on Σ^L (almost everywhere) or, equivalently, a section S_0 of the principal bundle of Lorentzian frames over Σ^L (a.e.). The section S_0 determines some (singular) Riemannian structure, say Σ_0, on the underlying manifold Σ of Σ^L. However, simultaneously, we obtain 1-parameter family of conformally unequivalent Riemannian structures Σ_α on Σ given by sections $S_\alpha = S\circ\alpha$; $\alpha\in SO(1,1)$. This collection $\left\{ \Sigma_\alpha \right\}_{\alpha\in SO(1,1)}$ forms some curve in the appropriate Teichmüller space $\mathcal{T}_{g,n}$ which we call *P*-line. The easiest way to see that any *P*-line is an infinite geodesic in the Teichmüller metric on $\mathcal{T}_{g,n}$ is to pass to a measured foliations approach. (From now, for simplicity and without the lost of generality, we will assume that the manifold Σ is compact). A time-like vector field $\bar{v}$ on Σ^L introduced above determines a pair $\bar{n}_1, \bar{n}_2$ of light-like vector fields. Now the pair of vector fields $\{\bar{n}_1, \bar{n}_2\}$ can be used not only to define a complex structure J on Σ by $J\cdot\bar{n}_1(m)=\bar{n}_2(m))$ but also to define a locally flat (singular) Riemannian metric g on Σ and a pair of measured transversal foliations. By the Hubbard and Masur result any pair of transversal measured foliations determines both: a conformal structure Σ_0 on Σ and some concrete holomorphic quadratic differential q on Σ_0. Since any holomorphic quadratic differential defines unique infinite Teichmüller geodesic we can easily check that this geodesic is exactly our P-line $\{\Sigma_\alpha\}_{\alpha\in SO(1,1)}$ [1].

The horizontal and vertical distributions of any holomorphic quadratic differential $q=\phi(z)dz^2$ are determined by a pair $\{\phi_1, \phi_2\}$ of local 1-forms $\phi_1 = Re\ \phi^{1/2}dz$, $\phi_2 = Im\ \phi^{1/2}dz$ which satisfy

$$\phi_i = \pm\ \phi_i' \qquad i=1,2 \tag{1}$$

on the overlap $U\cap U'$ of any two charts U, U' on Σ_0. If the cocycle defined by *(1)* determines a trivial line bundle over Σ_0 then the differential q is called orientable; if the corresponding bundle is not trivial then q is nonorientable. In the former case q is the square of some holomorphic 1-form i.e. $q=\omega^2$ and $\omega=\phi_1+i\phi_2$. In this case the holonomy group of a metric (of zero curvature) which arizes from q is trivial as well as we can construct (singular) global vector fields $\bar{n}_1, \bar{n}_2$ dual to ϕ_1, ϕ_2 respectively.

Since the Euler class of the underlying manifold Σ does not vanish ($g\geq 2$) we cannot construct a tangent line bundle over Σ. It means that Lorentzian structure Σ^L has to be a singular one or equivalently that any time-like vector field $\bar{v}$ on Σ^L has to be singular. It implies that the two light-like vector fields $\bar{n}_1(m)$ and $\bar{n}_2(m)$ determined by $\bar{v}(m)$ have singularity in the same points of Σ as $\bar{v}(m)$ as well as that these singularities are of the same kind. For any vector field on Σ_0 the Poincare-Hopf theorem tells us that the sum of its indices at zeroes is equal to the homological Euler characteristic $\chi(\Sigma)=2-2g$. On the other side, by the Riemann-Roth theorem we know that the degree of the divisor of the distributions ϕ_i, $i=1,2$ of holomorphic one forms is equal to $2g-2$. The singularities of vector fields $\bar{n}_i$, $i=1,2$ are at the same points as the zeroes of holomorphic one forms ϕ_i, $i=1,2$ and they have the same degree.

II. JENKIS-STREBEL RAYS AND DECAY.

Let $\{\delta_0, \delta_1\}$ be local coordinates on our world sheet Σ^L which are determined by the light-lines of an "observer" $e_0\in\mathbb{R}^{1,D}$. They induce Riemann surface structure, say Σ_0) and (locally) they form leaves of a pair of a transverse measured foliations. These leaves are horizontal and vertical trajectories of some concrete quadratic differential q respectively and $z=i\delta_0+\delta_1$ is its natural parameter i.e. $q=dz^2$. Local natural parameters on Riemann surface Σ_k belonging to the Teichmüller P-line l_q are

$$\delta_0' = \sqrt{\operatorname{ctg}\alpha}\;\delta_0 \qquad\qquad \alpha \in (\,0,\tfrac{\pi}{2}\,)$$

$$\delta_1' = \sqrt{\operatorname{tg}\alpha}\;\delta_1 \qquad\qquad z' = i\delta_0' + \delta_1'$$

(here we have identification of k with $\operatorname{tg}\beta$; $\beta= \frac{\pi}{2} - \alpha \in (\,-\frac{\pi}{4},\frac{\pi}{4}))$.

Let us consider a situation when Riemannian structure on our world sheet related to a concrete "observer" $e_0 \in \mathbb{R}^{1,D}$ posesses some concrete properties. The most regular situation would be when $X\colon \Sigma \longrightarrow \mathbb{R}^{1+D}$ realizes a minimal immersion into $\mathbb{R}^{1+D}$ (uniquely determined by $X\colon \Sigma^L \longrightarrow \mathbb{R}^{1,D}$ and by $e_0 \in \mathbb{R}^{1+D}$). However, it is known that although any noncompact Riemannian 2-manifold admits a proper embedding into $\mathbb{R}^k$, $k>5$, by a harmonic map it is not necessary a conformal one. Moreover there are no compact minimal submanifolds in $\mathbb{R}^n$, $n>3$. So we see that minimal immersion into $\mathbb{R}^{1+D}$ is not the case with high probability. The next, also very regular situation appears when an immersion X into $\mathbb{R}^{1+D}$ realizes a minimal immersion into the hypersphere S^D of $\mathbb{R}^{1+D}$. In this case the Gauss map associated to X is a harmonic and homothetic one [2].

Moreover we can check that the map $X\colon \Sigma \longrightarrow \mathbb{R}^{1+D}$ satisfies wave equation $\dfrac{\partial^2 X}{\partial\delta_0^2} - \dfrac{\partial^2 X}{\partial\delta_1^2} = 0$ if and only if $X\colon \Sigma_k \longrightarrow S^D_{r_k}$ is harmonic for every $\Sigma_k \in l_q$ [3]. If we have this case we call such P-line a harmonic one. If P-line is harmonic then we have (almost everywhere)

$$\frac{\partial^2 X^\mu}{\partial\delta_0^2} = \frac{\lambda}{2}\,X^\mu \qquad\qquad \frac{\partial^2 X^\mu}{\partial\delta_1^2} = \frac{\lambda}{2}\,X^\mu \qquad\qquad \mu=1,\dots,1+D$$

i.e. X^μ are periodic functions of δ_0 and δ_1 (a.e.). It means that our "physical" differential q has to have closed horizontal and closed vertical trajectories. Holomorphic quadratic differentials which satisfy this property are called Jenkis-Strebel differentials.

Jenkis-Strebel differentials have special features. Namely their critical graf Γ_q i. e. the set of critical trajectories with their singular endpoints (zeros) is compact. It implies that in this case we have defined a partition of $\Sigma_0-\Gamma_q$ onto ring domains R_i, $i=1\dots N \le 3g-3$, each of which is swept out by freely homotopic closed horizontal trajectories of q.

To investigate the problem of endpoints of Jenkis-Strebel rays we should pass to the Bers embedding Φ of $\mathcal{T}_g$ into the finite dimensional complex Banach space $B_2(\Gamma,L)$ of bounded differentials ($\Sigma_0=U/_\Gamma$; U is the upper half plane of $\mathbb{C}$ and L is the lower half plane).

The image of the Bers map Φ is bounded in $B_2(\Gamma,L)$ and the identification of $\mathcal{T}_g$ with $\Phi(\mathcal{T}_g) \subset B_2(\Gamma,L)$ determines a boundary $\partial\mathcal{T}_g$ which is called the complex boundary and which depends on the choice of the origin Σ_0 (or equivalently on the choice of $\Gamma \longrightarrow Möb_{\mathbb{R}}$). For each $\phi \in \partial\mathcal{T}_g \longrightarrow B_2(\Gamma,L)$ the group Γ^{ϕ} is always Kleinian and has only one invariant component $\Delta_1 = W_{\phi}(L)$ of Ω (here $\Gamma^{\phi} = W_{\phi}\Gamma W_{\phi}^{-1} \subset Möb$ and W_{ϕ} is appropriate normalized solution of the Schwarzian equation $\{W_{\phi},z\} = \phi \in B_2(\Gamma,L)$; Ω denotes the discontinuity region of Γ^{ϕ}). Such groups Γ^{ϕ} are called b-groups. Any other component Δ of Ω of Γ^{ϕ} is simply connected and not invariant. If Γ^{ϕ}_{Δ} denotes the stabilizer of Δ in Γ^{ϕ} then $\Delta/_{\Gamma^{\phi}}$ is a finite Riemann surface of type (p',n'). So for $\phi \in \partial\mathcal{T}_g$ the component $\Delta_2 = W_{\phi}(U)$ is (perhaps empty) union of all noninvariant components of discontinuity. In this case we write

$$\Omega/_{\Gamma^{\phi}} = \Delta_1/_{\Gamma^{\phi}} + \Delta'/_{\Gamma^{\phi}_{\Delta'}} + \Delta''/_{\Gamma^{\phi}_{\Delta''}} + \dots + \Delta^k/_{\Gamma^{\phi}_{\Delta}}$$

or

$$\Omega/_{\Gamma^{\phi}} = \bar{\Sigma}_0 + S_1 + S_2 + \dots + S_k$$

It turns out that "almost all" b-groups are totally degenerated, that is satisfy $\Delta_1 = \Omega$. A regular boundary group Γ^{ϕ} represents a Riemann surface $\bar{\Sigma}_0$ and one or more surfaces $S_1, S_2, \dots, S_k$ which may be thought to have been obtained drawing allowable Jordan curves on Σ_0 and then contracting each to a point on Σ_0. So we can say that for regular boundary point (b-group) $\Delta_2/_{\Gamma^{\phi}}$ is a finite union of Riemann surfaces which topologically may be derived from Σ_0 by cutting along an admissible system of Jordan curves $\gamma_1, \dots, \gamma_k$ and by gluing a punctured disc to each side of each cut [4]. (All of these considerations can be generalized to any Teichmüller space $\mathcal{T}_{g,n}$, where n is the number of punctures. It means that we will get exactly the same result for world-sheets which can be related to Riemann surfaces with n punctures.)

Let Σ^L be the world sheet of same string object and let l_q be a Teichmüller P-line through Σ_0 determined by a concrete observer. We know [5] that if Σ_0 is minimally immersed into a hypersphere $S^D \longrightarrow \mathbb{R}^{1+D}$ and if our P-line is a harmonic one then the quadratic differential q has to be Jenkis-Strebel differential. We will assume that each P-line is determined by Jenkis-Strebel differentials but that it is not necessarily a harmonic one. (The harmonicity is not a necessary condition to have P-line related to Jenkis-Strebel differential.)

Let $(k,-q)$ denote a Strebel ray through the point (Σ_0, id). Masur has shown [6] that its endpoint is given by the punctured model $\tilde{\Sigma}_0$ of Σ_0 and that there exists a regular boundary point $\tilde{\phi} \in \partial\mathcal{T}_p \subset B_2(\Gamma,L)$ such that $\tilde{\Sigma}_0 \cong \Delta_2/_{\Gamma^{\tilde{\phi}}}$ (As described above $\tilde{\Sigma}_0$ denotes the corresponding union of appropriate Riemann surfaces.)

From physical point of view the existance of such P-lines, determined by a quadratic differential with closed trajectories. (harmonic or not), seems to be the most plausible. In this case we have that any physical object which is related to a Lorentzian world sheet Σ^L cannot be stable. It has to be created - what is described by the so called opening procedure for horizontal cylinder of Jenkis-Strebel ray (k,q), and it has to decay - what is described by the endpoint of J-S ray $(k,-q)$. Since for any P-line l_q we have the identification of k with $k=tg\ \beta$ where $\beta=\frac{\pi}{4}-\alpha \in(-\frac{\pi}{4},\frac{\pi}{4})$ and α has well defined physical interpretation [1], the time orientability of Σ^L guarantees that the notions of "creation" and "decay" are definitely distinguished and well defined. The endpoints of the ray (k,q) can be interpreted as objects which take part in some collision process. Similarly any element S_1 of $\tilde{\Sigma}_0$ i.e. any element of decay can take part in some other collision process i.e. be one of the elements of some other opening procedure.

III DECAY ONCE MORE.

For 2-dimensional orientable manifold M and only for such manifolds, we have the following situation : Let ξ_G be a principal G-bundle over M with connection A. If M is given a complex structure then we have uniquely defined holomorphic $G^{\mathbb{C}}$ bundle over M, where $G^{\mathbb{C}}$ is the complexification of the Lie group G. To see this, let us pass to the extension $\xi_{G^{\mathbb{C}}}$ of the bundle ξ_G. The complex structure of $G^{\mathbb{C}}$ determines a complex structure of the fibres of $\xi_{G^{\mathbb{C}}}$ and a complex structure of M defines one in the horizontal directions given by A. Such defined almost complex structure of $\xi_{G^{\mathbb{C}}}$ satisfies the Newlander-Nirenberg integrability condition, i.e. defines a complex structure which, in our case, is additionally $G^{\mathbb{C}}$-invariant. And conversely, a connection A on ξ_G can be uniquely defined from the holomorphic structure of $G^{\mathbb{C}}$ bundle $\xi_{G^{\mathbb{C}}}$ and its reduction to ξ_G. Thus, on complex 1-dimensional manifolds (Riemann surfaces) G-bundles with connection are equivalent to holomorphic $G^{\mathbb{C}}$-bundles with a reduction to G [7].

On any Riemann surface Σ there exists unique, maximally unstable holomorphic projective bundle $\Phi \in H^1(\Sigma,\mathcal{PL}(1,\mathcal{O}))$. Now by choosing any concrete spinor bundle $\xi \in H^1(\Sigma,\mathcal{O}^*)$ we can construct holomorphic, maximally unstable $SL(2,\mathbb{C})$ bundle over Σ. Its flat representatives correspond to different projective structures on Σ which are associated to different holomorphic quadratic differentials. In this section we will show that for P-lines satisfying P-condition we can construct reductions of appropriate holomorphic, maximally unstable $SL(2,\mathbb{C})$ bundles to the $SU(2)$ group and we will give a physical interpretation of this fact (notice that $SU(2)^{\mathbb{C}}= SL(2,\mathbb{C})$).

A. SPINOR STRUCTURES.

A holomorphic square root of the canonical line bundle $\mathcal{K}$ of holomorphic 1-forms is called a spinor bundle. In other words if K denotes the canonical divisor class then any solution of the equation $2\mathcal{D} = K$ in the divisor class group corresponds to concrete spinor structure of Σ. We have 2^{2g} solutions of this equation i.e. there are 2^{2g} distinct spin structures on each surface Σ. Each of them corresponds to one of the 2^{2g} half points in the Jacobi variety $J(\Sigma)$ and the set $\mathcal{S}(\Sigma)$ of spinor structures has a natural structure of an affine space over $\mathbb{Z}_2$ with $H^1(\Sigma,\mathbb{Z}_2)$ as its group of translations. Moreover there is a natural quadratic function $f:\mathcal{S}(\Sigma)\longrightarrow\mathbb{Z}_2$ whose associated bilinear form is the cup product on $H^1(\Sigma,\mathbb{Z}_2)$. This function is defined as $f(\xi)=\dim\Gamma(\xi) \bmod 2$; $\xi \in \mathcal{S}(\Sigma)$ and does not depend on the complex structure of a surface. It can be seen that function f has exactly $2^{g-1}(2^g+1)$ zeroes what implies that there are precisely $2^{g-1}(2^g+1)$ isomorphic classes of spinor bundles which are spin-boundaries and which are called even. The remaining spinor structures ($2^{g-1}(2^g-1)$ in number) have to admit at least one holomorphic section and are called odd.

The better understanding of spinor structures can be obtained by using theta function and its zero-divisor known as as θ-divisor. Theta function is not srictly a function on the Jacobi variety $J(\Sigma)$. It is rather unique holomorphic section of a holomorphic line bundle on $J(\Sigma)$ called the ν-line bundle $\mathcal{L}$. The relation between spinor bundles and theta function is based on the Riemann's vanishing theorem which implies that to any spin structure we may associate a symmetric translate of θ. In other words if $(\varepsilon_1,\varepsilon_2)\in(\frac{1}{2}\mathbb{Z}/\mathbb{Z})^{2g}$ is a half point then the divisor of theta function with characteristic $[\varepsilon_1,\varepsilon_2]$ is a symmetric translate of θ. For even spin structures $4\varepsilon_1\cdot\varepsilon_2$ is even and for odd is odd.

Now let us consider a situation when we have a family of marked Riemann surfaces given by a corresponding Teichmüller space $\mathcal{T}_g$. The Bers embedding theorem provides holomorphic fibre space V_g (which is not a holomorphic fibre bundle) over $\mathcal{T}_g$ which is called the universal family. The fibres of $\pi:V_g\longrightarrow\mathcal{T}_g$ are compact genus g surfaces $\Sigma_\tau=\pi^{-1}(\tau)$, $\tau\in\mathcal{T}_g$. We can define a spin structure for a family of surfaces V_g as a holomorphic line bundle on V_g which restricts on each fibre surface Σ_τ to a spin bundle ξ_τ. The set $\mathcal{S}(g)$ of spin structures on V_g has the same description as the set of spin structures on each surface Σ_τ via a suitable process of holomorphic continuation. So there are 2^{2g} distinct spin structures on V_g.

B. PROJECTIVE STRUCTURES.

On any compact Riemann surface Σ of genus $g\geq2$ we always can introduce a special covering $\{U_i,\alpha_i\}$, $\alpha_i:U_i\longrightarrow \mathbb{C}$ with the property that coordinate functions $\alpha_i = \dfrac{a_{ij}\alpha_j + b_{ij}}{c_{ij}\alpha_j + d_{ij}}$ are projective transformations. Such covering forms, the so called, projective atlas and projective structure on Σ is an equivalence class of projective atlasses. In this case the coordinate transition functions $\psi_{ij}=\alpha_i\circ\alpha_j^{-1}$ associated to intersection $U_i \cap U_j$ satisfy $\psi_{ij}=\psi_{ji}^{-1}$ and $\psi_{ij}\psi_{jk}=\psi_{ik}$, (i.e. they define a flat complex projective bundle ψ over Σ; $\psi\in H^1(\Sigma, PL(1,\mathbb{C}))$).

Any representation of Riemann surface Σ as the quotient space of bounded Jordan region $D \subset \hat{\mathbb{C}}$ by a discontinuous group $\tilde{\Gamma}$ of projective transformations (quasifuchsian group with invariant domain D) provides a projective atlas on Σ and since there is a great many ways for such representations there is a lot of projective structures. For example, if U is the upper half plane then a covering map $p: U \longrightarrow \Sigma$ (whose existance is guaranteed by the uniformization theorem) gives rise to a projective structure on Σ given by sections of p over simply connected open subsets U_i of Σ.

In terms of a universal covering $\Pi: D \longrightarrow \Sigma$ any projective structure of Σ can be described by a complex analytic local homeomorphism f from D into the Riemann sphere $\hat{\mathbb{C}}$ satisfying

$$f(\gamma z) = \rho_\gamma \circ f(z) \qquad \text{for all } z\in D,\ \gamma\in\tilde{\Gamma} \qquad (1)$$

Here $\tilde{\Gamma}$ is (as above) a quasifuchsian group of covering translations and $\rho_\gamma \in PL(1,\mathbb{C})$. A projective structure determined by mapping f (which is called a developing map) is related to projective charts on each contractible open set U_i of Σ given by $f\cdot\Pi^{-1}$. From *(1)* we see that the mapping $\gamma \longrightarrow \rho_\gamma$ is a homomorphism from $\tilde{\Gamma}$ into the projective group $PL(1,\mathbb{C})$. We call this map the monodromy homomorphism of f or the representation of the projective structure described by developing map f [8].

Suppose we have two analytic projective structures α and β on Σ defined by atlasses (U_i,α_i) and (V_j,β_j); $\alpha_i : U_i \longrightarrow \hat{\mathbb{C}}$, $\beta_j : V_j \longrightarrow \hat{\mathbb{C}}$. Since the Schwarzian derivative $\{z(t),t\} = (\frac{z''}{z'})' - \frac{1}{2}(z''/_{z'})^2$ has the property $\{z,t\}=\{z,\delta\circ t\}=\{\delta\circ z,t\}$ for all $\delta\in PL(1,\mathbb{C})$ the quadratic forms $\{\beta_j,\alpha_i\}$ defined in $U_i \cap V_j$ coincide on open sets of form $U_{i1}\cap U_{i2}\cap V$ so they are induced by a quadratic differential $q=\alpha-\beta$ on Σ. Conversely, if α is any complex projective structure and $q=Q(\Sigma)=H^0(\Sigma,\mathcal{K}^{\otimes 2})$ then there is a unique projective structure β on Σ such that $\beta-\alpha=q$. So the space $\mathcal{P}(\Sigma)$ of all analytic projective structures on Σ is an affine space under $Q(\Sigma)$.

As we have mentioned above, coordinate transition functions of projective atlas (U_i,α_i) on Σ define a flat complex projective bundle, say $\psi\in H^1(\Sigma, PL(1,\mathbb{C}))$. The bundle ψ can also be reviewed as determined by the monodromy map $\rho: \tilde{\Gamma} \longrightarrow PL(,\mathbb{C})$ of a developing map f corresponding to α. Now the composition $f'=\delta\circ f$ is a developing map describing projective structure equivalent to α for all $\delta\in PL(1,\mathbb{C})$. Since its monodromy homomorphism

$\rho'=\delta\circ\rho\circ\delta^{-1}$ form representation conjugate to ρ we obtain well defined map from projective structures $\mathcal{P}(\Sigma)$ into the space $Hom(\pi_1\Sigma, SL(1,\mathbb{C}))/SL(1,\mathbb{C})$.

A flat

projective bundle ψ (corresponding to projective structure α on a marked Riemann surface Σ) is associated to a flat complex vector bundle ϕ. Namely the projective linear group $PL(1,\mathbb{C})$ is the quotient of the special linear group $SL(2,\mathbb{C})$ by its center. The commutative diagram of sheaves over Σ

$$\begin{array}{ccccccc} 0 \longrightarrow \mathbb{Z}_2 & \xrightarrow{e} & SL(2.\mathbb{C}) & \xrightarrow{\mu} & PL(1,\mathbb{C}) & \longrightarrow 1 \\ & & \downarrow & & \downarrow & \\ & & \mathcal{SL}(2,\mathcal{O}) & \longrightarrow & \mathcal{PL}(1,\mathcal{O}) & \end{array}$$

implies that on the level of cohomology we get

$$\begin{array}{ccccc} 0 \longrightarrow H^1(\Sigma,\mathbb{Z}_2) & \xrightarrow{e^*} & H^1(\Sigma,SL(2,\mathbb{C})) & \xrightarrow{\mu^*} & H^1(\Sigma,PL(1,\mathbb{C})) \\ & & \downarrow & & \downarrow \\ & & H^1(\Sigma,\mathcal{SL}(2,\mathcal{O})) & \longrightarrow & H^1(\Sigma,\mathcal{PL}(1,\mathcal{O})) \end{array}$$

If projective flat bundle $\psi\in H^1(\Sigma,PL(1,\mathbb{C}))$ is determined by the monodromy homomorphism related to some projective structure α on Σ then ψ is associated to a unique, up to factor in the finite group $H^1(\Sigma,\mathbb{Z}_2)$, flat $SL(2,\mathbb{C})$ bundle ϕ which divisor class $|div\phi|=g-1$. And conversely, any $SL(2,\mathbb{C})$ flat bundle ϕ with $|div\phi|=g-1$ is related (by μ^*) to the unique monodromy bundle for the developing map of some proojective structure on Σ. However, let us notice that for any flat bundle $\phi\in H^1(\Sigma,SL(2,\mathbb{C}))$ and for each line bundle $\xi\in H^1(\Sigma,\mathbb{Z}_2)$ we have $\mu^*(\phi) = \mu^*(\xi\otimes\phi)$

Two flat bundles are analytically equivalent when they determine the same complex analytic vector bundle. Since different conformal structures on the underlying surface introduce different relations of analytic equivalence on the set of flat vector bundles we obtain different foliations of $6g-6$ dimensional complex manifold $S \subset Hom(\pi_1\Sigma,SL(2,\mathbb{C}))/SL(2,\mathbb{C})$ onto $3g-3$ dimmensional leaves of analytic equivalent flat $SL(2,\mathbb{C})$ bundles. (Manifold S is formed by those representations of $\pi_1\Sigma \rightarrow SL(2,\mathbb{C})$ which do not have scalar commutants.)

All flat projective bundles related to projective structures on Σ are analytically equivalent i.e. determine a single element Ψ in $H^1(\Sigma,\mathcal{PL}(1,\mathbb{C}))$.[9] Similarly, all $SL(2,\mathbb{C})$ bundles with the same divisor (i.e. bundle ξ) of order $g-1$ (i.e. $c_1(\xi)=g-1$) such that $\xi^2=\mathcal{K}$ form one leaf of the manifold S (foliated by the analytic equivalence relation). The explicit form of the holomorphic bundle Φ corresponding to this leaf was found by Gunning and is the following:

$$\bar{\Phi}_{ij}(z_j) = \begin{bmatrix} \xi_{ij}(z_j) & \frac{d}{dz_j}\xi_{ij}(z_j) \\ 0 & \xi^{-1}_{ij}(z_j) \end{bmatrix} \tag{2}$$

So we can say that for each concrete spinor bundle $\xi=(\xi_{ij})\in H^1(\Sigma,\mathcal{O}^*)$ the space of all projective structures on Σ provides all flat representatives of the holomorphic bundle $\Phi\in H^1(\Sigma,\mathcal{SL}(2,\mathcal{O}))$ given by *(2)*.

C. P-LINE AND REDUCTION OF THE HOLOMORPHIC $SL(2,\mathbb{C})$ BUNDLE TO $SU(2)$ BUNDLE.

Let us consider once more a holomorphic fiber space $\pi\colon V_g \longrightarrow \mathcal{T}_g$ whose fibres are compact, genus g, surfaces. Similarly, as for spinor structures, the generalization of the uniformization theorem given by Bers does give relative projective structures on the universal curve V_g. (A relative projective atlas on V_g consists of an open cover $U=\{U_k\}$ of V_g and analytic maps $\alpha_k\colon U_k \longrightarrow \hat{\mathbb{C}}$ where restrictions to any fiber of π are isomorphisms onto their images and form projective atlas on this fiber.) More precisely, the Bers embedding theorem provides a holomorphic family $\tilde{\Pi}\colon F_g \longrightarrow \mathcal{T}_g$ of Jordan domains $\mathcal{D}_\tau = \tilde{\Pi}^{-1}(\tau)$; $\tau\in\mathcal{T}_g$ and discrete groups $\Gamma_\tau \longrightarrow PL(1,\mathbb{C})$ operating on $\mathcal{D}_\tau$ discontinuously to produce surfaces $\pi^{-1}(\tau) = \mathcal{D}_\tau/\Gamma_\tau$. Thus, according to our considerations in section B, we obtain a holomorphic family of projective structures on the family V_g of marked Riemann surfaces.

Now let us fix a spinor structure on V_g which corresponds to an odd theta characteristic. As a matter of fact we will fix such holomorphic line bundle over V_g which corresponds to the "typical" odd spinor structure on Σ_0 that admits only one holomorphic section [10]. Here we consider $\Sigma_0 = U/\Gamma$ as an origin of the Teichmüller space (i.e. $\Sigma_0 = \tau_0$) and projective structure α on Σ determined by $p\colon U \longrightarrow U/\Gamma = \Sigma_0 \cong \tau_0$ as an origin of $\mathcal{P}(\Sigma_0)$. Now, any holomorphic quadratic differential $q \in H^0(\Sigma_0, \mathcal{K}^{\otimes 2})$ defines a projective structure $\beta=\alpha+q$ on Σ_0 as well as a Teichmüller geodesic l_q through Σ_0. Let $\alpha(\tau)$ denote a projective structure on $\pi^{-1}(\tau)$ related to $\mathcal{D}(\tau) \longrightarrow \mathcal{D}(\tau)/\Gamma_\tau = \pi^{-1}(\tau)$, $\tau\in l_q$, and let $\beta(\tau)$ denote a projective structure given as $\beta(\tau) = \alpha(\tau) + q(\tau)$, where $q(\tau)$ is the terminal quadratic differential on $\pi^{-1}(\tau)$ uniquely determined by $q \in H^0(\Sigma_0, \mathcal{K}^{\otimes 2})$ [11].

Since we have fixed a spinor structure on V_g we have concrete spinor bundle $\xi(\tau)$ for all Riemann surfaces $\pi^{-1}(\tau)$, $\tau\in l_q$. So, we can construct the unique, maximally unstable holomorphic $SL(2,\mathbb{C})$ bundles $\Phi(\tau)$ as well as their flat representations $\phi(\alpha(\tau))$ and $\phi(\beta(\tau))$ respectively; $\tau\in l_q$. Moreover, solving the linear form of the Schwarzian equation $\{\beta(\tau),\alpha(\tau)\}=q(\tau)$ we can construct well defined section $\underline{\eta}(\tau)$ of the bundle $\xi^{-1}(\tau)\phi(\beta(\tau))$ for each $\tau\in l_q$. (The fact that $\xi(\tau_0)$ is a "typical" spinor bundle allows us to construct unique section $s(\tau_0)$ of the holomorphic bundle $\Phi(\tau_0)$.) To see how situation looks like for other $\tau\in l_q$, $\tau\neq\tau_0$, we have to consider the action of the diffeomorphisms group $Diff^{+}(\Sigma)$ (or equivalently the modular group $Modq$) on spinor structures.

In a general case the group of diffeomorphisms $Diff^{+}(\Sigma_0)$ preserves the parity of spinor bundles of any Riemann surface but it acts transitively on the subsets of odd and even ones. Moreover it is known [12] that any orientation preserving diffeomorphism of a compact oriented surface leaves fixed some spin structure. It is also known that there exists a subgroup $Modq^{spin}$ of finite index in $Modq$ that fixes all spinor structures [13]. So we see that if l_q contains a marked Riemann surface conformally equivalent to $\Sigma_0=U/\Gamma$ (i.e. related to Σ_0 by some element of the modular group $Modq$) then we could not be able to avoid a situation when we have two or more different, i.e. holomorphically unequivalent SL(2,$\mathbb{C}$) bundles over the same Riemann surface (they all correspond to the same, unique maximally unstable holomorphic projective bundle $\Psi\in H^1(\Sigma_0,\mathcal{PL}(1,\mathcal{O}))$). It suggests that we should require the following property for our P-lines:

Condition P: If two elementl τ and τ' laying on P-line l_q are conformally equivalent than they have to be related at most by element of the $Modq^{spin}$ group.

Which holomorphic quadratic differentials on Σ_0 determine Teichmüller lines with condition P satisfied is an open question. There is also an open question which quadratic differentials determine Teichmüller geodesics whose all "points" are conformally unequivalent. Nevertheless we will assume that our "physical observer" (i.e. time-like vector $e_0\in\mathbb{R}^{1,D}$) determines a pair of measured foliations on Lorentzian worldsheet Σ^L which define Riemann surface structure Σ_0 together with holomorphic quadratic differential q satisfying condition P.

We can construct a reduction of the holomorphic $SL(2,\mathbb{C})$ bundles $\Phi(\tau)$ over $\pi^{-1}(\tau)$: $\tau\in l_q$ to the 2-dimensional subgroup $N \longrightarrow SL(2,\mathbb{C})$ in any case (i.e. when q does or does not satisfy the condition P). This reduction is determined by concrete "typical" odd spinor structure and unique, mentioned above, holomorphic section of $\Phi(\tau)$. However, if we pass to the Riemann modular space $M_g=\mathcal{T}_g/Modq$ then we see that we obtain well defined reductions (to the group N) of the maximally unstable holomorphic $SL(2,\mathbb{C})$ bundles only for P-lines which satisfy condition P.

The group N, the so called spinorality group appears in the Iwasara decomposition of $SL(2,\mathbb{C})$ as its nilpotent subgroup ($SL(2,\mathbb{C})=SU(2)\cdot A\cdot N$) generated by two elements $A_1,A_2\in$ sL(2,$\mathbb{C}$) which have the following properties. Let L_{ij} be the standard generators of $SL(2,\mathbb{C})$ i.e. L_{23},L_{31},L_{12} are infinitesimal operators of rotations and L_{01},L_{02},L_{03} are the so called generators of proper Lorentz transformations. Every element X of the Lie algebra sL(2,$\mathbb{C}$) can be expressed as $X=F^{ik}L$ with F^{ik} as asymmetric tensor.

We can establish a connection between a complex 3-vector $\underline{F}=\underline{B}+i\underline{E}$ and *skev* tensor F^{ik} by $\underline{F} = \left[F^{23} + iF^{01},\ F^{31} + iF^{02},\ F^{12} + iF^{03} \right]$.

Now each generator A_α of N can be written as $A_\alpha = \underline{F}_\alpha$ with $\underline{F}_\alpha \cdot \underline{F}_\beta = \underline{F}_\alpha \cdot \underline{F}_\alpha = 0$, $\alpha,\beta=1,2$ It means that

$$\underline{B}_\alpha^2 - \underline{E}_\alpha^2 = 0 \quad \text{and} \quad \underline{B}_\alpha \cdot \underline{E}_\alpha = 0, \qquad \alpha=1,2 \qquad (3)$$

i.e. that each generator of N is represented by a Cantor-Whittaker *skev* tensor F^{ik}.

Let us recall the 4-dimensional geometry of the Minkowski vector space E. Every vector $\underline{x}$ in E can be described by the hermitian matrix $\underline{x} \cong \hat{x} = x^i \delta_i$, $i=0,\dots,3$, where $\delta_0 = \begin{bmatrix} 1 & 0 \\ 0 & 1 \end{bmatrix}$ and δ_i are the Pauli matrices. Each element $g \in SL(2,\mathbb{C})$ acts on $\underline{x}$ as $\underline{x} \longrightarrow g \circ \underline{x} = g\hat{x}g^{-1}$. The group $SL(2,\mathbb{C})$ acts transitively on the 2-dimensional spinor space $\mathbb{C}^2-\{0\}$ (with N as the stabilizer group) as well as it acts transitively on the homogeneous space of horospheres on the Lobatshewsky space $H \subset E$ of time like unit vectors (with $U(1)N$ as the stabilizer group). We have one-one correspondence between the set of horospheres $\{\omega(n)\}$ on H and the set of orbits $\{e^{i\phi}u\}$ of the $U(1)$ group in $\mathbb{C}^2\{0\}$; $u \in \mathbb{C}^2$ (i.e. each non-zero element $u \in \mathbb{C}^2$ determines some 2-dimensional horosphere $\omega(u)$ of the unit time like vectors). If additionally we will assume that we have some fixed basis in $\mathbb{C}^2$ then each element of N (or equivalently, each element of its Lie algebra) corresponds to some concrete time-like vector (with $SU(2)$ as its stabilizer group) on the horosphere $\omega(u) \subset H$.

Coming back to our holomorphic, maximally unstable $SL(2,\mathbb{C})$ bundles $\Phi(\tau)$ and their flat representatives $\Phi(\alpha(\tau))$, $\Phi(\beta(\tau))$, $\tau \in l_q$, $\beta(\tau)=\alpha(\tau)+q$, we observe that solutions of linearized form at the Schwarzian equations $\{\alpha(\tau),\beta(\tau)\}=q(\tau)$ define not only a section of $\Phi(\beta(\tau))$ but they also define concrete local trivialization of $\Phi(\tau)$. So, according to our considerations above, any section of the reduced N-bundle over $\pi^{-1}(\tau)$, $\tau \in l_q$, corresponds to the concrete reduction of the holomorphic $SL(2,\mathbb{C})$ bundle $\Phi(\tau)$ to the $SU(2)$ group.

D. PHYSICAL INTERPRETATION.

In section II. we have seen that P-lines which are related to Jenkis-Strebel differential (for example "harmonic" P-lines) describe world-sheet which has to be created and which has to decay. In this section we obtain that P-line satisfying P-condition is "associated" to reductions of appropriate holomorphic $SL(2,\mathbb{C})$ bundles(over Riemann surfaces determined by this line) to the $SU(2)$ group. It means (see that beginning of this section) that we have to do with $SU(2)$ bundles over Riemann surfaces equipped with a concrete connection A. If we interpret this connection as a gauge field of weak interaction (which is responsible for a process of decay) then we see that these completely different approaches yield to the same physical situation, namely to decay and creation. Moreover, holomorphic quadratic differentials which satisfy *P*-condition seems to be just Jenkis-Strebel differentials, or at least most of them (it is still open question). So we have to do with the following sheme

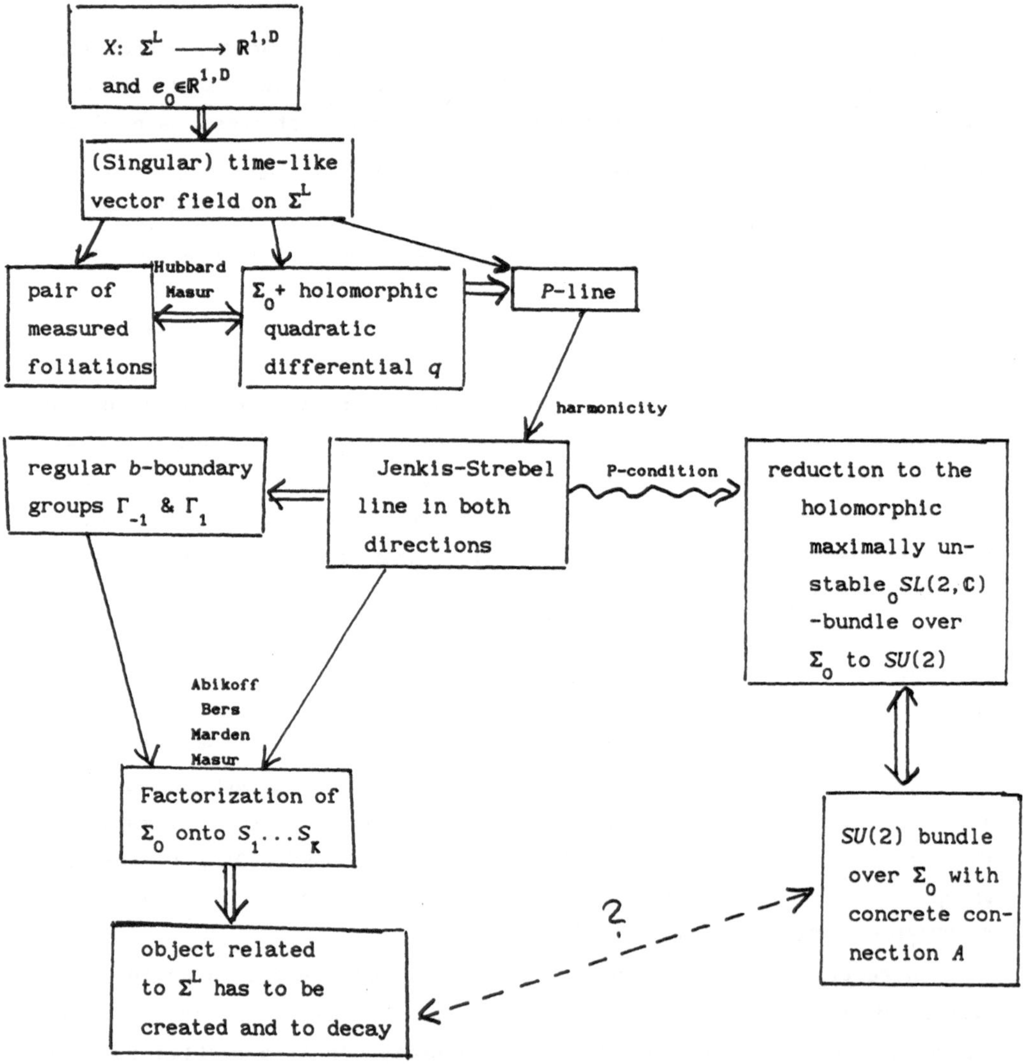

Is A a gauge field of weak interactions? Besides, let us notice that to get our reduction of holomorphic $SL(2,\mathbb{C})$ bundle to $SU(2)$ we have to fix a section, say f, of the N-bundle over Σ_0. So, formulas *(3)* suggest that our relation between A and f could correspond to electro-weak interaction problem.

[1] - K.Bugajska "Strings, observer ant Teichmüller line" (in print)
[2] - E.A.Ruth, J.Vilms, Trans.AMS **149** (1970) 569
T.Takahashi, J.Math.Soc. Japan **18** (1966) 380
Z.Muto, J.Math.Soc. Japan **32** (1980) 531
[3] - K.Bugajska "Harmonic, homothetic Gauss map of a world sheet" (in print)
[4] - L.Bers, Bull.AMS **5** (1981) 131-172
W.Abikoff, Acta Math. **134** (1975) 211-237
A.Marden, Ann.of Math. **99** (1974) 342-383
B.Maskit, Ann.of Math. **117** (1983) 569-668
[5] - K.Bugajska "Jenkis-Strebel rays and decay of elementary particles" (in print)
[6] - H.Masur, Ann.of Math. **102** (1975) 205-221
[7] - R.Bott, M.F.Atiyah, Phil.Trans.R.Soc. London **A308** (1982) 623-665
[8] - C.J.Earle, Ann.of Math. Studies **97** (1981) 87-99
J.Hubbard, Ann.of Math. Studies **97** (1981) 257-275
[9] - R.C.Gunning, Math.Annalen **170** (1967) 87-96
[10] - A.A.Belinson, Y.I.Manin, CMP **107** (1986) 359-376
[11] - L.Bers, Acta Math.**134** (1975) 73-98
[12] - M.F.Atiyah, Ann.Scient.Éc.Norm.Sup., **4** (1971) 47-62
[13] - W.J.Harvey, P.Teofilatto, "Projective and Superconformal Structures on Surfaces" King's College preprint
P.Sipe, Math.Ann. **260** (1982) 67-92

Breaking of Supersymmetry through Anomalies in Composite Spinor Operators

J. A. Dixon
Theory Group, Physics Department
University of Texas, Austin, TX 78712 USA

Rapallo, June, 1990

Abstract

The BRS cohomology of Super-Yang-Mills coupled to chiral matter has non-trivial BRS cohomology, which leads to the conjecture that the local gauge-invariant and supersymmetric composite operators have anomalies in their renormalization. Since these operators are interpolating operators for bound states which are gauge-invariant and which ought naively to fall into supersymmetry multiplets, it follows that one expects supersymmetry to be broken in bound states by anomalies in a way that can be calculated in perturbation theory.

Contents

1 Introduction

The present talk is designed to introduce some simple examples of potentially anomalous operators in supersymmetric Yang-Mills theories with chiral matter. Then various issues connected with these possible anomalies are discussed. The background for this discussion can be found in [1] which in turn is based on [2][3]. Referecnce [1] discusses the BRS cohomology of supersymmetric Yang-Mills theory coupled to chiral matter.

The potential anomalies to be discussed here are made possible by the recently discovered nontriviality of the cohomology of the BRS boundary operator for generalized actions. The generalized actions arise from the addition to the usual action of higher dimension gauge and supersymmetry invariant terms coupled to suitable external sources.

It is necessary to pick a specific gauge group and representation to discuss because the existence of potential anomalies is highly sensitive to these. The present examples are chosen because they are quite simple and yet show the general features that can be expected in other theories.

The examples are chosen from the the SU(2) supersymmetric gauge theory coupled to a chiral multiplet of isospin $\frac{1}{2}$ under the SU(2) group.

The lowest dimension operator which has a potential anomaly in this theory is a composite gauge invariant spinor operator (denoted Ψ^α) of dimension $4\frac{1}{2}$.

This operator is the F component of a (spinor) chiral superfield, so that its integral is also supersymmetry invariant. Integration of the operator is equivalent to contracting it with a constant (space-time independent) spinor source ϕ_α and adding the term $\phi^\alpha \Psi_\alpha$ to the action.

In this paper we also examine the feasibility of constructing anomalies for insertion of operators at non-zero momentum. Two ways to do this are discussed. The first way is a straightforward generalization of the zero-momentum insertion. The second involves the use of a gauge-invariant spinor superfield source. The latter raises a number of issues whose solution is unclear at present.

That the anomalies are not restricted to zero momentum is of some interest if one believes that the existence of anomalies in the renormalization of operators like Ψ^{α} may have some physical meaning.

A possible physical interpretation suggested here goes as follows. In non-Abelian gauge theories, one expects that there will be gauge-invariant spinor bound states. For example, in a supersymmetric version of QCD (assuming that such a theory gives confinement), the proton would be a bound state formed (mainly) from 3 quarks bound by non-Abelian vector gluons. The reason for examining a supersymmetric version of QCD is that it appears that supersymmetry (as it appears in the superstring) may form an essential part of a complete theory.

Unless something breaks supersymmetry, the bound states should not only be gauge-invariant (the usual result in QCD) but should also fall into supersymmetry multiplets. The particles in these multiplets should all have the same mass, as is usual for supersymmetric multiplets.

Now what is the relation between gauge-invariant bound states and local gauge-invariant composite operators (like Ψ_{α}) that have the same quantum numbers?

The answer to this question is of course that the operator may be an interpolating field for the bound state. This means that if $|p,s\rangle$ is a momentum eigenstate of the bound state spinor (e.g.proton) with spin s, then one would expect the matrix element:

$$\langle 0|\Psi_{\alpha}(x)|p,s\rangle = U_{\alpha}(p,s)e^{ip\cdot x} \tag{1}$$

to be nonzero and proportional (in some sense) to the admixture of the fields in Ψ_{α} in the bound state. If Ψ_{α} has the wrong quantum numbers then this matrix element should be zero. If Ψ_{α} is present in the bound state to a very small extent, then the matrix element should be small, etc. Thus, for example, the matrix element for the proton of a 5-quark Ψ_{α} should be 'smaller' than the matrix element of a 3-quark Ψ_{α} for normal momenta (those at which the usual constituent quark model gives reasonable results).

The computation of these matrix elements requires one to know the non-perturbative solution of the relevant Bethe-Salpeter equation for the bound state. This we do not know.

However the present results predict a rather independent phenomenon. They predict that these bound states will violate the BRS identity in a specific way that can be calculated in perturbation theory (assuming some value for the matrix elements above). This in turn seems to mean that supersymmetry is broken in some way by the anomaly in the bound states. How to make this into a calculable effect is a question for the future. Presumable something along the lines of PCAC might be a reasonable way to proceed.

The prediction that there can be an anomaly in the renormalization of Ψ^a comes from an examination of the BRS cohomology of the theory, which has been done elsewhere. Here we merely summarize the relevant part of those results.

As will be evident in the following, the incompleteness of the present knowledge of the BRS cohomology of supersymmetric theories prevents a satisfactory formulation of the problem to be analyzed. However the fact that it is known to be non-trivial at least for chiral fields shows that the problem is an important one.

2 Action in Components

In the present notation, the action for the super Yang-Mills theory coupled in a gauge-invariant way to chiral supersymmetric matter is:

$$S = \int d^4x \{ [-\frac{1}{4} G^a_{\mu\nu} G^{a\mu\nu} - \frac{1}{2} \lambda^{a\alpha} \sigma^{\mu}_{\alpha\dot\beta} D^{ab}_{\mu} \overline{\lambda}^{b\dot\beta} + \frac{1}{2} D^a D^a]$$

$$-\frac{1}{2} [D^{ij}_{\mu} A^j \overline{D}^{\mu ij} \overline{A}^j + \psi^{\alpha i} \sigma^{\mu}_{\alpha\dot\beta} \overline{D}^{\mu ij} \overline{\psi}^{j\dot\beta} - F^i \overline{F}^i]$$

$$+ T^{aij} D^a A^i \overline{A}^j + T^{aij} A^i \overline{\lambda}^{a\dot\alpha} \overline{\psi}^j_{\dot\alpha} + T^{aij} \overline{A}^i \lambda^{a\alpha} \psi^j_{\alpha} \tag{2}$$

This action is invariant under the following supersymmetry transformations:

$$\delta A^a_\mu = D^{ab}_\mu \omega^b + \frac{1}{2} c^\alpha \sigma^\mu_{\alpha\dot\beta} \overline{\lambda}^{a\dot\beta} + \frac{1}{2} \lambda^{a\alpha} \sigma^\mu_{\alpha\dot\beta} \overline{c}^{\dot\beta} + \epsilon^\nu \partial_\nu A^a_\mu \tag{3}$$

$$\delta \lambda^a_\alpha = +\frac{1}{2} G^a_{\mu\nu} \sigma^{\mu\nu}_{\alpha\beta} c^\beta - f^{abc} \lambda^b_\alpha \omega^c + i D^a c_\alpha + \epsilon^\nu \partial_\nu \lambda^a_\alpha \tag{4}$$

$$\delta \overline{\lambda}^a_{\dot\alpha} = +\frac{1}{2} G^a_{\mu\nu} \overline{\sigma}^{\mu\nu}_{\dot\alpha\dot\beta} \overline{c}^{\dot\beta} - f^{abc} \overline{\lambda}^b_{\dot\alpha} \omega^c - i D^a \overline{c}_{\dot\alpha} + \epsilon^\nu \partial_\nu \overline{\lambda}^a_{\dot\alpha} \tag{5}$$

$$\delta D^a = \frac{-i}{2} c^\alpha \sigma^\mu_{\alpha\dot\beta} D^{ab}_\mu \overline{\lambda}^{b\dot\beta} + \frac{i}{2} D^{ab}_\mu \lambda^{b\alpha} \sigma^\mu_{\alpha\dot\beta} \overline{c}^{\dot\beta} + f^{abc} D^b \omega^c + \epsilon^\nu \partial_\nu D^a \tag{6}$$

$$\delta \omega^a = -\frac{1}{2} f^{abc} \omega^b \omega^c + c^\alpha \sigma^\mu_{\alpha\dot\beta} \overline{c}^{\dot\beta} A^a_\mu + \epsilon^\nu \partial_\nu \omega^a \tag{7}$$

$$\delta\xi^a = Z^a + \epsilon^\mu\partial_\mu\xi^a \tag{8}$$

$$\delta Z^a = c^\alpha\sigma^\mu_{\alpha\dot\beta}\overline{c}^{\dot\beta}\partial_\mu\xi^a + \epsilon^\mu\partial_\mu Z^a \tag{9}$$

$$\delta A^i = c^\alpha\psi^i_\alpha + i\omega^a T^{aij}A^j + \epsilon^\mu\partial_\mu A^i \tag{10}$$

$$\delta\overline{A}^i = \overline{c}^{\dot\alpha}\overline{\psi}^i_{\dot\alpha} - i\omega^a T^{aij}\overline{A}^j + \epsilon^\mu\partial_\mu\overline{A}^i \tag{11}$$

$$\delta\psi^i_\alpha = D^{ij}_\mu A^j\sigma^\mu_{\alpha\dot\beta}\overline{c}^{\dot\beta} + F^i c_\alpha + i\omega^a T^{aij}\psi^j_\alpha + \epsilon^\mu\partial_\mu\psi^i_\alpha \tag{12}$$

$$\delta\overline{\psi}^i_{\dot\alpha} = D^{ij}_\mu\overline{A}^j\overline{\sigma}^\mu_{\dot\alpha\beta}c^\beta + \overline{F}^i\overline{c}_{\dot\alpha} - i\omega^a T^{aij}\overline{\psi}^j_{\dot\alpha} + \epsilon^\mu\partial_\mu\overline{\psi}^j_{\dot\alpha} \tag{13}$$

$$\delta F^i = D^{ij}_\mu\psi^{\alpha j}\sigma^\mu_{\alpha\dot\beta}\overline{c}^{\dot\beta} + \overline{c}^{\dot\alpha}\overline{\lambda}^a_{\dot\alpha}T^{aij}A^j + i\omega^a T^{aij}F^j + \epsilon^\mu\partial_\mu F^i \tag{14}$$

$$\delta\overline{F}^i = D^{ij}_\mu\overline{\psi}^{j\dot\alpha}\overline{\sigma}^\mu_{\dot\alpha\beta}c^\beta + c^\alpha\lambda_{\alpha a}T^{aij}\overline{A}^j - i\omega^a T^{aij}\overline{F}^j + \epsilon^\mu\partial_\mu\overline{F}^i\Big\} \tag{15}$$

If we add the variations:

$$\delta\epsilon^\mu = -c^\alpha\sigma^\mu_{\alpha\dot\beta}\overline{c}^{\dot\beta} = -c\cdot\sigma^\mu\cdot\overline{c} \tag{16}$$

$$\delta c^\alpha = 0, \tag{17}$$

$$\delta\overline{c}^{\dot\alpha} = 0, \tag{18}$$

then the variations acting on any field (counting ϵ_μ and c^α as constant fields) satisfy the relation:

$$\delta^2 = 0. \tag{19}$$

Note that

$$(c^\alpha\sigma^\mu_{\alpha\dot\beta}\overline{c}^{\dot\beta})^* = c^\beta\sigma^\mu_{\beta\dot\alpha}\overline{c}^{\dot\alpha} \tag{20}$$

is a real quantity.

It is worth emphasizing that even though c is a constant spinor, it is a mistake to think that the supersymmetry here is a 'global symmetry' if that is taken to mean that there can be no nontrivial cohomology or anomalies for this symmetry. The reason is that this constant parameter c sometimes accompanies transformations which convert a field into the derivative of another field–which is the sort of thing that no self-respecting 'global' transformation ought to do. For example λ is converted by δ to $\gamma^{\mu\nu}\partial_\mu A^a_\nu c$ (plus other things). This is just the sort of thing that gives rise to the well-known chiral anomalies–there we have A^a_μ converted by δ to $\partial_\mu\omega^a$.

3 BRS Identity and Cohomology

In [1] and [3] we derived the result that polynomials of the form:

$$\mathcal{I}_\alpha = \int d^4x\, \mathcal{A}_\alpha = \int d^4x d^2\overline{\theta}\Big\{c_\alpha[\overline{S}^{i_1}\overline{S}^{i_2}\cdots\overline{S}^{i_n}T^{i_1 i_2\cdots i_n}]\Big\}$$

$$= \int d^4x\Big\{c_\alpha[\overline{S}^{i_1}\overline{S}^{i_2}\cdots\overline{S}^{i_n}T^{i_1 i_2\cdots i_n}]_{\overline{F}}\Big\} \tag{21}$$

and their complex conjugates are in the cohomology space. Here the last formula means 'take the $\overline{F}$ component', and $T^{i_1 i_2\cdots i_n}$ is an invariant tensor under the gauge group, and S is a chiral superfield.

4 Insertion at Zero Momentum

To begin with we choose a simple example and try to find an expression in $\mathcal{H}$ with ghost charge one that matches an expression in $\mathcal{H}$ with ghost charge zero. They must have the same dimension to begin with, and this is already a tricky task in general. A simple theory is the SU(2) supersymmetric theory coupled to chiral matter in an isospin $\frac{1}{2}$ representation of the gauge group.

The lowest dimension match for this example appears to be available for dimension $4\frac{1}{2}$. The operator is:

$$\Psi_\beta = \left[S^i(\sigma^2\sigma^a)_{ij}S^j W^a_\beta\right]_F \tag{22}$$

where $T^a_{ij} = \sigma^a_{ij}$ are the Pauli matrices for this case.

Corresponding to this operator one has

$$\mathcal{A}_\alpha = [(\overline{S}^i(\sigma^2\sigma^a)_{ij}\overline{S}^j)(\overline{S}^k(\sigma^2\sigma^a)_{kl}\overline{S}^l)]_{\overline{F}}c_\alpha \tag{23}$$

which has the correct dimension and quantum numbers to appear as an anomaly in the renormalization of Ψ_α.

If we follow the lead in [1], then we would just couple the term

$$\int d^4x\ \{\phi^\alpha\Psi_\alpha\} \tag{24}$$

to the action, where ϕ^α is a constant (space-time independent) source term. Computation of the loop diagrams for this insertion could then lead to an anomaly of the form:

$$\delta\Gamma = \text{constant} \int d^4x \, \{\phi^\alpha A_\alpha\} \tag{25}$$

in the one particle-irreducible generating functional Γ.

5 Physical Interpretation

Suppose that one carries out the above calculation and finds that there is indeed an anomaly with some non-zero coefficient. How is this related to physics?

The operator discussed above, Ψ_α is 'F' component of a chiral superfield. If we try to take seriously the concept that any particle for which it is an interpolating field is in a supermultiplet, we will need to address the question 'what supermultiplet'.

The easiest answer to this question would be to suppose that the relevant supermultiplet contains the same fields as the chiral superfield that contains the operator Ψ_α. However, this is not an irreducible superfield. It has a similar structure to the field W_α that is encountered in an Abelian supersymmetric gauge theory. As is well known, one can impose a 'reality' constraint on a chiral superfield with an extra index, of the form:

$$D_\alpha W^\alpha = \overline{D}_{\dot\alpha} \overline{W}^{\dot\alpha} \tag{26}$$

whose solution yields relations and constraints among the otherwise independent fields in the superfield.

Somehow one has to disentangle the irreducible parts of the superfield and reconcile that with what one would expect for the supermultiplet of bound states with the appropriate quantum numbers.

It appears that this would also mix in the BRS cohomology of superfields that possess extra indices. More about this question below. But at any rate, it appears that the theory ought to naively predict degeneracy, and then that degeneracy ought to be broken by anomalies.

6 Insertion at non-Zero Momentum

Since we are interested in the possibility of interpretating the new anomalies (assuming their coefficients are non-zero) as anomalies in bound states, one of the first problems to confront is to see how to make composite operators that can be inserted at non-zero momentum while maintaining a suitable BRS identity. The reason is that in general bound states have (of course) non-zero momenta.

There are several ways to do this. The first one we shall consider is to add the following terms to the action:

$$\int d^4x\{\phi^\alpha \Psi_\alpha + f^{\alpha\mu} Y_{\alpha\mu} + \phi^{\alpha\mu\nu}\Psi_{\alpha\mu\nu} + f^{\alpha\mu\nu\sigma} Y_{\alpha\mu\nu\sigma} + \phi^{\alpha\mu\nu\sigma\lambda}\Psi_{\alpha\mu\nu\sigma\lambda}\} \tag{27}$$

We will not need the explicit form of the operators $Y_{\alpha\mu}, \Psi_{\alpha\mu\nu}, Y_{\alpha\mu\nu\lambda}, \Psi_{\alpha\mu\nu\lambda\sigma}$, but their construction is described below.

The point here is that we know that Ψ_α is invariant under the action of δ only up to the total derivative of some other local operator, since it is the F component of a superfield (It might be the D component in some cases):

$$\delta\Psi_\alpha = \partial_\mu Y^\mu_\alpha \tag{28}$$

Then since $\delta^2 = 0$, it follows that

$$\partial^\mu \delta\Psi_{\mu\alpha} = o \tag{29}$$

from which one can deduce that there exists a $\Psi_{\alpha\mu\nu}$ satisfying the relation:

$$\delta Y_{\alpha\mu} = \partial^\nu \Psi_{\alpha\mu\nu} \tag{30}$$

with

$$\Psi_{\alpha\mu\nu} = -\Psi_{\alpha\nu\mu} \tag{31}$$

In this way one builds up the series of operators. That there exist local operators satisfying these equations is a consequence of the triviality of the local cohomology of the operator $\delta = \epsilon^\mu \partial_\mu$ which has been demonstrated in [2] among other places.

Note that each operator has ghost charge equal to the number of free Lorentz indices on it, and statistics indicated by whether it is a greek or latin letter. Correspondingly the sources for these operators have ghost charge equal to minus the number of free Lorentz indices on them, and statistics indicated by greek or latin letters.

The action will then be invariant if we add the following transformations for the sources:

$$\delta\phi_\alpha = 0 \tag{32}$$

$$\delta f_{\alpha\mu} = \partial_\mu \phi_\alpha \tag{33}$$

$$\delta \phi_{\alpha\mu\nu} = \partial_\mu f_{\alpha\nu} - \partial_\nu f_{\alpha\mu} \tag{34}$$

$$\delta f_{\alpha\mu\nu\lambda} = \partial_\mu \phi_{\alpha\nu\lambda} + \partial_\nu \phi_{\alpha\lambda\mu} + \partial_\lambda \phi_{\alpha\mu\nu} \tag{35}$$

$$\delta \phi_{\alpha\mu\nu\lambda\sigma} = \partial_\mu f_{\alpha\nu\lambda\sigma} + \partial_\nu f_{\alpha\mu\lambda\sigma} + \partial_\lambda f_{\alpha\sigma\mu\nu} + \partial_\sigma f_{\alpha\mu\nu\lambda} \tag{36}$$

Now we have a new cohomology problem to solve, but it is easy to do. The result is that the cohomology space is isomorphic to the space of expressions of the form

$$\mathcal{I} = \int d^4x\, \mathcal{A} = \int d^4x d^2\overline{\theta} \{ \phi^\alpha c_\alpha [\overline{S}^{i_1} \overline{S}^{i_2} \cdots \overline{S}^{i_n} T^{i_1 i_2 \cdots i_n}] \} \tag{37}$$

and hence the result is very similar to that in the case where we inserted the operator at zero momentum. The isomorphism referred to means that the actual invariant $\mathcal{I}$ contains also terms involving the other sources in the case where ϕ_α is taken to depend on the spacetime argument. We shall not need the form of these terms here.

7 Supersymmetric Source

The spacetime-independent source ϕ used above is a superfield which is invariant under the superspace transformation. Hence it seems reasonable to extend it to a full chiral superfield of the form:

$$\Phi_\alpha(x) = \phi_\alpha(y) + \theta_\alpha K(y) + (\sigma^{\mu\nu})_{\alpha\beta}\theta^\beta H_{\mu\nu}(y) + \theta^2 \xi_\alpha(x) \tag{38}$$

where

$$y^\mu = x^\mu + \theta^\alpha \sigma^\mu_{\alpha\dot\beta} \overline{\theta}^{\dot\beta} \tag{39}$$

satisfies

$$\overline{D}_{\dot\alpha} y^\mu = 0 \tag{40}$$

Now the composite superfield Ξ of which Ψ is the 'F' component has a form similar to the above form of Φ. Let us write it in the form:

$$\Xi_\alpha(x) = \chi_\alpha(y) + \theta_\alpha M(y) + (\sigma^{\mu\nu})_{\alpha\beta}\theta^\beta N_{\mu\nu}(y) + \theta^2 \Psi_\alpha(x) \tag{41}$$

where each of the fields $\chi, M, N_{\mu\nu}$ and Ψ are composite fields made from the fields in the action and their derivatives.

Both Φ and Ξ are similar to the 'gauge spinor' superfield W_α that contains the gauge fields λ, D and A_μ in an Abelian supersymmetric gauge theory. However the composite superfield Ξ_α does not appear likely to satisfy any identities of the form

$$D^\alpha W_\alpha = \overline{D}^{\dot{\alpha}} \overline{W}_{\dot{\alpha}} \tag{42}$$

whereas these equations are satisfied by the Abelian gauge spinor superfield.

Consequently it would not be reasonable to impose these conditions on the source field Φ_α either.

The introduction of this unconstrained chiral spinor superfield raises the necessity of solving its BRS cohomology. We already know something about this problem. Namely, we know (or have assumed) that:

$$\phi_\alpha c^\alpha \tag{43}$$

is in the BRS cohomology space of this theory in the limit where the superfield source is taken as a constant. This appears to be correct because the supersymmetry transformations of ϕ is zero. It would seem likely that the result extends to non-zero momentum, but the details look complicated.

In addition to the problem of irreducibility under supersymmetry, there are also problems of mixing and of new (supersymmetry) gauge invariances that will come into this cohomology problem. It is also not clear to me at present how to relate the composite operators to bound states with the same quantum numbers. There is also the problem of determining whether bound states actually form in unbroken supersymmetric theories–for example what happens with respect to chiral symmetry breaking. There is a vast literature on these questions, but it does not seem easy to relate the literature to the present problem. The easiest thing to do (and it is not very easy) is to try to find the cohomology of the extended problem posed above.

8 Conclusion

Evidently, one could evaluate the relevant Feynman diagrams in these various theories to determine whether the potential anomalies do in fact occur. However it seems a bit early to do so without knowing whether one has understood the general structure and the significance of the calculations, particularly since the calculations look rather difficult.

Can one go from this vague formulation to a specific calculation of a physically interesting effect in a given theory? For example, do the anomalies break the usual supersymmetric mass degeneracy between (bound state) bosons and fermions in a calculable way? How does it relate to spontaneous symmetry breaking through vacuum expectation values of scalar fields?

The formulation that seems most likely to lead to some understanding of these questions is the one where the spinor superfield was introduced. In some sense this source could be related to a supermultiplet of bound states in the theory. The existence of the anomaly (assuming its coefficient is not zero) presumably leads to problems in the calculation of the functional $\Gamma(\Phi)$ where Φ is the superfield source.

Could this kind of calculation yield a sensible way to split the masses of the bound states using perturbation theory? Or is it impossible to disentangle the unitarity violation (and non-localities?) introduced by the anomalies?

Alternatively, an important problem would be to resolve the question of the extension to non-zero momentum in a supersymmetric way. If this can be done, then it would signal the presence of Lorentz-invariant supersymmetry anomalies, which would have implications for the corresponding theories of a more familiar kind.

In conclusion, the examination of the BRS cohomology of simple supersymmetric theories leads to many questions. Perhaps these may in turn lead to an explanation of why supersymmetry is not observed even though it may be present in a hidden way in the world around us.

Aknowledgments: I thank B. Campbell, G. Leibbrandt, O. Piguet, K. Stelle, R. Stora, J.C. Taylor and P.C. West for their interest and insights.

References

[1] Dixon, J. A.: Class. Quant. Grav. 7 (1990) 1511.

[2] Ibid. BRS Cohomology of Yang Mills theory. UVic Preprint, July 1989 to be published in Comm. Math. Phys.

[3] Ibid. Local BRS Cohomology of the Supersymmetric Chiral Multiplet. UVic Preprint, Oct. 1989.

Conformal field theory and moduli spaces of vector bundles over variable Riemann surfaces

Gregorio Falqui and Cesare Reina
SISSA, Strada Costiera 11 - TRIESTE (Italy)

Abstract. We give a geometric description of some representations of the semidirect sum of the Virasoro and Kac-Moody algebras in terms of line bundles over the moduli stacks of stable vector bundles over smooth Riemann surfaces.

Introduction.

Representation theory of infinite dimensional symmetry groups entering field theories (such as the gauge group or the group of diffeomorphisms) is expected to play a central role in the study of physical systems. Unfortunately little in known in general, with the exception of (spatial) dimension zero (i.e. mechanics) and of Conformal Field Theories, where Kac–Moody and Virasoro algebras and their representations have been the basic mathematical tool for physical applications (see, e.g. [6]).

Some of these have been recently given a geometric description (see [2] and references quoted therein). Indeed a subspace of the "classical" Virasoro algebra (i.e. of the Virasoro algebra without central extension) can be realized in terms of
a) vector fields over the moduli spaces $\hat{\mathcal{M}}_g$ of triples (C, p, z) where C is a Riemann surface of genus g, p a point on it and z a local coordinate vanishing at p,
b) vector fields over a suitably defined infinite dimensional Grassmannian manifold.
The set up b) is intimately related to the representation theory of the Virasoro algebra, while a) is directly related to algebraic geometry. The interplay between these two realizations is the reason why well known results in representation theory can be given a geometric explanation. For instance, the Mumford formula gives the central extension of the Virasoro algebra for b-c systems of spin j as the multiple $(6j^2 - 6j + 1)\lambda$ of the Hodge class λ on the moduli space of curves $\mathcal{M}_g$. The basic fact here is that the determinant index bundle $\det \bar{\partial}_j$ is non trivial over $\mathcal{M}_g$ and forces a central extension of the Virasoro algebra.

In this paper we will extend the geometrical approach to the study of some representations of $\mathfrak{sl}(r, \mathbf{C})$ Kac–Moody and Virasoro algebras. The first step is to notice that at the classical level, i.e. setting to zero the central extensions, suitable closures of these algebras have simple realizations in terms of Čech cocycles on Riemann surfaces. Actually, let C be a smooth algebraic curve (i.e. a compact Riemann surface), K^{-1} and $\mathrm{End}_0 E$ be the tangent sheaf of C and the sheaf of traceless endomorphisms of a holomorphic vector bundle $E \to C$. Fixing a point $p \in C$ we have a Stein covering $\{U_i\}$, $(i = 0, 1)$ of C with $p \in U_0$ a disk around p and $U_1 = C \setminus p$. If we choose a local coordinate z on U_0 with $z(p) = 0$ and a local trivialization ϕ for E, the group of one-cocycles with values in $K^{-1} \oplus \mathrm{End}_0 E$ can be represented on this covering as

$$\mathcal{C}^1(K^{-1} \oplus \mathrm{End}_0 E) = \mathbf{C}\{l_n, \tau^a_m\}, \; m, n \in \mathbf{Z}$$

with

$$l_n = -z^{n+1}\frac{d}{dz},\ \tau^a_m = \tau^a z^m$$

where τ^a, $(a = 1, ..., r^2-1)$ is a basis in the Lie algebra $\mathfrak{sl}(r, \mathbb{C})$ and $\mathbb{C}\{...\}$ denotes linear combinations of the elements in parentheses converging in $U_0 \setminus p$. The commutation relations one naturally gets in this way coincide with those of the semidirect sum of the Kac-Moody and Virasoro algebras:

$$[l_m, l_n] = (m-n)l_{m+n} + \frac{c_1}{12}m(m^2-1)\delta_{m,-n}$$

$$[\tau^a_m, \tau^b_n] = f^{ab}_c \tau^c_{m+n} + c_2 m\delta^{ab}\delta_{m,-n}$$

$$[l_m, \tau^a_n] = -n\tau^a_{m+n}$$

$$[l_n, c_i] = [\tau^a_n, c_i] = 0,\ (i = 1, 2)$$

with all the central terms $c_i = 0$, i.e. of the algebra $\mathcal{D} =: \mathrm{diff}^{\mathbb{C}}\, S^1 \ltimes L\mathfrak{sl}(r, \mathbb{C})$.

The basic hint coming from several recent papers is that both the central extensions of this algebra and its finite dimensional representations (i.e. the conformal blocks of rational Conformal Field Theories) can be understood via geometric quantization on certain (extensions of) algebraic varieties [4]; namely the moduli spaces of suitable algebraic objects over variable curves. The rationale for this is that if $\hat{V}$ is the space of isomorphism classes of quintuples $x = (C, p, z, E, \phi)$ the map

$$\hat{\phi}_x : \mathcal{D} \to T_x\hat{V} \simeq \mathcal{C}^1(C, K^{-1} \oplus \mathrm{End}_0 E)/\Gamma(U_1, K^{-1} \oplus \mathrm{End}_0 E)$$

given by the isomorphism above modulo $\Gamma(U_1, K^{-1} \oplus \mathrm{End}_0 E)$ is surjective and holomorphically depending on x. Of course restrictions on E should be required to get sensible moduli spaces.

At this stage we have simply realized the generators of $\mathcal{D}$ as holomorphic vector fields (among which "almost a half" identically vanishing) over an infinite dimensional "complex manifold" (actually a stack). There is a standard way [9] to geometrically quantize them by means a suitable infinite Grassmannian $Gr(H^r)$. We have indeed a (non–abelian) Krichever map $\mathcal{K} : \hat{V} \to Gr(H^r)$ which pulls back the determinant dual bundle Det* over $Gr(H^r)$ to a line bundle L over $\hat{V}$, together with the pull back $\hat{\nabla}$ of the universal connection ∇ on Det*. Now l_n and τ^a_n act on the space of holomorphic sections of L via covariant derivatives, yielding a projective representation since

$$\nabla_{[X,Y]} = [\nabla_X, \nabla_Y] + \hat{\omega}(X, Y)$$

where $\hat{\omega}$ is the pullback of the Kähler form ω on $Gr(H^r)$. We get in this way a concrete representation with $c_2 = 1$, $c_1 = r$. In particular the l_n in this representation correspond to the modes of the energy momentum tensor for a multiplet of scalars coupled to E. To get a level h representation, we start with the direct sum $E' = \oplus_1^h E$ getting $L' = L^h$ and central extensions $c_2 = h$, $c_1 = hr$.

The same can be done with fermions coupled to E. We need in this case the moduli space $\hat{F}$ of sextuples (C, p, z, ν, E, e) with ν a θ-characteristics on C and work with the Krichever map associated to $E' = \oplus_1^h E \otimes \nu$. This gives $c_2 = h$ and $c_1 = hr/2$.

Other representations clearly exist and can be in principle described in the same way, since one expects that $H^2(\hat{V}, \mathbf{C}) = \mathbf{C} \oplus \mathbf{C}$. Among these there is that arising from the Sugawara construction with $c_2 = h$, $c_1 = hN/(h+c)$ where $N = dimG$ and c is the dual Coxeter number. The corresponding l_n are the modes for the energy momentum tensor of a WZW σ-model with values in G. Although we do not have a direct way of understanding this representation whenever the WZW model cannot be fermionized, there is evidence that the Sugawara construction for $G = SU(2)$ is indeed related to geometrical data on moduli spaces. Some hints in this direction come from the differential geometrical description by Hitchin [7]. The central datum here is a vector bundle over Teichmüller space on which the Virasoro algebra acts via a projectively flat connection; an evidence of the existence of a non trivial line bundle over moduli spaces.

1. Deformations of vector bundles on curves.

Let E be a stable rank r vector bundle of degree d over a curve C. In the following the genus $g > 1$ of C will be fixed and left implicit. To keep things as simple as possible, i.e. to get rid of the abelian piece of the algebra $\operatorname{End} E$, we want to deform E with the constraint that the determinant bundle $\det E$ is kept fixed, whenever C is fixed. Of course, this has no meaning over a moving curve. However, over a pointed curve (C, p) we have a distinguished degree d line bundle; namely $\mathcal{O}(d \cdot p)$, and we will ask that $\det E = \mathcal{O}(d \cdot p)$.

As usual in deformation theory, one starts with infinitesimal deformations. The first step is to identify the sheaf of the infinitesimal automorphisms of the structure one wants to deform. An automorphism of a vector bundle $E \to C$ is a fibrewise linear biholomorphic map $\tilde{\mu}$ and a biholomorphic map μ such that the diagram

$$\begin{array}{ccc} E & \xrightarrow{\tilde{\mu}} & E \\ \downarrow & & \downarrow \\ C & \xrightarrow{\mu} & C \end{array}$$

is commutative. An automorphism of the form $(\tilde{\mu}, id)$ is called vertical. The sheaf $\operatorname{Aut}_v E$ of germs of vertical automorphisms of E is a subsheaf of $\operatorname{Aut} E$, i.e. we have an exact sequence

$$1 \to \operatorname{Aut}_v E \to \operatorname{Aut} E \to \operatorname{Aut} E / \operatorname{Aut}_v E \to 1$$

where the last sheaf is isomorphic to the sheaf of germs of biholomorphic maps $\mu : C \to C$. The infinitesimal version of the sequence above reads

$$0 \to \operatorname{End} E \to \Sigma_E \to K^{-1} \to 0$$

where Σ_E is the sheaf of germs of differential operators of order less or equal to one on E of the form $a(z)\frac{d}{dz} + \tau(z)$, with $a(z)\frac{d}{dz}$ is a local holomorphic vector field and $\tau(z)$ is a local endomorphism of E. Being an extension of the tangent sheaf K^{-1} by $\operatorname{End} E$, the sheaf Σ_E corresponds to a class in $H^1(C, K \otimes \operatorname{End} E) \simeq H^0(C, (\operatorname{End} E)^*)$. Now stability implies that $H^0(C, (\operatorname{End} E)^*) \simeq H^0(C, E^* \otimes E) \simeq \mathbf{C}$, i. e. global homomorphisms are of the form $\zeta\, id_E$, with $\zeta \in \mathbf{C}$. We shall be mainly interested in the case of traceless endomorphisms $\mathrm{End}_0 E$ fitting the sequence

$$0 \to \mathrm{End}_0 E \to \mathrm{End}\, E \stackrel{tr}{\to} \mathcal{O} \to 0$$

for which the extension

$$0 \to \mathrm{End}_0 E \to \Sigma^0_E \to K^{-1} \to 0$$

is trivial for E stable because $H^0(C, \mathrm{End}_0 E) = 0$. This corresponds to deforming the projective bundle $P(E)$ associated to E. One can actually show that on a curve C any bundle $P \to C$ in projective spaces is of the form $P = P(E)$ for some vector bundle E. Obviously, $P(E) = P(E')$ iff $E = E' \otimes L$ for some line bundle L. Fixing the determinant to be $\mathcal{O}(d \cdot p)$ implies that E can be recovered from its projectivization up to a point L of order $r = rankE$ in the Jacobian of C. Solving this ambiguity is the same as considering moduli spaces of curves with a level r structure. Although this may be relevant in connections with parafermionic theories, we will not dwell on this subtelty any further.

We want to study the set V of isomorphism classes of (projective) stable vector bundles of rank r and degree d over C of genus $g > 1$ with r, d coprime .

Our first goal is to construct a stack structure over V, by covering it with deformations i.e. for each point $[E \to C] \in V$ we want to construct a fibered manifold $\mathcal{E} \stackrel{\pi}{\to} B$, with B a ball in $\mathbf{C}^n$, together with an isomorphism between E and $\pi^{-1}(0)$, such that the Kodaira-Spencer map $KS : T_0 B \to H^1(C, \Sigma^0_E)$ is an isomorphism. Clearly $n = (r^2-1)(g-1) + 3g - 3$.

Lemma. Let $E \to C$ be a stable vector bundle of rank r over a curve C of genus g. For a generic point $p \in C$ there are isomorphisms

$$H^1(C, \mathrm{End}_0 E) \simeq H^0(C, \mathrm{End}_0 E((g-1)p)/\mathrm{End}_0 E)$$

and

$$H^1(C, K^{-1}) \simeq H^0(C, K^{-1}((3g-3)p)/K^{-1}).$$

Proof. This is an easy generalization of the proof of the second isomorphism which can be found e.g. in [5].

Summing up, we see that there is a subspace of the algebra $\mathcal{D}$ which parameterizes infinitesimal deformations of the couple (E, C). We can easily construct finite deformations out of them. We think of E as the disjoint union of $(U_0 \times \mathbf{C}^r) \coprod (U_1 \times \mathbf{C}^r)$ identifying (q_0, ξ_0) with (q_1, ξ_1) whenever $q_0 = q_1 \in C$ and $\xi_0 = g_{01}(z)\xi_1$, where z is a local coordinate on U_0 and $g_{01}(z)$ is a $Gl(r, \mathbf{C})$-valued function. Given $\Delta \times \mathbf{C}^r$ (Δ a small disk in $\mathbf{C}$) with coordinates (x, η), we glue it to $U_0 \times \mathbf{C}^r$ via the map

$$x = z + \sum_{k=1}^{3g-3} t_k z^{-k}$$

$$\eta = (1 + \sum_{l=1}^{g-1} s_{al} z^{-l} \tau^a)\xi_0.$$

This gives rise to a family of vector bundles $\mathcal{E}\xrightarrow{\pi'}B$ with B a ball in $\mathbf{C}^{3g-3+(r^2-1)(g-1)}$ such that $\pi'^{-1}(0)=E$. Taking derivatives of the transition functions of $\mathcal{E}$ at the origin we get

$$\frac{\partial}{\partial t_k} \rightsquigarrow z^{-k}\frac{\partial}{\partial z} \qquad k=1,...,3g-3$$

$$\frac{\partial}{\partial s_{ak}} \rightsquigarrow z^{-k}\tau^a \qquad l=1,...,(r^2-1)(g-1)$$

which shows that the deformation is universal by the lemma above.

In this way we get a stack V parameterizing isomorphism classes of couples (E,C), with E stable. We have clearly a projection $\pi : V \to \mathcal{M}$ over the moduli stack of curves, such that the preimage $\pi^{-1}(t)=V_t$ is the moduli stack of stable bundles over the fixed curve C_t. Both these stacks come together the "universal objects" they are parameterizing, i.e. we have a diagram

$$\begin{array}{ccccc} V & \xleftarrow{\tilde{f}} & V^{(1)} & \leftarrow & \mathcal{E} \\ \downarrow\pi & & \downarrow\tilde{\pi} & & \\ \mathcal{M} & \xleftarrow{f} & \mathcal{M}^{(1)}, & & \end{array}$$

where $\mathcal{M}^{(1)}$ is the universal curve over $\mathcal{M}$ and $V^{(1)}=\pi^*\mathcal{M}^{(1)}$. The vector bundle $\mathcal{E}$ is a universal bundle, in the sense that its restriction to $\tilde{f}^{-1}(v)$ is in the isomorphism class parameterized by $v\in V$. Clearly, this is not unique since $\mathcal{E}\otimes\tilde{f}^*N$, with N a line bundle on V, will do the job as well. We can get rid of this ambiguity by normalizing $\mathcal{E}$ by means of a simple generalization of what is done on a fixed curve (see e.g. [3]). First we pull back the diagram above over the moduli space $\mathcal{M}^{(1)}$ getting

$$\begin{array}{ccccc} V^{(1)} & \xleftarrow{\tilde{f}} & \tilde{\mathcal{C}} & \leftarrow & \mathcal{E} \\ \downarrow\pi^{(1)} & & \downarrow\tilde{\pi}^{(1)} & & \\ \mathcal{M}^{(1)} & \xleftarrow{f} & \mathcal{C}. & & \end{array}$$

For simplicity we still denote by $\mathcal{E}$, $\tilde{f}$ and by f the pull back bundle and projections. The rationale for this is that we need a section $\sigma:\mathcal{M}^{(1)}\to\tilde{\mathcal{C}}$ of the universal curve, which canonically exists over $\mathcal{M}^{(1)}$ but not over $\mathcal{M}$. This section pulls back under $\pi^{(1)}$ to a section still denoted by $\sigma: V^{(1)}\to\tilde{\mathcal{C}}$. Now, over $V^{(1)}$ we have two line bundles $A=\det\sigma^*\mathcal{E}$ and $B=\det\tilde{f}_!\mathcal{E}$ and over $\tilde{\mathcal{C}}$ we have their pull back under $\tilde{f}$ which will be denoted again by A and B. The vector bundle $\tilde{\mathcal{E}}=\mathcal{E}\otimes A^{-a}\otimes B^{-b}$ is fixed under tensoring, in the sense that $(\widetilde{\mathcal{E}\otimes N})=\tilde{\mathcal{E}}$ for $ar+b\chi(E)=1$, with $\chi(E)=d-r(g-1)$, i.e. for $bd\equiv 1(mod\, r)$. Since d and r are coprime, there is a minimum integer k_0 such that $b_0=(1+k_0r)/d\in\mathbf{Z}$. Then b_0 and $a_0=b_0(g-1)-k_0$ is a solution of the condition above.

Obviously enough, $\tilde{\mathcal{E}}=\tilde{\tilde{\mathcal{E}}}=\tilde{\mathcal{E}}\otimes\tilde{A}^{-a_0}\tilde{B}^{-b_0}$, with $\tilde{A}=\det\sigma^*\tilde{\mathcal{E}}$ and $\tilde{B}=\det\tilde{f}_!\tilde{\mathcal{E}}$. Therefore we get the relation $a_0c_1(\tilde{A})+b_0c_1(\tilde{B})=0$. For $d=2r(g-1)+1$ we have $\chi(E)=r(g-1)+1$, yielding $b_0=1$, $a_0=-(g-1)$ and therefore $\tilde{B}=\tilde{A}^{(g-1)}$ as in [3].

As a final remark, we notice that the bundle $\mathrm{End}_0\mathcal{E}$ does not depend on the normalization. The first Chern class of the relative tangent bundle $T_{V/\mathcal{M}}=\tilde{f}_!\mathrm{End}_0\mathcal{E}$ by the Grothendieck Riemann Roch theorem reads $c_1(T_{V/\mathcal{M}})=N\lambda+2\alpha$ where λ is the Hodge class on V, $\alpha=\tilde{f}_*[(r-1)ch_2(\mathcal{E})-c_2(\mathcal{E})]$ and $N=r^2-1$. In particular the relative canonical bundle reads $K_{V/\mathcal{M}} = \lambda^{-N}\alpha^{-2}$. From [11] we learn that α restricted to a fibre of $V \to \mathcal{M}$ generates the Picard group.

2. The non–abelian Krichever map.

We will now describe the construction of some infinite dimensional complex varieties which will be embedded by a non–abelian Krichever map into a suitable infinite Grassmannian. If $H = H_+ \oplus H_-$ is a polarized vector space, its grassmannian $Gr(H)$ is the set of closed subspaces $W \subset H$ such that the projection $p_W : W \to H_-$ is Fredholm. Its connected components are labelled by the index of p_W. In the following H will be identified with a suitable completion [2] of $\mathbf{C}[[z, z^{-1}]]$. Recall also[2] that one can construct a natural infinite dimensional variety $\hat{\mathcal{M}}$ modelled on $H_+ \times \mathbf{C}^{3g-2}$ whose points parameterize triples made of (curve C, points $p \in C$, local parameters z near p). Observe that, if $\mathcal{M}''$ parameterizes pointed curves and a non–zero tangent vector at the point, we have a natural projection

$$\hat{\mathcal{M}} \xrightarrow{q''} \mathcal{M}'' \to \mathcal{M}^{(1)}$$

sending (C, p, z) into $(C, p, \frac{\partial}{\partial z})$ with fibres isomorphic to the vector space H_+. Accordingly, q'' induces an isomorphism in cohomology $H^*(\hat{\mathcal{M}}, \mathbf{Z}) \equiv H^*(\mathcal{M}'')$.

Let us now consider vector bundles. Let (C, p) be a pointed curve, and E be a rank r stable vector bundle over C. A local trivialization ϕ of E at p is an isomorphism of the stalk of the sheaf of local sections of E at p with $\oplus_{i=1,\dots r}\mathcal{O}_p$. Two locally trivialized vector bundles (E_1, ϕ_1) and (E_2, ϕ_2) are equivalent if there is an isomorphism $E_1 \xrightarrow{T} E_2$ such that

$$\oplus_{i=1,\dots r}\mathcal{O}_p \overset{\phi_2 T \phi_1^{-1}}{\longrightarrow} \oplus_{i=1,\dots r}\, \mathcal{O}_p.$$

is the identity. It is clear that the set of equivalence classes of such pairs maps onto the moduli space of stable rank r vector bundles on C. The fiber of this map over (C, E) is the group $\mathrm{Aut}\,_V(E)\restriction_p$ modulo the group of global invertible elements in $H^0(C, \mathrm{End}\,(E))$. Hence, after choosing a local parameter z at p, this in turn can be identified with the space of elements of the form $g_p \cdot (1 + z \cdot A(z))$, with $A(z)$ a matrix valued holomorphic function of z and g_p is in $Gl(r, \mathbf{C})/H^0(C, \mathrm{End}\,(E))^*$, where * denotes the group of invertible elements. Thanks to stability $H^0(C, \mathrm{End}\,(E))^* \simeq \mathbf{C}^*$ and the fiber is $PGL(r, \mathbf{C}) \times \mathrm{End}\,(E)_+$, the latter being the space of local holomorphic matrix valued functions. Pasting together Kuranishi deformations of quintuples $(C, p, z, [\mathcal{E}, \phi])$ one defines the infinite dimensional moduli stack of such objects, $\hat{V}$. Notice also that, if V'' denotes the stack parameterizing vector bundles curves, points and a non–zero tangent vector, we have a natural map $\tilde{p}'' : \hat{V} \to V''$ which induces an isomorphism up to second rational cohomology as the fibers have the same homotopy type as $SU(r)/\mathbf{Z}_r$.

The correspondence between the datum of $x = (C, p, z, [E, \phi]) \in \hat{V}$ and the space $\Gamma(C \setminus p, E)$ of sections of E over $C \setminus p$ gives the following

Proposition. There exists a natural map of $\hat{V}$ into the infinite Grassmannian $Gr(H^r)$.

Proof. The Mayer–Vietoris sequence associated to the covering U_0, $U_1 = C \setminus p$ of C reads:

$$0 \to H^0(C, E) \to \Gamma(U_0, E) \oplus \Gamma(C \setminus p, E) \to \Gamma(U_0 \setminus p, E) \to H^1(C, E) \to 0.$$

Factoring out $\Gamma(U_0, E)$, one gets the exact sequence

$$0 \to H^0(C,E) \to \Gamma(C \setminus p, E) \to \Gamma(U_0 \setminus p, E)/\Gamma(U_0, E) \to H^1(C,E) \to 0.$$

Now, the local trivialization ϕ and the local parameter z gives identifications $\Gamma(U_0 \setminus p, E) \equiv H^r$ and $\Gamma(U_0, E) \equiv H^r_+$ so that the sequence can be interpreted as

$$0 \to H^0(C,E) \to \Gamma(C \setminus p, E) \xrightarrow{\phi} H^r/H^r_+ \equiv H^r_- \to H^1(C,E) \to 0,$$

which, by the completeness of the curve C exhibits $\Gamma(C \setminus p, E)$ as a Fredholm subspace of H^r of index $\chi(E)$. ■

The map

$$\begin{array}{ccc} \hat{V} & \xrightarrow{\mathcal{K}} & Gr(H^r) \\ x & \rightsquigarrow & \phi\Gamma(C \setminus p, E) \end{array}$$

is called "generalized" or "non-abelian" Krichever map [8][10]. Let us study this map from the infinitesimal point of view. If we consider a local universal family of curves $\mathcal{C} \to \mathcal{M}$ and the relative variety of moduli of vector bundles with "fixed determinant" (in the sense as explained in §1)

$$\begin{array}{ccc} \mathcal{E} & & \\ \downarrow & & \\ V & \xrightarrow{\pi} & \mathcal{M}. \end{array}$$

The tangent space exact sequence at any point $[E, C] \in V$ reads

$$0 \to H^1(C, \mathrm{End}_0(E)) \to H^1(C, \Sigma_E) \to H^1(C, K^{-1}) \to 0,$$

where Σ_E is the sheaf of differential operators of order less or equal to 1 with scalar symbol acting on sections of E. Given points $x = (C, p, z, [E, \phi]) \in \hat{V}$ and $y = \hat{\pi}(x) = (C, p, z) \in \hat{M}$, we have natural identifications

$$\Sigma_E{\restriction}p = H_+^{n^2-1} \oplus H_+ \cdot \partial \subset \mathcal{D}$$
$$K^{-1}{\restriction}p = H_+ \cdot \partial \subset \mathrm{diff}^{\mathbf{C}}\, S^1$$

and Lie algebras inclusions

$$\mathcal{D}_x \equiv \Gamma(C \setminus p, \Sigma_E) \hookrightarrow \mathcal{D}, \qquad \mathrm{diff}^{\mathbf{C}}\, S^1{}_y \equiv \Gamma(C \setminus p, K^{-1}) \hookrightarrow \mathrm{diff}^{\mathbf{C}}\, S^1.$$

In analogy with the abelian case one has the following

Proposition. For every $x \in \hat{V}$ one has the following commutative diagram

$$\begin{array}{ccccccccc} 0 & \to & \mathcal{D}_x & \to & \mathcal{D} & \to & T_x(\hat{V}) & \to & 0 \\ & & {\scriptstyle\sigma}\downarrow\uparrow{\scriptstyle i} & & {\scriptstyle\sigma}\downarrow\uparrow{\scriptstyle i} & & \downarrow{\scriptstyle d\hat{\pi}} & & 0 \\ 0 & \to & \mathrm{diff}^{\mathbf{C}}\, S^1{}_x & \to & \mathrm{diff}^{\mathbf{C}}\, S^1 & \to & T_y(\hat{\mathcal{M}}_g) & \to & 0, \end{array}$$

where σ is the symbol map and the horizontal sequences are exact.

The tangent to the Krichever map

$$T_x(\hat{V}) \equiv \mathcal{D}/\mathcal{D}_x \xrightarrow{d\mathcal{K}} T_{\mathcal{K}(x)}(Gr(H^r)) \equiv \mathrm{End}\,(\mathcal{K}(x), H^r/\mathcal{K}(x))$$

can be represented, if $O(z) = A(z) + b(z)\frac{\partial}{\partial z}$ maps to $v \in T_x(\hat{V})$ and $f(z) \in \mathcal{K}(x)$, as

$$d\mathcal{K}(v)(f) = O \cdot f \bmod \mathcal{K}(x).$$

Denoting with a_∞ the algebra of infinite size matrices and with ψ the natural representation of $\mathcal{D}$ in a_∞, the diagram

$$\begin{array}{ccc} \mathcal{D} \times \hat{V} & \overset{\psi \times \mathcal{K}}{\longrightarrow} & a_\infty \times Gr(H^r) \\ \downarrow & & \downarrow \\ T(\hat{V}) & \overset{d\mathcal{K}}{\longrightarrow} & T(Gr(H^r)) \end{array}$$

commutes. Pulling back the universal central extension of a_∞ via the ψ's given by the different choices of E listed in the introduction, one easily gets the corresponding central extensions.

The geometrical counterpart of this algebraic fact is that central extensions of the Lie algebra $\mathcal{D}$ are related to the Picard group of $\hat{V}$, the link being given by the following

Lemma. ([2]) Let X be a complex manifold and $\mathfrak{g}$ be an algebra acting on X such that
i) $\forall x \in X$ the evaluation map $\mathfrak{g} \xrightarrow{ev_x} T_x(X)$ is surjective;
ii) $\mathfrak{g}_x \equiv \ker ev_x$ is such that $[\mathfrak{g}_x, \mathfrak{g}_x] = \mathfrak{g}_x$
Then for any Lie algebra continuous extension

$$0 \to \mathbf{C} \to \hat{\mathfrak{g}} \to \mathfrak{g} \to 0$$

which is trivial on $\mathfrak{g}_x$ there is an associated continuous extension

$$0 \to X \times \mathbf{C} \to F \to T_x(X) \to 0$$

This defines a homomorphism $\cap_{x\in X} \ker r_x \to Ext^1(\mathcal{T}_X, \mathcal{O}_X)$ where $r_x : H^2(\mathfrak{g}, \mathbf{C}) \to H^2(\mathfrak{g}_x, \mathbf{C})$ is induced by the restriction.

We can apply this lemma to our situation thanks to the observation that, if $\mathcal{D}_x$ is the algebra of global differential operators of order ≤ 1 with scalar symbol acting on the sections of a vector bundle E over the affine curve $C \setminus p$, then $[\mathcal{D}_x, \mathcal{D}_x] = \mathcal{D}_x$ and, furthermore, every Lie algebra extension of $\mathcal{D}$ is trivial on $\mathcal{D}_x$. Summing up, we have the

Theorem. There exists a homomorphism

$$H^2(\mathcal{D}, \mathbf{C}) \xrightarrow{\rho} Ext^1(\mathcal{T}_{\hat{V}}, \mathcal{O}_{\hat{V}}) \equiv H^1(\Omega^1_{\hat{V}}).$$

The same argument yields a homomorphism

$$H^2(a_\infty) \to Ext^1(\mathcal{T}_{Gr(H^r)}, \mathcal{O}_{Gr(H^r)}).$$

This latter is spelled out by the following

Proposition. ([2]) The extension Σ induced by the standard extension of a_∞ is the sheaf Σ_{Det} of differential operators of order ≤ 1 acting on sections of the determinant bundle Det $\rightarrow Gr(H^r)$.

We next want to show that the extensions of $\mathcal{D}$ induced in this way actually come from line bundles. There is a canonical homomorphism

$$c : H^1(\mathcal{O}^*_{\hat{V}}) \rightarrow H^1(\Omega^1_{\hat{V}})$$

which associates to a line bundle L the isomorphism class of extensions represented by the sheaf of differential operators of order ≤ 1 acting on sections of L.

Proposition. The image of the homomorphism $H^2(\mathcal{D}) \xrightarrow{\rho} Ext^1(\mathcal{T}_{\hat{V}}, \mathcal{O}_{\hat{V}})$ is contained in the image of the homomorphism $H^1(\mathcal{O}^*_{\hat{V}}) \xrightarrow{c} H^1(\Omega^1_{\hat{V}})$

Proof. The second cohomology of $\mathcal{D}$ is generated by the Virasoro cocycle α_1 and the Kac–Moody cocycle α_2. Pulling back the generator ω of $H^2(a_\infty, \mathbf{C})$ under the representation $\mathcal{D} \xrightarrow{\psi} a_\infty$ a simple computation shows that $\alpha_0 = \psi^*(\omega) = \alpha_2 + r \cdot \alpha_1$ so that $\{\alpha_0, \alpha_1\}$ is still a basis for $H^2(\mathcal{D}, \mathbf{C})$. As in §1, we can define a normalized universal vector bundle $\tilde{\mathcal{E}}$ over the universal curve $\tilde{f} : \tilde{\mathcal{C}} \rightarrow \hat{V}$. On the other hand, we have a natural projection $\hat{\pi} : \hat{V} \rightarrow \hat{\mathcal{M}}_g$ under which we can pull-back the Hodge bundle λ. Let us consider the diagram

$$\begin{array}{ccc} \mathcal{D} \times \hat{V} & \longrightarrow & a_\infty \times Gr(H^r) \\ \downarrow & & \downarrow \\ T(\hat{V}) & \longrightarrow & T(Gr(H^r)) \end{array}$$

By construction, if $\mathcal{K} : \hat{V} \rightarrow Gr(H^r)$ is the non–abelian Krichever map, $\mathcal{K}^*(\text{Det}^*) = \det \tilde{f}_! \tilde{\mathcal{E}}$ so that, by the proposition above, the pull–back of Σ_{Det^*} is exactly $\Sigma_{\det \tilde{f}_! \tilde{\mathcal{E}}}$ and $\psi^*(\omega) = \alpha_0$. On the other hand, the standard Krichever map $\hat{M}_g \xrightarrow{W} Gr(H)$ gives us the analogous diagram

$$\begin{array}{ccc} \text{diff}^{\mathbf{C}} S^1 \times \hat{M}_g & \longrightarrow & a_\infty \times Gr(H) \\ \downarrow & & \downarrow \\ T(\hat{M}_g & \longrightarrow & T(Gr(H))) \end{array}$$

and $W^*(\text{Det}^*)$ is the Hodge class λ, which gives the cocycle α_1. ∎

The paper of Arbarello, De Concini, Kac and Procesi ended with the remarkable proof of the existence of an isomorphism between the group of line–bundles on the moduli space of pointed curves with a non zero cotangent vector and a degree $g-1$ line bundle and the second cohomology of the semidirect product of the Virasoro and Heisenberg algebra. What we can prove is a weaker result, and namely that there is an isomorphism between $H^2(\mathcal{D}, \mathbf{C})$ and the second rational cohomology of V'' in the case $r = 2, d = 1$.

Recall that $\hat{V}$ maps onto V'' and the map induces an isomorphism in second rational cohomology. Consider the universal vector bundle $\tilde{\mathcal{E}}$ over the universal curve $\tilde{\mathcal{C}}$. Then,

$\sigma^*(\det\,\tilde{\mathcal{E}})$ and $\det\,\tilde{f}_!\tilde{\mathcal{E}}$ when restricted to a fiber of π'' are non–zero multiples of the generator of the second cohomology of the fibre to the fibration

$$V'' \xrightarrow{\pi''} M''.$$

This means that $\dim H^2(V'',\mathbf{Q})/\pi''^*(H^2(M'',\mathbf{Q})) \geq 1$.

When dealing with pointed curves and for $r=2$, the fibration $V'' \xrightarrow{p} M''$ admits a section. In fact, given $p \in C$ one can consider the extension

$$0 \to \mathcal{O}(-p) \to E_{\omega_0} \to \mathcal{O}(2p) \to 0$$

where here ω_0 is the dual to the generator of $H^0(C,K_C(3p))/H^0(C,K_C(2p))$ and, thanks to the fact that the linear system $K_C(3p)$ is base point free, E is stable. The existence of such a section ensures that $\pi''^* : H^2(M'',\mathbf{Q}) \to H^2(V'',\mathbf{Q})$ is injective. Since [1] $\dim H^2(M'',\mathbf{Q}) = 1$, we have that $\dim H^2(V'',\mathbf{Q}) \geq 2$. By the Kunneth formula, $\dim H^2(V'',\mathbf{Q}) \leq 2$ and hence we have

Proposition. For $r=2$, $d=1$ there is an isomorphism $H^2(V'',\mathbf{C}) \simeq \mathbf{C} \oplus \mathbf{C}$.

Acknowledgements. We thank F. Bardelli, C. De Concini and M. Rothstein for enlightening discussions and remarks.

References.

[1] E. Arbarello, M. Cornalba "The Picard groups of the moduli space of curves", Topology **26**, 153 (1987).

[2] E. Arbarello, C. De Concini, V. G. Kac and C. Procesi "Moduli spaces of curves and representation theory" Commun. Math. Phys. **117** 1, (1988).

[3] M. F. Atiyah and R. Bott "The Yang-Mills equations over Riemann surfaces" Phil. Trans. R. Soc. Lond. **A308**, 523 (1982).

[4] S. Axelrod, S. Della Pietra, E. Witten "Geometric quantization of Chern–Simons Gauge theory", preprint IASSNS–HEP 89/57 (October 1989).

[5] G. Falqui, C. Reina "Superstrings and Supermoduli", to be published in Proceedings of the 1988 C.I.M.E. summer course 'Global Geometry and Mathematical Physics', Montecatini 1988.

[6] P. Goddard, D. Olive "Kac-Moody and Virasoro algebras in relation to quantum physics" Int. J. Mod. Phys. **A1** 303 (1986).

[7] N.J.Hitchin "Flat connections and geometric quantization" Commun. Math. Phys. **131** 347 (1990).

[8] M. Mulase "A correspondence between an infinite Grassmannian and arbitrary vector bundles on algebraic curves" in Theta Functions Bowdoin 1987, American Math. Soc. (1989)

[9] A.Pressley and G.Segal "Loop groups" Claredon Press, Oxford (1986)

[10] E.Previato and G.Wilson "Vector bundles over curves and solutions of the KP equations" in Theta Functions Bowdoin 1987, American Math. Soc. (1989)

[11] S. Ramanan "The moduli spaces of vector bundles over an algebraic curve" Math. Ann **200**, 69 (1973).

Instanton homology

A. Floer

Fakultät für Mathematik — Ruhr-Universität Bochum

In general relativity, one considers space-time manifolds X (of dimension 4) equipped with a pseudo Riemannian (Minkowski-type) metric. This specifies a "light cone" in each tangent space and hence, up to questions of completeness, a global product structure $X = R \times M$. If the pseudo-Riemannian metric is replaced by a Riemannian metric, space-time will generally be a cobordism between possible topologically distinct "space" manifolds M and N. Let us denote the space dimension by d, and assume that $M, N \ldots$ are closed oriented (smooth) manifolds, while X is a compact oriented cobordism, i.e. a compact $d+1$-dimensional manifold with boundary $\partial X = (-M) \cup N$. We will think of cobordisms as playing a role similar to maps between spaces, and write $X : M \longrightarrow N$. A "topological quantum field theory" (TQFT) assigns to every "space" M a Hilbert space $Q(M)$, and to every "space-time" $X : M \longrightarrow N$ a homomorphism $X_\star : Q(M) \longrightarrow Q(N)$. The basic axiom of any TQFT is that $X_\star$ depends only on the smooth structure of X (relative to M and N), and that for any two cobordisms $X : M \longrightarrow N$ and $Y : N \longrightarrow L$

$$(X \sharp Y)_\star = X_\star \circ Y_\star : Q(M) \longrightarrow Q(L)\,,$$

where $X \sharp Y$ is the "composite cobordism". This implies for example that $Q(M)$ is a diffeomorphism invariant of M, since for every diffeomorphism $\varphi : M \longrightarrow M'$ there exists an obvious "space-time" $X(\varphi) : M \longrightarrow N$ (a "product cobordism" such that $X(\varphi)X(\varphi^{-1})$ is diffeomorphic to the cylinder over M. If however X is not a product, then $X_\star$ will generally not be an isomorphism, i.e. "states" in the TQFT will "annihilated" or "created" by X.

Besides functoriality, a TQFT (in its stricter sense) should satisfy the unitarity axioms: $Q(M)$ should be a Hilbert space, a change in orientation should lead to $Q(\bar{M}) = (Q(M))^*$ and $(\bar{X}_\star)^*$. Thus any diffeomorphism defines a unitary equivalence). Such a "unitary TQFT" was indeed constructed for $d = 2$, i.e. for closed oriented surfaces as spaces and compact three-manifolds as space-times. In fact, functoriality and unitarity essentially determine all possible theories, leading to combinatorial constructions. The origin of these 2-dimensional TQFT's, however, lies in Witten's "path integral formulation" of V. Jones's new knot polynomials, and rather fascinating geometric and analytic concepts are used.

It is remarkable that functors on such a well known (and doubtlessly mathematically interesting) category as the cobordism category have emerged before in mathematics, or seem to have as straightforward a definition as say ordinary homology (which is a functor on the category of continous maps). Moreover, neither the original construction of the Witten-Jones TQFT, nor the combinatorial approaches generalize to other space dimensions. However, in space dimension 3, an entirely different approach based on instantons (self-dual Yang-Mills connections) on a space-time X leads to a cobordism functor $F_\star$. It is not unitary, in fact it actually takes values in the category of abelian groups. For details of the construction, we refer to [F1] and [A1], but roughly speaking, one applies Witten's alternative construction of homology on manifolds to the infinite dimensional manifold of connections on the space M. For these reasons, one should consider $F_\star$ as a homology theory which happens to be functorial with respect to cobordisms instead of continous maps. (In fact, the asterisk refers to grading by integers modulo 8).

There are two main lines of current interest in $F_\star(M)$: one is to formulate a "De Rham" or "Hodge" theory version. While the current definition uses exclusively the solutions of the "classical" instanton equation, a De Rham version leads to yet unsolved problems of 4-dimensional quantum field theory (see [W2]).

The other is to find further axioms in addition to functorial methods of computations as well as "uniqueness proofs" for $F_\star$. This has recently been accomplished in a way which is reminiscent of the axiomatic formulation of ordinary homology. The central axiom is a "long exact sequence": with any knot in a 3-manifold, one associates a certain natural sequence of surgery operations with corresponding surgery cobordisms

$$M \longrightarrow M' \longrightarrow M'' \longrightarrow M''' \longrightarrow \dots,$$

which is periodic of period 3, i.e. $M''' = M$. The homomorphisms induced by the cobordisms form an exact sequence, i.e. the set of "states" annihilated by one cobordism is precisely the image of the preceding cobordism. Though we are unable to make any "physical" sense out of this axiom, the mathematical implications are well known from ordinary homology theories, where exact sequences are the principal tools of calculation. (Consider for example the Mayer-Vietoris sequence in homology). There is even a notion of "completeness" of a system of axioms in homology. It neither implies uniqueness of the theory nor gives a straightforward algorithm for a calculation, but in practice comes reasonably close to both.

The most mysterious problem in instanton homology is its relation to the equally new invariants obtained for closed 3-manifolds as the "vacuum expectation values" of the $(2+1)$-dimensional Witten-Jones theory. Both theories use gauge theory and ideas of quantum field theory (more rigorously in $(2+1)$ dimensions, less rigorously in $(3+1)$ dimensions). However, there does not seem to be a mechanism in physics which could relate quantum field theories in different dimensions.

References

[A] M. Atiyah: New invariants of 3- and 4-dimensional manifolds, in: The mathematical heritage of Hermann Weyl, Am. Math. Soc., Proc. Symp. Pur. Maths. 48 (1988), 285-299.

[FI] A. Floer: An Instanton-invariant for 3-manifolds, Commun. Math. Phys. 118 (1988), 215-240.

[F2] A. Floer: Instanton homology and Dehn surgery, Preprint 1990.

[WI] E. Witten: Supersymmetry an Morse theory, J. Diff. Geom. 17 (1982), 661-692.

[W2] E. Witten: Topological quantum field theory, Commun. Math. Phys. 117 (1988), 353-356.

[W3] E. Witten: Quantum Field Theory and the Jones Polynomial, Comm. Math. Phys. 121 (1989), 351-399.

W- GEOMETRY

C. Itzykson
Service de Physique Théorique de Saclay
Laboratoire de la Direction des Sciences de la Matière
du Commissariat à l'Energie Atomique
F-91191 Gif-sur-Yvette Cedex France

1. During the meeting I presented work done with J.-B. Zuber on matrix combinatorics applied to triangulated surfaces. Rather than paraphrasing our paper, I thought it would be more appropriate to discuss here some investigations carried in collaboration with P. Di Francesco and J.-B. Zuber on the relation between geometry, differential operators and classical W- algebras. The latter were introduced by Zamolodchikov as a generalization of the Virasoro algebra in the context of conformal quantum field theory. Their status remains however still dubious (at least for this author) and it is difficult to pinpoint a precise geometric definition. Some still very preliminary and incomplete indications at the end of this paper were worked out in collaboration with M. Bauer.

Except for a few remarks, I will concentrate on W-algebra pertaining to the linear group, although following the work of Drinfeld and Sokolov -our basic reference- it is possible to generalize the notion to other simple Lie groups. Consequently the underlying geometry will be the projective or affine one.

Such a classical subject as linear differential operators in one variable is a very rich one. It embraces uniformization of algebraic curves, differential Galois theory, Kortweg de Vries integrable systems and their generalizations, including applications to the Schottky problem (of characterizing Jacobian varieties) or recent findings in two-dimensional quantum gravity, Painlevé equations... In a nutshell it may be interpreted as a linearization of many interesting non linear problems. One could also mention the background of quantum mechanics with its use of differential algebras as a natural realization of operators.

We will focus here on one main theme: to investigate those relations which have a covariant character. In other words we wish to emphasize properties which are independent of a specific choice of coordinate, thus treating differential operators as intrinsic geometric objects.

At first we need not be specific about the basic field, either the real or complex numbers, most relations involving in fact rational coefficients. We deal with generic linear differential operators with regular coefficients, i.e. admitting as many derivatives as necessary. According to the context those are defined in a fixed neighborhood either in the real or complex domain. In some case we need a non trivial integration cycle C. This could extend over the entire real axis (assuming coefficients vanishing fast enough at infinity) or over an interval for periodic coefficients or even on a closed curve encircling some singularity in the complex case. One could also treat the whole subject by purely algebraic means and formal series.

2. Linear differential operators give local maps of functions on functions (this will be qualified later). They are intrinsically determined (up to multiplication by a fixed function) by their kernel, a finite dimensional vector space. By quotienting this space through dilatations one can achieve two goals. On the one hand one can normalize the operator of order n to the form

$$Q_n = d^n + \sum_{p=2}^{n} a_p d^{n-p}$$

where d stands for the derivative $\frac{d}{dx}$ if x is the variable, by multiplying any function in the source space by an appropriate (x-dependent) factor. On the other hand, we retain only projective properties of the kernel (up to a finite cyclic extension which accounts for the difference between SL_n and PSL_n which does not affect infinitesimal properties). Hence above a fixed neighborhood U we can see the operator as defining a finite vector subspace E_n in the infinite dimensional vector space of regular functions, then its projectivization PE_{n-1}.

The normalizing procedure can also be understood as a means of selecting definite transformation properties of the source functions under diffeormorphisms (of U) and therefore of the coefficients of the operators as follows. Assume at first the source functions to have conformal weight λ, i.e. for a map $x_1 \longleftrightarrow x_2$ we have $f_1 \longleftrightarrow f_2$ such that $f_1(x_1)\,dx_1^{\lambda} = f_2(x_2)\,dx_2^{\lambda}$, it is natural to assume that Q_n increases the weight by n. Call $\mathcal{F}_\lambda$ the set of regular λ-differentials in U. Then

$$Q_n : \mathcal{F}_\lambda \longrightarrow \mathcal{F}_{\lambda+n}$$

Let us insist however to have a normalized operator in any coordinate. This requires the logarithmic derivative of the Wronskian of n linearly independent solutions (which is minus the coefficient of the term in d^{n-1}) to vanish in any coordinate. The Wronskian gives a map

$$\overset{n}{\Lambda}\mathcal{F}_\lambda \longrightarrow \mathcal{F}_{n\lambda+\frac{n(n-1)}{2}}$$

and its constancy can only be achieved if the target weight is zero, i.e. if

$$\lambda = \frac{1-n}{2}$$

If n is even $\frac{1\mp n}{2}$ are half integral and a square root of the non vanishing jacobian has to be chosen in a coordinate change, this however does not affect projective properties based on ratios.

For p functions $f_0, \ldots, f_{p-1}$ write the Wronskian as

$$W_p(f_0, \ldots, f_{p-1}) = \det f_k^{(\ell)}, \qquad 0 \le k, \ell \le p-1$$

where $f^{(\ell)} = d^\ell f$. If $f_0, \ldots, f_{n-1}$ stand for a basis of $E_n = \ker Q_n$ we have

$$f \longrightarrow Q_n f = \sum_{0}^{n} a_p d^{n-p} f = \frac{W_{n+1}(f, f_0, \ldots, f_{n-1})}{W_n(f_0, \ldots, f_{n-1})}$$

with $a_0 = 1$, $a_1 = -W_n'(f_0, \ldots, f_{n-1})/W_n(f_0, \ldots, f_{n-1})$ and

$$f \in \mathcal{F}_{\frac{1-n}{2}} \qquad Q_n f \in \mathcal{F}_{\frac{1+n}{2}}$$

Since W_n is a scalar $W_n'/W_n \in \mathcal{F}_1$. Its vanishing is therefore a coordinate free property. For simplicity we assume in the sequel W_n equal to unity. Said otherwise if we start with arbitrary a_1, $f \in \mathcal{F}_\lambda$ then $f \exp -\frac{1}{n}\int^x a_1 \in \mathcal{F}_{\frac{1-n}{2}}$ and after this change of function Q_n is normalized. This discussion assumes therefore $W_n \neq 0$. It could happen that for a specific choice of basis $f_0, \ldots, f_{n-1}, W_n$ vanishes at certain points, these would have to be deleted as they correspond to singularities in the coefficients of Q_n.

From the above formula for the coefficients a_k one can derive their transformation properties under diffeormorphisms, in particular if $x \longleftrightarrow \tilde{x}$, $a_2 \longleftrightarrow \tilde{a}_2(\tilde{x})$ with

$$\tilde{a}_2(\tilde{x}) = a_2(x)\left(\frac{dx}{d\tilde{x}}\right)^2 + \frac{n(n^2-1)}{12}\{x,\tilde{x}\}$$

where $\{x,\tilde{x}\}$ stands here for the Schwarzian derivative

$$\{x,\tilde{x}\} = \frac{x'''}{x'} - \frac{3}{2}\left(\frac{x''}{x'}\right)^2 \qquad x \equiv x(\tilde{x})$$

More generally in infinitesimal form, $x = \tilde{x} + \epsilon(\tilde{x})$, we have (with $a_0 - 1 = a_1 = 0$)

$$\delta a_k = \epsilon a'_k + k\epsilon' a_k + \sum_{\ell=0}^{k-1}\left[\frac{n-1}{2}\binom{n-\ell}{k-\ell} - \binom{n-\ell}{k-\ell+1}\right]\epsilon^{(k+1-\ell)}a_\ell$$

3. Rather than identifying the data coded in Q_n with the vector space $E_n = \ker(Q_n)$, we can equally well think of a map $U \longrightarrow V_n$ where V_n is a fixed vector space equipped with a basis $\underline{e}_0, \ldots, \underline{e}_n$ by treating n independent solutions $f_0, \ldots, f_{n-1}$, as the coordinates of a vector

$$x \in U \overset{Q_n}{\rightarrow} \underline{f}(x) = \sum_{p=0}^{n-1} f_p(x)\underline{e}_p \in V_n$$

Thus we have (a segment of) a curve in V_n and after projectivization in PV_{n-1}. Allowing invertible linear transformations among the f'_ps or $\underline{e}'_p$s (with constant coefficients) means that, in U, Q_n yields the neighborhood of a point on a curve in projective space of dimension $n-1$ (the lift in V_n being obtained by requiring $W_n = 1$) *up to an arbitrary projective transformation*. In other words Q_n codes the invariant (local and differential) aspects of this map. Note that one can draw a parallel with polynomial equations.

In the case of an algebraic curve we may remark that it not customary to embed it in projective space through $\frac{1-n}{2}$ (meromorphic) differentials. One uses rather 1 (or 3-) holomorphic differentials (the canonical embeddings) but in principle nothing prevents one from using other choices. For instance take the canonical embedding of the (genus 3) Klein curve giving the smooth quartic $uw^3 + vu^3 + wv^3 = 0$ in CP_2. For $n = 3$ using meromorphic vector fields it becomes a quintic $x_1^2x_2^3 + x_2^2x_3^3 + x_3^2x_1^3 = 0$ with 3 cusps at the base points.

The case $n = 2$ where PV_1 is one dimensional deserves special mention as it is related to the classical uniformization problem of algebraic curves studied in great depth by Poincaré and the late XIXth century mathematicians. In its normalized form Q_2 is a Schrödinger operator

$$Q_2 = d^2 + a_2(x)$$

acting on $-1/2$ differentials. Thanks to its "anomalous" transformation property, the "potential" a_2 can be made to vanish in an appropriate coordinate u so that $a_2(x) = \frac{1}{2}\{u,x\}$. In the uniformization coordinate u the general element of the kernel is a first degree polynomial, hence u is in general the ratio of two solutions defined up to a PSL_2 transformation and we have

$$a_2 = \frac{1}{2}\left\{\frac{f_1}{f_0}, x\right\}$$

which agrees with the above formula as a ratio of determinants for $W_2 = 1$. To describe an algebraic (complex) curve one has to look at the non trivial global homotopy group $\bar{\mathbb{C}} - \{\text{sing } a_2\}$ acting as a monodromy group on $u = f_1/f_0$, giving a discrete subgroup of PSL_2. In the best of all worlds this is a Fuchsian group with a fixed circle. The standard example corresponds to the hypergeometric equation with three (regular) singularities at 0, 1, ∞ as described in textbooks.

4. The transformation properties of the coefficients in Q_n are quite complicated except for a_2, which behaves like the energy momentum in conformal field theory. (This can be traced to the fact that we deal with projective representations of diffeormorphisms). This justifies to look for an invertible transformation from $a_2, a_3, ..., a_n$ to $w_2 \equiv a_2, w_3, ..., w_n$ where the w's are differential polynomials in the a's and vice versa (by differential polynomial we mean a polynomial both in the functions and finitely many of their derivatives) in such a way that w_k, $k \geq 3$ behaves like a k-differential ($w_k(x)dx^k = \tilde{w}_k(\tilde{x})\,d\tilde{x}^k$), the only anomaly appearing in $w_2 = a_2$ in the form of the only non trivial 2-cocycle, the Schwarzian derivative. As a result the w's will generate the same differential algebra as the a's.

The w's are not uniquely determined by the above properties even if we require that to leading order $w_k = a_k + ...$, for it is clear that if $k = k_1 + k_2$ with $k_1, k_2 \geq 3$ we can add to w_k a term proportional to $w_{k_1} w_{k_2}$ or if $k = k_1 + k_2 + 1$, $k_1 \neq k_2$, $k_1, k_2 \geq 3$ we can add to w_k a term in $k_1 w_{k_1} w'_{k_2} - k_2 w'_{k_1} w_{k_2}$ etc... . It turns out that $w_2 = a_2$, w_3, w_4, w_5 are in fact unique.

We can however make a canonical choice which excludes the former ambiguity by requiring that w_k be globally linear in the a'_ℓs and their derivatives ($\ell \geq 3$) with coefficients which are differential polynomials in a_2 up to an inhomogeneous term, itself a differential polynomial in a_2. We obtain this result as follows. We first pick a coordinate u in which a_2 vanishes. This requires solving a second order differential equation

$$\left(d^2 + \frac{6a_2}{n(n^2-1)}\right) f = 0 \qquad u = \frac{f_1}{f_0}$$

We then write the differential operator as

$$Q_n = \Delta_2^{(n)}(0) + \sum_{k=3}^{n} \Delta_k^{(n)}(w_k, 0)$$

where

$$\begin{aligned} \Delta_2^{(n)}(0) &= d_u^n \\ \Delta_k^{(n)}(w_k, 0) &= \sum_{\ell=0}^{n-k} \alpha_{k\ell} w_k^{(\ell)}(u) d_u^{n-k-\ell} \qquad k \geq 3 \end{aligned}$$

The coefficients

$$\alpha_{k,\ell} = \frac{\binom{k+\ell-1}{\ell}\binom{n-k}{\ell}}{\binom{2k+\ell-1}{\ell}}$$

are chosen in such a way that by returning to a general x-coordinate

$$\begin{aligned} Q_n(u) \longrightarrow Q_n(x) &= \varphi^{\frac{n+1}{2}} Q_n(u) \varphi^{\frac{n-1}{2}} \qquad \varphi(x) = \frac{du}{dx}(x) \\ Q_n(x) &= \Delta_2^{(n)}(a_2) + \sum_{k=3}^{n} \Delta_k^{(n)}(w_k, a_2) \end{aligned}$$

where the Δ's depend only on the Schwarzian derivative of the map, i.e. on $a_2 = \frac{1}{12}n\,(n^2-1)$ $\{u,x\}$, and its derivatives (as well as on the w's and their derivatives). Setting $b = \frac{1}{\varphi}\frac{d\varphi}{dx}$, the Schwarzian derivative is

$$\{u,x\} = \frac{db}{dx} - \frac{1}{2}b^2$$

so that with $j = \frac{n-1}{2}$

$$\Delta_2^{(n)}(x) = (d_x - jb)\,(d_x - (j-1)b)\cdots(d_x + jb)$$

and under a variation δb which leaves $\{u,x\}$ fixed i.e. such that $\delta b' - b\delta b = 0$ or

$$[d_x - (k+1)]\,\delta b = \delta b\,[d_x - kb]$$

we have

$$\begin{aligned}
\delta\Delta_2^{(n)}(x) &= -\sum_{-j}^{j}(d_x - jb)\cdots(d_x-(k+1)b)\,\delta b\,(d_x-(k-1)b)\cdots(d_x+jb)\\
&= -\left(\sum_{-j}^{+j}k\right)\delta b\,(d_x-(j-1)b)\cdots(d_x+jb) = 0
\end{aligned}$$

One uses the same reasoning for

$$\begin{aligned}
\Delta_k^{(n)}(x) &= \varphi^{\frac{n+1}{2}}\left(\sum_{\ell=0}^{n-k}\alpha_{k\ell}w_k^{(\ell)}(u)d_u^{n-k-\ell}\right)\varphi^{\frac{n-1}{2}}\\
&= \varphi^{\frac{n+1}{2}}\sum_{\ell=0}^{n-k}\alpha_{k\ell}\left[(\varphi^{-1}d_x)^{\ell}\varphi^{-k}w_k(x)\right](\varphi^{-1}d_x)^{n-k-\ell}\varphi^{\frac{n-1}{2}}\\
&= \sum_{\ell=0}^{n-k}\alpha_{k\ell}\left[\mathcal{D}^{\ell}w_k\right]\mathcal{D}^{n-k-\ell}
\end{aligned}$$

where $\mathcal{D}$ is a covariant derivative which acts from p differentials to $(p+1)$ ones as $d_x - pb$, i.e.

$$\left[\mathcal{D}^{\ell}w_k\right] = (d_x-(k+\ell-1)b)\cdots(d_x-kb)\,w_k$$

and

$$\mathcal{D}^{n-k-\ell} = (d_x-(n-k-\ell-j-1)b)\cdots(d_x+jb)$$

Requiring $\alpha_{k,0} = 1$ and $\delta\Delta_k^{(n)}(x) = 0$ under a variation δb with fixed a_2 leads to the recursion relation

$$\ell(\ell+2k-1)\alpha_{k,\ell} = (k+\ell-1)(n+1-k-\ell)\alpha_{k,\ell-1}$$

with the solution given above. To relate w's and a's one identifies w_3 as the coefficient of d^{n-3} in $Q_n - \Delta_2^{(n)}(a_2)$, w_4 as the coefficient of d^{n-4} in $Q_n - \Delta_2^{(n)}(a_2) - \Delta_3^{(n)}(w_3, a_2)$ and so on. This yields

$$\begin{aligned}
w_2 &= a_2\\
w_k &= \sum_{\ell=2}^{k}B_{k,\ell}a_\ell^{(k-\ell)} + \sum_{\substack{0\le p_1\le p_2\le\ldots\le p_r\\ \Sigma p_i+2r=k}} C_{p_1,\ldots,p_r}a_2^{(p_1)}\cdots a_2^{(p_r)}\\
&\quad+\sum_{\substack{0\le p_1\le p_2\le\ldots\le p_r\\ 3\le\ell\le k-\Sigma p_i-2r}} D_{p_1,\ldots,p_r}a_2^{(p_1)}\cdots a_2^{(p_r)}a_\ell^{(k-\ell-\Sigma p_i-2r)}
\end{aligned}$$

The coefficients $B_{k,\ell}$ (the only ones necessary in the u-coordinate) are given by

$$B_{k,\ell} = (-1)^{k-\ell} \frac{\binom{k-1}{k-\ell}\binom{n-\ell}{k-\ell}}{\binom{2k-2}{k-\ell}}$$

while conversely

$$\begin{aligned} a_k &= \sum_{\ell=2}^{k} A_{k\ell} w_\ell^{(k-\ell)} + \text{ non linear terms} \\ A_{k,\ell} &= \alpha_{\ell,k-\ell} = \frac{\binom{k-1}{k-\ell}\binom{n-\ell}{k-\ell}}{\binom{k+\ell-1}{k-1}} \end{aligned}$$

Both $A_{k,\ell}$ and $B_{k,\ell}$ vanish when $\ell > k$ while one checks that

$$\sum_\ell B_{k\ell} A_{\ell m} = \delta_{k,m}$$

Explicitly

$$\begin{aligned} w_2 &= a_2 \\ w_3 &= a_3 - \frac{n-2}{2} a_2' \\ w_4 &= a_4 - \frac{n-3}{2} a_3' + \frac{(n-2)(n-3)}{10} a_2'' - \frac{(n-2)(n-3)(5n+7)}{10n(n^2-1)} a_2^2 \\ w_5 &= a_5 - \frac{n-4}{2} a_4' + \frac{3(n-3)(n-4)}{28} a_3'' - \frac{(n-2)(n-3)(n-4)}{84} a_3''' \\ &+ \frac{(n-3)(n-4)(7n+13)}{14n(n^2-1)} [(n-2) a_2 a_2' - 2 a_3 a_2] \\ &\ldots \end{aligned}$$

and it is a tedious but straightforward matter to extend the list. Of course for a given n we need only $w_2, \ldots, w_n$.

Because for generic n we made a definite, but arbitrary, choice of w's, it is at first glance frivolous to try to find a direct geometric interpretation of the w's. However a statement such as $w_k = 0$, $k \geq 3$ has an intrinsic meaning and implies that the curve is rational in U. For small n the w's are unique and we shall try an interpretation in the last section.

Moreover it is an interesting matter to count the number of linearly independent k-differentials ($k \geq 3$) as differentials polynomials in w's (or a's). Such a formula has been obtained by N. Sochen and J.-B. Zuber.

5. Differential operators admit a natural involution

$$Q_n \longleftrightarrow Q_n^* = \sum a_p^* d^{n-p} = \sum (-d)^{n-p} a_p \quad ,$$

which act as antiautomorphism: $(Q_1 Q_2)^* = Q_2^* Q_1^*$. Note that we *do not* conjugate the coefficients would they happen to be complex. The above decomposition $Q_n = \sum_{k=2}^n \Delta_k^{(n)}$ enjoys the property $\Delta_k^{(n)*} = (-1)^{n-k} \Delta_k^{(n)}$.

The relation between the kernels of Q_n and Q_n^* (if Q_n is normalized to $a_0 - 1 = a_1 = 0$ so is $(-1)^n Q_n^*$) follows from a factorization of Q_n. Let $\{f_0\}, \{f_1, f_0\}, ..., \{f_{n-1}, ..., f_0\}$ be an increasing sequence of subspaces of dimensions $1, 2, ..., n$ in $E_n = \ker Q_n$ (a flag). Set

$$W_0 = 1 \qquad W_p \equiv W_p(f_{p-1}, ..., f_0) \qquad W_n = \text{cst}$$

then

$$Q_n = (d - b_n)(d - b_{n-1}) .... (d - b_1)$$

where

$$b_p = \left(\ell n \frac{W_p}{W_{p-1}}\right)' \quad ,$$

so that $\sum_1^n b_p = \frac{W_n'}{W_n} = 0$. We observe that W_p is a $\frac{p(p-n)}{2}$ differential but of course not a differential polynomial in the a's since it depends on the choice of a flag. This makes factorizations not unique as opposed to the polynomial case. (The map $b \longrightarrow a$ is sometimes called a Miura transformation, it might as well be called a Ricatti transformation). The stabilizer of a flag in GL_n is a Borel sub-group of upper triangular matrices B_n so that the manifold of complete factorizations in GL_n/B_n is of dimension $n^2 - \frac{1}{2}n(n+1) = \frac{1}{2}n(n-1)$.

From

$$(-1)^n Q_n^* = (d + b_1) ... (d + b_n)$$

it follows that $\ker Q_n^*$ is spanned by the Wronskians W_{n-1} (since W_n = cst) for sets of $(n-1)$ linearly independent elements in $\ker Q_n$ i.e. we have the duality formula

$$\ker Q_n^* = \overset{n-1}{\Lambda} \ker Q_n$$

For $n = 3$ when we interpret $\ker Q_3$ as a curve in P_2 (two-dimensional projective space) we recognize that $\ker Q_3^*$ is associated to the dual curve. In general this suggests the consideration of the family of adjoint curves, describing tangents, osculating 2-planes, 3-planes etc... in $P_{\binom{n}{k}-1}$ and raises the question of finding the corresponding differential operators (or sets of operators). Since the corresponding Grassmannians are intersections of quadrics, we deal in the simplest case with a pair of differential equations which can be reduced to a pseudo-differential operator!

As a first instructive and non trivial example take $n = 4$ and $k = 2$, i.e. the set of tangent lines to a curve in P_3. These are points on a quadric in P_5. If

$$Q_4 = d^4 + a_2 d^2 + a_3 d + a_4$$

corresponds to the original curve and if for two independent elements f and g in $\ker Q_4$, we use the Plucker coordinates

$$u = 2(f'g'' - f''g') \qquad v = (fg' - f'g)$$

one finds the coupled differential system

$$\begin{aligned} (d^3 + a_2 d + a_3)\, v &= du \\ (d^3 + d a_2 - a_3)\, u &= 2\,[a_4, d]_+\, v \end{aligned}$$

or equivalently the pseudo differential operator (obtained by eliminating u)

$$\begin{aligned} Q_4 \longrightarrow \hat{Q} &= d^5 + [a_2, d^3]_+ + \left[a_3' - a_2'' - 2a_4 + \frac{1}{2}a_2^2, d\right]_+ \\ &- (a_3 - a_2')\, d^{-1} (a_3 - a_2') \end{aligned}$$

As in the six-dimensional vector space a quadratic form was left invariant and as SL_4 acting on $V_4 \wedge V_4$ is isomorphic to SO_6, we have explicitly exhibited here the relation between the differential operator Q_4 associated to the Lie algebra A_3 and the pseudo differential operator $\hat{Q}$ ($\hat{Q}^* = -\hat{Q}$) pertaining to $D_3 \approx A_3$ (Drinfeld and Sokolov). In agreement with this interpretation we see that the above system admits a quadratic differential invariant

$$d\left\{uv'' - u'v' + u''v + a_2vu - 2a_4v^2 - \frac{u^2}{2}\right\} = 0$$

It is an interesting excercise to show that the bracket does transform as a scalar (v has weight -2 but u is a scalar only mod v').

6. If one starts with a curve in projective space P_{n-1} in the neighborhood of a point and introduces homogeneous coordinates, one has a natural ring of functions but it is not invariant point wise with respect to projective transformations in P_{n-1}. On the other hand if the curve is parametrized and the homogeneous coordinates are promoted to $\frac{1-n}{2}$ differentials, with constant Wronskian, we have the set of differential polynomials in the coefficients of the corresponding Q_n as projective invariants. We have seen that it requires some work to find the appropriate combinations which transform covariantly under diffeomorphisms. It is interesting to organize the differential polynomials using a natural basis suggested by the analogy between the expression of the coefficients a_k and the characters of the linear group although I must admit that this line of thought did not lead to any further understanding of the covariance properties. Let therefore a sequence of integers $0 \leq \ell_0 \leq \ell_1 \leq ... \leq \ell_{n-1}$ correspond to a Young tableau $Y\{\ell\}$ and let $\{f_{n-1}, ..., f_0\}$ be a basis of ker Q_n. The following "differential" characters

$$a_Y = \frac{1}{W_n\left(f_{n-1}, ..., f_0\right)} \det \begin{vmatrix} d^{n-1+\ell_{n-1}} f_{n-1} & & d^{n-1+\ell_{n-1}} f_0 \\ \vdots & & \vdots \\ d^{0+\ell_0} f_{n-1} & & d^{0+\ell_0} f_0 \end{vmatrix}$$

are independent of the choice of basis, and are such that

$$a_\phi = 1 \qquad a_{\,\boxminus\}p \text{ boxes}} = (-1)^p a_p$$

Clearly the a_Y's are differential polynomials in $a_0 = 1$, $a_1 = 0$, $a_2, ..., a_n$, and reduce to ordinary characters when $f_k = \exp q_k x$, in which case the a_k's become up to a sign the (constant) fundamental symmetric functions in the q's. The latter are nothing but the n (generally distinct) solutions of the characteristic equation $\sum_0^n a_k q^{n-k} = 0$ since the Wronskian is up to a factor $\exp \sum q_k$ the discriminant of this equation.

More generally on the n-th Cartesian product $\overset{n}{\times} U$ we can define the antisymetric Slater determinant (or fermionic wave function, or Plucker coordinate)

$$\psi_n\left(z_{n-1}, ..., z_0\right) = \det \begin{vmatrix} f_{n-1}\left(z_{n-1}\right) & \cdots & f_0\left(z_{n-1}\right) \\ \vdots & & \vdots \\ f_{n-1}\left(z_0\right) & \cdots & f_0\left(z_0\right) \end{vmatrix}$$

Changing $\{f_{n-1}, ..., f_0\}$ to linearly independent combinations (with constant coefficients) only multiplies ψ by a constant (the determinant of the transformation). This ψ_n is a generating function for the differential characters as follows from the Taylor expansion around a coinciding point. Set $z_p = x + y_p$, then

$$\begin{aligned} \psi_n\left(x + y_{n-1}, ..., x + y_0\right) &= W_n(x) \prod_{0 \leq i < j \leq n-1} \left(y_j - y_i\right) \\ &\times \sum_{Y\{\ell\}} \frac{1}{\prod_0^{n-1}\left(p + \ell_p\right)!} \mathrm{ch}_Y(y) a_Y(x) \end{aligned}$$

where $\mathrm{ch}_Y(y)$ are the polynomial characters of the linear group, i.e. with $r_p = p + \ell_p$ an increasing sequence of non negative integers

$$\mathrm{ch}_Y = \frac{\det \begin{vmatrix} y_{n-1}^{r_{n-1}} & \cdots & y_0^{r_{n-1}} \\ \vdots & & \vdots \\ y_{n-1}^{r_0} & \cdots & y_0^{r_0} \end{vmatrix}}{\det \begin{vmatrix} y_{n-1}^{n-1} & \cdots & y_0^{n-1} \\ \vdots & & \vdots \\ 1 & & 1 \end{vmatrix}}$$

In particular $da_Y = \Sigma a_{Y'}$ where Y' correspond to those irreducible representations of the linear group which occur in the tensor product of the one pertaining to Y by the fundamental (n-dimensional) representation. Symbolically

$$da_Y = \Sigma a_{Y'} \longleftrightarrow Y \otimes \square = \oplus Y'$$

If under a diffeormorphism $x = g(t)$, $f(x)\mathrm{d}x^{\frac{1-n}{2}} = \tilde{f}(t)\mathrm{d}t^{\frac{1-n}{2}}$ we write for the polar part

$$\frac{g'(u)^{\frac{n+1}{2}}}{[g(u) - g(t)]^{r+1}} = \sum_{k=0}^{r} \frac{A_{r,k}(t)}{(u-t)^{k+1}} + \text{reg.}$$

we can find the transformation properties of the differential characters in closed form as

$$\frac{a_Y(x)}{\prod_0^{n-1} r_p!} = \sum_{Y' \leq Y} A_{Y,Y'}(t) \frac{\tilde{a}_{Y'}(t)}{\prod_0^{r-1} k_p!}$$

with the partial ordering meaning that if $Y \longrightarrow \{0 \leq r_0 < r_1 < \ldots < r_{n-1}\}$ and $Y' \longrightarrow \{0 \leq k_0 < k_1 < \ldots < k_{n-1}\}$ then $Y' \leq Y$ means $k_0 \leq r_0, k_1 \leq r_1, \ldots, k_{n-1} \leq r_{n-1}$. Finally

$$A_{YY'}(t) = \det \left(A_{r_p, k_s} \right)$$

In infinitesimal form $x = t + \epsilon(t)$, to leading order

$$\begin{aligned} A_{r,r} &= 1 + \left(\frac{n-1}{2} - r \right) \epsilon'(t) \\ A_{r,k} &= \frac{1}{(r-k+1)!} \left[(r-k+1)\frac{n+1}{2} - r - 1 \right] \epsilon^{(r-k+1)}(t) \qquad 0 \leq k < r \end{aligned}$$

The rule $A_{YY'} \neq 0$ only for $Y' \leq Y$ is of course in agreement with the transformation proprieties of the quantities $a_2, a_3, \ldots, a_n$. Note also that $A_{YY} = g'(t)^{-\Sigma \ell_p} = \left(\frac{\mathrm{d}t}{\mathrm{d}x}\right)^{\Sigma \ell_p}$. Unfortunately we were not able, using this formalism, to find a simple algorithm to compute the set of k differential forms.

7. Except for the discussion of the behaviour of differential operators under diffeomorphisms, we have hardly justified up to now the title of this contribution. We want now to study general W-deformations of these operators. The symbol W seems quite appropriate given the prominent role of Wronskian. For this purpose we first recast the previous analysis as follows. In infinitesimal form the maps of $\mathcal{F}_{\frac{1-n}{2}}$ on themselves

$$f \longrightarrow F = (1 + X_1) f \qquad g \longrightarrow G = (1 + Y_1) g$$

with

$$X_1 = \epsilon d + \frac{1-n}{2}\epsilon' \qquad Y_1 = \epsilon d + \frac{1+n}{2} = -X_1^*$$

transform the map $f \longrightarrow g = Q_n f$ into $F \longrightarrow G = (Q_n + \delta Q_n) F$ where $Q_n + \delta Q_n$ is again an n-th order normalized operator with

$$\delta Q_n = Y_1 Q_n - Q_n X_1$$

We are therefore led to look for W-generalizations as pairs of differential operators (we drop an infinitesimal scale factor) such that the variation of a *generic* Q_n

$$\delta Q_n = Y Q_n - Q_n X \tag{1}$$

be a differential operator of order less than or equal to $n-2$, with X and Y defined up to the addition of a common (irrelevant) constant term. Indeed Q_n commutes with the multiplication by a constant.

It is good to remember that isospectral flows are just of this form with the requirement $X = Y$. Since we drop this constraint X and Y will now depend on some arbitrary functions. In a very loose sense this is a "gauged" form of the generalized KdV flows (pertaining to the linear group).

For the sequel it is convenient to use freely the algebraic concept of pseudo-differential operator. We rely on the exposition of Drinfeld and Sokolov where the reader can find some of the original references (some older ones seem to be to Pincherle and Schur). We already introduced the symbol d^{-1} such that

$$d^{-k} f = \sum_{i=0}^{\infty} (-1)^i \begin{pmatrix} k+i-1 \\ i \end{pmatrix} f^{(i)} d^{-i-k} \qquad (d^{-1})^* = -d^{-1}$$

Let R_+ stand for the differential part of a pseudo differential operator R and $R_- = R - R_+$. The coefficient in d^{-1} (written to the right or to the left, they are equal) is called the residue of R and enjoys the fundamental property

$$\text{Res}\,[R_1, R_2] = \text{total derivative}$$

which leads to the definition of a commutative trace on pseudo differential operators

$$\text{Tr}\, R = \int_C \text{Res}\,(R)$$

Any non zero pseudo differential operator has a finite formal inverse R^{-1} and the $*$ involution extends to pseudo differential operators. Since Y is a differential operator, from (1) written as

$$Y = Q_n X\, Q_n^{-1} + \delta Q_n Q_n^{-1}$$

it also follows that

$$Y = (Q_n X\, Q_n^{-1})_+$$

showing that Y is entirely determined by X and of the same order. Moreover since the order of $\delta Q_n \leq n-2$, the residue of $\delta Q_n Q_n^{-1}$ vanishes, hence X is constrained by the unique condition

$$\text{Res}\; Q_n X\, Q_n^{-1} = 0$$

If we write $X = \tilde{X} + c$ with c a function the above reads

$$\text{Res } Q_n \tilde{X}\, Q_n^{-1} + nc'(x) = 0$$

showing that we can pick a fiducial $\tilde{X}$, requiring for instance that it admits constants in its kernel: $\tilde{X} \cdot 1 = 0$ in terms of which

$$X = \tilde{X} - \frac{1}{n} \int_C^x \text{Res } \left(Q_n \tilde{X}\, Q_n^{-1} \right)$$

Finally we see that if we add to X any term of the form $P\, Q_n$ (P differential operator) then Y increases by $Q_n P$ while δQ_n is unaffected. By the Euclidean division algorithm applied to differential operators we can reduce X (and Y) to differential operators of order at most $n-1$. Given the above linear constraint on X this shows that we can break the most general W-deformation into a linear combination of $n-1$ independent ones, hence that it depends on $n-1$ arbitrary functions (of definite weight as we shall see). This number $n-1$ (as the number of w's) is, not surprisingly, the rank of SL_n.

To illustrate the above considerations on the first of these deformations, we have X_1, Y_1 of order 1 with

$$\tilde{X}_1 = \epsilon d \qquad X_1 = \epsilon d - \frac{1}{n} \int^x \text{Res } (Q_n(\epsilon d) Q_n^{-1})$$

Hence

$$X_1 = \epsilon d - \frac{1}{n} \int^n \frac{n(n-1)}{2} \epsilon'' = \epsilon d - \frac{n-1}{2} \epsilon'$$

We note that the constraint Res $Q_n X Q_n^{-1} = 0$ relates the coefficients in X to those of Q_n. As a result if δ_X stands for the deformation of Q_n (or rather its coefficients) the commutator of two such operations reads

$$[\delta_X, \delta_{X'}] = \delta_{[X',X]} + \delta_X X' - \delta_X X$$

Where consistently by $\delta_X X'$ we mean the δ_X deformation of its dependence on Q_n.

We now want to find a basis of these deformations $\delta_k(\eta) \equiv \delta_{X_k(\eta)}$ each one depending on a single function of weight $-k$, i.e. such that,

$$[\delta_k(\eta), \delta_1(\epsilon)] = \delta_k (\epsilon\eta' - k\eta\epsilon')$$

which diagonalizes the effect of diffeomorphisms. This is achieved most efficiently by using an appropriate Poisson bracket (or Hamiltonian) structure on the manifold of Q_n's in such a way (as it will turn out) that

$$\delta_k(\eta) \cdot Q_n(x) = \sum_{p=2}^{n} \int_C \text{d}y\; \eta(y) \left\{ w_{k+1}(y),\; a_p(x) \right\} d_x^{n-p}$$

Of course one could also directly try to solve the above condition.

8. The goal is now to describe a Hamiltonian structure on (integrals of) differential polynomials in the coefficients of the generic n-th order differential operator. We shall briefly reproduce the construction given by Drinfeld and Sokolov. Assuming $a_1 = 0$, it is sufficient to deal with linear forms

$$\ell_U(Q_n) = \int \text{d}x\; \sum_{i=2}^{n} a_i(x) u_i(x)$$

where the collection of cofficients can be associated to a pseudo differential operator

$$U = d^{1-n}u_2 + d^{2-n}u_3 + \cdots + d^{-1}u_n$$

in such a way that

$$\ell_U(Q_n) = \mathrm{Tr}\ Q_n U$$

Since a_1 vanishes one can freely add to U an arbitrary term $d^{-n}u_1$. This freedom will be used in the sequel.

The Poisson bracket structure to be defined is derived from a first order matrix formalism, substituting the differential operator Q_n, which then allows a neat generalization to other Lie groups. Suppose we are given a $n \times n$ matrix valued function on the cycle $C : x \longrightarrow A(x)$ and consider an (action) functional

$$S(A) = \int_C L(A)$$

where $L(A)$ (the Lagrange function) is a differential polynomial in the coefficients of A. For two such maps let

$$(B,C) = (C,B) = \int_C \mathrm{tr}\ BC$$

where tr is the ordinary matrix trace. With dots denoting derivatives $\dot{B} \equiv [d,B]$ we have obviously

$$\left(\dot{B},C\right) + \left(B,\dot{C}\right) = 0 \qquad ([A,B],C) + (B,[A,C]) = 0$$

Thus

$$([d+A,B],C) + (B,[d+A,C]) = 0$$

Now define the standard functional derivative $\frac{\delta S}{\delta A}(x) \equiv \mathrm{grad}_A S$

$$\frac{\mathrm{d}}{\mathrm{d}t} S(A+tH)\bigg|_{t=0} = \int_C \mathrm{tr}\,(\mathrm{grad}_A\ S, H)$$

for $H(x)$ an arbitrary matrix function on C. This means

$$(\mathrm{grad}_A S)_{ij} = \sum_{p \geq 0} (-d)^p \frac{\partial L}{\partial A_{ji}(x)^{(p)}}$$

where we observe the transposition of indices implied by the trace. Define the Poisson bracket of two such actions as

$$\{S_1, S_2\} = (\mathrm{grad}_A S_1,\ [d+A,\ \mathrm{grad}_A\ S_2])$$

which is a natural generalization of the standard one dimensional prescription. One readily verifies that it is antisymmetric and satisfies the Jacobi identity.

To the differential operator (we need not assume for the time being that $a_1\ (Q_n)$ vanishes)

$$Q_n = d^n + a_1 d^{n-1} + a_2 d^{n-2} + \ldots$$

one associates a first order matrix differential operator

$$d + A \equiv d + a - J = \begin{pmatrix} d+a_1 & a_2 & a_3 & \cdots & a_n \\ -1 & d & 0 & \cdots & 0 \\ & -1 & d & \cdots & 0 \\ & & \ddots & \ddots & \vdots \\ & & & -1 & d \end{pmatrix}$$

where d multiplies the identity matrix, J has 1 in the first diagonal below the main one and a has $a_1, ..., a_n$ in the first row. Clearly to $f \in \ker Q_n$ there corresponds a vector $\underline{f} = \left(f^{(n-1)}, f^{(n-2)}, ..., f\right)^T$ such that

$$(d+A)\underline{f} = (Q_n f, 0, ..., 0)^T = 0$$

and conversely. Hence the kernels are in one to one correspondence.

The above formalism endows the actions (integrals of differential polynomials in the a_k's) of a Poisson bracket structure which, as we shall verify, in the case of linear functionals turns out to be

$$\begin{aligned}\{\ell_U, \ell_V\} &= \mathrm{Tr}\left((Q_n U)_+ Q_n V - V Q_n (U Q_n)_+\right) \\ &= \ell_V \left((Q_n U)_+ Q_n - Q_n (U Q_n)_+\right)\end{aligned}$$

the so-called second Gelfand Diki Hamiltonian structure. The analogy of this formula with the deformation equation is striking and suggests a mean to construct pairs (X, Y) of differential operators corresponding to the W-algebra. Before setting this point it is however interesting to present the interpretation given by Drinfeld and Sokolov of the matrix formulation to enable one to generalize it to other geometries (or other Lie groups).

The combination $d + A$ leads to the interpretation of $x \longrightarrow A(x)$ as a matrix valued gauge potential with values in the Lie algebra $g\ell_n$ ($s\ell_n$ when a_1 vanishes). Now $A = a - J$, so one needs to discuss both $a(x)$ and J. Diagonal elements are generating a maximal abelian (i.e. Cartan) subalgebra (for the time being we ignore the distinction between $g\ell_n$ and $s\ell_n$). Each matrix with a unique 1 just below the diagonal may be interpreted as a generator corresponding to a simple (positive) root while matrices with non vanishing elements above the diagonal correspond to the nihilpotent Lie algebra generated by elements corresponding to the negative roots. If we include the diagonal ones we get a Borel subalgebra $\mathcal{B}$. Let us assume in general that $a \in \mathcal{B}$ and that J is the sum of Chevalley generators indexed by simple positive roots. Finally let $\mathcal{N}$ denote the group generated by the above nihilpotent algebra. In the above presentation they are matrices with zero below the diagonal and one along it. If $N \in \mathcal{N}$ then

$$N^{-1} J N = J + b \qquad b \in \mathcal{B}$$

From this it follows that under a gauge transformation under the subgroup $\mathcal{N}$, i.e. with $x \longrightarrow N(x) \in \mathcal{N}$

$$d + A \longrightarrow d + \tilde{A} = N^{-1}(d+A)N$$

we have

$$\begin{aligned}\tilde{A} &= N^{-1} A N + N^{-1}(dN) = \tilde{a} - J \\ \tilde{a} &= N^{-1} a N + N^{-1}(dN) - \left(N^{-1} J N - J\right)\end{aligned}$$

with each term in the r.h.s. belonging to $\mathcal{B}$ if a does. Therefore the condition $a \in \mathcal{B}$ is stable under $\mathcal{N}$-gauge transformation. It allows therefore to consider actions $S(a) \equiv S(A)$ with $a \in \mathcal{B}$ invariant under the above gauge transformations. We call them for short gauge invariant and write $S(a)$ or $S(A)$ since only a is allowed to vary.

The definition of $\mathrm{grad}_A S$ is now ambiguous since one is only allowed to consider variations $S(A + tH)$, $H \in \mathcal{B}$, so using the invariant quadratic form in the general case as a substitute for the trace in the definition of (A, B)

$$\frac{\mathrm{d}}{\mathrm{d}t} S(A + tH)\,|_{t=0} = (H, \mathrm{grad}_A S)$$

and taking into account that $(H, \mathrm{Lie}\,\mathcal{N}) = 0$ for $H \in \mathcal{B}$ we see that $\mathrm{grad}_A S$ is only defined up to an element in Lie $\mathcal{N}$ (here strictly upper triangular matrices). Nevertheless this indetermination does not affect the Poisson bracket between two such functionals. Indeed for

$N = 1 + \eta M$, $M \in \mathrm{Lie}\,\mathcal{N}$, η infinitesimal constant, we have to first order

$$d + A_\eta = d + A + \eta[d + A, M]$$

$$\frac{\mathrm{d}}{\mathrm{d}\eta} S(A_\eta)\,|_{\eta=0} = \frac{\mathrm{d}}{\mathrm{d}\eta} S(A)\,|_{\eta=0} = 0 = (\mathrm{grad}_A S, [d + A, M])$$

which insures that the definition of the bracket

$$\{S_1, S_2\} = (\mathrm{grad}_A S_1, [d + A, \mathrm{grad}_A S_2])$$

is insensitive to changing $\mathrm{grad}_A S_i \longrightarrow \mathrm{grad}_A S_i + M_i$. Furthermore one also wants to show that the bracket is itself ($\mathcal{N}$-) gauge invariant. Let $x \longrightarrow N(x) \in \mathcal{N}$ induce such a transformation, $A \longrightarrow \tilde{A}$. For B also in $\mathcal{B}$ we have (η constant)

$$S_i(A + \eta B) = S_i\left(\widetilde{A + \eta B}\right) = S_i\left(\tilde{A} + \eta N^{-1} B N\right)$$

Hence

$$\begin{aligned} \frac{\mathrm{d}}{\mathrm{d}\eta} S_i(A + \eta B)\,|_{\eta=0} &= (\mathrm{grad}_A S_i, B) = \left(\mathrm{grad}_{\tilde{A}} S_i, N^{-1} B N\right) \\ &= \left(N\ \mathrm{grad}_{\tilde{A}} S_i N^{-1}, B\right) \end{aligned}$$

and

$$\mathrm{grad}_A S_i = N\ \mathrm{grad}_{\tilde{A}} S_i N^{-1} \qquad \mathrm{mod\ Lie}\,\mathcal{N}$$

and since Lie $\mathcal{N}$ is invariant under the adjoint action of $\mathcal{N}$ we have

$$\mathrm{grad}_{\tilde{A}} S_i = N^{-1} \mathrm{grad}_A S_i N \qquad \mathrm{mod\ Lie}\,\mathcal{N}$$

Therefore

$$\begin{aligned} \{S_1, S_2\}\left(\tilde{A}\right) &= \left(N^{-1}\mathrm{grad}_A S_1 N, \left[N^{-1}(d + A)N, N^{-1}\mathrm{grad}_A S_2 N\right]\right) \\ &= \{S_1, S_2\}(A) \end{aligned}$$

Finally using $\mathcal{N}-$ gauge invariance one can bring A, hence $a \in \mathcal{B}$, into a unique canonical form. For instance in the linear case a recursive reasoning shows that one can take either

$$a_{\mathrm{can}} = \begin{pmatrix} a_1 a_2 \ldots a_n \\ \\ 0 \end{pmatrix} \quad \text{or} \quad a^{\mathrm{can}} = \begin{pmatrix} 0 & a_n^* \\ & -a_{n-1}^* \\ & \vdots \\ & (-1)^{n-1} a_1^* \end{pmatrix}$$

Note that switching from Q_n to $(-1)^n Q_n^*$ amounts to change $(-1)^{n-p} a_p^* \longrightarrow (-1)^p a_p$ in a^{can}.

In the case of an arbitrary simple Lie algebra $\mathcal{L}$ one can grade it as follows. Write the Chevalley-Serre basis indexed by simple roots as h_i, x_i, y_i with i ranging over the nodes of the Dynkin diagram

$$\begin{aligned} [h_i, h_j] &= 0 & [x_i, y_j] &= \delta_{ij} h_i \\ [h_i, x_j] &= n_{ij} x_j & [h_i, y_j] &= -n_{ij} y_j \end{aligned}$$

$$(\mathrm{ad}\ x_i)^{1-n_{ij}}\, x_j = (\mathrm{ad}\ y_i)^{1-n_{ij}}\, y_j = 0 \quad i \neq j$$

with integers n_{ij} given in terms of the diagram.

Then

$$\mathcal{L} = \bigoplus_{-k \le j \le k} \mathcal{L}^{(j)} \quad x_i \in \mathcal{L}^{(1)} \quad y_i \in \mathcal{L}^{(-1)} \quad h_i \in \mathcal{L}^{(0)}$$

and the grading is compatible with commutation. The integer $h = k + 1$ is the Coxeter number and the (positive) integer j is a Coxeter exponent if and only if the injective map ($J \equiv \sum_i x_i$)

$$\text{ad } J \quad \mathcal{L}^{(-j-1)} \longrightarrow \mathcal{L}^{(-j)}$$

is *not* surjective with mult(j) = dim$\mathcal{L}^{(-j)}$ − dim$\mathcal{L}^{(-j-1)}$ (hence $\sum$ mult(j) = dim$\mathcal{L}^{(0)}$ = rank$\mathcal{L}$).

From the above, a canonical form for a can be chosen such that it has non vanishing elements in a complementary of the images of ad J in each $\mathcal{L}^{(-j)}$, thus a number of elements equal to the rank, each one with a grade corresponding to a Coxeter exponent.

In the linear case let us briefly discuss the relation with the differential operator $Q_n\,(a_{\text{can}})$. To find the application $d + A \longrightarrow Q_n$, one considers that $d + A$ is a matrix in the differential ring generated by smooth functions and d, loosely speaking that matrix elements are quantum mechanic operators and one writes

$$d + A = \begin{pmatrix} \underline{\alpha} & \beta \\ \Delta & \underline{\gamma} \end{pmatrix}$$

with $\underline{\alpha}$ and ($\underline{\gamma}$) a row (column) vector with $n-1$ entries, β a 1×1 matrix, Δ a $(n-1)\times(n-1)$ matrix with -1 along the diagonal and zero below, hence invertible. One then sets

$$d + A \longrightarrow Q_n \equiv \beta - \underline{\alpha}\Delta^{-1}\underline{\gamma}$$

It follows that Q_n is an n-th order differential operator. This arises by comparing the kernels when writing a column vector on which $d + A$ acts as $(\underline{g}, f)^T$ where $\underline{g}$ has $n-1$ entries, then from the system

$$\begin{aligned} \underline{\alpha}\cdot\underline{g} + \beta f &= 0 \\ \Delta\underline{g} + \underline{\gamma} f &= 0 \end{aligned}$$

one obtains by elimination

$$\left(\beta - \underline{\alpha}\Delta^{-1}\underline{\gamma}\right) f = 0$$

One shows that Q_n is independent from ($\mathcal{N}$-) gauge transformation and that when $a = a_{\text{can}}$ it reduces to the required form.

The final (and painful) exercise is to express the Hamiltonian structure in terms of the coefficients a_k. We present the calculation again in the linear case starting with a linear functional $\ell_U\,(Q_n)$ corresponding to a unique gauge invariant functional $S_U(A)$. For $H(x) \in B$

$$S_U(A + tH) = \int \text{Res}\,(Q_n(t)U)$$

with

$$d + A + tH = \begin{pmatrix} \underline{\alpha}(t) & \beta(t) \\ \Delta(t) & \underline{\gamma}(t) \end{pmatrix}$$

If $H = \begin{pmatrix} \underline{\xi} & \zeta \\ \delta & \underline{\eta} \end{pmatrix}$

$$\underline{\alpha}(t) = \underline{\alpha} + \underline{\xi}t, \qquad \beta(t) = \beta + \zeta t, \qquad \underline{\gamma}(t) = \underline{\gamma} + \underline{\eta}t,$$

$$\Delta(t) = \Delta + \delta t, \qquad \Delta^{-1}(t) = \Delta^{-1} - t\Delta^{-1}\delta\Delta^{-1} + \dots$$

implying

$$(\mathrm{grad}_A S_U, H) = \int \mathrm{Res}\frac{\mathrm{d}}{\mathrm{d}t}Q_n(t)\,|_{t=0}\, U$$

where

$$\frac{\mathrm{d}Q_n(t)}{\mathrm{d}t}\,|_{t=0} = \zeta - \underline{\xi}\Delta^{-1}\underline{\gamma} - \underline{\alpha}\Delta^{-1}\underline{\eta} + \underline{\alpha}\Delta^{-1}\delta\Delta^{-1}\underline{\gamma}$$

i.e.

$$\mathrm{grad}_A S_U = \mathrm{Res}\begin{pmatrix} -\Delta^{-1}\underline{\gamma}U & \Delta^{-1}\underline{\gamma}\otimes U\underline{\alpha}\Delta^{-1} \\ U & -U\underline{\alpha}\Delta^{-1} \end{pmatrix}$$

where we note the transposition of the sizes of the blocks. Since when P is a pseudo differential operator

$$d\ \mathrm{Res}P = \mathrm{Res}[d, P]$$

we find for $S_V(A)$ corresponding to another linear functional

$$[d + A, \mathrm{grad}_A S_V] = \mathrm{Res}\left[\begin{pmatrix} \underline{\alpha} & \beta \\ \Delta & \underline{\gamma} \end{pmatrix}, \begin{pmatrix} -\Delta^{-1}\underline{\gamma}V & \Delta^{-1}\underline{\gamma}\otimes V\underline{\alpha}\Delta^{-1} \\ V & -V\underline{\alpha}\Delta^{-1} \end{pmatrix}\right]$$

Remembering that $Q_n = \beta - \underline{\alpha}\Delta^{-1}\underline{\gamma}$ one sees that

$$[d + A, \mathrm{grad}_A S_V] = \mathrm{Res}\left\{\begin{pmatrix} Q_n V & -Q_n V\underline{\alpha}\Delta^{-1} \\ 0 & 0 \end{pmatrix} - \begin{pmatrix} 0 & -\Delta^{-1}\underline{\gamma}VQ_n \\ 0 & VQ_n \end{pmatrix}\right\}$$

Explicitly with $a \equiv a_{\mathrm{can}}$

$$\underline{\alpha} = (d + a_1, a_2, \dots, a_{n-1}) \qquad \beta = a_n$$

$$\Delta^{-1} = -\begin{pmatrix} 1 & d & d^2 & \dots & d^{n-2} \\ & 1 & d & \dots & d^{n-3} \\ 0 & & & \ddots & 1 \end{pmatrix}$$

hence

$$\begin{aligned} \underline{\alpha}\Delta^{-1} &= -\left(d + a_1, d^2 + a_1 d + a_2, \dots, d^{n-1} + a_1 d^{n-2} + \dots + a_{n-1}\right) \\ &= -\left((Q_n d^{1-n})_+, \dots, (Q_n d^{-1})_+\right) \end{aligned}$$

$$\Delta^{-1}\underline{\gamma} = -\begin{pmatrix} d^{n-1} \\ \vdots \\ d \end{pmatrix}$$

Therefore

$$\begin{aligned} \{\ell_U, \ell_V\}(A) = \{S_U, S_V\}(A) &= \int \mathrm{tr}\ (\mathrm{grad}_A S_U, [d + A, \mathrm{grad}_A S_V]) \\ &= \int \mathrm{tr}\ \mathrm{Res}\Lambda_1\ \mathrm{Res}\Lambda_2 \end{aligned}$$

with

$$\Lambda_1 = \begin{pmatrix} d^{n-1}U & \begin{bmatrix} d^{n-1} \\ \vdots \\ d \end{bmatrix} \otimes U\left[\left(Q_n d^{1-n}\right)_+, \ldots, \left(Q_n d^{-1}\right)_+\right] \\ \vdots & \\ dU & \\ U & U\left[\left(Q_n d^{1-n}\right)_+, \ldots, \left(Q_n d^{-1}\right)_+\right] \end{pmatrix}$$

$$\Lambda_2 = \begin{pmatrix} \left[Q_nV, Q_nV\left(Q_nd^{1-n}\right)_+, \ldots, Q_nV\left(Q_nd^{-2}\right)_+\right] & Q_nV\left(Q_nd^{-1}\right)_+ - d^{n-1}VQ_n \\ 0 & -\begin{bmatrix} d^{n-2}VQ_n \\ \vdots \\ dVQ_n \\ VQ_n \end{bmatrix} \end{pmatrix}$$

This means that (with $\mathrm{Res}\left(Q_nd^{-1}\right)_+ = 1$)

$$\begin{aligned} \{\ell_U, \ell_V\}(A) &= \int \sum_{k=1}^{n} \Big[\mathrm{Res}\left(d^{k-1}U\right) \mathrm{Res} Q_nV\left(Q_nd^{-k}\right)_+ \\ &\quad - \mathrm{Res}\left(U\left(Q_nd^{-k}\right)_+ \mathrm{Res}\left(d^{k-1}VQ_n\right)\right)\Big] \end{aligned}$$

Since for α a function $\alpha\ \mathrm{Res}P = \mathrm{Res}\alpha P = \mathrm{Res}P\alpha$ the first sum under the integral reads

$$\begin{aligned} &\mathrm{Res}\left(Q_nV \sum_{k=1}^{n} \left(Q_nd^{-k}\right)_+ \mathrm{Res}\left(d^{k-1}U\right)\right) \\ &= \mathrm{Res}\left(Q_nV \sum_{k=1}^{n} \left(Q_nd^{-k}\right)_+ u_{n+1-k}\right) \\ &= \mathrm{Res}\left(Q_nV\left(Q_n \sum_{k=1}^{n} d^{-k}u_{n+1-k}\right)_+\right) = \mathrm{Res}\left(Q_nV\left(Q_nU\right)_+\right) \end{aligned}$$

where we have written $U = \sum_{k=1}^{n} \mathrm{d}^{-k}u_{n+1-k}$. Similarly the second sum reads

$$\mathrm{Res}\left(U\left(Q_n\left(VQ_n\right)_-\right)_+\right)$$

Thus one form of the bracket is

$$\{\ell_U, \ell_V\} = \mathrm{Tr}\left[Q_nV\left(Q_nU\right)_+ - U\left(Q_n\left(VQ_n\right)_-\right)_+\right]$$

But

$$\mathrm{Tr}\left[U\left(Q_n\left(VQ_n\right)_-\right)_+\right] = \mathrm{Tr}\left[UQ_n\left(VQ_n\right)_-\right] - \mathrm{Tr}\left[U\left(Q_n\left(VQ_n\right)_-\right)_-\right]$$

and the second term is readily seen to vanish. In the first we can replace similarly UQ_n by $(UQ_n)_+$ then $(VQ_n)_-$ by VQ_n and use $\mathrm{Tr}\,AB = \mathrm{Tr}\,BA$ to rewrite finally

$$\{\ell_U, \ell_V\} = \mathrm{Tr}\left[\left(Q_nU\right)_+ Q_nV - VQ_n\left(UQ_n\right)_+\right]$$

as claimed above.

9. We have obtained the Poisson structure corresponding to $g\ell_n$ rather than $S\ell_n$ which corresponds to $a_1(Q_n) = 0$. In order to apply the above expressions to find a pair, tentatively $X = (UQ_n)_+$, $Y = (Q_nU)_+ = \left(Q_nXQ_n^{-1}\right)_+$ for a deformation of Q_n, we want the condition $a_1 = 0$ to apply also to the deformed operator. Indeed such a pair X, Y would in general lead to $\delta a_1 \neq 0$ or equivalently $\mathrm{Res}\left(Q_nXQ_n^{-1}\right) \neq 0$.

But recall that a term $d^{-n}u$, in U does not affect the functional $\ell_U(Q_n)$ when a_1 vanishes but would affect the above brackets. One is therefore free to adjust u_1 to

$$u_1 = \frac{1}{n}\int^x \mathrm{Res}\,[U, Q_n]$$

to remove the variation δa_1 in $\delta Q_n = (Q_n U)_+ Q_n - Q_n (UQ_n)_+$. So that we use the above brackets with the understanding that v_1 vanishes and u_1 is given by the above formula.

We are now ready to express the W-deformations using Poisson brackets as follows. Let $\ell_V(Q_n)$ stand for an arbitrary linear functional of Q_n

$$\ell_V(Q_n) = \int \sum_2^n a_i v_i = \mathrm{Tr}\, Q_n V$$

$$V = d^{1-n} v_2 + \ldots + d^{-1} V_n$$

Choose the non-linear function ($\epsilon(x)$ of weight $-k$)

$$H_k = \int \epsilon w_{k+1}$$

Compute its gradient and associate to it a pseudo differential operator

$$U_k = \sum_{i=1}^{n-1} d^{-i} \frac{\delta H_k}{\delta a_{n+1-i}} + \frac{1}{n} d^{-n} \int^x \mathrm{Res}\left[\sum_{i=1}^{n-1} d^{-i} \frac{\delta H_k}{\delta a_{n+1-i}}, Q_n\right]$$

where the functional derivative (the gradient) $\frac{\delta H}{\delta a_p}$ is a short hand for

$$\sum_r (-1)^r \left(d^r \frac{\delta H}{\delta a_p^{(r)}}\right)$$

we have then the k-th "W-flow" (depending on ϵ)

$$\delta_k \ell_V(Q_n) = \{H_k, \ell_V(Q_n)\} = \ell_V(\delta_k Q_n)$$

$$\delta_k Q_n = (Q_n U_k)_+ Q_n - Q_n (U_k Q_n)_+$$

Since we have the expression of w_{k+1} we derive the pairs X_k, Y_k of differential operators explicitly as

$$\begin{aligned}
X_k &= \tilde{X}_k - \frac{1}{n}\int^x \mathrm{Res}\left(Q_n \tilde{X}_k Q_n^{-1}\right) \\
Y_k &= \tilde{Y}_k + \frac{1}{n}\int^x \mathrm{Res}\left(Q_n^{-1} \tilde{Y}_k Q_n\right) \\
\tilde{X}_k &= \left(\left(\sum_{i=1}^{n-1} d^{-i} \frac{\delta H_k}{d a_{n+1-i}}\right)\left(d^n + \sum_{j=2}^n a_j d^{n-j}\right)\right)_{++} \\
\tilde{Y}_k &= \left(\left(d^n + \sum_{j=2}^n a_j d^{n-j}\right)\left(\sum_{i=1}^{n-1} d^{-i} \frac{\delta H_k}{\delta a_{n+1-i}}\right)\right)_{++}
\end{aligned}$$

where the symbol $++$ means that we keep the differential part omitting the constant term $(\tilde{X}_k \cdot 1 = \tilde{Y}_k \cdot 1 = 0)$.

Using the notations of section 4 the terms in $\tilde{X}_k$, $\tilde{Y}_k$ that are independent of a's or w's take a simple form

$$\begin{aligned}\tilde{X}_k(0) &= \sum_{s=1}^{k} B_{k+1,s+1}\epsilon^{(k-s)}d^s \\ \tilde{Y}_k(0) &= (-1)^k \tilde{X}_k^*(0) = \sum_{s=1}^{k} C_{k+1,s+1}\epsilon^{(k-s)}d^s\end{aligned}$$

where

$$\begin{aligned}C_{k+1,s+1} &= (-1)^k \sum_{\ell=0}^{j-1}(-1)^{\ell+s}\begin{pmatrix}\ell+s\\ \ell\end{pmatrix} B_{k+1,\ell+s+1} \\ &= B_{k+1,s+1}(n \longrightarrow -n)\end{aligned}$$

where use has been made of the identity

$$B_{k+1,s+1} = (-1)^{k-s}\sum_{\ell=0}^{k-s}\begin{pmatrix}n-s-1\\ \ell\end{pmatrix} B_{k+1,\ell+s+1}$$

The first few pairs X_k, Y_k are displayed in the following table.

$$\begin{aligned}X_1 &= \epsilon d - \frac{n-1}{2}\epsilon' \\ Y_1 &= \epsilon d + \frac{n+1}{2}\epsilon' \\ X_2 &= \epsilon d^2 - \frac{n-2}{2}\epsilon' d + \left\{\frac{2}{n}\epsilon a_2 + \frac{1}{12}(n-1)(n-2)\epsilon''\right\} \\ Y_2 &= \epsilon d^2 + \frac{n+2}{2}\epsilon' d + \left\{\frac{2}{n}\epsilon a_2 + \frac{1}{12}(n+1)(n+2)\epsilon''\right\} \\ X_3 &= \epsilon d^3 - \frac{n-3}{2}\epsilon' d^2 + \left\{\frac{(n-2)(n-3)}{10}\epsilon'' + \frac{6}{5}\frac{3n^2-7}{5n\,(n^2-1)}a_2\epsilon\right\} d \\ &+ \left\{\frac{3}{n}w_3\epsilon - \frac{3(n+2)(n-7)}{10n(n+1)}a_2'\epsilon - \frac{(n-3)(4n-7)}{5n(n+1)}a_2\epsilon' - \frac{(n-1)(n-2)(n-3)}{5!}\epsilon'''\right\} \\ Y_3 &= \epsilon d^3 + \frac{n+3}{2}\epsilon' d^2 + \left\{\frac{(n+2)(n+3)}{10}\epsilon'' + \frac{6}{5}\frac{3n^2-7}{5n\,(n^2-1)}a_2\epsilon\right\} d \\ &+ \left\{\frac{3}{n}w_3\epsilon + \frac{3(n-2)(n+7)}{10n(n-1)}a_2'\epsilon + \frac{(n+3)(4n-7)}{5n(n-1)}a_2\epsilon' + \frac{(n+1)(n+2)(n+3)}{5!}\epsilon'''\right\}\end{aligned}$$

With the explicit expressions of the w's, X's and Y's we can now from the Poisson brackets of w's among themselves obtaining thus the classical W-algebra (for SL_n). For small n it already appears in the literature. Of course it always contains the Virasoro algebra (as a Lie subalgebra) together with the brackets expressing that w_k has weight k. Thus using distributions

$$\begin{aligned}\{w_2(y), w_2(x)\} &= \left(w_2'(x) + 2w_2(x)d_x + \frac{n\,(n^2-1)}{12}d_x^3\right)\delta(x,y) \\ \{w_2(y), w_2(x)\} &= \left(w_k'(x) + kw_k(x)d_x\right)\delta(x,y)\end{aligned}$$

The other brackets are of course easier to compute in the coordinate u where $a_2 \equiv w_2$ vanishes. If

$$\{w_k(v), w_\ell(u)\}_{w_2=0} = \Delta\left(w_j(u), d_u\right)\delta(u,v)$$

with Δ an ordered polynomial in d_u, $w_j(u)$ and their derivatives, then in the generic coordinate

$$\{w_k(y), w_\ell(x)\} = \varphi^k \Delta\left(\varphi^{-j} w_k(x), \varphi^{-1} d_x\right) \varphi^{\ell-1} \delta(x,y)$$

with $\varphi = \left(\frac{du}{dx}\right)$ and the δ-distribution has contributed an extra factor φ^{-1}. The differential operator Δ has of course to satisfy specific constraints in order that the final expression depend only on the Schwarzian derivative of the map hence on w_2. As an illustration for generic n in the u-coordinate we have

$$\{w_3(v), w_3(u)\}|_{w_2=0} = \left(2\,[w_4, d]_+ - \frac{(n-2)(n-1)n(n+1)(n+2)}{6!} d^5\right) \delta(u,v)$$

$$\{w_3(v), w_4(u)\}|_{w_2=0} = \left(5w_5 d + 2w_5' - \frac{(n-3)(n+3)}{70}\left(14 w_3 d^3 + 14 w_3' d^2 + 6 w_3'' d + w_3'''\right)\right) \delta(u,v)$$

$$\{w_4(v), w_4(u)\}|_{w_2=0} = \left(3\,[w_6, d]_+ - \frac{n^2-19}{30}\left(3\,[w_4, d^3]_+ - 2\,[w_4'', d]_+\right)\right.$$
$$\left. -3\frac{n-3}{n} w_3 d w_3 + \frac{(n-3)(n-2)(n-1)n(n+1)(n+2)(n+3)}{20.7!} d^7\right) \delta(u,v)$$

where on the r.h.s. all w's are evaluated at u and $d \equiv d_u$. By restoring the w_2 dependence and truncating to n (i.e. $w_\ell = 0$ for $\ell > n$) one obtains the w-algebra.

In particular one can show that the "central term" i.e. the term independent of w's is diagonal in the indices k, ℓ and of the form

$$(-)^k \frac{\begin{pmatrix} n+k-1 \\ n-k \end{pmatrix}}{\begin{pmatrix} 2k-2 \\ k-1 \end{pmatrix}} \delta_{k\ell} d_x^{k+\ell-1} \delta(x,y)$$

10. We will not extend the algebraic machinery beyond the previous indications, referring to our paper for the discussion of the orthogonal case. Rather in this last and still very speculative part I would like to look more closely at the simplest case of W_3 (for SL_3) and ask what is really this kind of structure.

Apart from diffeormorphisms generated by $w_2 = a_2$ corresponding to the pair

$$X_1 = \epsilon d - \epsilon' \qquad Y_1 = \epsilon d + 2\epsilon'$$

we have in this case transformations generated by

$$w_3 = a_3 - \frac{1}{2} a_2'$$

with

$$\begin{aligned} X_2 &= \epsilon d^2 - \frac{1}{2}\epsilon' d + \frac{2}{3}\epsilon a_2 + \frac{1}{6}\epsilon'' \\ Y_2 &= \epsilon d^2 + \frac{5}{2}\epsilon' d + \frac{2}{3}\epsilon a_2 + \frac{5}{3}\epsilon'' \end{aligned}$$

while

$$\begin{aligned} \{w_2(y), w_2(x)\} &= \left(w_2'(x) + 2w_2(x) d_x + 2d_x^3\right) \delta(x,y) \\ \{w_2(y), w_3(x)\} &= \left(w_3'(x) + 3w_3(x) d_x\right) \delta(x,y) \\ \{w_3(y), w_3(x)\} &= -\frac{1}{6}\left(d_x^5 + \frac{5}{2}\left[w_2(x), d_x^3\right]_+ - \frac{3}{2}\left[w_2''(x) d_x\right]_+ + 2\left[w_2^2, d\right]_+\right) \delta(x,y) \end{aligned}$$

While this is *not* the general situation, let us assume nevertheless that ker Q_3 describes a smooth projective curve in CP_2 encoded by the vanishing of an irreducible homogeneous polynomial $P(u_1, u_2, u_3)$ of degree n in three variables. Hence the curve has genus $(n-1)(n-2)/2$ (we assume no double points, no cusps,..., an example would be the Fermat curve $u^n + v^n + w^n = 0$).

Thus w_3 is a meromorphic 3-differential characterized by its zeroes and poles. Once the curve is parametrized (by patches), the poles only occur where the Wronskian vanishes i.e. at the $3n(n-2)$ inflexion points and it is readily seen that they all occur as triple poles (this arises from the necessity of absorbing a_1 into a multiplicative factor on the coordinates). The poles are given by the vanishing of the Hessian

$$H_1(u_1, u_2, u_3) = \det \frac{\partial^2 P}{\partial u_1 \partial u_j}$$

a homogeneous polynomial of degree $3(n-2)$ which will appear cubed in the denominator of w_3. On the other hand we have to understand what the zeroes stand for. For this we first get rid of a_1, then go to a coordinate where a_2 vanishes (call it t), so that locally the operator is $Q_3 = d_t^3 + w_3$. If we assume that w_3 has a single zero at $t = 0$ in the vicinity of the origin, i.e. $w_3 \sim t$ it is readily seen that we can take three solutions of $Q_3 f = 0$ of the form

$$\begin{aligned} f_0 &\sim 1 + 0(t^4) \\ f_1 &\sim t + 0(t^5) \\ f_2 &\sim t^2 + 0(t^6) \end{aligned}$$

Hence

$$f_2/f_0 \sim \left(\frac{f_1}{f_0}\right)^2 + 0\left(\frac{f_1}{f_0}\right)^6$$

which means that the osculating *non degenerate* conic (the would be conic with 5 points of contact at the origin; recall that a conic is defined by requiring that it passes through 5 points) has in fact six points of contact with the curve at the origin. These are analogous to Weirstrass points (the inflexion points). Let H_2 be a homogeneous polynomial vanishing only on these points. Then

$$w_3 = \frac{H_2}{H_1^3}$$

The homogeneity degree of w_3 is then deg $H_2 - 3$ deg H_1. Now it is known that for a smooth curve of degree $n > 3$ the coordinates can be taken of weight $\frac{1}{n-3}$. This is one for a quartic (genus 3, corresponding to the canonical embedding) and it can be verified in general by finding three holomorphic differentials with zeroes of order $n-3$. For instance for a Fermat curve $u_1^n + u_2^n + u_3^n = 0$ seen as a n-fold branch cover over the $x = \frac{u_2}{u_1}$ plane with $y = \frac{u_3}{u_1}$ (or vice versa) take

$$\begin{aligned} \omega_1 &= \frac{dx}{y^{n-1}} && n-3 \text{ fold zeroes over the } n-\text{points at infinity} \\ \omega_2 &= -\frac{dy}{x^2} && n-3 \text{ fold zeroes at } n-\text{th roots of } y=-1,\ (x=0) \\ \omega_3 &= \frac{dx}{y^2} && n-3 \text{ fold zeroes at } n-\text{th roots of } x=-1,\ (y=0) \end{aligned}$$

then $\omega_1 : \omega_2 : \omega_3 = u_1^{n-3} : u_2^{n-3} : u_3^{n-3}$ (of course $n(n-3) = 2g-2$).

In order that w_3 be indeed a 3-differential it is thus required that

$$3(n-3) = \deg H_2 - 3 \deg H_1$$

or

$$\deg H_2 = 3[n-3+3(n-2)] = 3[4n-9]$$

giving a total of $3n[4n-9]$ points where a conic touches a smooth curve with a contact of order six. I dont known whether this is a classical result (it surely is) but at any rate for $n=3$ (27) and $n=4$ (84) it agrees with values which are quoted in textbooks. Thus at least when Q_3 describes a smooth curve in $\mathbf{C}P_2$ we have a global view of the meromorphic 3-differential w_3 and with more elaborate geometry one could find analogs in higher dimension.

Finally let us inquire about the deformations

$$\left\{\int \eta w_3,\ Q_3\right\}$$

with $Q_3 = d^3 + \frac{1}{2}[w_2, d]_+ + w_3$. If we ignore covariance and take $\eta = \text{cst}$, this is the isospectral KdV flow

$$\begin{aligned} \dot{w}_2 &= 2w_3' \\ \dot{w}_3 &= -\frac{1}{6}\left[w_2''' + 4w_2 w_2'\right] \end{aligned}$$

where the dots stand for derivatives with respect to the parameter along the flow. Taking an extra derivative we obtain for w_2 the Boussinesq equation

$$\ddot{w}_2 = -\frac{1}{3}\left[w_2'' + 2w_2^2\right]''$$

On the other hand if η varies (as it should, having weight -2) we get

$$\begin{aligned} \delta_\eta w_2 &= 3\eta' w_3 + 2\eta w_3' \\ \delta_\eta w_3 &= -\frac{1}{6}\left[\eta^{(5)} + 5\eta''' w_2 + \frac{15}{2}\eta'' w_2' + \frac{9}{2}\eta' w_2'' + \eta w_2''' + 4\eta' w_2^2 + 4\eta w_2 w_2'\right] \end{aligned}$$

Even it we have chosen a coordinate where w_2 vanishes, but not w_3, then a general w_3-flow will have $\delta_\eta w_2 \neq 0$. But this can be compensated by changing the new coordinate, so let us look at the case $w_2 = 0$, $\delta_\eta w_2 = 0$. It therefore follows that $\eta^3 w_3^2 = \text{cste}$, hence η can be taken as one of the cube roots of w_3^{-2} (this has the correct weight at least away from the zeroes or poles of w_3). Choosing one of these cube roots we get a W_3-flow

$$\frac{\mathrm{d}w_3}{\mathrm{d}t} = \left(\frac{\mathrm{d}}{\mathrm{d}x}\right)^5 \frac{1}{w_3^{2/3}}$$

with coordinates (of weight -1) changing as

$$\frac{\mathrm{d}}{\mathrm{d}t} f = -\left[4\eta'' - 6\eta'\delta + 6\eta d^2\right] f, \qquad \eta = \frac{1}{w_3^{2/3}}$$

Up to this 3-fold ambiguity this flow should have a geometric interpretation in terms of the embedded curve in $\mathbf{C}P_2$ involving osculating conics. I find it very suggestive that if we look at two such neighboring conics, three of their common points have limits distinct from the point on the curve and by drawing lines to the osculating point define three deformation directions. This construction becomes singular at inflexion points or when there is a sixfold contact!

This report is primarily based on the work of

V.G. Drinfeld and V.V. Sokolov
"Lie algebras and equations of Kortweg-de Vries type"
Journ. Sov. Math. **30** (1985) 1975-2036

W-algebras were introduced by

A.B. Zamolodchikov
"Infinite additional symmetries in two-dimensional conformal quantum field theory"
Theor. Math. Phys. **65** (1985) 1205-1213

Our geometric inspiration came from

G. Sotkov and M. Stanishkov
"Affine geometry and W_n-gravities"
Preprint ISAS-70/90/EP (June 1990)

Further references can be found in

P. Di Francesco, C. Itzykson, J.-B. Zuber
"Classical W-algebras"
Preprint SPhT/90-149 and PUPT-1211 (Oct. 1990) submitted to Communications in Mathematical Physics.

The paper quoted in the introduction is

C. Itzykson, J.-B. Zuber
"Matrix integration and combinatorics of modular groups"
Communications in Mathematical Physics, **134** (1990) 197-207.

Connections between CFT and topology via Knot Theory

R.J. Lawrence[1]

Department of Mathematics
Harvard University
Cambridge, Massachusetts

Abstract. In this paper we shall discuss some of the isomorphisms established between the approach to conformal field theory on $\mathbf{P}^1$ of [TK], and the topological construction of braid group representations of [L 1]. These approaches both lead, in the simplest cases, to the one-variable Jones polynomial invariant of links, but can be generalised to give other invariants. The case of higher spin representations of $\mathfrak{sl}_2$ is discussed from the point of view of both approaches, and is used to re-interpret the well known connection with cabled links. The structure of the braid group representation obtained is also discussed in both the spin-$1/2$ and higher spin cases, and is extended to give a representation of the category of tangles.

1: Introduction

Artin's *braid group* [Ar], B_n, can be defined as the fundamental group of the configuration space $\widetilde{X}_n$ of n (unordered) distinct points in $\mathbf{C}$. Indeed, if X_n denotes the configuration space of n (ordered) distinct points in $\mathbf{C}$, then its fundamental group, P_n, the pure braid group, is the normal subgroup of B_n given by the kernel of the natural map:

$$\sigma: B_n \longrightarrow S_n \, .$$

Here, $\sigma(g)$ describes the permutation effected under a loop g in $\widetilde{X}_n$ on the n points of $\mathbf{C}$ constituting the base point of the loop. There is also a natural action of S_n on X_n, permuting the n points, and with respect to this, $\widetilde{X}_n = X_n/S_n$. An element of B_n, an *n-string braid*, can be described pictorially by n curves in $\mathbf{C} \times [0,1]$ joining two sets of n points in $\mathbf{C} \times \{0\}$, $\mathbf{C} \times \{1\}$. Alternatively, by projection, this becomes equivalent to a two-dimensional picture between parallel lines, joining two sets of n points on these lines, with at most two strands crossing at any point, and, at any such crossing, over and under strands being marked. Such a representation is shown in Fig. 1. This leads to the familiar description in terms of the generators $\sigma_1, \ldots, \sigma_{n-1}$ and relations,

$$\left.\begin{aligned} \sigma_i\sigma_{i+1}\sigma_i &= \sigma_{i+1}\sigma_i\sigma_{i+1} && \text{for } i = 1,2,\ldots,n-2, \\ \sigma_i\sigma_j &= \sigma_j\sigma_i && \text{for } |i-j| > 1 . \end{aligned}\right\}$$

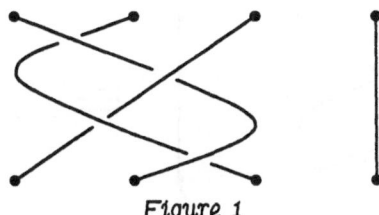

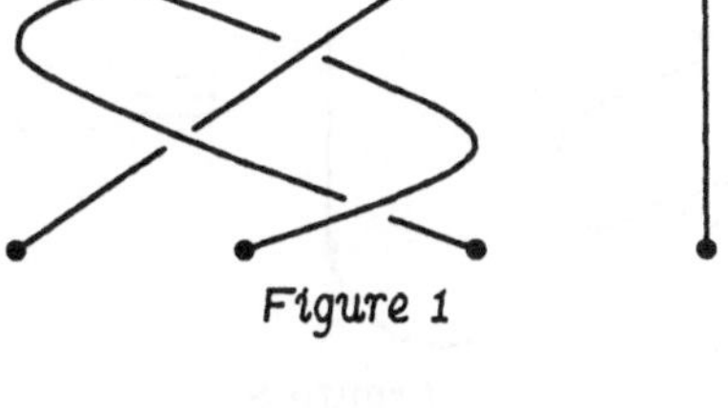

Figure 1

[1] The author is a Junior Fellow of the Society of Fellows.

Such braids may be related to links in two important ways. In this context, a link refers to an embedding of a finite disjoint union of circles, S^1, in S^3. The two kinds of closure we shall discuss are *braid closure* and *plait closure*. The braid closure, $\widehat{\beta}$, of a braid β, is formed from the pictorial presentation by joining corresponding points in $\mathbf{C} \times \{0\}$ and $\mathbf{C} \times \{1\}$, thus providing n further curves; the resultant object is a link, and in the case of the example of Fig. 1 it is illustrated in Fig. 2.

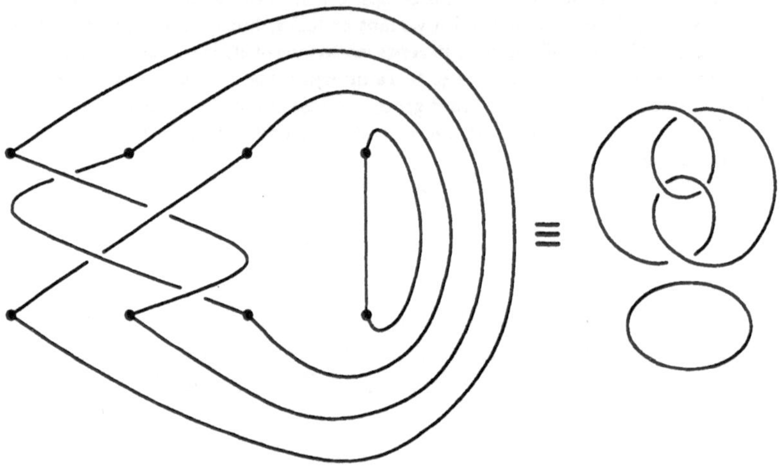

Figure 2

The operation of plait closure can only be performed on braids of even string number. Adjacent points in each of $\mathbf{C} \times \{0\}$ and $\mathbf{C} \times \{1\}$ are joined in pairs. Once again, this provides an additional n arcs, which close up the braid β to give a link $\widetilde{\beta}$. However, the link obtained is usually distinct from that found via braid closure; see Fig. 3 for an illustration of the process applied to the braid of Fig. 1.

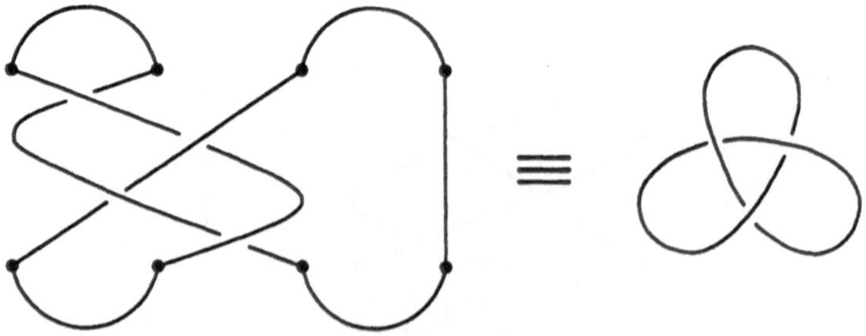

Figure 3

The importance of these operations is that the maps,

$$\widehat{\ }: \coprod_n B_n \longrightarrow \{\text{links}\}/(\text{ambient isotopy equivalence}),$$

$$\widetilde{\ }: \coprod_n B_{2n} \longrightarrow \{\text{links}\}/(\text{ambient isotopy equivalence}),$$

are both surjective (see [Al] and note that the surjectivity of $\tilde{}$ follows from that of $\hat{}$), while the structures of the inverse images of any fixed link $\{L\}$ are known (see [M] for $\hat{}$ and [B] for $\tilde{}$). That is, the operations of closure are not injective, so that there exist an infinite number of braids whose closure give a given link L; however, all such braids are related by certain simple moves. Invariants of links may therefore be specified by suitable families of representations of the braid groups.

In this paper, three general families of representations of $\{B_n \mid n \in \mathbf{N}\}$ will be compared. They arise from, (i) the R-matrix approach, (ii) correlation functions in conformal field theory, and (iii) a topological approach. These methods are briefly reviewed in the succeeding sections, and their comparison leads to the existence of certain isomorphisms, discussed in §5. All these methods lead to representations of B_n, which in some sense 'fit together' for different values of n, so as to give invariants of links.

2: R-MATRIX APPROACH

It was shown by Drinfel'd [D] and Jimbo [Ji], how a quantum group $U_q\mathfrak{g}$ could be constructed, from an arbitrary semisimple Lie algebra $\mathfrak{g}$. This is a non-cocommutative algebra over $\mathbf{C}[[h]]$ (where $q = e^h$), deforming the universal enveloping algebra $U\mathfrak{g}$, equipped with Hopf algebra structure as well as quasi-triangularity. That is, there is a naturally defined element $R \in U_q\mathfrak{g} \otimes U_q\mathfrak{g}$ satisfying the *quantum Yang-Baxter equation* (QYBE),

$$R_{12}R_{13}R_{23} = R_{23}R_{13}R_{12}| >,$$

where $R_{12} = R \otimes 1$, $R_{23} = 1 \otimes R$ and $R_{13} = P_{12}(R_{23})$ are elements of $(U_q\mathfrak{g})^{\otimes 3}$ and P_{12} denotes the operation of interchanging the first two components (see [D]). It is easily seen that R satisfying QYBE is equivalent to $\widetilde{R} = P \circ R$ satisfying the braid relation,

$$\widetilde{R}_{12}\widetilde{R}_{23}\widetilde{R}_{12} = \widetilde{R}_{23}\widetilde{R}_{12}\widetilde{R}_{23}$$

and thereby leads to a braid group representation in which $\sigma_i \mapsto \widetilde{R}_{i\,i+1}$. A map $B_n \to (U_q\mathfrak{g})^{\otimes n}$ is thus produced, and for any representation ρ on V of $U_q\mathfrak{g}$, it reduces to a representation π_n of B_n on $V^{\otimes n}$. The associated link invariants produced can be written in the form,

$$P(\widehat{\gamma}) = \alpha^{-n}\beta^{-e}\,\mathrm{tr}\big(\mu^{\otimes n} \circ \pi_n(\gamma)\big)\,,$$

where $\alpha, \beta \in \mathbf{C}$, $\mu: V \to V$ and $\rho(\widetilde{R}) \in \mathrm{End}(V \otimes V)$ are said to form a quadruple known as an *enhanced Yang-Baxter operator* (EYBO), see [T]. In the above, γ denotes a braid in B_n whose braid closure $\widehat{\gamma}$ is the link concerned, and whose exponent sum is e. It can be shown that for any $\mathfrak{g}$, α, β and μ can be suitably chosen so that the quadruple gives a well-defined invariant, P, as written above (see [BGZ], [R] and [T]).

The above outlines a procedure associating a link invariant with every Lie group and associated representation. To $\mathfrak{sl}_2$ and the standard vector (spin-$1/2$) representation, the invariant associated is the one-variable Jones polynomial V_L (see [Jo 1]). To $\mathfrak{sl}_m$ and the m-dimensional vector representation, is associated a slice of the two-variable Jones polynomial X_L (see [FYHLMO], [Jo 2]). When $\mathfrak{sl}_2$ with a higher spin representation is used, no essentially new invariant is obtained, simply V_M for M cablings of the original link L; see [KR], [R], [MS].

The two-variable Jones polynomial $X_L(q,\lambda)$ satisfies an important property, known as the *skein relation*, which can in fact be used to define X_L; see [LM] where a change of variables is employed to give $P_L(l,m)$. This is a linear relation existing between invariants of links differing within a small sphere, and evaluated at the same values of q and λ; namely,

$$q^{1/2}\lambda^{1/2}X(\times)+(q^{-1/2}-q^{1/2})X()()-q^{-1/2}\lambda^{-1/2}X(\times)=0\,,$$

i.e., by the condition that any representation σ_i has two eigenvalues $\lambda^{-1/2}$ and $-q^{-1}\lambda^{-1/2}$. The quadratic algebra so obtained (at least when σ is replaces by $-q\lambda^{1/2}\sigma$ so as to give eigenvalues 1 and $-q$) is known as the Iwahori-Hecke algebra $H_n(q)$, and its representation theory, for q not a root of unity, is similar to that of S_n. Thus to any Young diagram Λ of n squares, an irreducible representation π_Λ may be associated; see [We].

3: Conformal field theory approach

In [BPZ], the foundations of conformal field theory were laid, and the idea of vertex operators introduced. Braid group representations arise in this context as the monodromy of a certain integrable system of differential equations satisfied by n-point correlation functions. Consider the affine Lie algebra $\widehat{\mathfrak{g}}=\mathfrak{sl}_2\otimes\mathbf{C}[t,t^{-1}]\oplus\mathbf{C}c$, with integrable highest weight modules $\mathcal{H}_j$. For any triple of non-negative half-integers $\left(\begin{smallmatrix} & j & \\ j_2 & & j_1\end{smallmatrix}\right)=\mathbf{j}$ satisfying the Clebsch-Gordon condition (CG),

$$|j_1-j_2|\le j\le j_1+j_2\,,\qquad j+j_1+j_2\in\mathbf{Z}\,,$$

a *vertex operator* $\Phi_{\mathbf{v}}(z)$, of type $\mathbf{v}$, may be defined as a map,

$$V_j\otimes\mathcal{H}_{j_1}\longrightarrow\widehat{\mathcal{H}}_{j_2}\,,$$

satisfying certain conditions: holomorphicity with respect to the parameter $z\in\mathbf{C}^*$, and two other conditions specifying gauge invariance (under the action of $\mathfrak{sl}_2$) and the equation of motion. Here $\widehat{\mathcal{H}}_j$ is a suitable completion of $\mathcal{H}_j$. There is only one vertex operator for any particular type $\mathbf{v}$, up to scalar multiple. Such vertex operators can be composed and one can then compute an expectation value, to give a correlation function,

$$\begin{aligned} f\colon X_n &\longrightarrow (V_{1/2}^{\otimes n}\otimes V_t)\\ (w_1,\dots,w_n)&\longmapsto\langle u\mid\Phi_n(w_n;v_n)\dots\Phi_1(w_1;v_1)\mid\mathrm{vac}\rangle\end{aligned}$$

where $w_i\in\mathbf{C}$, $v_i\in V_{1/2}$ and Φ_i is a vertex operator of type $\left(\begin{smallmatrix} & 1/2 & \\ p_i & & p_{i-1}\end{smallmatrix}\right)=\mathbf{v}_i$ and $p_0=0$, $p_n=t$, while $|\,\mathrm{vac}\rangle$ generates V_0 and $\langle u|\in V_t$. Note that here V_t denotes the spin-t representation of $\mathfrak{sl}_2$, of dimension $2t+1$. Such a correlation function may be shown to be well-defined for $0<|w_1|<\cdots<|w_n|$. However it can be analytically continued to a multi-valued meromorphic function on the whole of X_n, and as such satisfies various properties (see [TK], [KZ]). For us, we only need to note that f actually maps into the $\mathfrak{sl}_2$-invariant part, and satisfies the *Knizhnik-Zamalodchikov* equation (KZ),

$$\kappa\frac{\partial f}{\partial w_i}=\sum_{j\ne i}\frac{\Omega_{ij}}{w_i-w_j}f\,,$$

where Ω_{ij} denotes the action of the polarised $\mathfrak{sl}_2$ Casimir operator, $1/2 H \otimes H + E \otimes F + F \otimes E$ upon the ith and j^{th} factors of $(V_{1/2}^{\otimes n} \otimes V_t)_0$. This equation is integrable and therefore is equivalent to the flatness of the section specified by f, of a vector bundle, with base $\widetilde{X}_n$, on which a certain flat connection has been placed [Ko]. There is thus a well-defined monodromy action of $\pi_1(\widetilde{X}_n)$ on the fibres.

Theorem 1 [TK] *The monodromy representation of $B_n = \pi_1(\widetilde{X}_n)$ upon the space of n-point correlation functions specified by all possible half-integers p_i, with $p_0 = 0$ and $p_n = t$, for which $\mathbf{v}_i$ satisfies (CG), is $q^{3/4}\pi_{\Lambda^n_m}$, a multiple of a representation of $H_n(q^{-1})$, associated with the 2-row Young diagram Λ^n_m with rows of length $n-m$ and m. Here m, q, t and l are related by,*

$$m = 1/2 n - t, \qquad q = \exp\left(2\pi i/(l+2)\right).$$

This result can be extended to give representations of B_n from any general $\mathfrak{g}$ and representation V in the natural way. The shift of 2 in the expression for q becomes the dual Coxeter number, while the factor of 1/2 in the expression for m is associated with the weight of the representation chosen.

4: Topological approach

The approaches of §§2,3 both employed Lie algebras and their representations in an essential way, and it was therefore natural that they gave rise to braid group representations with these as data. In the approach described in this section, although this data does enter, it is not in such an explicit fashion.

Consider the fibration of X_{m+n} over X_n given by the first n points. Use local coordinates for X_n (base) given by $w_1, \ldots, w_n$ and for the fibre $Y_{\mathbf{w},m}$ over a point $\mathbf{w} \in X_n$, given by $z_1, \ldots, z_m$. There is a natural action of $S_m \times S_n$ on this fibre bundle, given by permuting the z's and w's amongst themselves, and when one divides out by this action, one obtains the fibration below.

$$\begin{array}{ccc} X_{m+n}/(S_m \times S_n) & \supseteq & \widetilde{Y}_{\mathbf{w},m} = Y_{\mathbf{w},m}/S_m \\ \downarrow & & \downarrow \\ \widetilde{X}_n & \ni & \mathbf{w} \end{array}$$

The first homology of the fibre is $\mathbf{Z} \times \mathbf{Z}$, generated by the two types of cycles in which z_i goes around a loop around either another z_j, or a w_k. The characters of $\mathbf{Z}^2$ are specified by two, independent, non-zero complex parameters α and q, and thus one can define a local coefficient system $\chi(\alpha, q)\colon \pi_1(Y_{\mathbf{w},m}) \to \mathbf{C}^*$ on $Y_{\mathbf{w},m}$ giving twistings of α and q around the two types of generators of $\pi_1(Y_{\mathbf{w},m})$. The vector bundle E^m over the base $\widetilde{X}_n$, whose fibre over $\mathbf{w} \in \widetilde{X}_n$ is the middle cohomology,

$$E^m_{\mathbf{w}} = H^m(\widetilde{Y}_{\mathbf{w},m}, \chi(\alpha, q)),$$

has a natural flat connection, due to the homotopy invariance of homology. There is thus obtained a natural monodromy representation of $\pi_1(\widetilde{X}_n) = B_n$. Since $Y_{\mathbf{w},m}$ is obtained from $\mathbf{C}^m$, by removing hyperplanes, it is a Stein manifold, and so H^m can be computed from $H^{m,0}$. Thus the elements of $E^m_{\mathbf{w}}$ may be represented in Dolbeault cohomology by classes $[f\, dz_1 \wedge \cdots \wedge dz_m]$. Here f is a multi-valued holomorphic function on $\widetilde{Y}_{\mathbf{w},m}$ twisting according to $\chi(\alpha, q)$; that is, it has form $f_0.g$, where,

$$f_0 = \prod_{1 \le i < j \le m} (z_i - z_j)^a \prod_{\substack{1 \le i \le m \\ 1 \le k \le n}} (z_i - w_k)^b$$

and g is a single valued holomorphic function on $\widetilde{Y}_{\mathbf{w},m}$ with $\alpha = e^{2\pi i a}$ and $q = e^{2\pi i b}$. Let $\mathcal{T}_1$ denote the set of m-tuples of distinct elements of $\{w_1, \ldots, w_n\}$. This is a set of order $\binom{n}{m}$, and to each element $\underline{\alpha} \in \mathcal{T}_1$ we may associate,

$$f_{\underline{\alpha}} = f_0 \sum_{\sigma \in S_m} \prod_{i=1}^{m} (z_i - \alpha_{\sigma(i)})^{-1} .$$

Theorem 2 [L 1] *The subspace of $E^m_{\mathbf{w}}$ spanned by $\{[f_{\underline{\alpha}}] \mid \underline{\alpha} \in \mathcal{T}_1\}$ when $\alpha = q^{-2}$ (i.e. $a + 2b = 0$), is preserved by parallel transport under the natural connection. Moreover, $\sum_{\underline{\alpha} \in \mathcal{T}_1} A_{\underline{\alpha}}(\mathbf{w}) f_{\underline{\alpha}}$ defines a flat section if, and only if,*

$$\frac{\partial \mathbf{A}}{\partial w_i} = \sum_{j \ne i} \frac{\mathbf{C}_{ij}}{w_i - w_j} \mathbf{A} ,$$

where $\mathbf{C}_{ij}$ are suitable constant matrices.

The matrices $\mathbf{C}_{ij}$ can be computed and are found to have eigenvalues $-2b$ and 0 with multiplicities $\binom{n-2}{m-1} - \binom{n-2}{m-2}$ and $\binom{n-2}{m} - \binom{n-2}{m-3}$, respectively.

Theorem 3 [L 1] *The monodromy representation of B_n obtained from the natural flat connection on E^m contains, when $\alpha = q^{-2}$, a large irreducible part, on $W_{n,m}$, which factors through $H_n(q)$ and is $\pi_{\Lambda^n_m}$.*

The subspace $W_{n,m}$ referred to in Theorem 3 can be described either as the invariant space in Theorem 2, or as the set,

$$\left\{ [f] \,\middle|\, \int_{\gamma_{\underline{\alpha}}} f = 0 \qquad \forall \underline{\alpha} \in \mathcal{S} \backslash \mathcal{T}_1 \right\} .$$

Here $\int_{\gamma_{\underline{\alpha}}} f$ denotes the pairing of f with a chain $\gamma_{\underline{\alpha}}$, that is, an embedding of a torus T^m in $Y_{\mathbf{w},m}$ defined by iterated loops, with the ith generator on the torus given by a loop, in which z_i goes around α_i. The chains $\gamma_{\underline{\alpha}}$ may be constructed for any $\underline{\alpha} \in \mathcal{S}$, that is, for any m-tuple of elements of $\{z_1, \ldots, z_m\} \cup \{w_1, \ldots, w_n\}$ for which $\alpha_i \ne \{z_1, \ldots, z_i\}$ $\forall i$. See [L 2] for more details.

The one-variable Jones polynomial $V_L(q)$ may be represented as a linear combination of Hecke algebra characters,

$$V_L(q) = \frac{(\sqrt{q})^{e-n-1}(-1)^{n-1}}{1 - q^2} \sum_{m=0}^{[n/2]} (q^m - q^{n-m+1}) \chi_{\Lambda^n_m}(\gamma) ,$$

where $L = \widehat{\gamma}$ (see [Jo 2]). Theorem 2 provides a geometric construction of $\pi_{\Lambda^n_m}$, and thus also of $V_L(q)$. However this is better expressed in terms of a functorial presentation. In other words, the spaces $W_{n,m}$ can be used as images of objects in the category of tangles, and when suitable images are defined for the morphisms, a representation of the category of tangles is defined. The space associated in [At], [S] with a sphere and $2n$ marked points, with $\mathfrak{sl}_2$, spin-$1/2$ representation attached to each marked point is isomorphic to $W_{2n,n}$. For more details, see [L 3]. This is also intimately related to the Pimsner-Popa-Temperley-Lieb representation of B_n or $H_n(q)$ or the Jones algebra A_n, and helps to connect the original picture of V_L in [Jo 1] with the R-matrix approach (see [Jo 2]).

Everything in this section so far could be said to be associated with $\mathfrak{sl}_2$ and spin-$1/2$, although they have not entered explicitly anywhere in the construction. The analogous construction for spin-j simply replaces $\alpha = q^{-2}$ by $\alpha = q^{-1/j}$ and $\mathcal{T}_1$ by $\mathcal{T}_{2j}$, the set of m-tuples of elements of $\{w_1, \ldots, w_n\}$, no element of which is mentioned more than $2j$ times. Theorem 2, so transformed, remains valid. However, the monodromy representation obtained, no longer factors through $H_n(q)$, having instead $2j+1$ possible eigenvalues for the images of the generators σ_i. The alternative description of $W_{n,m}$ referred to in the spin-$1/2$ case remains valid with $\mathcal{T}_{2j}$ replacing $\mathcal{T}_1$. The eigenvalues of the matrices $b^{-1}\mathbf{C}_{ij}$ in Theorem 2 now become: $\nu(\nu-1)/2j$, $\nu = 0, 1, \ldots, 2j$ with associated multiplicities $\binom{n-2}{m-\nu}_j - \binom{n-2}{m-4j-\nu-1}_j$, where $\binom{s}{t}_j$ denotes the coefficient of x^t in $(1+x+\cdots+x^{2j})^s$.

A link invariant may be computed associated with this more general picture; and indeed a representation of the category of tangles may be obtained. In this description, the space associated with a plane section of a link, which is $W_{2n,n}$ when $j = 1/2$, is replaced by $W^j_{2n,2nj}$ for general j. The local coeffient system $\chi(q^{2j}, q^{-2})$ on $Y_{\mathbf{w},2nj}$ is 'equivalent' to the system $\chi(q, q^{-2})$ on $Y_{\mathbf{w}^*,2nj}$, where $\mathbf{w}^* \in X_{4nj}$ is formed from $\mathbf{w} \in X_{2n}$ by replacing each point with a clump of $2j$ points The structure of the construction then makes it clear that one would expect such a link invariant, evaluated on a link L, to be related to $V_{L^{(2j)}}$, the one-variable Jones polynomial evaluated on the $2j^{\text{th}}$ cabling of L. This is an already well-known result—see [KR], [R]. See also [L 4] for more details.

5: Connections between the approaches

Kohno has shown that the representations of Theorem 1 are precisely those of §2 given by R-matrices constructed for the appropriate quantum group. Theorem 1 was proved essentially by an analysis of the differential equations (KZ) satisfied by the correlation functions. Similar differential equations appear in Theorem 2, §4, and by a theorem of Kohno's [Ko], the following can be deduced.

Theorem 4 [L 4] *There exists an isomorphism,*

$$\alpha\colon W^j_{n,m} \longrightarrow (V_j^{\otimes n} \otimes V_t)_0$$

under which $b^{-1}\mathbf{C}_{ij}$ *corresponds to* $\kappa^{-1}\mathbf{\Omega}_{ij} - 2j(j+1)\mathbf{I}$, *where* $m = nj - t$, $q = \exp\left(2\pi i/\kappa\right)$ *and* $aj + b = 0$ *i.e.* $\alpha = q^{-1/j}$.

This theorem was proved in [L 1] for the case of $j = 1/2$ (see also [L 2] Theorem 5.20). The results of §4 were all proved without reference to the structure of any system of differential equations, and solely used the dual picture in terms of chains in homology.

For the case of $\mathfrak{sl}_2$ and spin-j, the connections found are summarised in the table below. The standard approach to CFT can be viewed as the *local* picture, as opposed to the topological approach outlined in §4, which considers the *global* structure. In the former picture, the analytic continuations of the correlation functions form a natural basis for the representation; whereas it is the functions $\{f_{\underline{\alpha}} \mid \underline{\alpha} \in \mathcal{T}_{2j}\}$ which play this role in §4. The transformation between these bases involves hypergeometric functions.

CFT ($\mathfrak{sl}_2$, spin-j)	**Topology**
Knizhnik-Zamolodchikov equation	Condition for flat section
Kohno connection	Gauss-Manin connection
$\kappa^{-1}\mathbf{\Omega}_{ij}$	$b^{-1}\mathbf{C}_{ij} + 2j(j+1)\mathbf{I}$
Correlation functions	$\{f_{\underline{\alpha}}\}$
Conformal blocks $(V_j^{\otimes n} \otimes V_t)_0$	$W^j_{n,m}$

The construction of $W^j_{n,m}$ can be carried out analogously for any Lie algebra $\mathfrak{g}$ and representation V and similar isomorphisms still exist. The data that enters into the construction of §4 is the Cartan matrix; see [SV] for the appropriate twistings and definitions of the associated functions $f_{\underline{\alpha}}$ in this general case. The monodromy representations of B_n obtained will then be related to those discussed in §2.

It has been shown by Kohno that the monodromy representations obtained in §3 from the solutions to the Knizhnik-Zamolodchikov equation, for a general Lie algebra $\mathfrak{g}$, are isomorphic to those arising directly from quantum groups as discussed in §2.

It should be noted that this discussion relies upon q not being a root of unity. When $q^N = 1$, some N, $H_n(q)$, for $n > N$, is not semi-simple and its representation theory becomes more complex. However, an irreducible representation π_Λ can still be obtained for suitable Young diagrams Λ, although its dimension may be smaller than that of the associated symmetric group representation. Accordingly, a subspace of $W_{n,m}$ must be chosen (see [FSV]), in order that it correspond to the conformal blocks. It is only when q *is* a root of unity, that the functorial approach outlined in [At], [Wi], via functional integrals, makes sense; the conformal blocks mentioned above here are thought of as Hilbert spaces.

The three approaches detailed in §§2,3 and 4 should be viewed as different facets of the same construction. Note that whereas the approach of §2 may be viewed as the most non-commutative, involving quantum groups, the approach of §4 is at the opposite extreme. In §3, Lie groups enter, although only in the classical sense, and so in this respect, it may be regarded as an intermediate approach.

It is the belief of the author that a fuller understanding of the relationships between the three approaches, and indeed of the associated knot polynomials themselves, will come by combining the geometrical power of topology with the concise algebraic form of quantum groups.

References

[Al] J.W. ALEXANDER, 'A lemma on a system of knotted curves', *Proc. Nat. Acad. Sci. USA* **9** (1923) p.93–95.

[Ar] E. ARTIN, 'Theorie der Zöpfe', *Abh. Math. Sem. Univ. Hamburg* **4** (1925) p.47–72.

[At] M.F. ATIYAH, 'The geometry and physics of knots', *Lezioni Lincee* Cambridge University Press (1990).

[B] J. BIRMAN, 'Braids, links and mapping class groups', *Ann. Math. Stud.* **82** (1974).

[BGZ] A.J. BRACKEN, M.D. GOULD, R.B. ZHANG, 'Quantum group invariants and link polynomials', *To appear in Commun. Math. Physics.*

[BPZ] A.A. BELAVIN, A.M. POLYAKOV, A.B. ZAMOLODCHIKOV, 'Infinite conformal symmetry in two-dimensional quantum field theory', *Nuc. Phys.* **B241** (1984) p.333–380.

[D] V.G. DRINFEL'D, 'Quantum groups', *Proc. International Congress of Mathematicians, Berkeley* (1986) p.798–820.

[FSV] B.L. FEIGEN, V.V. SCHECHTMAN, A.N. VARCHENKO, 'On algebraic equations satisfied by correlations in Wess-Zumino-Witten models', *Preprint* (1990).

[FYHLMO] P. FREYD, D. YETTER, J. HOSTE, W. LICKORISH, K. MILLET, A. OCNEANU, 'A new polynomial invariant of knots and links', *Bull. A.M.S.* **12** (1985) p.239–246.

[Ji] M. JIMBO, 'A q-analogue of $U(\mathfrak{gl}(N+1))$, Hecke algebras and the Yang-Baxter Equations', *Lett. Math. Phys.* **11** (1986) p.247–252.

[Jo 1] V.F.R. JONES, 'A polynomial invariant for knots via von Neumann algebras', *Bull. A.M.S.* **12** (1985) p.103–111.

[Jo 2] V.F.R. JONES, 'Hecke algebra representations of braid groups and link polynomials', *Ann. Math.* **126** (1987) p.335–388.

[Ko] T. KOHNO, 'Hecke algebra representations of braid groups and classical Yang-Baxter equations', *Adv. Studies in Pure Maths.* **16** (1988) p.255–269.

[KR] A.N. KIRILLOV, N.YU. RESHETIKHIN, 'Representations of the algebra $U_q(\mathfrak{sl}_2)$, q-orthogonal polynomials and invariants of links', *Infinite dimensional Lie algebras and groups* World Scientific (1988) p.285–342.

[KZ] V.G. KNIZHNIK, A.B. ZAMOLODCHIKOV, 'Current algebras and Wess-Zumino models in two-dimensions', *Nuc. Phys.* **B247** (1984) p.83–103.

[L 1] R.J. LAWRENCE, 'Homology representations of braid groups', *D.Phil. Thesis, Oxford* (June 1989).

[L 2] R.J. LAWRENCE, 'Homological representations of the Hecke algebra', *To appear in Commun. Math. Physics.*

[L 3] R.J. LAWRENCE, 'A functorial approach to the one-variable Jones polynomial', *Harvard University preprint* (1990).

[L 4] R.J. LAWRENCE, 'The homological approach applied to higher representations', *Harvard University preprint* (1990).

[LM] W.B.R. LICKORISH, K.C. MILLETT, 'A polynomial invariant of oriented links', *Topology* **26** (1987) p.107–141.

[M] A.A. MARKOV, 'Über die freie Aquivalenz geschloßener Zöpfe', *Recueil Math. Moscow* **1** (1935) p.73–78.

[MS] H.R. MORTON, P. STRICKLAND, 'Jones polynomial invariants for knots and satellites', *Preprint* (1990).

[R] N.YU. RESHETIKHIN, 'Quantised universal enveloping algebras, the Yang-Baxter equation, and invariants of links I, II', *LOMI Preprints E-4-87, E-17-87* **I, II** (1988).

[S] G.B. SEGAL, 'Two-dimensional conformal field theories and modular functors', *Proc. IXth Int. Congr. on Mathematical Physics* (1989) p.22–37.

[SV] V.V. SCHECHTMAN, A.N. VARCHENKO, 'Integral representations of n-point conformal correlators in the WZW model', *Preprint* (1989).

[T] V.G. TURAEV, 'The Yang-Baxter equation and invariants of links', *Invent. Math.* **92** (1988) p.527–553.

[TK] A. TSUCHIYA, Y. KANIE, 'Vertex operators in conformal field theory on $\mathbf{P}^1$ and monodromy representations of braid groups', *Adv. Studies in Pure Math.* **16** (1988) p.297–372, Erratum ibid **19** (1990) p.675–682.

[We] H. WENZL, 'Hecke algebra representations of type A_n and subfactors', *Invent. Math.* **92** (1988) p.349–383.

[Wi] E. WITTEN, 'Quantum field theory and the Jones polynomia;', *Commun. Math. Phys.* **121** (1989) p.351–399.

Stochastic Calculus in Superspace and Supersymmetric Hamiltonians

Alice Rogers

Department of Mathematics, King's College, Strand, London WC2R 2LS, Great Britain

The intention of this talk is to describe some techniques in fermionic path integration for supersymmetric systems. These techniques, which are a generalisation of conventional stochastic methods to include spaces of anticommuting variables, exploit the supersymmetry of the system to the full. A specicific motivation for this work is the desire to make rigorous the path-integral steps in the supersymmetric proofs of the index theorem given by Alvarez-Gaumé [1]and Friedan and Windey [2]. More generally, the recent spate of results achieved by non-rigorous path integral methods in quantum field theory suggests that further study of these methods is valuable; and that particularly the use of anticommuting degrees of freedom to represent fermions and ghosts merits a rigorous treatment.

There are two novel features of the path-integral treatment of supersymmetric quantum mechanics in curved superspace which this paper describes. One is the use of stochastic calculus, and in particular of solutions of stochastic differential equations, which removes the operator-ordering ambiguities which plague more heuristic approaches. The other is the use of superpaths, that is of paths parametrised by a commuting and an anticommuting variable. (A heuristic notion of superpaths in flat superspace was introduced in [3].) The emphasis of the talk is on explanation of the stochastic techniques, which even in their purely bosonic, classical form are not always familiar. The discussion is at a fairly informal level, so that general features of the method should be clear. A more rigorous presentation containing the technical details may be found in a longer paper [4].

A supersymmetric system is characterised by the fact that its Hamiltonian H is the square of a supercharge Q which is a Dirac-type operator of the form

$$Q = \psi^a e_a^i(x)\partial_i + \phi(\psi, x), \tag{1}$$

where the ψ^a are γ-matrices or Clifford algebra operators. (Throughout this paper the summation convention, that repeated indices are to be summed over their range, will be used.) The key idea of this talk is to consider super Brownian paths $b(t) + \tau\xi(t)$ and their curved space analogues (the notation is explained below) to study the evolution operator $e^{-Ht-Q\tau}$ of a supersymmetric system, and particularly to establish the supersymmetric Feynman-Kac formula (28). The simplest Feynman-Kac formula is for a quantum-mechanical system with Hamiltonian

$H = -\frac{1}{2}\partial_i\partial_i + V(x)$. In this case, using physicist's terminology, the matrix elements of the evolution operator e^{-Ht} are given by the Feynman-Kac formula

$$e^{-Ht}(a,b) = \int \mathcal{D}x \left(\exp - \int_0^t \tfrac{1}{2}\dot{x}^2 + V(x)\,ds\right) \tag{2}$$

where the notation $\mathcal{D}x$ means summation over all paths $x(s)$ beginning at a when $s = 0$ and ending at b when $s = t$. The more precise mathematical expression for this formula is

$$e^{-Ht} f(x) = \int d\mu_b \left(\exp - \int_0^t V\left(x + b(s)\right) ds\right) f\left(x + b(t)\right) \tag{3}$$

where $d\mu_b$ denotes standard (purely bosonic) Wiener measure, which is the analytic equivalent of the heuristic $\mathcal{D}x \exp - \int_0^t \frac{1}{2}\dot{x}^2 ds$. This formula will be generalised in two ways.

(i) The Brownian paths $b(s)$ will be replaced by more general paths $x(s)$ which are solutions to a carefully chosen stochastic differential equation. (This allows the inclusion of more general differential operators in the Hamiltonian H.)

(ii) Superpaths $x(s,\sigma)$ and $\zeta(s,\sigma)$ which depend on an anticommuting parameter σ will be used. (This allows the Feynman-Kac formula to be given in terms of the supercharge Q directly, rather than in terms of the Hamiltonian which is a more complicated and less fundamental object in a supersymmetric theory.)

The first step in this process is to generalise the Itô method for integrating functions along Brownian paths $b(s), 0 \leq s \leq t$ to include functions depending on fermionic paths, and to obtain the appropriate Itô formula for functions of Itô integrals. Brownian paths are almost everywhere too irregular for it to be possible to integrate along them in the standard manner of differential calculus. In classical stochastic calculus the Itô method of integration is one method of avoiding these difficulties; the Itô integral of a suitably regular function f along $b(s)$ is defined by

$$\int_0^t f(b(s))\, db(s) = \lim_{N\to\infty} \sum_{r=0}^{2^N-1} f(b(s_r))\left[b(s_{r+1}) - b(s_r)\right] \tag{4}$$

where $s_r = \frac{rt}{2^N}$. In fact these integrals are defined not only for $f\left(b(t)\right)$, but for a more general class of stochastic process, the class of adapted stochastic processes. Loosely speaking one may think of an adapted process $f_s, 0 \leq s \leq t$ as a class of functions of Brownian paths $b(s)$ where f_s only depends on $\{b(u)|u \leq s\}$. With this definition of integration along Brownian paths a chain rule, known as the Itô formula, can be established. The irregularities of Brownian motion mean that this chain rule includes a second order correction term which would not be present for deterministic calculus. Essentially this is because $\int d\mu_b \left(b(s_1) - b(s_2)\right)^2 = |s_1 - s_2|$, and thus second order variations may contribute to the integral. Considering now Brownian motion in m-dimensional space, suppose that for $i = 1,\ldots,p$ and $a = 1,\ldots,m$ g^i_s and $f^i_{s,a}$ are adapted stochastic processes and that for $i = 1,\ldots,m$

$$z_s^i = \int_0^t f_{s,a}^i db^a(s) + \int_0^t g_s^i ds. \tag{5}$$

Then, if h is a suitably smooth function of $\mathbf{R}^p$,

$$\begin{aligned} h(z_t) - h(z_0) = \int_0^t \partial_i h(z_s) f_{s,a}^i db^a(s) \\ + \int_0^t \partial_i h(z_s) g_s^i ds + \tfrac{1}{2} \int_0^t \partial_i \partial_j h(z_s) f_{s,a}^i f_{s,a}^j ds. \end{aligned} \tag{6}$$

In order to study sytems with fermions (or, more mathematically, to study operators on spin bundles) fermionic paths are introduced. Following the original idea of Martin [5], fermionic paths are paths in a space of anti-commuting variables. In [6] the author has described how an analytic theory of fermionic measure can be developed. The approach may be understood by analogy with conventional, bosonic Wiener measure $d\mu_b$, which is the rigorous version of the heuristic expression $\mathcal{D}x \exp - \int \frac{1}{2}\dot{x}^2 ds$, based on the action of a free particle. In the fermionic case it is useful to remain in phase space and consider pairs of paths $(\theta(t), \rho(t))$, with fermionic Wiener measure $d\mu_f$ defined as the rigorous version of $\mathcal{D}\theta\mathcal{D}\rho \exp - \int \rho\dot{\theta} ds$. The useful property of this fermionic measure is that it can be used to evaluate the kernels of Clifford algebra operators in the following manner. For $a = 1, \ldots, n$ (where n is even) let ψ^a be operators satisfying

$$\psi^a \psi^b + \psi^b \psi^a = 2\delta^{ab}. \tag{7}$$

These operators can be represented on the space of functions of n anticommuting variables by

$$\psi^a = \theta^a + \frac{\partial}{\partial \theta^a}. \tag{8}$$

Then, if $\psi^\mu = \psi^{\mu_1} \ldots \psi^{\mu_k}$ is a product of such operators with $1 \leq \mu_1 < \ldots < \mu_k \leq n$, and f is a function of n anticommuting variables $\theta^1, \ldots, \theta^n$,

$$\psi^\mu f(\theta) = \int d\mu_f (\theta^{\mu_1} + \xi^{\mu_1}(s)) \ldots (\theta^{\mu_k} + \xi^{\mu_k}(s)) f(\theta + \theta(s)), \tag{9}$$

where $\xi(s) = \theta(s) + i\rho(s)$. When considering bosons and fermions together, one may use the super Wiener space of paths $(b(s), \theta(s), \rho(s))$ in (m,2n)-dimensional superspace, with super Wiener measure $d\mu_s$ being the product of bosonic Wiener measure $d\mu_b$ and fermionic Wiener measure $d\mu_f$. The Itô formula equation (6) survives if the following two modifications are made:

(1) the $f_{s,a}^i$ and g_s^i are adapted processes on super-Wiener space, so that each $f_{s,a}^i = f_{s,a}^i(b(u), \theta(u), \rho(u) | 0 \leq u \leq s)$, and each g_s^i is similarly restricted.

(2) The function h depends explicitly on $\theta(s)$. (This is necessary because one cannot express $\theta(t)$ as $\int_0^t d\theta(s)$, fermionic Brownian paths being too irregular for some analogue of the Itô integral to exist. As a compensation, it is analytically straightforward to remain in fermionic phase space, and thus handle a wide

class of fermionic operators without needing to use more general fermionic stochastic processes.)

The next ingredient required is the solution to some stochastic differential equations. A stochastic differential equation (here generalised to include fermionic paths) is an equation to be satisfied by some adapted stochastic process z_s^i, $i = 1,\ldots,p$ which takes the form

$$\begin{aligned} z_t^i &= \int_0^t \sum_{a=1}^m f_a^i\left(z_s,\theta(s),\rho(s)\right) db^a(s) + \int_0^t g^i\left(z_s,\theta(s),\rho(s)\right) ds \quad i=1,\ldots,p \\ z_0 &= A, \end{aligned} \tag{10}$$

where A is some fixed elemnt of $\mathbf{R}^p$ or $\mathbf{C}^p$. Provided that the functions f_a^i and g^i are sufficiently regular, it can be shown that such an equation has a unique solution [4], the method of proof being very similar to the classical proof, with a solution being constructed by successive approximations [7]. It is solutions to carefully chosen stoachastic differential equations which provide the appropriate Brownian paths for curved space and superspace.

In order to construct a supersymmetric Feynman-Kac formula, some elementary calculus in a (1,1)-dimensional superspace is required [3]. With the convention that lower-case Roman letters denote commuting quantities, lower-case Greek letters denote anticommuting quantities and upper-case Roman letters denote combinations of both kinds of quantities or quantities of unspecified parity, let us suppose that (1,1)-dimensional superspace is parametrised by $T = (t,\tau)$. The superderivative D_T is defined to be

$$D_T = \partial/\partial\tau + \tau\partial/\partial t. \tag{11}$$

This operator has the property ${D_T}^2 = \partial/\partial t$. One may check by explicit calculation that

$$\begin{aligned} D_T\left(e^{-Ht-Q\tau}\right) &= Qe^{-Ht-Q\tau} \\ &= e^{-Ht-Q\tau}\left(Q - 2\tau H\right). \end{aligned} \tag{12}$$

Integration is defined by

$$\int_0^\tau d\sigma \int_0^t ds\, F(s,\sigma) = \int d\sigma \int_0^{t+\sigma\tau} ds\, F(s,\sigma) \tag{13}$$

where the integral on the right hand side is carried out as follows. The first integration, with respect to the commuting variable s, is carried out by regarding $\int_0^u du\, F(s,\sigma)$ as a function of u, and evaluating this function at $u = t + \sigma\tau$ by Taylor expansion about $u = t$. Subsequently the second integration, with respect to the anticommuting variable σ, is carried out by the usual Berezin prescription. One then finds by explicit calculation that

$$\int_0^\tau d\sigma \int_0^t ds(A(s) + \sigma B(s)) = \int_0^t B(s)ds + \tau A(t) \tag{14}$$

and also that if one takes the superderivative of the integral with respect to its upper limits then one obtains the integrand, that is,

$$D_T\left(\int_0^T d\sigma \int_0^t ds F(s,\sigma)\right) = F(t,\tau). \tag{15}$$

One may use equation (14) to define stochastic integrals in (1,1)- dimensional superspace, in the following manner. Suppose that G_s and H_s are adapted stochastic processes, and that $F_S = G_s + \sigma H_s$. Then

$$\int_0^T d\sigma \int_0^t ds\, F_S =_{def} \int_0^t ds\, H_s + \tau G_t. \tag{16}$$

One then finds that the following supersymmetric Itô formulae are satisfied.

Theorem 1 *For $i = 1,\ldots,p$ and $a = 1,\ldots,m$, let x_s^i and $e^i_{(s)a}$ be adapted stochastic processes on (m,n)-dimensional super-Wiener space, and ϕ be a function of p commuting and m anticommuting variables which is linear in the anticommuting argument. Also, for any smooth function g of a single complex variable, let*

$$g_T =_{def} g\left(\int_0^T d\sigma \int_0^t ds\, \phi(x + z_S + \frac{i}{\sqrt{2}}\sigma\eta_s, \theta + \zeta_S)\right) \tag{17}$$

$$\begin{aligned} \text{where} \quad & z_S^i = x^i(s) + \frac{i}{\sqrt{2}}\sigma\xi^a(s)e^i_{(s)a} \\ & \zeta_S^a = \xi^a(s) + \sqrt{2}i\sigma\dot{b}^a(s) \\ \text{and} \quad & \eta_s^i = \theta^a e^i_{(s)a}. \end{aligned} \tag{18}$$

(The expression $\dot{b}$ is formal, because Brownian paths are almost everywhere not differentiable; a factor of $\dot{b}$ will only occur in conjunction with a factor ds, and the product $\dot{b}\,ds$ is to be interpreted as $db(s)$. Because of the factor σ in ζ_S^a no higher powers of $\dot{b}$ occur.) Then

$$\begin{aligned} g_T - g_0 = \int_0^T d\sigma \int_0^t ds \Bigg[& g'_{s,\sigma}\phi(x + z_S^i + \frac{i}{\sqrt{2}}\sigma\eta_s, \theta + \zeta_S) \\ - \Big(2\sigma g''_{s,\sigma}\delta_a\phi(x + z_S^i + & \frac{i}{\sqrt{2}}\sigma\eta_s, \theta + \zeta_S)\delta_a\phi(x + z_S^i + \frac{i}{\sqrt{2}}\sigma\eta_s, \theta + \zeta_S)\Big)\Bigg]. \end{aligned} \tag{19}$$

A second theorem, which is a supersymmetric version of the restricted Itô theorem, can also be proved. This theorem applies to Brownian motion in (m,m)-dimensional superspace.

Theorem 2 *Suppose that f is a function of m commuting and m anticommuting variables and that e^i_a, $i,a = 1,\ldots,m$ are smooth functions on $\mathbf{R}^m$. Then, with the* *notation of the above theorem, and the addtional requirement that x^i_s satisfies the stochastic differential equation*

$$dx^i_s = e^i_a(x_s)db^a + L^i_s ds, \tag{20}$$

$$\begin{aligned} &f(z_t + \frac{i}{\sqrt{2}}\tau\eta, \theta + \theta_t) - f(x,\theta) \\ &=_I \int_0^\tau d\sigma \int_0^t ds \left[\sqrt{2}i\psi^a e^i_a(x + x_s)\partial_i f(z_S + x + \frac{i}{\sqrt{2}}\sigma\eta, \theta + \theta(s))\right. \\ &\left. + \sigma e^i_a(x_s + x)e^j_a(x_s + x)\partial_i\partial_j f(z_S + x + \frac{i}{\sqrt{2}}\sigma\eta, \theta + \theta(s))\right] \\ &\int_0^\tau d\sigma \int_0^t dx^i_s\, \sigma\partial_i f(z_S + x + \frac{i}{\sqrt{2}}\sigma\eta, \theta + \theta(s)). \end{aligned} \tag{21}$$

These two theorems can both be proved by expanding in powers of τ using (14), and then using the Itô formula (6) [4].

After this brief interlude in the world of (1,1)-dimensional calculus, the main theorem of this talk can be derived. Both the Itô formula and the existence of solutions to certain stochastic differential equations are essential. The key step is the following definition of the operator $U_{t,\tau}$ acting on functions f on an (m,m)-dimensional superspace with coordinates (x,θ):

$$\begin{aligned} &U_{t,\tau} f(x,\theta) \\ &= \int\int d\mu_s \left\{ \left[\exp - \int_0^\tau d\sigma \int_0^t ds\, \phi\left(x + z_S + \sigma\theta, \theta + \zeta_S\right)\right] \right. \\ &\left. \times f\left(x + \tau\theta + z_T, \theta + \theta(t)\right) \right\}. \end{aligned} \tag{22}$$

Here for $a, i = 1, \ldots, m$, e^i_a is a smooth function on $\mathbf{R}^m$ and

$$z^i_S = x^i_s + \sigma\xi^a(s)e^i_a(x_s) \tag{23}$$

with x_s satisfying the stochastic differential equation

$$\begin{aligned} x^i_s - x^i_0 &= \int_0^t e^i_a(x_s)db^a(s) \\ &\quad + \int_0^t -\tfrac{1}{2}e^j_a(x_s)\partial_j e^i_b(x_s)\left(\delta^{ab} + \xi^a_s\xi^b_s\right) ds, \\ x^i_0 &= 0, \\ \xi^a(s) &= \theta^a(s) + i\rho^a(s) \\ \text{and}\quad \zeta^a_S &= \xi^a(s) + 2\sigma\dot{b}^a(s). \end{aligned} \tag{24}$$

The final step is to apply stochastic calculus to (22) and show that in fact $U_{t,\tau}$ is the desired evolution operator $e^{-Ht-Q\tau}$. If one applies the Itô formula to

$$\int\int d\mu_s\Bigg\{\left[\exp-\int_0^\tau d\sigma\int_0^t ds\phi\left(x+z_S+\sigma\theta,\theta+\zeta_S\right)\right]$$
$$\times f\left(x+\tau\theta+z_T,\theta+\theta(t)\right)\Bigg\}.$$

one finds that

$$\begin{aligned}
&U_{t,\tau}-U_{0,0}\\
&=\int\int d\mu_s\Bigg\{\int_0^\tau d\sigma\int_0^t ds\left[\exp-\int_0^\sigma d\eta\int_0^s du\phi\left(x+x_U+\eta\theta,\theta+\zeta_U\right)\right]\\
&\quad\times\left(Q-2\sigma H\right)f\left(x+\sigma\theta+z_S,\theta+\theta(s)\right)\Bigg\}\\
&=\int_0^\tau d\sigma\int_0^t dsU_{s,\sigma}(Q-2\sigma H)f(x,\theta).
\end{aligned}\tag{25}$$

(Here we have used the additional running super variable $U=(u,\eta)$.)

By taking the superderivative of this result with respect to $T=(t,\tau)$, using (15), we find that

$$D_TU_{t,\tau}f(x,\theta)=U_{t,\tau}(Q-2\tau H)f(x,\theta)\tag{26}$$

and hence, recalling (12), that

$$U_{t,\tau}=e^{-Ht-Q\tau}.\tag{27}$$

Thus we have obtained the desired supersymmetric Feynman-Kac formula

$$\begin{aligned}
&e^{-Ht-Q\tau}f(x,\theta)\\
&=\int\int d\mu_s\Bigg\{\left[\exp-\int_0^\tau d\sigma\int_0^t ds\phi\left(x+z_S+\sigma\theta,\theta+\zeta_S\right)\right]\\
&\quad\times f\left(x+\tau\theta+z_T,\theta+\theta(t)\right)\Bigg\}.
\end{aligned}\tag{28}$$

This formula expresses the evolution operator $e^{-Ht-Q\tau}$ directly in terms of the potential ϕ of the supercharge, and may consequently be regarded as a supersymmetric Feynman-Kac formula. It is also possible to construct a more conventional Feynman-Kac formula, without using superpaths, in terms of the Hamiltonian. But this is more complicated and less fundamental because in a supersymmetric theory the supercharge Q is the simpler, fundamental operator from which H is derived.

A further paper is in preparation which will show how these methods give a natural approach to differential operators in spin bundles, and in particular show how the path-integral step in the supersymmetric proof of the index theorems [1][2]may be made rigorous.

Acknowledgement The author is grateful to the Royal Society for financial support.

References

1. L. Alvarez-Gaumé: Comm. Math. Phys. **90** 161 (1983)
2. D. Friedan and P. Windey: Nuc. Phys. B **235** 395 (1984)
3. A. Rogers: Phys. Lett. B **193** 48 (1987)
4. A. Rogers: King's College Preprint, 1990
5. J. Martin: Proc. Roy. Soc. A **251** 543 (1959)
6. A. Rogers: Comm. Math. Phys. **113** 353 (1987)
7. A. Friedman: *Stochastic Differential equations and applications Vol. 1*, Academic Press, New York 1975

Geometric models and the moduli spaces for string theories

Arkady Vaintrob

Seminar on supermanifolds, Department of Mathematics, Stockholm University, Box. 6701, S-11385, Stockholm, Sweden

Introduction

In [M1] Manin conjectured that the two seemingly quite remote branches of mathematics: the hamiltonian and lagrangean appproach to the representation of infinite dimensional Lie algebras on th one hand and the geometry of the moduli spaces of algebraic curves on the other hand are directly connected. The quantum theory of strings has recently tied up these branches; the connection has been simultaneously established in [AKCP], [BS], [K]. It turns out that the Lie algebra $\mathfrak{w}$ (usually called the Witt algebra or centerless Virasoro algebra) of vector fields on the punctured disk $S=\{z: 0<|z|<1\}$ acts on the space of triples (a compact Riemann surface C, a point $p \in C$, a local parameter z at p). This action changes the complex structure on C and determines an infinitesimal transitive action of $\mathfrak{w}$ on the space M_g of compact Riemann surfaces of genus g. The Virasoro algebra $\mathfrak{v}$, the nontrivial central extension of $\mathfrak{w}$, acts on the space of the determinant bundle over M_g. In terms of this action one can easily write the differential equations for the Polyakov-Mumford measure on the moduli space, for the correlator functions, etc..

In the theory of fermionic string the operator formalism and Polyakov's path integration relate, respectively, the representation theory of Neveu-Schwarz superalgebra (shortly, NS(1)) and the moduli spaces of (1|1)-dimensional supermanifolds with a contact structure (falsy but dynamically called *super Riemann surfaces* or even *SUSY-curves*). The mathematical side of this relationship is not yet studied as thoroughly as for the bosonic string, but even the obtained results (mainly due Deligne [D], see also [Wo], [Ma2], [UY]) leave no room for doubt that everything here is absolutely analogous to the bosonic case.

Since the early 70s it became known that beside the Neveu-Schwarz superalgebra there exist other simple superextensions of the Witt algebra, some of which have nontrivial central extensions similar to the Virasoro algebra, cf. [A], [RS], where some real forms of some of such superalgebras are described.

The problem of classification of simple "stringy" superalgebras and their central extensions with the subsequent classification of their real forms was initiated by Leites, see [LF] and [S1], [S2].

Remark. It is worth mentioning that Serganova proved in [S2] that there are three real forms of the complex Virasoro algebra (two of which have unitary representations; only one of these two forms seem to be under the physicists' study). This result also enabled her to give a correct description of real forms of simple Kac-Moody superalgebras, in particular of Kac-Moody algebras, see the two papers that follow [LF] and Serganova's thesis translated in an enlarged form in [L1], #22.

Regrettably, the list of cocycles and even of the algebras themselves in [LF] is incomplete; the gaps were discovered in [SS], where some deformations of a divergence-free series are found, and [Sch], where additional central extensions are discovered. A systematic proof was renewed by Kac and van de Leur [KL], who conjectured that now we know the complete list (to prove this it is necessary to describe all the deformations of these superalgebras; alas, this is insufficient, we need a result akin to that of Mathieu [M] on Lie algebras).

The representations of the stringy superalgebras other then NS(1) are being under pretty active study but the geometric side of these theories is practically not developed. Only quite recently there started to appear papers devoted to the study of the moduli spaces of SUSY-2 supermanifolds ([M4], [FR]) but the relation with the corresponding Lie superalgebra is not touched there.

The main purpose of this talk is to define for each of the distinguished stringy superalgebra defined in Table 1 a class of geometric objects purpoted to replace the algebraic curves for the Virasoro case and SUSY-curves in the NS(1) case.

The main result of the paper is theorem on existence of a local superspace of moduli for each of these objects. The construction is based on the application of the results of the theory of deformations of superanalytic structures similar to that used in [V2] for the construction of the moduli space of SUSY-curves. All the necessary background on supermainifold theory can be found in the books [L1], [M3]. (I also recommend to read the classic [B] and for some ramifications look in [L4].) The details will be published in [L3], v.3.

Acknowledgements. I am thankful to D.Leites who put the problem and to B.Feigin for the helpful discussions. Financially I was partly supported by Swedish NFR while a guest at Stockholm University which I gratefully acknowledge.

1. Superizations of the Virasoro algebra

A *stringy* (super)algebra is a (super)algebra from the class distinguished in [A], [RS] and [FL], namely a $\mathbb{Z}$-graded Lie (super)algebra of growth 1, simple, containing W, and which, as W-module, is a direct sum of modules $F_\lambda = \mathbb{C}[z^{-1}, z]dt^\lambda$ ("strings"). It is sometimes more convenient to ascribe to stringy superalgebras certain "derived" algebras of the above: the result of their nontrivial central extension or the algebra of derivatives. A distinguished stringy superalgebra is the one which has a nontrivial central extention or, by abuse of language, the result of such an extension.

All the known stringy superalgebras are subalgebras of the Lie superalgebra

$\mathfrak{vect}(1|N) = \mathfrak{der}\, \mathbb{C}[z^{-1}, z, \xi_1, ..., \xi_N]$

of vector fields on the complex superdisc $S^{1|N} = (S, \mathcal{O}(S) \otimes \Lambda(N))$, where $\Lambda(N)$ is the Grassmann superalgebra in N indeterminates. They are listed in Table 1 together with the structures they preserve.

Table 1. Stringy superalgebras
In this Table $\mathfrak{g}' = [\mathfrak{g}, \mathfrak{g}]$.

$\mathfrak{g}$	the structure preserved on $S^{1\|N}$	conditions for simplicity	when $\mathfrak{g}$ or $\mathfrak{g}'$ is distinguished
$\mathfrak{vect}(1\|N)$	-	-	$N = 0, 1, 2$
$\mathfrak{svect}(1\|N)$	the volume element *vol* with constant coeff.	for $N \geq 2$; $\mathfrak{g}'$ is simple codim $\mathfrak{g}' = \epsilon^N$	$N = 2$
$\mathfrak{svect}(1\|N; \alpha)$	the volume element $vol = z^\alpha vol, \alpha \in \mathbb{C}$	for $N \geq 2$ and $\alpha \notin \mathbb{Z}$; $\mathfrak{g}'$ is simple for $\alpha \in \mathbb{Z}$	$N = 2$
$\mathfrak{k}(1\|N)$	the contact form $\alpha = dz - \sum_{i=1}^N \theta_i d\theta_i$	$N \neq 4$; $\mathfrak{g}'$ is simple for $N = 4$	$N \leq 4$
$\mathfrak{k}_+(1\|N)$	the contact form $\alpha_+ = dz - \sum_{i=1}^{N-1} \theta_i d\theta_i - z\theta_N d\theta_N$	-	$N \leq 4$
$\mathfrak{m}(1)$ -	the odd contact form $\beta = d\tau - q d\pi - \pi dq$	-	always

Comments to Table 1. The Lie algebra $\mathfrak{vect}(1|0)$ is the above mentioned Witt algebra $\mathfrak{w}$; its nontrivial central extention is the Virasoro algebra $\mathfrak{v}$. The nontrivial central extensions of Lie superalgebras $\mathfrak{k}(1|1)$ and $\mathfrak{k}_+(1|1)$ are called the Neveu-Schwarz and Ramond superalgebras and denoted NS(1) and R(1), respectively. Clearly, $\mathfrak{svect}(1|N;\alpha) = \mathfrak{svect}(1|N;\beta)$ if $\alpha-\beta\in\mathbb{Z}$.

2. SUSY-curves corresponding to stringy superalgebras

Let us assign to every stringy Lie superalgebra $\mathfrak{g}$ a $\mathfrak{g}$-*curve*, which is a (1|N)-dimensional complex superspace C with an additional structure such that $\mathfrak{g}$ is the Lie superalgebra of the supergroup of automorphisms of this structure in the local model, i.e. when C is $S^{1|N}$. For convenience, the definition is given in the form of Table 2.

Table 2. $\mathfrak{g}$-supercurves

$\mathfrak{g}$	the local model: $S^{1\|N}$ with an additional structure	$\mathfrak{g}$-supercurve C: a complex supervariety $C^{1\|N}$ with fixed points $x_1,\dots,x_k$ and additional conditions
$\mathfrak{vect}(1\|N)$		$k\geq 0$
$\mathfrak{svect}(1\|N)$	a volume element reducible to the form *vol*	as above and with a volume element of the form *vol*
$\mathfrak{svect}(1\|N;\alpha)$	a volume element reducible to the form $vol = z^\alpha vol, \alpha\in\mathbb{C}$	$k\geq 1$ with a volume form on $C^{1\|N}\setminus\{x_1,\dots,x_k\}$ with singularity of the form $z^\alpha vol$ at each of the x_i
$\mathfrak{k}(1\|N)$	the contact form $\alpha = dz - \sum_{i=1}^N \theta_i d\theta_i$	with the contact structure
$\mathfrak{k}_+(1\|N)$	the contact form $\alpha_+ = dz - \sum_{i=1}^{N-1}\theta_i d\theta_i - z\theta_N d\theta_N$	a ramified two-sheeted covering of the above
$\mathfrak{m}(1)$	the odd contact form $\beta = d\tau - q d\pi - \pi dq$	with the odd contact structure

$\mathfrak{k}(1|1)$-curves are just the usual SUSY-curves. The superspace of their moduli has been studied in numerous mathematical and physical papers. A detailed description of the complex structure on this superspace see in [LR] and [CR] (and [V4]).

$\mathfrak{k}(1|2)$-curves are usually called SUSY(2)-curves. The superspace of their moduli is considered in [FR]. The isomorphism $\mathfrak{k}(1|2) = \mathfrak{vect}(1|1) = \mathfrak{m}(1)$ (first mentioned in [ALS], see [L2]) suggests a relation between $\mathfrak{k}(1|2)$-curves or $\mathfrak{m}(1)$-curves and (1|1)-dimensional supermanifolds without any structure. Such a relation exists all right and was noted in 1986 by J.Bernstein and later by P.Deligne [D] who showed that the category of the (1|1)-dimensional supermanifolds is equivalent to that of oriented or nontwisted $\mathfrak{k}(1|2)$-curves. (Leites told me that by the same arguments Deligne got similar description for the category of $\mathfrak{m}(1)$-curves. Unpublished, as usual.)

A local complex analytic structure on the superspace of $\mathfrak{g}$-curves is introduced with the help of deformation theory [V2]-[V4] as follows.

Theorem. *Let the supermanifold C of a $\mathfrak{g}$-curve be compact and $\mathfrak{g}$ any stringy superalgebra except $\mathfrak{svect}(1|N;\ \alpha)$ with $\alpha \notin \mathbb{Q}$. Then the $\mathfrak{g}$-curve C possesses a universal deformation whose base is a finite-dimensionsional complex supermanifold B.*

The Lie superalgebra $\mathfrak{g}$ acts locally transitively on B.

The construction of the universal deformation is done separately for each series of stringy superalgebras. For $\mathfrak{g} = \mathfrak{vect}(1|N)$ the supercurves have no additional structure and for them the universal deformation is constructed in [V1]. For the other series the construction is performed with the help of the theorems on existence of versal deformations of supermanifolds, bundles over them and cohomology classes proved in [V2]. The example $\mathfrak{g} = \mathfrak{k}(1|1)$ is considered in [V3]. The other cases are considered with all the details of the proof in [V4].

3. Concluding remarks and open problems

1) If a stringy superalgebra is distinguished and $\hat{\mathfrak{g}}$ is the result of its nontrivial central extension, then it seems very plausible that $\hat{\mathfrak{g}}$ acts in the space of a rank 1 (or ε) vector bundle over the superspace of moduli of $\mathfrak{g}$-curves. For example if $\mathfrak{g}$ is $\mathfrak{vect}(1|0)$ or $\mathfrak{k}(1|1)$ this bundle is the determinant bundle on the moduli space of (super)curves ([M1]). It would be interesting to give a construction of such a bundle for other stringy superalgebras and find analogues of Mumford's formula (see [M1], [ACKP], [F])

$$\det_j = (\det)^{6j^2-6j+1}$$

For the distinguished contact Lie superalgebras of the both Neveu-Schwarz and Ramond type this had been done by Serganova, prompted by Leites, in 1985 (unpublished) and for NS(1) by A.Raina (sec. 5 in [KL]). The answer for all distinguished superalgebras with the construction of the bundles (they are the bundles of certain semiinfinite forms) is contained, among other things, in [CLS]. For example, for $\mathfrak{k}(1|2)$ we have $\det_j = (\det)^2$; for $\mathfrak{g}=\mathfrak{vect}(1|2)$ and $\mathfrak{svect}(1|2;\ \alpha)$ the exponent in the right hand side is str(V), where V is the fiber of a vector bundle over the supercurve (an irreducible $\mathfrak{g}_0$-module) with the help of which the module of semiinfinite forms is constructed.

2) What do the analogues of Mumford's form, equations for correlator functions, etc. look like for an arbitrary stringy superalgebra? Is there a canonical flat connection on the superspace of moduli of $\mathfrak{g}$-curves, cf. [BS], [Vo], [M2]?

3) What is the relation between the represertations of $\mathfrak{g}$ and and the geometry of the corresponding moduli superspace similar to such a relation in the case of $\mathfrak{g} = \mathfrak{vect}(1|0)$ or $\mathfrak{k}(1|1)$, cf. [ACKP] and [UY]? To an extent the answer is given in [CLS].

4) The real (i.e. over $\mathbb{R}$) variant of the above set of ideas is related with the description of contact-projective structures on real supercircles, see [OOT], [F] and papers by O. and V. Ovsienkos and T. Khovanova in [L1], #23. The real forms of the stringy superalgebras are described [S2] (note that the quaternionic forms whose definition see in [S1] are not described yet) and in some cases when the Schwarz derivative can exist it and its global version, the Bott cocycle, are written [R2], [Sch]. It would be interesting to trace the relation further, at least write the Bott cocycle for all the supergroups corresponding to the distinguished stringy superalgebras.

5) What are analogues of the ∞-dimensional Grassmannian and Krichiver's construction for an arbitrary $\mathfrak{g}$-curve? See [R1] - [R3], where the cases $\mathfrak{g} = \mathfrak{k}(1|1)$ and $\mathfrak{vect}(1|1)$ are considered. To answer this question the knowledge of a "right" super version of

$\mathfrak{gl}(\infty)$ is handy, and this has been recently obtained by a participant of our Seminar on supermanifolds [E].

References

[A] Ademollo M. , Brink L., D'Adda A., D'Auria R., Napolitano E., Sciuto S., Del Guidice E., Di Veccia P., Ferrara S., Gliozzi F., Musto R., Pettorino R., Supersymmetric strings and colour confinement. Phys. Lett. B 62, 1976, 105-110; Nucl. Phys. B 111, 1976, 77; id., ibid., 114, 1976, 297

[ALS] Alekseevsky D., Leites D., Shchepochkina I., New examples of simple Lie superalgebras of vector fields. C.r. Acad. Bulg. Sci., v. 34, N9, 1980, 1187-1190 (in Russian)

[ACKP] Arbarello E., De Concini C., Kac V., Procesi C., Moduli spaces of curves and representation theory. Commun. Math. Phys. 117, 1988, 1-36

[B] Berezin F. Analysis with anticommuting variables. Kluwer, 1987

[BS] Beilinson A., Schekhtman V., Determinant bundles and Virasoro algebras. Commun. Math. Phys. 118, 1988, 651-701

[CLS] Chaichian M., Leites D., Serganova V., On ghost, semiinfinite and highly diagonalizable representations. CERN preprint, 1990

[CR] Crane L., Rabin J.M., Super Riemann surfaces: uniformisation and Teichmüller theory. Commun. Math. Phys. 113, 1988, 601

[D] Deligne P., Letters to Yu. Manin (fall, 1988)

[E] Egorov G., How to superize $\mathfrak{gl}(\infty)$. Sec. 5 in [L5]

[F] Feigin B., Notes on the Virasoro algebra, Schrödinger operator and projective structures on curves. Preprint, 1986 (to appear in [L3])

[FR] Falqui G., Reina C., N = 2 super Riemann surfaces and algebraic geometry. J. Math. Phys. 31(4), 1990, 948-952

[KL] Kac V., van de Leur J., On classification of superconformal algebras. In: Strings-88, World Sci., 1989, 77-106

[K] Kontsevich M.L., Virasoro algebra and Teichmüller spaces. Sov. J. Func. Anal. Appl. 21(2), 1987, 156-157

[LR] Le Brun C., Rothstein M., Moduli of super Riemann surfaces. Commun. Math. Phys. 117, 1988, 159

[L1] Leites D., Supermanifold theory. Petrozavodsk, Karelia branch of the USSR Acad of Sci., 1983, 200p. (in Russian; an expanded English version is [L3]; meanwhile see a preprinted version in 34 issues, Reports of Dept. of Math. Stockholm Univ., 1987-1990, 2800p.)

[L2] Leites D., Lie superalgebras. In: Itogi nauki i tehniki. Ser. Sovr. probl. matem. Novejshie dostizheniya. v.25, VINITI, 1984, 1-50 (Russian; Engl. transl. in Sov. J. Math. (JOSMAR) 30

[L3] Leites D.(ed.) Seminar on supermanifolds. v.1-4, Kluwer(?), 1991

[L4] Leites D. Supermanifolds and quantization. Supplement 3. In: Berezin F., Shubin M., Schrödinger equation, Kluwer, 1991

[L5] Leites D. On superized Leznov-Saveliev equations and their their relation with superized KdV and KP. Appendix in: A.Leznov, M.Saveliev. A group-theoretical method for integrating nonlinear dynamical systems. Birkhauser, 1991

[LF] Leites D., Feigin B., New Lie superalgebras of string theories. In: Group-theoretical methods in physics, Zvenigorod, 1983. Nauka, Moscow, v. 1, 1984, (in Russian; Engl. transl. published by Harwood Publ. Co., 1986)

[M1] Manin Yu., Critical dimensions of string theories and the dualizing sheaf on the moduli space of (super)curves. Sov. J. Funct. Anal. Appl. 20 (3), 1986, 244-245

[M2] Manin Yu., Neveu-Schwarz sheaves and differential equations for Mumford superforms. In: Geometry and Analysis.

[M3] Manin Yu., Gauge fields and complex geometry. Springer, 1988

[M4] Manin Yu., Superalgebraic curves and quantum strings. Compositio Math.

[M] Mathieu O. Talk at ICM-90.
[OOT]Ovsienko O., Ovsienko V., Tchekanov Yu., Classification of the contact-projective structures on supercircle. Russian Math. Surveys, 44 (3), 1989,
[R1] Rabin J., The geometry of the super KP flows. Preprint UC San Diego, June, 1990
[R2] Radul A., Superizing Schwarz derivative and Bott's cocycle. In: [L1], #1.
[R3] Radul A., Algebro-geometric solution to the super Kadomsev-Petviashvily hierarchy . In: [L1], #28, 1988-10, 1-10
[RS] Ramond P., Schwarz J., Phys. Lett. B 64, 1976, 75; J. Math. Phys., v.21, #4, 1980
[Sch] Schoutens K., A nonlinear representation of the d = 2 so(4)-extended superconformal algebra. Phys. Lett. B 194, 1987, 75-80; id. O(N)-extended superconformal field theory in superspace. Nucl. Phys. B 295, 1988, 634-652
[S1] Serganova V. Classification of simple real Lie superalgebras and of classical superdomains.Sov. J. Func. Anal. Appl. 18(2), 1984, 59-60
[S2] Serganova V. Real forms of stringy Lie superalgebras. Sov. J. Func. Anal. Appl. 18(2), 1984, 59-60
[SS] Schwimmer A., Seiberg N., Comments on the N = 2, 3, 4 superconformal algebras in two dimensions. Phys. Lett. B 184, 1987, 191-196
[UY] Ueno K., Yamada H., Some observations on geometric representations of the superconformal algebras and a superanalogue of the Mumford sheaves. In: Prospects of Algebraic Analysis, Acad. Press, 1988
[V1] Vaintrob A. Deformations of complex supermanifolds. Sov. J. Func. Anal. Appl. 18(2), 1984, 59-60; id. Deformations of complex supermanifolds. In: Group-theoretical methods in physics, Yurmala, 1985. Nauka, Moscow, v. 1, 1985, (in Russian; Engl. transl. published by VNU Sci. Press, 1987)
[V2] Vaintrob A. Deformations of complex structures on supermanifolds. In: [L1], #24.
[V3] Vaintrob A. Deformations of complex superspaces and coherent sheaves on them. In: Itogi nauki i tehniki. Ser. Sovr. probl. matem. Novejshie dostizheniya. v.32, VINITI, 1988, 125-201 (Russian; Engl. transl. in JOSMAR)
[V4] Vaintrob A. Complex structures on supermanifolds and their deformations. In: [L3], v.3.
[Vo] Voronov A., A formula for the Mumford measure in the superstring theory. Sov. J. Funct. Anal Appl. 22(2), 1988, 67-68

4. Superalgebras and supermanifolds

Supersymmetric products of SUSY-curves †

J.A. Domínguez Pérez,

D. Hernández Ruipérez and C. Sancho de Salas

Departamento de Matemáticas, Universidad de Salamanca, Plaza de la Merced 1-4, 37008, Salamanca (Spain)

Abstract: The supersymmetric product of a SUSY-curve over a field is constructed with the aid of a theorem of invariants, and the notion of relative superdivisor is introduced. A universal superdivisor is defined in the supersymmetric product by means of Manin's superdiagonal, and it is proven that every superdivisor can be obtained in a unique way as a pullback of the universal superdivisor.

1. Introduction

1.1 Preliminaries on SUSY-curves

Super Riemann surfaces or *supersymmetric curves* (SUSY-curves) were introduced by Baranov, Schwarz [3] and Friedan [12] as an important tool for understanding the geometric structure of superstrings. The moduli space of these objects is necessary for the development of a Polyakov formalism for superstring theory.

As often happens in supermanifold theory, there are two different approaches to SUSY-curves; the one based on Rogers [30] and DeWitt's [11] formulation of supermanifolds, and the other based on Berezin-Leĭtes-Kostant's [8], [22], [20] formulation of graded manifolds.

In the first approach, super Riemann surfaces, or superconformal manifolds as they are also often called, are — quoting directly from [31] — "(1, 1)-dimensional complex supermanifolds composed of superdomains glued together by superconformal transformations (that is to say, by transformations preserving the operator $D = \frac{\partial}{\partial\theta} + \theta\frac{\partial}{\partial z}$ up to a factor)". In addition to the pioneering work by Baranov, Schwarz and Friedan, remarkable contributions to the geometry of super Riemann surfaces and their moduli space following this approach have been supplied, among others, by Crane, Rabin, Freund and Topiwala [10], [28], [29], and Baranov, Rosly, Schwarz and Voronov [3], [31], [32].

† Talk given by the second author.

In the second approach, closer to the spirit of Algebraic Geometry, graded Riemann surfaces are regarded as $(1,1)$-dimensional graded locally ringed spaces $(X, \mathcal{A})$ such that $(X, \mathcal{O} = \mathcal{A}/\mathcal{N})$ is a smooth proper algebraic curve ($\mathcal{N}$ denoting the nilpotent ideal of $\mathcal{A}$), and where the odd part $\mathcal{A}_1$ of $\mathcal{A}$ is a spin structure on $(X, \mathcal{O})$, i.e., a line bundle whose square is isomorphic with the canonical bundle. SUSY-families or (relative) SUSY-curves were defined as proper flat morphisms of relative dimension $(1,1)$ endowed with a relative spin structure. In that context, a graded Riemann surfaces is a single SUSY-curve or a SUSY-family over a point (the spectrum of a field) [23]. There are many works in this direction, of very different depth and scope, considering either families or single SUSY-curves. Among others, those of Manin [23] [24], Batchelor and Bryant [6], Giddings and Nelson [13], [14], Bartocci, Bruzzo and Hernández Ruipérez [5], and LeBrun and Rothstein [21] — devoted to the construction of the moduli space of SUSY-curves as a superorbifold — might be cited.

This paper follows the second approach and should be considered as a first step towards an algebraic construction of a Jacobian supervariety of a single SUSY-curve $\mathcal{X}$ fulfilling the usual functorial requirements. Thus, in this paper by a SUSY-*curve* we shall understand a SUSY-*family over a point*, that is to say, a *graded Riemann surface*.

In order to establish these requirements more precisely, a survey of the most outstanding aspects of the corresponding theory for ordinary (non-graded) algebraic curve seems to be necessary. Although very interesting reviews of this topic can be found in [26] or [19], let us briefly recall in the following section some aspects of the diverse theories about the Jacobian variety of an algebraic curve, stressing those more relevant to SUSY-curves.

1.2. The Jacobian variety of an algebraic curve. A historical survey

One on the most prominent features the theory of algebraic curves is the study of the Jacobian variety, i.e. the variety that parametrizes (modulo isomorphisms) the line bundles of degree g on the curve. There are many ways for constructing the Jacobian variety, differing either in method or in scope.

The first definition of the Jacobian variety $J(X)$ of a complex algebraic curve is of an analytic nature and is based on the period matrix. Once a 'canonical' basis $(\delta_1, \dots, \delta_{2g})$ of 1-cycles and a basis $(\omega_1, \dots, \omega_g)$ of holomorphic 1-forms on X have been chosen, $J(X)$ is defined as the complex torus $H^0(X, \Omega^1)/\Lambda$, where Λ is the lattice generated by the columns $\Pi_h = (\int_{\delta_h} \omega_1, \dots, \int_{\delta_h} \omega_g)$ of the period matrix. Then, if $\mathrm{Div}^0(X)$ denotes the group of divisors of degree 0 of X, we have the so-called Abel map:

$$\begin{aligned} &\mu \colon \mathrm{Div}^0(X) \to J(X) \\ &\sum_{i=1}^{g} (q_i - p_i) \mapsto \Big(\int_{p_1}^{q_1} \omega_1, \dots, \int_{p_g}^{q_g} \omega_g\Big)/\Lambda . \end{aligned}$$

Two classical theorems, namely Abel's theorem and the Jacobi inversion theorem, ensure that the Abel map induces an isomorphism

$$\bar{\mu}\colon \mathrm{Div}^0(X)/\text{linear equivalence} \simeq J(X)$$

which is generically one-to-one, i.e., it is a birational map.

The second classical construction of the Jacobian variety is due to A. Weil [33], who defined $J(X)$ as a set by letting $J(X) = \mathrm{Div}^0(X)/(\text{linear equivalence})$, thus shifting the problem to the determination of the structure of $J(X)$ as an abelian variety. This was done as follows: let $\mathrm{Div}^g(X) = S^g X$ be the set of positive divisors of degree g of X, that is, unordered families of g points. If a point $x \in X$ is fixed, the Abel map

$$\mu^{(g)}\colon \mathrm{Div}^g(X) \to J(X)$$
$$x_1 + \ldots + x_g \mapsto [x_1 + \ldots + x_g - gx]\,,$$

where g is the genus of X and [ ] means the linear equivalence class, is shown to be one-to-one for generic x. Moreover, if a point $x \in X$ has been fixed, it is proved that, given two generic positive divisors D, D' of degree g, the divisor $D + D' - gx$ is linearly equivalent to a unique positive divisor $\bar{D}$ of degree g, thus allowing us to define a local group law in the Jacobian variety by letting

$$[D - gx] + [D' - gx] = [\bar{D} - gx]\,.$$

The third procedure for the construction of the Jacobian variety is due to Chow [9], who noticed that for a great enough p the Abel map $\mu^{(p)}\colon \mathrm{Div}^p(X) \to J(X)$ is a projective bundle, so that $J(X)$ can be endowed with a structure of projective algebraic group.

Inspired by this procedure, but with an enormous change in mentality, Grothendieck constructed the Picard schemes (the Jacobian of a curve is a particular case of a Picard scheme) as quotients of Hilbert schemes by the action of a flat equivalence relation [15], [1]. The philosophy of this method was the theory of the *functor of points*, which is based on the following principle: classically, the knowledge of the points of a variety allows us to determine the variety only as a set, and the algebraic or analytic structure has to be assigned separately. However, if we consider not only ordinary points, i.e., morphisms of a single point into the variety, but rather S-valued 'points', i.e. morphisms of a variety S into our variety, then the knowledge of all these 'points' *characterizes* the algebraic variety. The techniques for determining a scheme from the functor of its points are called 'construction techniques' and the theorems stating that a certain functor is actually the functor of points of a scheme are known as 'representability theorems'.

Let us explain briefly how these techniques apply to construction of the Picard scheme of an algebraic variety X. As a first step, we wish the points of the Picard scheme to be the line bundles on X. But, as we have seen, we need all the S-valued 'points' for every scheme S and not only the ordinary points. Since these 'points' should be, roughly speaking, 'families of line bundles' of X parametrized by S, Grothendieck defined the Picard functor of X by taking $\mathrm{Pic}_X(S)$ as the set of line bundles on $X \times S$ modulo those obtained by pulling back line bundles on the parameter space S. In this way we get the candidate $\mathrm{Pic}_X(S)$ for the set of S-valued 'points' of the Picard scheme. Under certain conditions, the Picard functor is representable; that is, it is the functor of points of a well determined scheme, namely, the Picard scheme of X.

The method of the functor of points is not only as elegant as Grothendieck's ideas used to be, but is also extremely precise. It was exploited to show how even for smooth algebraic varieties, the Picard scheme may fail to be reduced. In fact the Picard scheme of the so-called Igusa surface is a scheme whose ring sheaf has nilpotent elements (see [15]).

Weil's construction of the Jacobian of a smooth algebraic curve was drawn to this framework by Iversen [17] and Artin [2]. Iversen showed that the functor of relative divisors of degree g of $X \times S \to S$, once properly defined, is the functor of points of the symmetric product $S^g X$ considered as the quotient of the cartesian product $X \times \overset{g)}{\dots} \times X$ by the action of the symmetric group. Weil's techniques for defining an algebraic group structure on an algebraic variety which is birationally equivalent to an algebraic group (afterwards generalized to arbitrary schemes by Artin [2]) applies to this case to perform the Abel morphism, taking the quotient with respect to the linear equivalence relation in order to represent the Picard functor of the curve and thus constructing its Jacobian as an abelian variety.

1.3. Towards the Jacobian of a SUSY-curve. The supersymmetric product

Within the context of supermanifolds *à la* Rogers, the analytic construction of the Jacobian (the one based on the period matrix) has been generalized for super Riemann surfaces in [31] and [32]. This Jacobian is an ordinary (non-super) manifold, but the superanalogous of Abel's and Jacobi inversion theorems have been proved. In the same framework, a cohomological construction of the Picard variety for a particular kind of Roger supermanifolds, the DeWitt ones, has been made in [4].

However, as far as we know, there are no constructions of the Jacobian scheme of SUSY-curves within the algebraic approach, i.e. via graded ringed spaces. Our plan is to extend the Weil-Iversen construction to this case, and the first step to do so is to construct the supersymmetric product of a SUSY-curve and to determine its functor of points.

There are further reasons to adopt this procedure, based on the intrinsic interest of the supersymmetric product. In fact, regardless of the method used to construct the Jacobian, the structure of the symmetric products $S^p X$ and the diverse Abel morphisms $S^p X \to J(X)$ turn out to be a key point in the theory of Jacobian varieties (see, for instance, [18], [25]), and have proved to be a very important tool for the solution of the Schottky problem ([27]).

This paper is organized as follows. The supersymmetric product $S^g \mathcal{X}$ of a superscheme $\mathcal{X}$ of dimension $(1,1)$ is constructed in Part 2 as the orbit ringed space obtained through the action of the symmetric group on the cartesian product of g copies of the supercurve, g being its genus. Actually, this supersymmetric product is a superscheme of dimension (g,g); this statement is far from trivial, and is equivalent to a theorem on invariants, namely, the fact that the invariant subalgebra of a certain exterior algebra is an exterior algebra as well. It should be stressed that this result is no longer true for superschemes of higher odd dimension ($n > 1$).

Part 3 deals with the determination of the functor of points of the supersymmetric product. The main result here is that this functor is the functor of the so-called *relative superdivisors*. This means that there exists a universal relative superdivisor in $\mathcal{X} \times S^g\mathcal{X} \to S^g\mathcal{X}$ such that every relative superdivisor of $\mathcal{X} \times \mathcal{S} \to \mathcal{S}$ can be obtained in a unique way as the pullback of the universal one through a morphism $\mathcal{S} \to S^g\mathcal{X}$.

The construction of the universal superdivisor is inspired by Manin's definition of superdiagonal ([23], [7]) and by the fact — also pointed out by him — that for SUSY-curves relative superdivisors of $\mathcal{X} \times S^g\mathcal{X} \to S^g\mathcal{X}$ are not $\mathcal{S}$-valued 'points', because $\mathcal{S}$-points have codimension $(1,1)$ and not codimension $(1,0)$ as superdivisors do.

However, given an $\mathcal{S}$-point, defined locally in relative conformal coordinates by the equations $z - z_0 = \theta - \theta_0 = 0$, one can associate to this point the relative superdivisor of degree 1 with the same support defined by the equation $z - z_0 - \theta\theta_0 = 0$, which is in fact indepedent of the coordinate system, and is therefore globally defined.

The $\mathcal{S}$-point σ can be recovered from the function $f = z - z_0 - \theta\theta_0$, because the defining ideal $(z - z_0, \theta - \theta_0)$ for the point is exactly the ideal $(f, D(f))$, where $D = \frac{\partial}{\partial\theta} + \theta\frac{\partial}{\partial z}$ is the local generator of the conformal structure.

This shows the nature of the equations defining relative divisors of degree 1. Relative divisors of degree g should be then defined 'generically' by products of equations of this kind. This can be stated in a precise way only for a suitable universal superdivisor and for those superdivisors obtained from it by pullback. This is the reason underlying our definition of relative superdivisor and our construction of the universal relative superdivisor.

This notion is related to some definitions given in [31]. In that reference, a principal zero of a meromorphic section s of a line bundle on a superconformal manifold is defined as a point P fulfilling $s(P) = 0$ and $Ds(P) = 0$, and the relationship between Cartier divisors, points, and principal zeroes of meromorphic sections in that context is discussed. Our definition will lead to an equivalent notion of a Cartier divisor if the base scheme $\mathcal{S}$ is taken as a (non-graded) single point.

As a final comment, let us say that the construction of the supersymmetric product of a SUSY-curve and the determination of its functor of points can be extended straightforwardly to SUSY-families parametrized by an *ordinary* scheme. The case of SUSY-families over an arbitrary superscheme will be considered in a forthcoming paper.

2. Supersymmetric products

2.1. Definitions

Let us start with the definition of *graded scheme* or *superscheme*. Let $\mathcal{X} = (X, \mathcal{A})$ be a graded ringed space, consisting of a topological space X and a sheaf $\mathcal{A}$ of $\mathbb{Z}_2$-graded algebras.

Definition 1. A superscheme of dimension (m, n) over a field k is a graded ringed space $\mathcal{X} = (X, \mathcal{A})$ where $\mathcal{A}$ is a sheaf of graded k-algebras such that:
1) $(X, \mathcal{O})$ is an algebraic scheme of dimension m over k, where $\mathcal{O} = \mathcal{A}/\mathcal{J}$, $\mathcal{J}$ being the ideal $A_1 + A_1^2$.
2) $\mathcal{J}/\mathcal{J}^2$ is a locally free rank n $\mathcal{O}$-module and $\mathcal{A}$ is locally isomorphic to $\bigwedge_{\mathcal{O}}(\mathcal{J}/\mathcal{J}^2)$.

$(X, \mathcal{O})$ will be called the underlying ordinary scheme.

Let us consider the product

$$\mathcal{X}^g = (X^g, \mathcal{A}^{\otimes g}) \tag{1}$$

where X^g denotes the cartesian product $X \times \overset{g)}{\ldots} \times X$, and $\mathcal{A}^{\otimes g} = \mathcal{A} \otimes \overset{g)}{\ldots} \otimes \mathcal{A}$.

The symmetric group S_g acts on $\mathcal{X}^g$ by graded automorphisms of superschemes according to the rule:

$$\begin{aligned} \sigma\colon X^g &\to X^g \\ (x_1, \ldots, x_g) &\mapsto (x_{\sigma(1)}, \ldots, x_{\sigma(g)}) \end{aligned} \tag{2}$$

on the space and:

$$\begin{aligned} \sigma^*\colon \mathcal{A}^{\otimes g} &\to \sigma_* \mathcal{A}^{\otimes g} \\ f_1 \otimes \ldots \otimes f_g &\mapsto \prod_{\substack{i<j \\ \sigma(i)>\sigma(j)}} (-1)^{|f_i||f_j|} f_{\sigma(1)} \otimes \ldots \otimes f_{\sigma(g)} \end{aligned} \tag{3}$$

on the sheaves. This action reduces to the ordinary action of S_g on the scheme $(X^g, \mathcal{O}^{\otimes g})$. Then, we have the orbit space $S^g X$, a natural projection $p\colon X^g \to S^g X$, and an invariant sheaf $\mathcal{O}_g = (\mathcal{O}^{\otimes g})^{S_g}$ on $S^g X$, whose sections on an open subset $V \subset S^g X$ are

$$\mathcal{O}_g(V) = \{f \in \mathcal{O}^{\otimes g}(p^{-1}(V)) \,|\, \sigma^* f = f \text{ for every } \sigma \in S_g\}\,. \tag{4}$$

It is well-known that if $(X, \mathcal{O})$ is a projective scheme, the ringed space $(S^g X, \mathcal{O}_g)$ is a scheme, the *g-th symmetric product* of $(X, \mathcal{O})$.

Let us consider the sheaf $\mathcal{A}_g = (\mathcal{A}^{\otimes g})^{S_g}$ of graded invariants on $S^g X$ defined as above by letting:

$$\mathcal{A}_g(V) = \{f \in \mathcal{A}^{\otimes g}(p^{-1}(V)) \,|\, \sigma^* f = f \text{ for every } \sigma \in S_g\} \tag{5}$$

for every open subset $V \subset S^g X$.

2.2. The case of supercurves

Definition 2. A proper smooth supercurve is a superscheme of dimension $(1,n)$, such that the underlying ordinary scheme is a proper smooth algebraic curve over a field k.

Let $\mathcal{X}$ be a smooth proper supercurve. As $(X,\mathcal{O})$ is projective (because it has very ample sheaves), the symmetric product $(S^gX,\mathcal{O}_g)$ is a scheme (in fact, it is smooth, which is no longer true for a higher dimensional X).

Theorem 1. *Let $\mathcal{X}=(X,\mathcal{A})$ be a smooth proper supercurve. The graded ringed space $S^g\mathcal{X}=(S^gX,\mathcal{A}_g)$ is a superscheme if and only if $n=1$, that is, if and only if $\mathcal{X}$ is a superscheme of dimension $(1,1)$.*

In that case, $S^g\mathcal{X}$ will be called the g-th supersymmetric product of $\mathcal{X}$.

2.2.1. Sketch of the proof

There is a natural projection $\mathcal{A}_g\to\mathcal{O}_g$, and we have only to ascertain whether $\mathcal{A}_g$ is locally the exterior algebra of a certain locally free $\mathcal{O}_g$-module. We can assume that $\mathcal{A}=\bigwedge_{\mathcal{O}}(\mathcal{N})$, $\mathcal{N}$ being a free rank n $\mathcal{O}$-module.

Let us write $\mathcal{N}_i=\mathcal{O}\otimes\ldots\otimes\overset{\downarrow i)}{\mathcal{N}}\otimes\ldots\otimes\mathcal{O}$ and $\mathcal{M}=\mathcal{N}_1\oplus\ldots\oplus\mathcal{N}_g$. Now, if we write $\bar{\mathcal{O}}=\mathcal{O}^{\otimes g}$ and $\bar{\mathcal{A}}=\mathcal{A}^{\otimes g}$, we have:

$$\bar{\mathcal{A}}=\textstyle\bigwedge_{\mathcal{O}}(\mathcal{N})\otimes_{\mathcal{O}}\overset{g)}{\ldots}\otimes_{\mathcal{O}}\bigwedge_{\mathcal{O}}(\mathcal{N})\simeq\bigwedge_{\bar{\mathcal{O}}}(\mathcal{M})\,. \tag{6}$$

The symmetric group S_g acts on $\mathcal{M}$ by $\sigma(n_1+\ldots+n_g)=n_{\sigma(1)}+\cdots+n_{\sigma(g)}$ and this action provides an action $\sigma\colon\bar{\mathcal{A}}\to\bar{\mathcal{A}}$ on the exterior algebra $\bar{\mathcal{A}}=\bigwedge_{\bar{\mathcal{O}}}(\mathcal{M})$, given by $\sigma(m_1\wedge\ldots\wedge m_p)=\sigma(m_1)\wedge\ldots\wedge\sigma(m_p)$

Thus, we have two actions of S_g on $\bar{\mathcal{A}}$, the one just defined, and that previously introduced in (3). However, it turns out that the two actions actions are the same, because they coincide on $\bigwedge^1_{\bar{\mathcal{O}}}(\mathcal{M})=\mathcal{M}$ and are morphisms of graded algebras.

Let us denote by $\mathcal{M}^{S_g}$ the $\mathcal{O}_g$-module consisting of the invariant sections of $\mathcal{M}$. Our main theorem now follows from:

Lemma 1. *The natural morphism of sheaves of graded $\mathcal{O}_g$-algebras over S^gX,*

$$\phi\colon\textstyle\bigwedge_{\mathcal{O}_g}(\mathcal{M}^{S_g})\to(\bigwedge_{\bar{\mathcal{O}}}(\mathcal{M}))^{S_g}=\mathcal{A}_g\,, \tag{7}$$

is an isomorphism if and only if $n=1$.

Proof. The proof is a calculus of invariants in the exterior algebra of a free module over a *commutative* ring, which allows us to use standard methods of Commutative Algebra. We shall proceed by steps.

Let us first notice that the statement is invariant by completion at the maximal points of $\mathcal{O}_g$. This reduces the problem to the case of the polynomial ring $\mathcal{O}=k[x]$.

Now, for every $0 \leq p \leq g$ we have:

$$\bigwedge^p(\mathcal{M}) = \bigoplus_{i_1<\ldots<i_p} \mathcal{N}_{i_1} \wedge \ldots \wedge \mathcal{N}_{i_p} \tag{8}$$

and S_g acts transitively by permutation of summands.

In fact, $\mathcal{N}_{i_1} \wedge \ldots \wedge \mathcal{N}_{i_p} = \sigma_{i_1 \ldots i_p}(\mathcal{N}_1 \wedge \ldots \wedge \mathcal{N}_p)$, $\sigma_{i_1 \ldots i_p} \in S_g$ being any permutation of type $\sigma_{i_1 \ldots i_p} = \begin{pmatrix} 1 & \ldots & p & \ldots \\ i_1 & \ldots & i_p & \ldots \end{pmatrix}$.

Then, an invariant element $m = \sum_{i_1<\ldots<i_p} n_{i_1} \wedge \ldots \wedge n_{i_p}$ is characterized by $n_1 \wedge \ldots \wedge n_p$, and we have an isomorphism:

$$\begin{aligned} (\mathcal{N}_1 \wedge \ldots \wedge \mathcal{N}_p)^{S_p \times S_{g-p}} &\simeq (\bigwedge^p(\mathcal{M}))^{S_g} \\ n_1 \wedge \ldots \wedge n_p &\mapsto \sum_{\sigma \in S_g/(S_p \times S_{g-p})} \sigma(n_1 \wedge \ldots \wedge n_p) \end{aligned} \tag{9}$$

where $S_p \times S_{g-p}$ denotes the subgroup of S_g consisting of the permutations leaving the subset $\{1, \ldots, p\}$ invariant. In particular, $(\mathcal{N}_1)^{S_{g-1}} \simeq \mathcal{M}^{S_g}$.

Let us now assume that $n = 1$. Then

$$\mathcal{N}_1 \wedge \ldots \wedge \mathcal{N}_p = \bar{\mathcal{O}} \cdot e_1 \wedge \ldots \wedge e_p \,, \tag{10}$$

and we have:

$$(\mathcal{N}_1 \wedge \ldots \wedge \mathcal{N}_p)^{S_p \times S_{g-p}} \simeq \bar{\mathcal{O}}^{-S_p \times S_{g-p}} \tag{11}$$

where $\bar{\mathcal{O}}^{-S_p \times S_{g-p}}$ stands for the subset of those $f \in \bar{\mathcal{O}}$ such that $(\sigma \times \mu)(f) = \operatorname{sign}(\sigma) \cdot f$ for every $(\sigma \times \mu) \in S_p \times S_{g-p}$. In particular,

$$\mathcal{N}_1^{S_{g-1}} \simeq \bar{\mathcal{O}}^{S_{g-1}} \,. \tag{12}$$

The original morphism

$$\phi_p \colon \bigwedge^p_{\mathcal{O}_g}(\mathcal{M}^{S_g}) \to (\bigwedge^p_{\bar{\mathcal{O}}}(\mathcal{M}))^{S_g} = \mathcal{A}_g \tag{13}$$

is now a morphism

$$\bar{\phi}_p \colon \bigwedge^p \bar{\mathcal{O}}^{S_{g-1}} \to \bar{\mathcal{O}}^{-S_p \times S_{g-p}} \tag{14}$$

described by

$$\bar{\phi}_p(f_1 \wedge \ldots \wedge f_p) = \sum_{\mu \in S_p} \operatorname{sign}(\mu) \sigma_{\mu(1)}(f_1) \ldots \sigma_{\mu(p)}(f_p) \tag{15}$$

where σ_i is the transposition of 1 and i.

A lengthy calculation involving symmetric functions allows us to prove that $\bar{\phi}_p$ is an isomorphism. We can also prove, basically because there exist invariant elements in the tensor product that cannot be written as tensor products of invariant elements, that ϕ_p is not an isomorphism when $n > 1$, thus finishing the sketch of the proof. □

Moreover, if $\mathcal{X}$ is a proper smooth supercurve of odd dimension 1, the ring sheaf of the symmetric product $S^g \mathcal{X}$ is globally an exterior algebra. This is due to the following fact. The natural epimorphism $\mathcal{A} \to \mathcal{O}$ induces an isomorphism $\mathcal{A}_0 \stackrel{\sim}{\to} \mathcal{O}$ between the even part of $\mathcal{A}$ and $\mathcal{O}$. The odd part is an invertible $\mathcal{O}$-module $\mathcal{L}$, so that $\mathcal{A} = \mathcal{O} \oplus \mathcal{L} \stackrel{\sim}{\to} \bigwedge_{\mathcal{O}}(\mathcal{L})$. Then there is a canonical isomorphism:

$$\bigwedge\nolimits_{\mathcal{O}_g}(\mathcal{M}^{S_g}) \stackrel{\sim}{\to} \mathcal{A}_g \tag{16}$$

where $\mathcal{M} = \oplus_{i=1}^{g}(\mathcal{O} \otimes \ldots \otimes \overset{\downarrow i)}{\mathcal{L}} \otimes \ldots \otimes \mathcal{O})$.

3. The functor of points of the Supersymmetric product

3.1. SUSY-curves

Henceforth calligraphic types will be reserved to graded ringed spaces and the structure ring sheaf of any ringed space will be denoted by $\mathcal{O}$ with the name of the ringed space as a subscript. Thus, for instance, $\mathcal{X} = (X, \mathcal{O}_{\mathcal{X}})$ or simply $\mathcal{X}$ will mean a graded ringed space, whereas $(X, \mathcal{O}_X)$ or X will stand for the underlying ordinary ringed space.

Definition 3. A *supersymmetric curve* or SUSY-curve over a field k is a smooth proper supercurve $\mathcal{X} = (X, \mathcal{O}_{\mathcal{X}})$ of odd dimension 1, endowed with a locally free submodule $\mathcal{D}$ of rank $(0,1)$ of the tangent sheaf $T_{\mathcal{X}} = \mathcal{D}er(\mathcal{O}_{\mathcal{X}})$ such that the composition of maps

$$\mathcal{D} \otimes_{\mathcal{O}_{\mathcal{X}}} \mathcal{D} \xrightarrow{[\,,\,]} \mathcal{D}er(\mathcal{O}_{\mathcal{X}}) \to \mathcal{D}er(\mathcal{O}_{\mathcal{X}})/\mathcal{D} \tag{17}$$

is an isomorphism of $\mathcal{O}_{\mathcal{X}}$-modules (see for instance [21]). $\mathcal{D}$ is called the conformal structure of the SUSY-curve

If $\mathcal{X} = (X, \mathcal{O}_{\mathcal{X}}, \mathcal{D})$ is a SUSY-curve, X can be covered by affine open subsets $U \subset X$ with local relative coordinates (z, θ) such that $\mathcal{D}$ is locally generated by $D = \dfrac{\partial}{\partial \theta} + \theta \dfrac{\partial}{\partial z}$. These coordinates are called *conformal.*

There is a natural isomorphism $\mathcal{D}^* \stackrel{\sim}{\to} \mathcal{B}er(\mathcal{O}_{\mathcal{X}})$ and a 'Berezinian differential':

$$\delta\colon \Omega^1_{\mathcal{X}} \to \mathcal{B}er(\mathcal{O}_{\mathcal{X}}) \simeq \mathcal{D}^* \tag{18}$$

which is nothing but the natural projection induced by the immersion $\mathcal{D} \to T_{\mathcal{X}}$. In conformal coordinates δ is described by $\delta(df) = \left[dz \otimes \dfrac{\partial}{\partial \theta}\right] \cdot D(f)$ where $\left[dz \otimes \dfrac{\partial}{\partial \theta}\right]$ denotes the local basis of $\mathcal{B}er\mathcal{O}_{\mathcal{X}}$ determined by (z, θ) (see [16]).

3.2. The universal divisor for an algebraic curve

This section is devoted to summarize the theory of the variety of divisors and the universal divisor for an (ordinary) algebraic curve X, also explaining how the relevant constructions and theorems can be recast if the variety of parameters is a superscheme. Suitable references are [15] or [17].

In this case, divisors of degree g are unordered families of g points, so that they are parametrized by the symmetric product S^gX. This can be made precise by means of the notion of relative divisor. If S is another scheme, relative divisors of $X \times S \to S$ of degree g, or families of divisors of X depending on a scheme of parameters S, are simply subschemes $Z \to X$ such that $\mathcal{O}_Z$ is a locally free $\mathcal{O}_S$-module of rank g. There is a nice relative divisor Z_U of degree g of $X \times S^gX \to S^gX$, whose fibre at a point $(x_1, \dots, x_g) \in S^gX$ is the divisor $x_1 + \dots + x_g$ of X defined by it. Z_U is called the ***universal divisor*** because the map

$$\begin{aligned} \operatorname{Hom}(S,\, S^gX) &\to \operatorname{Div}^g_S(X \times S) \\ \phi &\mapsto (1 \times \phi)^{-1}(Z_U)\,, \end{aligned} \tag{19}$$

where $\operatorname{Div}^g_S(X \times S)$ denotes the set of relative divisors of degree g, is one-to-one. This means that each divisor can be obtained as a pullback of the universal divisor, which statement is known as the ***representability theorem*** for the symmetric product.

This theory is still viable when a superscheme is allowed as the space of parameters. The corresponding notion of relative divisor is:

Definition 4. Let X be an ordinary smooth curve and $(\mathcal{S}, \mathcal{O}_\mathcal{S})$ a superscheme. A relative divisor of degree g of $X \times \mathcal{S} \to \mathcal{S}$ is a closed sub-superscheme $\mathcal{Z}$ of $X \times \mathcal{S}$ of codimension $(1,0)$ defined by a homogeneous ideal J of $\mathcal{O}_{X \times \mathcal{S}}$ such that $\mathcal{O}_{X \times \mathcal{S}}/J$ is a locally free $\mathcal{O}_\mathcal{S}$-module of rank $(g, 0)$.

The representability theorem now reads:

Theorem 2. *Let X be a smooth proper curve over a field k and Z_U the universal divisor. The map*

$$\begin{aligned} \operatorname{Hom}(\mathcal{S},\, S^gX) &\to \operatorname{Div}^g_\mathcal{S}(X \times \mathcal{S}) \\ \phi &\mapsto (1 \times \phi)^{-1}(Z_U)\,, \end{aligned} \tag{20}$$

where $\operatorname{Div}^g_\mathcal{S}(X \times \mathcal{S})$ denotes the set of relative divisors of degree g, is one-to-one for every superscheme $\mathcal{S}$.

The argument proving the representability theorem for ordinary schemes applies to this case only with minor changes. Two key points are then involved in the proof; the first is the construction of the universal divisor. If $\pi_i \colon X^g \to X$ is the ith projection and Δ_i is the relative divisor of $X \times X^g \to X^g$ obtained by pulling back the diagonal $\Delta \subset X \times X$ through $1 \times \pi_i \colon X \times X^g \to X \times X$, one can prove that there exists a unique relative divisor Z_U of $X \times S^gX \to S^gX$ such that

$$(1 \times p)^* Z_U = \Delta_1 + \ldots + \Delta_g \,, \tag{21}$$

where $p: X^g \to S^g X$ is the natural projection. This divisor Z_U is the universal divisor and is locally described by the equation

$$(z - z_1) \cdot \ldots \cdot (z - z_g) = z^g - s_1 z^{g-1} + \ldots + (-1)^g s_g = 0 \,, \tag{22}$$

where z is a local parameter on X, $z_i = 1 \otimes \ldots \otimes \overset{\downarrow i)}{z} \otimes \ldots \otimes 1$, and $s_1, \ldots, s_g$ are the symmetric functions of $z_1, \ldots, z_g$.

The second key point is the 'determinant morphism' $\mathcal{S} \to S^g \mathcal{Z}$, $\mathcal{Z}$ being a relative divisor of degree g, which deserves this name because its composition with $S^g \mathcal{Z} \to S^g X$ provides the inverse mapping of (20) (See [17]). The determinant morphism for the locally free $\mathcal{O}_\mathcal{S}$-module $\mathcal{O}_\mathcal{Z}$ of rank $(g, 0)$ is defined as follows. Each element b in the invariant sheaf $(\mathcal{O}_\mathcal{Z})_g = (\mathcal{O}_\mathcal{Z}^{\otimes g})^{S_g}$ acts on the $\mathcal{O}_\mathcal{S}$-module $\bigwedge_{\mathcal{O}_\mathcal{S}} \mathcal{O}_\mathcal{Z}$ of rank $(1,0)$ as the multiplication by a well-determined element $\det(b)$ in $\mathcal{O}_\mathcal{S}$. This gives rise to a morphism of sheaves $(\mathcal{O}_\mathcal{Z})_g \to \mathcal{O}_\mathcal{S}$, and thus to a morphism of schemes $\mathcal{S} \to S^g \mathcal{Z}$.

The determinant morphism provides the inverse mapping of (20) because if b is an even element in $\mathcal{O}_\mathcal{Z}$, $b_i = 1 \otimes \ldots \otimes \overset{\downarrow i)}{b} \otimes \ldots \otimes 1 \in \mathcal{O}_\mathcal{Z}^{\otimes g}$, and we denote by $s_i(b)$ the symmetric functions of $b_1, \ldots, b_g$, we have that

$$a_i = \det(s_i(b)) \quad (i = 1, \ldots, g) \,, \tag{23}$$

where $z^g - a_1 z^{g-1} + \ldots + (-1)^g a_g$ is the characteristic polynomial of b acting on $\mathcal{O}_\mathcal{Z}$ by multiplication.

3.3. Superdivisors on a SUSY-curve and universal superdivisor

Let $(\mathcal{X}, \mathcal{O}_\mathcal{X}, \mathcal{D})$ be a SUSY-curve, $(X, \mathcal{O}_X)$ the ordinary underlying curve and $(\mathcal{S}, \mathcal{O}_\mathcal{S})$ a superscheme.

Every closed sub-superscheme $\mathcal{Z}$ of $\mathcal{X} \times \mathcal{S} \to \mathcal{S}$ induces a closed sub-superscheme $\hat{\mathcal{Z}} = \mathcal{Z} \times_\mathcal{X} X$ of $X \times \mathcal{S} \to \mathcal{S}$ by reduction through the canonical immersion $X \to \mathcal{X}$.

Definition 5. A relative superdivisor of degree g of $\mathcal{X} \times \mathcal{S} \to \mathcal{S}$ is a closed sub-superscheme $\mathcal{Z}$ of $\mathcal{X} \times \mathcal{S}$ of codimension $(1,0)$ whose reduction $\hat{\mathcal{Z}} = \mathcal{Z} \times_\mathcal{X} X$ is a relative divisor of degree g of $X \times \mathcal{S} \to \mathcal{S}$ (definition 4).

Let us describe relative superdivisors in local terms.

Lemma 3. *A closed sub-superscheme $\mathcal{Z}$ of $\mathcal{X} \times \mathcal{S}$ of codimension $(1,0)$ defined by an homogeneous ideal J of $\mathcal{O}_{\mathcal{X} \times \mathcal{S}}$ is a relative superdivisor of degree g if and only if $\mathcal{O}_\mathcal{Z} = \mathcal{O}_{\mathcal{X} \times \mathcal{S}} / J$ is a locally free $\mathcal{O}_\mathcal{S}$-module of dimension (g, g) and, once a system of conformal coordinates (z, θ) has been chosen, J can be locally generated by an element of the form*

$$f = z^g - (a_1 + \theta b_1)z^{g-1} + \ldots + (-1)^g(a_g + \theta b_g), \tag{24}$$

where the a_i's are even and the b_j's odd elements in $\mathcal{O}_S$.

Now, in order to parametrize superdivisors by means of the supersymmetric product $S^g\mathcal{X} = (S^gX, \mathcal{O}_{S^g\mathcal{X}})$, we need a suitable definition of universal superdivisor, which is obtained in terms of Manin's superdiagonal. If Δ denotes the ideal of the diagonal immersion $\delta\colon \mathcal{X} \to \mathcal{X} \times \mathcal{X}$, we have that the kernel of the composition $\delta_*\Omega^1 \to \delta_*\Omega^1_{\mathcal{X}} \simeq \Delta/\Delta^2 \xrightarrow{\partial} Ber(\mathcal{O}_{\mathcal{X}})$ (see (18)) is a homogeneous ideal I of $\mathcal{O}_{\mathcal{X}\times\mathcal{X}}$, thus defining a sub-superscheme Δ^s called the ***superdiagonal***.

Lemma 4. (Manin, [23]) *The superdiagonal $\Delta^s = (X, \mathcal{O}_{\mathcal{X}\times\mathcal{X}}/I)$ is a closed sub-superscheme of codimension $(1,0)$ (a superdivisor). In conformal coordinates (z,θ), it can be described by the equation:*

$$z_1 - z_2 - \theta_1\theta_2 = 0 \tag{25}$$

where, as usual, $z_1 = z \otimes 1$ and $z_2 = 1 \otimes z$.

The universal superdivisor for SUSY-curves is now constructed as in the non-graded case. Let $p_i\colon \mathcal{X}^g \to \mathcal{X}$ be the i-th projection and let Δ_i^s be the sub-superscheme of $\mathcal{X} \times \mathcal{X}^g$ obtained by pulling the superdiagonal Δ^s back through $(1 \times p_i)\colon \mathcal{X} \times \mathcal{X}^g \to \mathcal{X} \times \mathcal{X}$. Let us consider the sub-superscheme $\Delta_1^s + \ldots + \Delta_g^s$ defined by the intersection ideal.

Theorem 3. *There is a unique closed sub-superscheme of codimension $(1,0)$, $\mathcal{Z}^s$ of $\mathcal{X} \times S^g\mathcal{X} \to S^g\mathcal{X}$ such that*

$$(1 \times p)^*\mathcal{Z}^s = \Delta_1^s + \ldots + \Delta_g^s, \tag{26}$$

where $p\colon \mathcal{X}^g \to S^g\mathcal{X}$ is the natural projection. Moreover, $\mathcal{Z}^s$ is a relative superdivisor of degree g, that will be called the Universal Superdivisor.

Proof. The first part of the theorem holds because in local conformal coordinates the equation of $\mathcal{Z}^s$ is

$$(z - z_1 - \theta\theta_1) \cdot \ldots \cdot (z - z_g - \theta\theta_g) = 0, \tag{27}$$

which is a polynomial in (z,θ) with coefficients in the invariant sheaf $(\mathcal{O}_{\mathcal{X}})_g$, since:

$$\begin{aligned}(z - z_1 - \theta\theta_1) \cdot \ldots \cdot (z - z_g - \theta\theta_g) &= \prod_{i=1}^{g}(z - z_i) - \theta\sum_{i=1}^{g}\theta_i\prod_{j\neq i}(z - z_i) \\ &= z^g + \sum_{h=1}^{g}(-1)^h(s_h + \theta\varsigma_h)z^{g-h};\end{aligned} \tag{28}$$

here the s_h's are the symmetric functions of $(z_1, \ldots, z_g)$, and $\varsigma_h = \sum_{i=1}^{g}\sigma_i(\theta_1\bar{s}_{h-1})$ are the odd symmetric functions, $\bar{s}_h$ being the symmetric functions of $(z_2, \ldots, z_g)$ and σ_i any permutation of $(1, \ldots, g)$ sending 1 to i, as in (9).

The second part follows directly from lemma 3. □

Remark. The coefficients of the universal divisors enjoy an important property: if $\bar{D} = \sum_{i=1}^{g}(\theta_i \frac{\partial}{\partial z_i} + \frac{\partial}{\partial \theta_i})$ is the local derivation of $S^g\mathcal{X}$ induced by $D = \theta\frac{\partial}{\partial z} + \frac{\partial}{\partial \theta}$, then

$$\varsigma_i = \bar{D}(s_i)\,. \tag{29}$$

3.4 The Representability Theorem for superdivisors

Let $(\mathcal{X}, \mathcal{D})$ be a SUSYcurve. For every superscheme $\mathcal{S}$ let us denote by

$$\mathcal{S} \to \operatorname{Div}^g_{\mathcal{S}}(\mathcal{X} \times \mathcal{S}) \tag{30}$$

the functor of relative superdivisors of degree g of $\mathcal{X} \times \mathcal{S} \to \mathcal{S}$.

Let us also consider the functor of points of the supersymmetric product, defined by:

$$\mathcal{S} \to \operatorname{Hom}(\mathcal{S}, S^g\mathcal{X})\,. \tag{31}$$

Since $\mathcal{O}_{S^g\mathcal{X}} \simeq \bigwedge_{\mathcal{O}_{S^g X}}(\mathcal{M}^{S_g})$ (17), a morphism $\varphi: \mathcal{S} \to S^g\mathcal{X}$ is characterized by the induced morphism $\bar{\varphi}: S \to S^g X$ and a morphism $\psi: \mathcal{M}^{S_g} \to \mathcal{O}_S$ of $\mathcal{O}_{S^g X}$-modules, where $\mathcal{O}_S$ has the $\mathcal{O}_{S^g X}$-module given by $\bar{\varphi}$. This will be very useful in the next section.

Our next step is to prove the main result of this paper, the ***Representability Theorem for the Supersymmetric product.***

Theorem 4. *The natural morphism of functors:*

$$\begin{aligned} \Theta: \operatorname{Hom}(\mathcal{S}, S^g\mathcal{X}) &\to \operatorname{Div}^g_{\mathcal{S}}(\mathcal{X} \times \mathcal{S}) \\ \varphi &\mapsto (1 \times \varphi)^{-1}(\mathcal{Z}^s) \end{aligned} \tag{32}$$

is a functorial isomorphism between the functor of points of the supersymmetric product $S^g\mathcal{X}$ and the functor of relative superdivisors of degree g.

Proof. Let us outline a sketch of the proof. The first question is the construction of the inverse map of Θ. Let $\mathcal{Z}$ be a relative superdivisor of degree g of $\mathcal{X} \times \mathcal{S} \to \mathcal{S}$ defined by an ideal J of $\mathcal{O}_{\mathcal{X}\times\mathcal{S}}$. We need a morphism $\varphi : \mathcal{S} \to S^g\mathcal{X}$ such that $(1 \times \varphi)^{-1}(\mathcal{Z}^s) = \mathcal{Z}$. However, as we have already seen, this is equivalent to giving a morphism $\bar{\varphi}: S \to S^g X$ and a morphism $\psi: \mathcal{M}^{S_g} \to \mathcal{O}_S$ of $\mathcal{O}_{S^g X}$-modules.

The morphism $\bar{\varphi}: S \to S^g X$ is what one obtains through Theorem 2 from the reduction $\bar{\mathcal{Z}}$, which is an ordinary relative divisor of degree g of $X \times S \to S$.

The morphism of $\mathcal{O}_{S^g X}$-modules $\psi: \mathcal{M}^{S_g} \to \mathcal{O}_S$ can now be defined locally. We can then take conformal coordinates (z, θ) on $\mathcal{X}$, such that the ideal J of $\mathcal{Z}$ is generated by an element:

$$f = z^g - (a_1 + \theta b_1)z^{g-1} + \ldots + (-1)^g(\bar{a}_g + \theta b_g) \tag{33}$$

where the a_i's are even and the b_j's are odd elements in $\mathcal{O}_S$. Since the module $\mathcal{M}^{S_\bullet}$ is generated by the odd symmetric functions ς_i, we can define a morphism $\psi: \mathcal{M}^{S_\bullet} \to \mathcal{O}_S$ of $\mathcal{O}_{S_\bullet X}$-modules by imposing the condition that $\varsigma_i \mapsto b_i$ for $i = 1, \dots, g$.

On the other hand, since the ring morphism induced by $\bar{\varphi}: S \to S^g X$ maps the symmetric functions s_i to the coefficients a_i of the local generator f of J (see (23)), in our coordinate open subset we have $(1 \times \varphi)^{-1}(\mathcal{Z}^s) = \mathcal{Z}$.

Using the conformal structure and, more specifically, the remark following Theorem 3, we can prove that this construction of ψ is independent of the choice of the conformal coordinates, thus providing the global morphism we are seeking. In this way, we have a morphism of functors:

$$\Xi: \mathrm{Div}\,^g_{\mathcal{S}}(\mathcal{X} \times \mathcal{S}) \to \mathrm{Hom}\,(\mathcal{S}, S^g\mathcal{X})\,. \tag{34}$$

From the above local description we can easily see that Ξ and Θ are one the inverse of the other. □

Remark. As in the non-graded case, the meaning of Theorem 4 is that the supersymmetric product $S^g\mathcal{X}$ of a SUSY-curve and its universal superdivisor $\mathcal{Z}^s$ parametrize relative superdivisors of degree g for every SUSY-family $\mathcal{X} \times \mathcal{S} \to \mathcal{S}$, i.e., every relative superdivisor can be obtained in a unique way as a pullback of the universal relative superdivisor of the supersymmetric product.

According to the remark following Theorem 3, the odd coefficients of the equation of a relative superdivisor can be obtained 'generically' by derivation of the even ones with the aid of the conformal structure. Such a statement, which is certainly meaningless for a particular divisor because we cannot derive its coefficients (which are constant) is nevertheless behind our definition of relative superdivisor.

Acknowledgments.

We thank J.M. Muñoz Porras for many enlightening comments about the theory of Jacobian varietes, and J.M. Rabin for drawing references [31] and [32] to our attention. We also owe many thanks to the organizers of the Conference for their warm hospitality in Rapallo.

References

1. A. Altman, S. Kleiman: Adv. in Math. **35** 5 (1980)
2. M. Artin: In Lecture Notes in Math. **152** 632 (1970), Springer-Verlag, Berlin
3. M.A. Baranov, A.S. Schwarz: JETP Lett. **42** 419 (1986)
4. C. Bartocci, U. Bruzzo: C. R. Acad. Sci. Paris Sèr. I Math. **309** 75 (1989)
5. C. Bartocci, U. Bruzzo, D. Hernández Ruipérez: In Differential-Geometric Methods in Theoretical Physics (1990), World Scientific, Singapore
6. M. Batchelor, P. Bryant: Commun. Math. Phys. **114** 243 (1988)

7. A.A. Beilinson, V.V. Schechtman: Commun. Math. Phys. **118** 651 (1988)
8. F.A. Berezin, D.A. Leĭtes: Soviet Math. Dokl. **16** 1218 (1975)
9. W.L. Chow: Amer. J. Math. **76** 453 (1954)
10. L. Crane, J.M. Rabin: Commun. Math. Phys. **113** 601 (1988)
11. B. DeWitt: Supermanifolds (1984), Cambridge Univ. Press, London
12. D. Friedan: In Workshop on Unified String Theories, p.162 (1986) World Scientific, Singapore
13. S.B. Giddings, P. Nelson: Commun. Math. Phys. **116** 607 (1988)
14. S.B. Giddings, P. Nelson: Commun. Math. Phys. **118** 289 (1988)
15. A. Grothendieck: Seminaire Bourbaki 1961/62, Exp. 232
16. D. Hernández Ruipérez, J. Muñoz Masqué: C. R. Acad. Sci. Paris Sèr. I Math. **301** 915 (1985)
17. B. Iversen: Lecture Notes in Math. **174** (1970), Springer-Verlag, Berlin
18. G. Kempf: Ann. of Math. **98** 178 (1973)
19. G. Kempf: Abelian Integrals. Monografías del Inst. Mat. Univ. Aut. Mexico (1984)
20. B. Kostant: In Lecture Notes in Math. **570** 177 (1977), Springer-Verlag, Berlin
21. C. LeBrun, M. Rothstein: Commun. Math. Phys. **117** 159 (1988)
22. D.A. Leĭtes: Russian Math. Surveys **35** 1 (1980)
23. Yu.I. Manin: Funct. Anal. Appl. **20** 244 (1987)
24. Yu.I. Manin: J. Geom. Phys. **5** 161 (1988)
25. A. Mattuck: Amer. J. of Math. **87** 779 (1965)
26. D. Mumford: Curves and their Jacobians (1975) Univ. of Michigan Press
27. J.M. Muñoz Porras: Preprint Princeton Univ. (1990)
28. J.M. Rabin, P.G.O. Freund: Commun. Math. Phys. **114** 131 (1988)
29. J.M. Rabin, P. Topiwala: Preprint Univ. of California (1988)
30. A. Rogers: J. Math. Phys. **21** 1352 (1980)
31. A.A. Rosly, A.S. Schwarz, A.A. Voronov: Commun. Math. Phys. **119** 129 (1988)
32. A.A. Rosly, A.S. Schwarz, A.A. Voronov: Commun. Math. Phys. **120** 437 (1989)
33. A. Weil: Variétés abéliennes et courbes algébriques, Hermann (1948), Paris

Classical superspaces and related structures

D.Leites* (speaker), V. Serganova**, G.Vinel***
SFB-170, Dept of Math. Göttingen (*on leave of absence from Dept. of Math., Stockholm Univ., Stockholm, 11385, Sweden; **Department of Mathematics, Yale Univ. New Haven, CT, USA; ***Dept. of Math. UConn., Storrs, CT, USA)

Introduction

The main object in the study of Riemannian geometry is (properties of) the Riemann tensor which, in turn, splits into the Weyl tensor, Ricci tensor and scalar curvature. The word "splits" above means that at every point of the Riemannian manifold M^n the space of values of the Riemann tensor constitutes an O(n)-module which is the sum of three irreducible components (unless $n = 4$ when the Weyl tensor aditionally splits into 2 components).

More genearlly, let G be any group, not necessarily O(n). In what follows we recall definition of *G-structure* on a manifold and of (the space of) its *structure functions* (SFs) which are obstructions to integrability or, in other words, to possibility of flattening the G-structure. Riemannian tensor is an example of SF. Among the most known (or popular of recent) examples of G-structures are:

- an almost *conformal structure*, $G = O(n)\times\mathbb{R}^*$, SF are called the *Weyl tensors*;
- Penrose' *twistor* theory, $G = SU(2)\times SU(2)\times\mathbb{C}^*$, SF -- the *Penrose tensor* -- splits into 2 components whose sections are called "*α-forms*" and "*β-forms*";
- an almost *complex structure*, $G = GL(n;\mathbb{C})\subset GL(2n;\mathbb{R})$, SF is called the *Nijenhuis tensor*;
- an almost *symplectic structure*, $G = Sp(2n)$, (no accepted name for SF).

The first two examples are examples of a "conformal" structure which preserves a tensor up to a scalar. In several versions of a very lucid paper [G] Goncharov calculated (among other things) all SF for all structures with a simple group of conformal transformations, whose subgroup of *linear* transformations is the reductive part of the stabilizer of a point of the space and is the "G" which determines the G-structure on the manifold. Remarkably, Goncharov's examples correspond precizely to the classical spaces, i.e. irreducible compact Hermitian symmetric spaces (CHSS). Goncharov did not, however, write down the highest weights of irreducible components of SFs; this is done in [LPS1] and some of these calculations are interpreted as leading to generalized Einstein equation.

In this talk we advertize results (mostly due to E.Poletaeva) of calculating SF (and interpretation of them) for classical superspaces who are defined and partly listed in [S] and [L2] (see also [V], containing interesting papers on supergravity and where curved supergrassmannians are introduced). The problem was raised in [L2], cf. [L4], and the above examples are now superized in [P] and [LPS]. The passage to supermanifolds naturally hints to widen the usual approach to SFs in order to embrace at least the following cases:

- 2 types of infinite dimensional generalizations of Riemannian geometry connected with: (1) string theories of physics (these infinite dimensional examples have no analogues on manifolds because they require no less then **three** odd coordinates of the superstring; the list of corresponding hermitian superspaces deduced from [S] is given in [L2]; dual pairs, etc. will be considered elsewhere) and (2) Kac-Moody (super) algebras (see Table 5);
- the G-structures of the N-extended Minkowski superspace: the tangent space to the Minkovski superspace for $N\neq 0$ is naturally endowed with a 2-step nilpotent Lie superalgebra structure that highly resembles the contact structure on a manifold. We start studying such structures in earnest in [LPS2], compair our approach with that of the GIKOS group lead by V.I. Ogievetsky. More generally, we shall calculate SF for the G-structures of the type corresponding to any "flag variety", not just Grassmannians, particular at that, see Table 1.

Elsewhere we will generalize the machinery of Jordan algebras, so useful in the study of geometry of CHSSs [Mc], to the cases we consider (this is Vinel's thesis).

Can programmers help? A good part of the calculations we need are very simple (to calculate cohomology is to solve systems of linear equations [F]). Still, though the number of papers on supergravity is counted by thousands (see reviews in our bibliography, of which [OS3], [WB], [We] are easy to understand) there is remarkably small progress in actual calculations (cf. mathematical papers [Sch], [RSh], [Me]). It is yet unclear what are all supergravities for N>1. The reason to that: the calculations *are* voluminous besides, these calculations also have to be "glued" in an answer and there are no rules for doing so, cf. [P4]. Thus the problem is a challenge for a computer scientist, our calculations, together with [LP1] and [P1-4], illustrate [LP2]. For our cohomology of our infinite dimensional Lie (super)algebras there are NO recipes at all (not even from Feigin-Fuchs nor Roger [FF]).

In this text we deal with linear algebra: at a point. The global geometry, practically not investigated, is nontrivial, cf. [M], [MV].

Acknowledgements. We are thankful to D. Alekseevsky, J. Bernstein, P.Deligne, A. Goncharov, V.Ogievetsky, A. Onishchik and I. Shchepochkina for help. During the preparation of the manuscript D.L. was supported by I.Bendixson and NFR grants, Sweden; MPI, Bonn; and NSF grants: via Harvard and DMS-8610730 via IAS; SFB-170 supported D.L. and V.S. at the final stage.

Preliminaries

Terminological conventions. 1) A $\mathfrak{g}$ - module V with highest weight ξ and *even* highest vector will be denoted by V_ξ or R(ξ). An irreducible module with highest weight $\Sigma a_i\pi_i$, where π_i is the i-th fundamental weight, will be denoted sometimes by its numerical labels R(Σa_i; a) the highest weight with respect to the center of $\mathfrak{g}$ stands after semicolon, cf.[OV], Reference Chapter.

2) Let $\mathfrak{c}\mathfrak{g}$ denote the trivial central "extent" (the result of the extention) of a Lie (super)algebra $\mathfrak{g}$; let $\mathfrak{p}$ stand for projectivization (as in $\mathfrak{psl}$, $\mathfrak{pq}$) and $\mathfrak{s}$ for "trace"-less part (as in $\mathfrak{sl}$, $\mathfrak{sq}$, $\mathfrak{sh}$).

0.1. Structure functions. Let us retell some of Goncharov's results ([G]) and recall definitions ([St]).

Let M be a manifold of dimension n over a field $\mathbb{K}$; think $\mathbb{K} = \mathbb{C}$ (or $\mathbb{R}$). Let F(M) be the frame bundle over M, i.e. the canonical principal GL(n; $\mathbb{K}$)-bundle. Let $G \subset GL(n; \mathbb{K})$ be a Lie group. A *G-structure* on M is reduction of the frame bundle to the principal G-bundle corresponding to inclusion $G \subset GL(n; \mathbb{K})$, i.e. a *G-structure* is the possibility to select transition functions so that their values belong to G.

The simplest G-structure is the *flat* G-structure defined as follows. Let V be $\mathbb{K}^n$ with a fixed frame. Consider the bundle over V whose fiber over $v \in V$ consists of all frames obtained from the fixed one under the G-action, V being identified with T_vV.

Obstructions to identification of the k-th infinitesimal neighbourhood of a point $m \in M$ on a manifold M with G-structure and that of a point of the flat manifold V with the above G-structure are called *structure functions of order k*. Such an identification is possible provided all structure functions of lesser orders vanish.

Proposition. ([St]). *SFs of order k are elements from the space of* (k,2)-*th Spencer cohomology.*

Recall definition of the Spencer cochain complex. Let S^i denote the operator of the i-th symmetric power. Set $\mathfrak{g}_{-1} = T_mM$, $\mathfrak{g}_0 = \mathfrak{g}$ = Lie(G) and for i > 0 put:

$$(\mathfrak{g}_{-1}, \mathfrak{g}_0)_* = \oplus_{i \geq -1}\, \mathfrak{g}_i, \text{ where } \mathfrak{g}_i = \{X \in Hom(\mathfrak{g}_{-1}, \mathfrak{g}_{i-1}): \ X(v)(w,...) = X(w)(v,...)$$

for any $v,w \in \mathfrak{g}_{-1}\} = S^i(\mathfrak{g}_{-1})^* \otimes \mathfrak{g}_0 \cap S^{i+1}(\mathfrak{g}_{-1})^* \otimes \mathfrak{g}_{-1}$.

Suppose that

the $\mathfrak{g}_0$*-module* $\mathfrak{g}_{-1}$ *is faithful.* (0.1)

Then, clearly, $(\mathfrak{g}_{-1}, \mathfrak{g}_0)_* \subset \mathfrak{vect}(n) = \mathfrak{der}\ \mathbb{K}[[x_1,..., x_n]]$, where n = dim $\mathfrak{g}_{-1}$. It is subject to an easy verification that the Lie algebra structure on $\mathfrak{vect}(n)$ induces a Lie algebra structure on

$(\mathfrak{g}_{-1}, \mathfrak{g}_0)_*$. The Lie algebra $(\mathfrak{g}_{-1}, \mathfrak{g}_0)_*$, usually abbreviated to $\mathfrak{g}_*$, will be called *Cartan's prolong* (the result of *Cartan prolongation*) of the pair $(\mathfrak{g}_{-1}, \mathfrak{g}_0)$.

Let E^i be the operator of the i-th exterior power; set $C^{k,s}_{\mathfrak{g}_*} = \mathfrak{g}_{k-s} \otimes E^s(\mathfrak{g}_{-1}^*)$; usually we drop the subscript or at least indicate only $\mathfrak{g}_0$. Define the differential $\partial_s: C^{k,s} \to C^{k-1,s+1}$ setting for any $v_1, \dots, v_{s+1} \in V$ (as always, the slot with the hatted variable is ignored):

$$(\partial_s f)(v_1, \dots, v_{s+1}) = \Sigma(-1)^i f(v_1, \dots, \hat{v}_{s+1-i}, \dots, v_{s+1})(v_{s+1-i})$$

As usual, $\partial_s \partial_{s+1} = 0$, the homology of this complex is called *Spencer cohomology* of $(\mathfrak{g}_{-1}, \mathfrak{g}_0)_*$.

0.2. Case of simple $\mathfrak{g}_*$ over $\mathbb{C}$. The following remarkable fact, though known to experts, is seldom formulated explicitely:

Proposition. *Let $\mathbb{K} = \mathbb{C}$, $\mathfrak{g}_* = (\mathfrak{g}_{-1}, \mathfrak{g}_0)_*$ be simple. Then only the following cases are possible:*

1) $\mathfrak{g}_2 \neq 0$ and then $\mathfrak{g}_*$ is either $\mathfrak{vect}(n)$ or its special subalgebra $\mathfrak{svect}(n)$ of divergence-free vector fields, or its subalgebra $\mathfrak{h}(2n)$ of hamiltonian fields;

2) $\mathfrak{g}_2 = 0$, $\mathfrak{g}_1 \neq 0$ then $\mathfrak{g}_*$ is the Lie algebra of the complex Lie group of automorphisms of a CHSS (see above).

Proposition explains the reason of imposing the restriction (0.1) if we wish $\mathfrak{g}_*$ to be simple. Otherwise, or on supermanifolds, where the analogue of Proposition does not imply similar restriction, we have to (and do) broaden the notion of Cartan prolong to be able to get rid of restriction (0.1).

When $\mathfrak{g}_*$ is a simple finite-dimensional Lie algebra over $\mathbb{C}$ computation of structure functions becomes an easy corollary of the Borel-Weyl-Bott-... (BWB) theorem, cf. [G]. Indeed, by definition $\oplus_k H^{k,2}_{\mathfrak{g}_*} = H^2(\mathfrak{g}_{-1}; \mathfrak{g}_*)$ and by the BWB theorem $H^2(\mathfrak{g}_{-1}; \mathfrak{g}_*)$, as $\mathfrak{g}$-module, has as many components as $H^2(\mathfrak{g}_{-1})$ which, thanks to commutativity of $\mathfrak{g}_{-1}$, is just $E^2(\mathfrak{g}_{-1})$; the highest weights of these modules, as explained in [G], are also deducible from the theorem. However, [G] pityfully lacks this deduction, see [LP1] and [LPS1] where it is given with interesting interpretations.

Let us also immediately calculate SF corresponding to case 1) of Proposition: we did not find these calculations in the literature. Note that vanishing of SF for $\mathfrak{g}_* = \mathfrak{vect}$ and $\mathfrak{k}$ (see 0.5) follows from the projectivity of $\mathfrak{g}_*$ as $\mathfrak{g}_0$-modules and properties of cohomology of coinduced modules [F]. In what follows $R(\Sigma a_i \pi_i)$ denotes the irreducible $\mathfrak{g}_0$-module. The classical spaces are listed in Table 1 and some of them are bapthized for convenience of further references.

Theorem. 1)(Serre [St]). *In case 1) of Proposition structure functions can only be of order 1.*

a) $H^2(\mathfrak{g}_{-1}; \mathfrak{g}_*) = 0$ *for* $\mathfrak{g}_* = \mathfrak{vect}(n)$ *and* $\mathfrak{svect}(m)$, $m>2$;

b) $H^2(\mathfrak{g}_{-1}; \mathfrak{g}_*) = R(\pi_3) \oplus R(\pi_1)$ *for* $\mathfrak{g}_* = \mathfrak{h}(2n)$, $n>1$;

$H^2(\mathfrak{g}_{-1}; \mathfrak{g}_*) = R(\pi_1)$ *for* $\mathfrak{g}_* = \mathfrak{h}(2)$.

2)(Goncharov [G]). *SFs of* Q_3 *are of order 3 and constitute* $R(4\pi_1)$. *SF for Grassmannian* Gr_m^{m+n} (when neither m nor n is 1, i.e. Gr is not a projective space) *is the direct sum of two components whose weights and orders are as follows:*

Let $A = R(2, 0, \dots, 0, -1) \otimes R(1, 0, \dots, 0, -1, -1)$, $B = R(1, 1, 0, \dots, 0, -1) \otimes R(1, 0, \dots, 0, -2)$.

Then if $mn \neq 4$ both A and B are of order 1;
if $m = 2$, $n \neq 2$ A is of order 2 and B of order 1;
if $n = 2$, $m \neq 2$ A is of order 1 and B of order 2;
if $n = m = 2$ both A and B are of order 2.

SF of G-structures of the rest of the classical CHSSs are the following irreducible $\mathfrak{g}_0$-modules whose order is 1 (recall that $Q_4 = Gr_2^4$)*:*

CHSS	$\mathbb{P}^n$	OGr_m	LGr_m	Q_n, n>4
weight of SF		$E^2(E^2(V^*))\otimes V$	$E^2(S^2(V^*))\otimes V$	$E^2(V^*)\otimes V$

$E_6/SO(10)\times U(1)$	$E_7/E_6\times U(1)$
$E^2(R(\pi_5)^*))\otimes R(\pi_5)$	$E^2(R(\pi_1)^*))\otimes R(\pi_1)$

0.3. SF for reduced structures. In [G] Goncharov considered conformal structures. SF for the corresponding generalizations of the Riemannian structure, i.e. when $\mathfrak{g}_0$ is the semisimple part $^\wedge\mathfrak{g}$ of $\mathfrak{g}$ = Lie (G), seem to be more difficult to compute because in these cases $(\mathfrak{g}_{-1}, \mathfrak{g}_0)_* = \mathfrak{g}_-\oplus\mathfrak{g}_0$ and the BWB-theorem does not work. Fortunately, the following statement, a direct corollary of definitions, holds.

Proposition ([G], Th.4.7). For $\mathfrak{g}_0 = {}^\wedge\mathfrak{g}$ and $\mathfrak{g}$ SF of order 1 are the same and SF of order 2 for $\mathfrak{g}_0 = {}^\wedge\mathfrak{g}$ are $S^2(\mathfrak{g}_1) = S^2(\mathfrak{g}_{-1}{}^*)$. (There are clearly no SF of order 3 for $\mathfrak{g}_0 = {}^\wedge\mathfrak{g}$).

Example: Riemannian geometry. Let G = O(n). In this case $\mathfrak{g}_1 = \mathfrak{g}_{-1}$ and in $S^2(\mathfrak{g}_{-1})$ a 1-dimensional subspace is distinguished; the sections through this subspace constitute a Riemannian metric g on M. (The habitual way to determine a metric on M is via a symmetric matrix, but actually this is just one scalar matrix-valued function.) The values of the Riemannian tensor at a point of M constitute an O(n)-module $H^2(\mathfrak{g}_{-1}; \mathfrak{g}_*)$ which contains a trivial component whose arbitrary section will be denoted by R. What is important, this trivial component is realised by Proposition as a submodule in $S^2(\mathfrak{g}_{-1})$. Thus, we have two matrix-valued functions: g and R each being a section of the trivial $\mathfrak{g}_0$-module. What is more natural than to require their ratio to be a constant (rather than a function)?

$$R = \lambda g, \text{ where } \lambda \in \mathbb{R}. \qquad (EE_0)$$

Recall that the Levi-Civita connection is the unique symmetric affine connection compatible with the metric. Let now t be the structure function (sum of its components belonging to the distinct irreducible O(n)-modules that constitute $H^2(\mathfrak{g}_{-1}; \mathfrak{g}_*)$) corresponding to the Levi-Civita connection; the process of restoring t from g involves differentiations thus making (EE_0) into a nonlinear pde. This pde is not Einstein Equation yet. Recall that in adition to the trivial component there is another O(n)-component in $S^2(\mathfrak{g}_{-1})$, the Ricci tensor Ri. ***Einstein equations*** (in vacuum and with cosmological term λ) are the *two* conditions: (EE_0) and

$$Ri = 0. \qquad (EE_{ric})$$

A generalization of this example to G-structures associated with certain other CHSSs, flag varieties, and to supermanifolds is considered in [LPS1] and [LP3].

0.4. SF of flag varieties. Contact structures. In heading a) of Proposition 0.2 there are listed all simple Lie algebras of (polynomial or formal) vector fields except those that preserve a contact structure. Recall that a *contact structure* is a maximally nonintegrable distribution of codimension 1, cf. [A].

To consider contact Lie algebra we have to generalize the notion of Cartan prolongation: the tangent space to a point of a manifold with a contact structure possesses a natural structure of the Heisenberg algebra. This is a 2-step nilpotent Lie algebra. Let us consider the general case corresponding to "flag varieties" -- quotients of a simple complex Lie group modulo a parabolic subgroup. (The necessity of such a generalization was very urgent in the classification of simple Lie superalgebra, see [Shch] and [L2], where it first appeared, already superized.)

Given an arbitrary (but $\mathbb{Z}$-graded) nilpotent Lie algebra $\mathfrak{g}_- = \oplus_{0>i\geq -d}\mathfrak{g}_i$ and a Lie subalgebra $\mathfrak{g}_0 \subset \mathfrak{der}\ \mathfrak{g}_-$ which preserves $\mathbb{Z}$-grading of $\mathfrak{g}_-$, define the i-th *prolong* of the pair $(\mathfrak{g}_-, \mathfrak{g}_0)$ for i > 0 to be:

$$\mathfrak{g}_i = (S^*(\mathfrak{g}_-)^* \otimes \mathfrak{g}_0 \cap S^*(\mathfrak{g}_-)^* \otimes \mathfrak{g}_-)_i,$$

where the subscript singles out the component of degree i.Similarly to the above, define $\mathfrak{g}_*$, or rather, $(\mathfrak{g}_-, \mathfrak{g}_0)_*$, as $\oplus_{i\geq -d}\mathfrak{g}_i$; then, by the same reasons as in 0.1, $\mathfrak{g}_*$ is a Lie algebra (subalgebra of $\mathfrak{k}(\dim \mathfrak{g}_-)$ for d = 2 and $\dim \mathfrak{g}_{-2} = 1$) and $H^i(\mathfrak{g}_-; \mathfrak{g}_*)$ is well-defined. $H^i(\mathfrak{g}_-; \mathfrak{g}_*)$ naturally splits into homogeneous components whose degree corresponds to what we will call the *order*. (For the particular case of Lie algebras of depth 2 the obtained bigraded complex was independently and much earlier defined by Tanaka [T] and used in [BS] and [O]. No cohomology was explicitely calculated, however; see calculations in [LPS2] and [LP3].)

The space $H^2(\mathfrak{g}_-; \mathfrak{g}_*)$ is the space of obstructions to flatness. In general case the minimal order of SF is 2-d. For d > 1 we did not establish correspondence between the order of SF and the number of the infinitesimal neighbourhood of a point of a supermanifold with the flat G-structure.

Examples. 1) G* is a simple Lie group, P its parabolic subgroup, G the Levi subgroup of P, $\mathfrak{g}_0$ = Lie(G), $\mathfrak{g}_-$ is the complementary subalgebra to Lie(P) in Lie(G*). The corresponding SF, calculable from the BWB-theorem if $\mathfrak{g}_*$ is finite-dimensional and simple describe for the first time the local geometry of flag varieties other than CHSSs, see [LP3] for details. Here is the simplest example.

2) Let $\mathfrak{g} = \mathfrak{csp}(2n)$, $\mathfrak{g}_{-1} = R(\pi_1; 1)$, $\mathfrak{g}_{-2} = R(0)$; then $\mathfrak{g}_* = \mathfrak{k}(2n+1)$ and

$$C^{k,s}_{\mathfrak{g}_*} = \mathfrak{g}_{k-s} \otimes E^s(\mathfrak{g}_{-1}^*) \oplus \mathfrak{g}_{k-s-1} \otimes E^{s-1}(\mathfrak{g}_{-1}^*) \otimes \mathfrak{g}_{-2}^*.$$

Theorem. *For* $\mathfrak{g}_* = \mathfrak{k}(2n+1)$ *all SF vanish.*

This is a reformulation of the Darboux theorem on a canonical 1-form, actually.

0.5. SF for projective structures. It is also interesting sometimes to calculate $H^2(\mathfrak{g}_-; \mathfrak{h})$ for some $\mathbb{Z}$-graded subalgebras $\mathfrak{h} \subset \mathfrak{g}_*$, such that $\mathfrak{h}_i = \mathfrak{g}_i$ for $i \leq 0$. For example, if $\mathfrak{g} = \mathfrak{gl}(n)$ and $\mathfrak{g}_{-1}$ is its standard (identity) representation we have $\mathfrak{g}_* = \mathfrak{vect}(n)$ and, as we have seen, all SF vanish; but if $\mathfrak{h} = \mathfrak{sl}(n+1) \subset \mathfrak{vect}(n)$ then the corresponding SF are nonzero and provide us with obstructions to integrability of what is called the *projective connection.*

Theorem. 1) *Let* $\mathfrak{g}_* = \mathfrak{vect}(n)$, $\mathfrak{h} = \mathfrak{sl}(n+1)$. *Then SF of order 1 and 2 vanish, SF of order 3 are* R(2,1, 0, ..., 0, -1)

2)*Let* $\mathfrak{g}_* = \mathfrak{k}(2n+1)$, $\mathfrak{h} = \mathfrak{sp}(2n+2)$. *Then SF are* $R(\pi_1+\pi_2; 3)$ *of order 3.*

0.6. Case of simple $\mathfrak{g}_*$ over $\mathbb{R}$.

Example: Nijenhuis tensor. Let $\mathfrak{g}_0 = \mathfrak{gl}(n) \subset \mathfrak{gl}(2n;\mathbb{R})$, $\mathfrak{g}_{-1}$ is the identity module. In this case $\mathfrak{g}_* = \mathfrak{vect}(n)$, however, in seeming contradiction with Theorem 0.1.2, the SF are nonzero. There is no contradiction: now we consider not $\mathbb{C}$-linear maps but $\mathbb{R}$-linear ones.

Theorem. *Nonvanishing SF are of order 1 and constitute the* $\mathfrak{g}_0$*-module* $\overline{\mathfrak{g}_{-1}} \otimes_{\mathbb{C}} E^2_{\mathbb{R}}(\mathfrak{g}_{-1}^*)$, *where* $g(cv) = \overline{c}V$ *for* $c \in \mathbb{C}$, $g \in \mathfrak{gl}(n)$, $v \in \overline{V}$ *and a* $\mathfrak{gl}(n)$*-module V.*

One of our mottos is: *simple $\mathbb{Z}$-graded Lie superagebras of finite growth* (SZGLSAFGs) *are as good as simple finite-dimensional Lie algebras*; the results obtained for the latter should hold, in some form, for the former. So we calculate

SF on supermanifolds: Plan of campaign

The necessary background on Lie superalgebras and supermanifolds is gathered in a condenced form in [L5], see also [L1, L2]. The above definitions are generalized to Lie superalgebras via Sign Rule.

On the strength of the above examples we must list $\mathbb{Z}$-gradings of SZGLSAFGs of finite depth (recall that a $\mathbb{Z}$-graded Lie (super)algebra of the form $\oplus_{-d\leq i\leq k}\mathfrak{g}_i$ is said to be of ***depth*** d and ***length*** k; here d, k >0), calculate projective-like and reduced structures for the above and then go through the list of real forms.

Our theorems are cast in Tables. In Table 1 we set notations. Tables 2 and 3 complement difficult tables of [S]. Table 4 lists all symmetric superspaces of depth 1 of the form G/P with a simple finite-dimensional G. Table 5 lists all hermitian superspaces corresponding to simple loop supergroups different from the obvious examples of loops with values in a hermitian superspace. *Notice that there are 3 series of nonsuper examples.*

We compensate superfluity of exposition by wast bibliography with further results. Let us list some other points of interest in the study of SF on superspaces.

- there is no complete reducibility of the space of SF as $\mathfrak{g}_0$-module;

- Serre's theorem reformulated for superalgebras shows that there are SFs of order >1, see [LPS1];

- faithfulness of $\mathfrak{g}_0$-actions on $\mathfrak{g}_{-1}$ is violated in natural examples of: (a) supergrassmannians of subsuperspaces in an (n,n)-dimensional superspace when the center $\mathfrak{z}$ of $\mathfrak{g}_0$ acts trivially; retain the same definition of Cartan prolongation; the prolong is then the semidirect sum $(\mathfrak{g}_{-1}, \mathfrak{g}_0/\mathfrak{z})_* \ltimes S^*(\mathfrak{g}_{-1}{}^*)$ with the natural $\mathbb{Z}$-grading and Lie superalgebra structure; notice that the prolong is ***not*** subalgebra of $\mathfrak{vect}(\dim \mathfrak{g}_{-1})$; (b) the exterior differential d preserving structure.

More precisely, recall that for supermaifolds the good counterpart of differential forms on manifolds are not differential but rather ***pseudodifferential and pseudointegrable forms. Pseudodifferential forms*** on a supermanifold X are functions on the supermanifold X' associated with the bundle τ^*X obtained from the cotangent one by fiber-wise change of parity. *Differential forms* on X are fiber-wise ***polynomial*** functions on X'. In particular, if X is a manifold there are no pseudodifferential forms. The *exterior differential* on X is now considered as an odd vector field d on X'. Let $x = (u_1, ..., u_p, \xi_1, ..., \xi_q)$ be local coordinates on X, $x_i{}' = \pi(x_i)$. Then $d = \Sigma x_i{}'\partial/\partial x_i$ is the familiar coordinate expression of d. The Lie superalgebra $\mathfrak{C}(d) \subset \mathfrak{vect}(m+n/m+n)$, where (m/n) = dim X, -- the Lie superalgebra of vector fields preserving the field d on X' (see definition of the Nijenhuis operator P_4 in [LKW]) -- is neither simple nor transitive and therefore did not draw much attention so far. Still, the corresponding G-structure ($\mathfrak{C}(d) = (\mathfrak{g}_{-1}, \mathfrak{g}_0)_*$, where $\mathfrak{g}_0 = \mathfrak{gl}(k) \ltimes \Pi(\mathfrak{gl}(k))$ and where $\Pi(\mathfrak{gl}(k))$ is abelean and constitutes the kernel of the $\mathfrak{g}_0$-action on $\mathfrak{g}_{-1}$ = id, the standard (identity) representation of $\mathfrak{gl}(k)$) is interesting and natural. Let us call it the *d-preserving* ***structure***. The following theorem justifies pseudocohomology introduced in [LKW].

Theorem. *SFs of the d-preserving structure are 0.*

An interesting counterpart of the d-preserving structure is the odd version of the hamiltonian structure. In order to describe it recall that ***pseudointegrable forms*** on a supermanifold X are functions on the supermanifold 'X associated with the bundle τX obtained from the ***tangent*** one by fiber-wise change of parity. Fiber-wise ***polynomial*** functions on 'X are called polyvector fields on X. (In particular, if X is a manifold there are no pseudointegrable forms.) The ***exterior*** *differential* on X is now considered as an odd nondegenerate (as a bilinear form) bivector field div on X'. Let $x = (u_1, ..., u_p, \xi_1, ..., \xi_q)$ be local coordinates on X, $'x_i = \pi(\partial/\partial x_i)$. Then div $= \Sigma\partial^2/\partial x_i{}'\partial x_i$ is the coordinate expression of the Fourier transform of the exterior differential d with respect to primed variables; the operator is called "div" because it sends a polyvector field on X, i.e. a function on 'X to its divergence. The Lie superalgebra $\mathfrak{aut}$(div) is isomorphic to the Lie superalgebra $\mathfrak{le}$(m+n) which is the simple subalgebra of $\mathfrak{vect}$(n+m|n+m) that preserves a nondegenerate odd differential 2-form $\omega = \Sigma dx_i{}'dx_i$; an interesting algebra is the superalgebra $\mathfrak{sle}$(m+n) which preserves both div and ω; for both of these Lie superalgebras and their deformations the corresponding SF are calculated in [PS] and [LPS1] .

References

[A] Arnold V., Mathematical methods of classical mechanics. Springer, 1980

[BL] Bernstein J., Leites D., Invariant differential operators and irreducible representations of the Lie superalgebra of vector fields, Sel. Math. Sov., v.1, 1982

[BS] Burns D. Jr., Sneider S., Real hypersurfaces in complex manifolds. In: Wells R.O. Jr. (ed.), Proc. Symp. in Pure Math. of the AMS, williams college, 1975. v.30, pt. 1-2, AMS, 1977

[F] Fuchs D., Cohomology of infinite dimensional Lie algebras, Consultunts Bureau, NY, 1987; Feigin B., Fuchs D. Cohomology of Lie groups and Lie algebras. Итоги науки. Совр. пробл. математики. Фунд. Напр. # 22, ВИНИТИ. 1988. (Russian, to ap. in English in Springer series Sov.Math. Encycl.)

[FT] Ferrara S., Taylor J. (eds.), Supergravity '81. Cambridge Univ. Press, 1982

[Fe] Freund P. Introduction to supersymmetry. Cambridge Univ. Press, 1986

[Fo] Fronsdal C., Essays on supersymmetry. D.Reidel, 1986

[G1] Goncharov A., Infinitesimal structures related to hermitian symmetric spaces, Funct. Anal. Appl, 15, n3 (1981), 23-24 (Russian); a detailed version see in [G2]

[G2] Goncharov A., Generalized conformal structures on manifolds, An enlarged English translation in: [L3], #11 and Selecta Math. Sov. 1987

[GI1] Galperin A., Ivanov E., Ogievetsky V., Sokachev E., N = 2 supergravity in superspace: different versions and matter couplings. Class. Quantum Grav. 4, 1987, 1255-1265

[GI2] Galperin A., Ivanov E., Ogievetsky V., Sokachev E., Gauge field geometry from complex and harmonic analiticities I, II. Ann. Phys. 185, #1, 1988, 1-21; 22-45

[GKS] Galperin A., Nguen Anh Ky, Sokachev E., N = 2 supergravity in superspace: solutions to the constraints and the invariant action. Class. Quantum Grav. 4, 1987, 1235-1253

[K] Kac V., Classification of simple $\mathbb{Z}$-graded Lie superalgebras and simple Jordan superalgebras. Commun. Alg. 5(13), 1977, 1375-1400

[L1] Leites D., Lie superalgebras, JOSMAR, v. 30, #6, 1984; id. Introduction to Supermanifold theory. Russian Math. Surveys. v.33, n1,1980, 1-55; an expanded version: [L2]

[L2] Leites D., Supermanifold theory. Karelia Branch of the USSR Acad. of Sci., Petrozavodsk, 1983, 200pp. (Russian) = in English a still more expanded version (in 7 volumes) is to be published by Kluwer in 1991-92; meanwhile see the preprinted part in: [L3]

[L3] Leites D. (ed.), Seminar on supermanifolds, Reports of Dept. of Math. of Stockholm Univ. n1-34, 2800 pp., 1986-89

[L4] Leites D., Selected problems of supermanifold theory. Duke Math. J. v. 54, #2,1987, 649-656

[L5] Leites D., Quantization and supermanifolds. Appendix 3. In: Berezin F., Shubin M., Schroedinger equation. Kluwer, Dordrecht, 1990

[LKW] Leites D., Kochetkov Yu., Weintrob A., New invariant differential operators and pseudocohomology of supermanifolds and Lie superalgebras. In: Proc. Topological Conf., Staten Island, 1989, Marcel Decker, 1991

[LP1] Leites D., Poletaeva E., Analogues of the Riemannian structure for classical superspaces. Proc. Intnl. Algebraic Conf. Novosibirsk, 1989. (to appear) (see [L3, #34])

[LP2] Leites D., Post G., Cohomology to compute. In: Kaltofen E, Watt S.M. eds., Computers and Mathematics, Springer, NY ea, 1989, 73-81

[LP3] Leites D., Premet A., Structure functions of flag varieties. Geom. Dedicata (to appear)

[LPS1] Leites D., Poletaeva E., Serganova V., Einstein equations on manifolds and supermanifolds (to appear)

[LPS2] Leites D., Poletaeva E., Serganova V., Structure functions on contact supermanifolds (to appear)

[Mc] McCrimmon K. Jordan algebras and their applications. Bull. Amer. Math. Soc. v.84, n4, 1978, 612-627

[M] Manin Yu., Topics in non-commutative geometry. M.B.Porter lectures. Rice Univ., Dept. of Math., 1989

[MV] Manin Yu., Voronov A., Supercell decompositions of flag supervariety. In: Итоги науки. Совр. пробл. матем. Новейшие дост. т.32, ВИНИТИ, Moscow, 1988, 125-211 (Russian) = Engl. translation in J. Sov. Math. (JOSMAR)

[Mk] Merkulov S., N = 2 superconformal superspaces. Class. Quantum Grav. 7, 1990, 439-444

[Mi] Miklashevsky I., Connections, conformal structures and Einstein equation. In: [L3, #27], 1988-9

[O] Ochiai T., Geometry associated with semisimple flat omogeneous spaces, Trans. Amer. Math. Soc., 152, 1970, 159-193

[OS1] Ogievetsky V.I., Sokachev E.S., The simplest group of Einstein supergravity. Sov. J. Nucl. Phys. v.31, #1, 1980, 264-279

[OS2] Ogievetsky V.I., Sokachev E.S., The axial gravity superfield and the formalism of the differenrial geometry. Sov. J. Nucl. Phys. v.31, #3, 1980, 821-840

[OS3] Ogievetsky V., Sokachev E., Supersymmetry and superspace.J.Sov. Math. (JOSMAR) 36, 1987, 721-744 (transl. from Itogi Nauki i Tekhn. Ser. Math. Anal. 22, 1984, 137-173, in Russian)

[OV] Onishchik A.L., Vinberg E.B., Seminar on algebraic groups and Lie groups. Springer, Berlin ea, 1990

[P1] Poletaeva E., Structure functions on (2, 2)-dimensional supermanifolds with either of the differential forms $\omega_+ = d\eta^{(2\lambda-1)/(1-\lambda)}((1-\lambda)dpdq + \lambda d\xi d\eta)$ or $\omega_- = d\xi^{1/\lambda-2}(dpdq + d\xi d\eta)$, $\lambda \in \mathbb{C}$. In: [L3, # 34]

[P2] Poletaeva E., Penrose tensors on supergrassmannians. Math. Scand., 1991(to appear)

[PS] Poletaeva E., Serganova V. Structure functions on the usual and exotic symplectic supermanifolds (to appear)

[RSh] Rosly A.A., Schwarz A.S., Geometry of N = 1 supergravity.I, II. Commun. Math.Phys. 95. 1984, 161-184

[Sch] Schwarz A.S. Supergravity, complex geometry and G-structures. Commun. Math.Phys. 87. 1982, 37-63

[S] Serganova V. Classification of real simple Lie superalgebras and symmetric superspaces. Funct. Anal. Appl. 17, #3, 1983, 46-54

[SH] Shander V., Analogues of the Frobenius and Darboux theorems for supermanifolds. C. R. de l'Acad. bulg. de Sci., t. 36, #3, 1983, 309-312

[Shch] Shchepochkina I., Exceptional infinite dimensional Lie superalgebras of vector fields. C. R. de l'Acad. bulg. de Sci., t. 36, #3, 1983, 313-314 (Russian)

[So] Sokachev E. Off-shell six-dimensional supergravity in harmonic superspace Class. Quantum Grav. 5, 1459-1471, 1988

[St] Sternberg S. Lectures on differential geometry, 2nd ed, Chelsey, 1985

[T] Tanaka N., On infinitesimal automorphisms of Siegel domains, J. Math. Soc. Japan 22, 1970, 180-212.

[V] Problems of nuclear physics and cosmic rays. v.24, Kharkov, Вища школа, 1985 (Russisan) Festschrift on the occation of D.Volkov's birthday.

[W] Weintrob A. Almost complex structures on supermanifolds. Questions of group theory and homological algebra.,Yaroslavl Univ. Press, Yaroslavl, 1985, 1 (Russian) = English translation in [L3, n24]

[We] West P. Introduction to supersymmetry and supergravity. World Sci. , Singapore, 1986

[WB] Wess J., Bagger J., Supersymmetry and supergravity. Princeton Univ. Press, 1983

Notations in tables. Everywhere we assume the notational conventions of [S] and definitions adopted there without mentioning this specifically. In Table 1 $\mathfrak{s} = (\mathrm{Lie}(S_c))\otimes\mathbb{C}$, NCHSS is an abbreviation for noncompact hermitian symmetric space, in the diagram of $\mathfrak{s}$ the vertex defining the minimal parabolic subalgebra $\mathfrak{p}$ = Lie (P), such that X can be presented as $(S_c)^{\mathbb{C}}/P$, is shaded. In Table 4 we call a homogeneous space G/P, where G is a simple Lie supergroup P its parabolic subsupergroup corresponding to several omitted generators of a Borel subalgebra (description of these generators can be found in [L3, # 31]), of *depth* d and *length* l if such are the depth and length of Lie (G) in the $\mathbb{Z}$-grading compatible with that of Lie(P). Note that all superspaces of Table 4 possess an hermitian structure (hence are of depth 1) except PeGr (no hermitian structure), PeQ (no structure, length 2), $CGr_{0,k}^{0,n}$ and $SCGr_{0,k}^{0,n}$ (no structure, lengths n-k and, resp. n-k-1)

Table 1. Hermitian symmetric spaces

Name of CHSS X	$X=S_c/G_c$	$\mathfrak{s}_0=(\mathfrak{g}_c)^{\mathbb{C}}$	The diagram of $\mathfrak{s}$	$\mathfrak{s}_{-1}\cong T_oX$	$(S_c)^*$	names of NCHSS
$\mathbb{C}P^n$	$SU(n+1)/U(n)$	$\mathfrak{gl}(n)$		id	$SU(1, n)$	$\mathbb{C}P^n$
Gr_p^{p+q}	$SU(p+q)/S(U(p)\times U(q))$	$\mathfrak{s}(\mathfrak{gl}(p)\oplus\mathfrak{gl}(q))$		$id\otimes id^*$	$SU(p, q)$	$^*Gr_p^{p+q}$
OGr_n	$SO(2n)/U(n)$	$\mathfrak{gl}(n)$	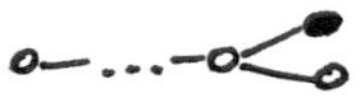	$\Lambda^2 id$	$SO(n, n)$	$^*OGr_n$
Q_n	$SO(n+2)/SO(2)\times SO(n)$	$\mathfrak{co}(n)$		id	$SO(n, 2)$	$^*Q_n$
LGr_n	$Sp(2n)/U(n)$	$\mathfrak{gl}(n)$		$S^2 id$	$Sp(2n;\mathbb{R})$	$^*LGr_n$
$(\mathbb{O}P^2)$	$E_6/SO(10)\times U(1)$	$\mathfrak{co}(10)$			E_6^*	
	$E_7/E_6\times U(1)$	$\mathfrak{ce}_6$			E_7^*	

Occasional isomorphisms: $Gr_p^{p+q} \cong Gr_q^{p+q}$, $Q_1 \cong \mathbb{C}P^1$, $Q_3 \cong LGr_2$, $Q_2 \cong S^2\times S^2$, $OGr_2 \cong LGr_1 \cong \mathbb{C}P^1$, $OGr_3 \cong Gr_3^4$, $Q_4 \cong Gr_2^4$.

Table 2. Dual pairs of homogeneous symmetric superspaces

$(\mathfrak{p})\mathfrak{sl}_r(m\|2n)/\mathfrak{osp}(m,p\|2n)$	$(\mathfrak{p})\mathfrak{su}(m,p\|2n,n)/\mathfrak{osp}(m,p\|2n)$
$(\mathfrak{p})\mathfrak{sl}_r(2m\|2n)/(\mathfrak{p}_r)\mathfrak{s}_r\mathfrak{l}(m\|n)$	$(\mathfrak{p})\mathfrak{su}^*(2m\|2n)/(\mathfrak{p}_r)\mathfrak{s}_r\mathfrak{l}(m\|n)$
$\mathfrak{psl}_r(n\|n)/\mathfrak{pq}_r(n)$	${}^o\mathfrak{pq}(n)/\mathfrak{pq}_r(n)$
$\mathfrak{psl}_r(n\|n)/\mathfrak{spe}_r(n)$	$\mathfrak{supe}(n)/\mathfrak{spe}_r(n)$
$(\mathfrak{p})\mathfrak{su}(m,p\|n,q)/$ $/(\mathfrak{p})\mathfrak{s}(\mathfrak{u}(r+s,r\|t+v,v)\oplus$ $\oplus\,\mathfrak{u}(m-r-s,p-r\|n-t-v,q-t)$	$(\mathfrak{p})\mathfrak{su}(m,p+s-r\|n,v+q)/$ $/(\mathfrak{p})\mathfrak{s}(\mathfrak{u}(r+s,r\|t+v,v)\oplus$ $\oplus\,\mathfrak{u}(m-r-s,p-r\|n-t-v,q-t)$
$(\mathfrak{p})\mathfrak{su}(2n,m\|2n,2q)/\mathfrak{osp}^*(2m\|2n,2q)$	$(\mathfrak{p})\mathfrak{su}^*(2m\|2n)/\mathfrak{osp}^*(2m\|2n,2q)$
$\mathfrak{psu}(m,p\|n,q)/\mathfrak{puq}(n,p)$	$\mathfrak{supe}(n)/\mathfrak{puq}(n,p)$
$\mathfrak{psu}^*(2n\|2n)/\mathfrak{pq}^*(2n)$	${}^o\mathfrak{pq}(n)/\mathfrak{pq}^*(2n)$
$\mathfrak{psu}^*(2n\|2n)/\mathfrak{su}^*(2n)$	$\mathfrak{supe}(2n)/\mathfrak{spe}^*(2n)$
$\mathfrak{osp}(m,p\|2n)/\mathfrak{osp}(s+r,r\|2q)\oplus$ $\oplus\,\mathfrak{osp}(m-r-s,p-s\|2n-2q)$	$\mathfrak{osp}(m,p+s-r\|2n)/\mathfrak{osp}(s+r,r\|2q)\oplus$ $\oplus\,\mathfrak{osp}(m-r-s,p-s\|2n-2q)$
$\mathfrak{osp}(m,p\|2n)/\mathfrak{u}(m/2,p/2\|n,q)$	$\mathfrak{osp}^*(m\|2n,2q)/\mathfrak{u}(m/2,p/2\|n,q)$
$\mathfrak{osp}^*(2m\|2n,2q)/\mathfrak{osp}^*(2p\|2s+2r,2r)\oplus$ $\oplus\,\mathfrak{osp}^*(2m-2p\|2n-2r-2s,2q-2r)$	$\mathfrak{osp}^*(2m\|2n,2q+2s-2r)/$ $/\mathfrak{osp}^*(2p\|2s+2r,2r)\oplus$ $\oplus\,\mathfrak{osp}^*(2m-2p\|2n-2r-2s,2q-2r)$
$\mathfrak{osp}^*(2m\|2n,n)/\mathfrak{osp}_{\mathbb{C}}(m\|n)$	$\mathfrak{osp}(2m\|2n,n)/\mathfrak{osp}_{\mathbb{C}}(m\|n)$
$\mathfrak{psq}_r(2n)/\mathfrak{p}_r\mathfrak{s}_r\mathfrak{q}(n)$	$\mathfrak{psq}^*(2n)/\mathfrak{p}_r\mathfrak{s}_r\mathfrak{q}(n)$
$\mathfrak{psq}_r(2n)/{}^0\mathfrak{p}_r\mathfrak{q}(n)$	$\mathfrak{psq}^*(2n)/{}^0\mathfrak{p}_r\mathfrak{q}(n)$
$\mathfrak{psuq}(m,p)/\mathfrak{ps}(\mathfrak{uq}(r+s,r)\oplus$ $\oplus\,\mathfrak{uq}(m-r-s,p-r))$	$\mathfrak{psuq}(m,p+s-r)/\mathfrak{ps}(\mathfrak{uq}(r+s,r)\oplus$ $\oplus\,\mathfrak{uq}(m-r-s,p-r))$
$\mathfrak{psuq}(m,p)/\mathfrak{pu}(r+s,r\|m-r-s,p-r)$	$\mathfrak{psuq}(m,p+s-r)/\mathfrak{pu}(r+s,r\|m-r-s,p-r)$
$\mathfrak{spe}_r(2n)/\mathfrak{upe}(n)$	$\mathfrak{spe}^*(2n)/\mathfrak{upe}(n)$
$\mathfrak{spe}_r(2n)/\mathfrak{s}_r\mathfrak{pe}(n)$	$\mathfrak{spe}^*(2n)/\mathfrak{s}_r\mathfrak{pe}(n)$
$\mathfrak{sh}(n,p)/\text{Ш}(k,m,p,n)$	$\mathfrak{sh}(n,p+l-k)/\text{Ш}(k,m,p,n)$

Table 3. Selfdual homogeneous symmetric superspaces

$(\mathfrak{p})\mathfrak{su}^*(2m|2n)/(\mathfrak{p})\mathfrak{s}(\mathfrak{u}^*(2p|2q)\oplus\mathfrak{u}^*(2m-2p|2n-2q))$;
$(\mathfrak{p})\mathfrak{sl}_r(m|2n)/(\mathfrak{p})\mathfrak{s}(\mathfrak{gl}_r(p|q)\oplus\mathfrak{gl}_r(n-p|n-q))$;
$(\mathfrak{p})\mathfrak{su}(2m,\ m|2n,\ n)/\mathfrak{p}_{im}\mathfrak{s}_{im}\mathfrak{l}(m|n)$; ${}^o\mathfrak{pg}(n)/\mathfrak{p}({}^o\mathfrak{g}(p)\oplus{}^o\mathfrak{g}(n-p))$;
${}^o\mathfrak{pg}(n)/\mathfrak{p}_r\mathfrak{s}_{im}\mathfrak{l}(p|n-p)$; $\mathfrak{supe}(n)/\mathfrak{s}(\mathfrak{upe}(p)\oplus\mathfrak{upe}(n-p))$; $\mathfrak{supe}(n)/\mathfrak{p}_{im}\mathfrak{s}_r\mathfrak{l}(p|n-p)$
$\mathfrak{osp}(2m,\ m|2n)/\mathfrak{gl}_r(m|n)$; $\mathfrak{osp}^*(2m|2n,\ n)/\mathfrak{u}^*(m|n)$; $\mathfrak{psg}_r(n)/\mathfrak{ps}(\mathfrak{g}_r(p)\oplus\mathfrak{g}_r(n-p))$
$\mathfrak{psg}_r(n)/\mathfrak{pgl}_r(p|l-p)$; $\mathfrak{psg}^*(2n)/\mathfrak{ps}(\mathfrak{g}^*(2p)\oplus\mathfrak{g}^*(2n-2p))$; $\mathfrak{psug}(2m,\ m)/\mathfrak{p}_{im}\mathfrak{s}_{im}\mathfrak{g}(m)$;
$\mathfrak{psug}(2m,\ m)/{}^o\mathfrak{p}_{im}\mathfrak{g}(m)$; $\mathfrak{spe}_r(n)/\mathfrak{s}(\mathfrak{pe}_r(n-p)\oplus\mathfrak{pe}_r(p))$; $\mathfrak{spe}_r(n)/\mathfrak{sl}_r(p|n-p)$
$\mathfrak{spe}^*(2n)/\mathfrak{s}(\mathfrak{pe}^*(2p)\oplus\mathfrak{pe}^*(2n-2p))$; $\mathfrak{spe}^*(2n)/\mathfrak{su}^*(2p|2n-2p)$; $\mathfrak{sh}(2n,\ n)/\text{Ц}_r(n)$

Table 4. Classical superspaces of depth 1

$\mathfrak{g}$	$\mathfrak{g}_0$	$\mathfrak{g}_{-1}$	Interpretation	Underlying domain	Name of the superdomain
$\mathfrak{sl}(m\|n)$	$\mathfrak{s}(\mathfrak{gl}(p\|q)\oplus\mathfrak{gl}(m-p\|n-q))$	$id\otimes id^*$	Supergrassmannian of the $(p\|q)$-dimensional subsuperspaces in $\mathbb{C}^{m\|n}$-dimensional one	$Gr_p^m\times Gr_q^n$	$Gr_{p,q}^{m,n}$
$\mathfrak{psl}(m\|n)$	$\mathfrak{ps}(\mathfrak{gl}(p\|p)\oplus\mathfrak{gl}(m-p\|n-q))$	$id\otimes id^*$	Same for m=n, p=q	$Gr_p^m\times Gr_p^m$	$Gr_{p,p}^{m,m}$
$\mathfrak{osp}(m\|2n)$	$\mathfrak{cosp}(m-2\|2n)$	id	Superquadric of $(1\|0)$-dimensional isotropic with respect to the nondege-nerate even form lines in $\mathbb{C}^{m\|n}$	Q_{m-2}	$Q_{m-2,n}$
$\mathfrak{osp}(2m\|2n)$	$\mathfrak{gl}(m\|n)$	E^2 id	Ortholagrangean supergrassmannian of $(m\|m)$-dimensional isotropic with respect to the nondegenerate even form subsuperspaces in $\mathbb{C}^{2m\|n}$	$OGr_m\times {}^*LGr_n$ ${}^*OGr_m\times LGr_n$	$OLGr_{m,n}$
$\mathfrak{sq}(n)$ $\mathfrak{psq}(n)$	$\mathfrak{s}(\mathfrak{q}(p)\oplus\mathfrak{q}(n-p))$ $\mathfrak{ps}(\mathfrak{q}(p)\oplus\mathfrak{q}(n-p))$	$irr(id\otimes id^*)$	Queergrassmannian of q-symmetric $(p\|p)$-dimensional subsuperspace in $\mathbb{C}^{n\|n}$	Gr_p^n	QGr_p^n
$\mathfrak{pe}(n)$ $\mathfrak{spe}(n)$	$\mathfrak{cpe}(n-1)$ $\mathfrak{cspe}(n-1)$	id	Odd superquadric of $(1\|0)$-dimensional isotropic with respect to the nondegenerate odd form lines in $\mathbb{C}^{n\|n}$	$\mathbb{CP}^{n-1}$	PeQ_{n-1}
$\mathfrak{pe}(n)$ $(\mathfrak{spe}(n))$	$\mathfrak{gl}(p\|n-p)$ $(\mathfrak{sl}(p\|n-p))$	$\pi(S^2(id))$ or $\pi(E^2(id))$	Odd lagrangean supergrassmannian of $(p\|n-p)$-dimensional (and with a fixed volume for $\mathfrak{spe}$) subsuperspaces in $\mathbb{C}^{n\|n}$ isotropic with respect to the odd symmetric or skewsymmetric form	Gr_p^n	$PeGr_p^n$
$\mathfrak{vect}(0\|n)$	$\mathfrak{vect}(0\|n-k)\oplus\mathfrak{gl}(k;\Lambda(n-k))$	$\Lambda(k)\otimes\pi(id)$	Curved supergrassmanian of $(0\|1)$-dimensional subsupermanifolds in $\mathbb{C}^{0\|n}$		$CGr_{0,k}^{0,n}$
$\mathfrak{svect}(0\|n)$	$\mathfrak{vect}(0\|n-k)\oplus\mathfrak{sl}(k;\Lambda(n-k))$	$\pi(Vol)$if k=1	Same with volume elements preserved in the sub- and ambient supermanifolds		$SCGr_{0,k}^{0,n}$
$\mathfrak{h}(0\|m)$ $\mathfrak{sh}(m)$	$\mathfrak{h}(0\|m-2)\oplus\Lambda(m-2)\cdot z$ $\mathfrak{sh}(m-2)\oplus\Lambda(m-2)\cdot z$	$\pi(id)$	Curved superquadeic of $(0\|1)$-dimensional isotropic with respect to the (partly) split symmetric form subsupermanifolds in $\mathbb{C}^{0\|m}$		$CQ_{m-2,0}$

$\mathfrak{d}(\alpha)=$ $=\mathfrak{osp}(4\|2)_\alpha$	$\mathfrak{cosp}(2\|2)\cong(\mathfrak{gl}(2\|1))$	id	$\mathbb{CP}^1\times\mathbb{CP}^1$
$\mathfrak{ab}(3)$	$\mathfrak{cosp}(2\|4)$ $\mathfrak{cd}(2)$	$L_{3\varepsilon_1}$	$\mathbb{CP}^1\times Q_5$

Table 5. Gradings of twisted loop (super)algebras corresponding to hermitian superdomains

$\mathfrak{g}_\varphi^{(m)}$	φ	*grading elements from* $\mathfrak{h}$	$(\mathfrak{g}_\varphi^{(m)})_0$
$\mathfrak{sl}(2m/2n)^{(2)}$	$(-st)\cdot\mathrm{Ad}\ \mathrm{diag}(\pi_{2m},J_{2n})$	$\mathrm{diag}(1_m,-1_m,1_n,-1_n)$	$\mathfrak{sl}(m/n)^{(1)}$
$\mathfrak{sl}(2m)^{(2)}$	$(-t)\cdot\mathrm{Ad}\ (\pi_{2m})$		$\mathfrak{sl}(m)^{(1)}$
$\mathfrak{sl}(2n)^{(2)}$	$(t)\cdot\mathrm{Ad}\ (J_{2n})$		$\mathfrak{sl}(n)^{(1)}$
$\mathfrak{sl}(n/n)^{(2)}$	π	$\mathrm{diag}(1_p,0_{n-p},1_p,0_{n-p})$	$\mathfrak{s}(\mathfrak{gl}(p/p)^{(2)}{}_\pi\oplus\mathfrak{gl}(n-p/n-p)^{(2)}{}_\pi)$
$\mathfrak{sl}(n/n)^{(2)}$	$\pi\cdot(-st)$	$\mathrm{diag}(1_p,-1_{n-p},-1_p,1_{n-p})$	$\mathfrak{s}(\mathfrak{gl}(p/p)^{(2)}{}_{\pi\cdot(-st)}\oplus\mathfrak{gl}(n-p/n-p)^{(2)}{}_{\pi\cdot(-st)})$
$\mathfrak{osp}(2m/2n)^{(2)}$	$\varphi_{m,n}\mathrm{Ad}\ \mathrm{diag}(1_{2m-1},1,1_{2n})$	$\mathrm{diag}(2J_2,\ O_{2(m+n-1)})$	$(\mathfrak{cosp}(2m-2/2n))^{(1)}{}_{\varphi_{m-1,n}}$
$\mathfrak{o}(2m)^{(2)}$			$(\mathfrak{co}(2m-2))^{(1)}$
$\mathfrak{psq}(2n)^{(4)}$	$(-st)\cdot\sigma_i$	$\mathrm{diag}(J_{2n},J_{2n})$	$\mathfrak{psq}(n)^{(2)}{}_{\delta_{-1}}$
$\mathfrak{sh}(2n)^{(2)}$	A	$H_{\xi_2\xi_3}$	$(\mathfrak{sh}(2n-2)\oplus\Lambda(2n-2))^{(2)}{}_A$
$\mathfrak{psq}(n)^{(2)}$	σ_{-1}	$\mathrm{diag}(1_p,0_{n-p},1_p,0_{n-p})$	$\mathfrak{ps}(\mathfrak{q}(p)^{(2)}{}_{\delta_{-1}}\oplus\mathfrak{q}(n-p)^{(2)}{}_{\delta_{-1}})$

REMARKS ON THE DIFFERENTIAL IDENTITIES IN SCHOUTEN-NIJENHUIS ALGEBRA

Zbigniew Oziewicz
Department of Mathematics, Gannon University
Erie, Pennsylvania 16541, USA
On leave on absence from:
Institute of Theoretical Physics, Wrocław University
pl. Dabrowszczaków 38, PL 50 204 Wrocław, Poland

There are two extensions of the Lie product of the derivations of the algebra F of smooth functions on manifolds, i.e. the two extensions of the Lie module structure of the vector fields DerF, to some tensor fields: the Frölicher and Nijenhuis Lie module and the Schouten and Nijenhuis Lie module. For somehow completeness we consider these Lie module structures on symmetric and on skew-symmetric tensor algebras. The first section fix our notation. Essentially this is the short presentation of the graded Lie algebra of derivations of the Grassmann algebra of differential forms, due to Frölicher and Nijenhuis (1956) and Michor (1985, 1987, 1988, 1989). These results are adopted here for our needs in slightly different context. Last section is devoted to the presentation of our main result, the differential identities relating the Grassman algebras homomorphisms to the Schouten and Nijenhuis bracket. They are interesting for applications in Hamiltonian mechanics (cf Lichnerowicz 1977, 1978 and 1988). We believe that similar differential identities should involve also the Frölicher and Nijenhuis bracket which could be useful in the theory of connections (cf Oziewicz and Gruhn 1983 and profound results by Michor 1988). It is interesting open problem how much of analogous relations remains to hold in noncommutative differential geometry.

I want thank those who have helped me by discussing, explaining and advising, Rafał Abłamowicz, Andrzej Borowiec, Michel Dubois-Violette, Giusepe Marmo, Marco Modugno and Jerzy Różański. In particular I want to acknowledge Peter Michor for sending me bunch of his unpublished important results. Special thanks are due to Konrad Bleuler and Ugo Bruzzo also for warm hospitality at Rapallo.

Lie algebras of derivations of associative algebras

Z-graded associative algebra G is (graded) Lie admissible. The Lie structure given by (graded) commutator {,} is operating as the inner (graded) derivation on the associative structure "°". For A, B and C in G and for $q \in \{-1, 0, +1\}$,

$$\{,\}: G \longrightarrow \mathrm{Der}(G,°), \qquad \deg\{,\} = 0,$$

$$\{A,B\} = A°B - q^{(\deg A)(\deg B)}\, B°A,$$

$$\{A,\{B,C\}\} = \{\{A,B\},C\} + q^{(\deg A)(\deg B)}\{B,\{A,C\}\},$$

$$\{A,B°C\} = \{A,B\}°C + q^{(\deg A)(\deg B)}B°\{A,C\}. \tag{1}$$

It is important to realize that these formulas hold true either for $q = -1$ and for $q = +1$ ($q = 0$ is trivial). These two values corresponds exactly to fermionic and to bosonic situations. To see this let G = v-End H, where H is a Fock space, i.e. associative Z-graded algebra, graded commutative (skew-symmetric) for multi-fermions or commutative (symmetric) for multi-bosons. Here v-End H denote the vector space endomorphisms of H to be distinguished from the algebra endomorphisms a-End H which we will need late on. Obviously any derivation in Der H $\subset$ v-End H, should be compatible with the graded-commutativity or ($q = +1$)--commutativity in algebra H, which imply the above (1) Lie structure {,}, with $q = -1$ for graded-commutative H and with $q = +1$ for commutative H. Let's mention that in connection with so called quantum deformations of Lie algebras (strictly speaking one deforms universal enveloping algebra) the q-deformed commutator {,} has been considered for arbitrary complex value of q, with the q-deformed versions of (1).

Let M denote an object in category of smooth manifolds and let F be the structural sheaf on M, i.e. the sheaf of the R-algebras of smooth real (or complex) valued functions on M. The (commutative) differential geometry of the manifold M can be described completely by means of the structural sheaf F of the commutative and associative algebras, in terms of the corresponding F-modules and F-algebras.

Let Ω and $\mathfrak{X}$ denote the pair of functors of the Z-graded symmetric or Grassmann tensor F-algebras, mutually F-dual. One can distinguish notationaly symmetric and skew-symmetric cases by introducing say Ω^+ for

$q = +1$ and Ω^- for $q = -1$, but we are not going to be so pedantic. Here Ω denote contravariant functor of the algebra of the differential forms. When the functor of the algebras of the multivector fields $\mathfrak{X}$ is restricted to the category mutually diffeomorfic manifolds then can be chosen as covariant or as contravariant one. We will denote a contravariant functor of the differential forms in value in the category of the cochain R-complexes also for short by Ω, i.e. Ω_M could denote de Rham complex on M with coboundary differential d.

The Z-graded associative algebras v-$\mathrm{End}_R\Omega$ and v-$\mathrm{End}_R\mathfrak{X}$ are our examples of (graded) Lie admissible algebras according to (1).
There are an exact sequences of (graded) Lie algebras

$$0 \longrightarrow \mathrm{End}_F\Omega \longrightarrow \mathrm{End}_R\Omega \longrightarrow \mathrm{End}_R\Omega/\mathrm{End}_F\Omega \longrightarrow 0 \tag{2}$$

and the same for v-$\mathrm{End}_R\mathfrak{X}$. In (2) $\mathrm{End}_F\Omega$ is short for m-$\mathrm{End}_F\Omega$, the F-module endomorphisms.

We will use the nontrivial F-module isomorphism $\Omega \otimes \mathfrak{X} \approx \mathrm{End}_F\Omega$, to be referred as the skew-isomorphism which we describe as follows.

Let e denote the associative binary operation (e for exterior), $e(\alpha,\beta) = e_\alpha\beta = (e\alpha)\beta$, for α and β in Ω_M (and the same for $\mathfrak{X}_M$, although it would be better to distinguish notationally the binary operations in two different associative F-algebras Ω_M and $\mathfrak{X}_M$). By lower case Greek letters, α, β, ω, .., we denote usually the differential forms from Ω_M, whereas the capital X, Y, Z, denote the multivector fields in $\mathfrak{X}_M$. Now A is in a F-module endomorphisms m-$\mathrm{End}_F\Omega$ (i.e. A is a tensor) iff $\{A,e_f\} = 0$ for every f in F_M. Annihilator of F in v-$\mathrm{End}_R\Omega$ will be denoted by Ann F. Similarly D is in a graded Lie algebra of derivations $\mathrm{Der}_R\Omega$ iff $\{D,e_\omega\} = e_{D\omega}$ for every ω in Ω_M. In particular the (graded) commutator $\{,\}$ of derivations is again a derivation which is equivalent to graded Jacobi identity (1). We have $\mathrm{Der}_F\Omega = (\mathrm{Der}_R\Omega) \cap (\text{m-}\mathrm{End}_F\Omega) = (\mathrm{Der}_R\Omega) \cap (\mathrm{Ann}\ F)$. Also for $q = -1$, $\{d,e_\omega\} = e_{d\omega}$ and $\{d,\{d,A\}\} = 0$, for every $A \in \mathrm{End}_R\Omega$. We will need late on the following trivial facts. Let A and B are in Ann F and $f \in F$, then $\{A,\{B,d\}\}f = (A\circ B)df$. If A and B are in m-$\mathrm{End}_F\Omega$ then $\{ef,\{\{d,A\},B\}\} = -\{\{edf,A\},B\}$ for $f \in F$.

From (1) it follows that $\{e_\alpha\circ T, e_\beta\} = e_\alpha\circ\{T, e_\beta\}$, which is obvious generalization of $\{e\alpha,e\beta\} = 0$ for $T = \mathrm{id}_\Omega$. This show that $\mathrm{Der}_R\Omega$ is a graded left Ω-module (Michor 1985).

The exact sequence (2) give rise to exact sequence of Lie subalgebras

$$0 \longrightarrow \mathrm{Der}_F\Omega \longrightarrow \mathrm{Der}_R\Omega \longrightarrow \mathrm{Der}_R\Omega/\mathrm{Der}_F\Omega \longrightarrow 0. \tag{3}$$

Let i denote pull-back of e (here this is a transposition), $i = e^* = \mathrm{Hom}_F(e,F)$. In fact this is too short for $(eX)^* = iX$, i.e. $\{(iX)\alpha\}Y = (i_X\alpha)Y = \alpha(e_XY)$ and $\beta\{(i\alpha)X\} = \beta(i_\alpha X) = (e_\alpha\beta)X$ for all X and Y in $\mathfrak{X}$ and all α and β in Ω. Evidently i is the co-multiplication, known as the internal product. Both e and i are monomorphisms of associative F-algebras, e.g. $e \in \mathrm{Mon}_F(\Omega, \mathrm{End}_F\Omega)$. Evident generalization of $\{iX, iY\} = 0$ (and of $\{i\alpha, i\beta\} = 0$) is $\{iX{\circ}T,\ iY\} = iX{\circ}\{T, iY\}$.

PROPOSITION. Let $X \in \mathfrak{X}$ and $\alpha \in \Omega$. Then for both values of q = +1 and -1, we have

$\{iX,\ e\alpha\} = e(iX)\alpha$ for all α in Ω, iff $\deg X = 1$.

$\{i\alpha,\ eX\} = e(i\alpha)X$ for all X in $\mathfrak{X}$, iff $\deg \alpha = 1$.

$\{i\alpha,\ eX\} = i(iX)\alpha$ for all α in Ω, iff $\deg X = 1$.

$\{iX,\ e\alpha\} = i(i\alpha)X$ for all X in $\mathfrak{X}$, iff $\deg \alpha = 1$.

This means that e.g. iX is in $\mathrm{Der}_F\Omega$ iff $\deg X = 1$. The last two statements follows from the first one through the pull-back, $i^* = e$. Note that the above relations are known for q = -1 as the relations for multi-fermion (and multi-anti-fermion) creation (e) and annihilation (i) operators and for q = +1 they corresponds to multi-boson case (cf Oziewicz 1986).

For $\omega \otimes X$ in $\Omega \otimes \mathfrak{X}$ we denote the skew-isomorphic image in $\mathrm{End}_F\Omega$ by $e_\omega \circ i_X$, or

$$(e{\circ}i)(\omega \otimes X) = e_\omega \circ i_X. \tag{4}$$

Notice the difference between natural F-module isomorphism $\Omega \otimes \mathfrak{X} \simeq \mathrm{End}_F\Omega$, denoted by ev (as short for the evaluation) and (4): $\{ev(\omega \otimes X)\}\alpha = 0 \in \Omega$ for $\deg X \neq \deg\alpha$, while this need not be the case for $e{\circ}i$! (Here α is in Ω.)

Obviously the above F-modules skew-isomorphism (4) extent to both: graded associative F-algebras isomorphism and graded Lie F-algebras isomorphism. The induced Lie F-algebra structure on the space of tensors $\Omega \otimes \mathfrak{X}$ is known as the Nijenhuis-Richardson algebra. For T and R in $\Omega \otimes \mathfrak{X}$, the Nijenhuis-Richardson bracket $[T,R]_{N-R} \in \Omega \otimes \mathfrak{X}$, is defined as

$$(e{\circ}i)([T,R]_{N-R}) = \{(e{\circ}i)T,\ (e{\circ}i)R\}. \tag{5}$$

From the above proposition it follows that $(e^{\circ}i)T$ is in $Der_F\Omega$ iff the "contravariant" valence of T is equal 1, i.e. iff T is a one-vector field-valued differential form or a differential form - valued one-vector field.

Frölicher-Nijenhuis Lie module

Consider the Lie F-algebra $Der_F\Omega$ wrt to (graded) bracket (1) for both, +1 and -1, values of q. The Frölicher and Nijenhuis binary Lie R-operation on F-algebra $Der_F\Omega$ will be denoted by N or by traditional bracket $N = [\,,\,]_{F-N}$, where $N_A B = (NA)B = = [A,B]_{F-N}$. By definition, which holds true for each q value, the Frölicher and Nijenhuis Lie structure N is operating as the derivation of the (graded) Lie structure $\{,\}$,

$$N:\ Der_F(\Omega,e) \longrightarrow Der_R(\ Der_F(\Omega,e),\ \{,\}), \qquad \deg N = +1,$$

$$\{N_A,\ \{B,\ldots\}\} = \{N_A B,\ldots\}$$

$$N_A B = (q)^{(\deg A)(\deg B)} N_B A \tag{6}$$

Therefore $Der_F\Omega$ is equipped with the pair of the Lie structures $(Der_F\Omega,\ \{,\},\ N)$. From the definition (6) the Jacobi identity follows (N is the self-derivation)

$$\{NA\ ,\ NB\} = N_{(NA)B},$$

which means that N is homomorphism of Lie R-algebras,

$$N\ \varepsilon\ Hom_R((Der_F(\Omega,e),N),(Der_R Der_F(\Omega,e),\{,\}),\{,\})).$$

On the contrary the classical approach for q = -1 case is based on the following

PROPOSITION (Frölicher and Nijenhuis 1956, cf also with Michor 1985).
Any $D\ \varepsilon\ Der_R\Omega$ can uniquely be decomposed into the form $D = \{K,\ d\} + L$, where both K and L are in $Der_F\Omega$.

Complete proof. Let $f\ \varepsilon\ F$. Then $K\ \varepsilon\ Der_F\Omega$ is defined by $Kdf = Df$.

Then $L = D - \{K, d\} \in \mathrm{Der}_F\Omega$. Moreover $\{D, ef\} = 0 \Rightarrow K = 0$ and $\{d, D\} = 0 \Rightarrow D = \{K, d\}$ (i.e. $L = 0$). qed

Any F-linear map A from one-forms to Ω give rise to unique F-linear derivation of Ω, denoted by $\delta_A \in \mathrm{Der}_F\Omega$, such that $\delta_A df = (\delta A)df = Adf$ for all f in F. Evidently deg $\delta = 0$, and $\delta|\mathrm{Der}_F\Omega$ = identity. For example for A and B in $\mathrm{Der}_F\Omega$, $\delta_{\{A,B\}} = \{A,B\} \in \mathrm{Der}_F\Omega$, which means that

$$\delta_{A\circ B} = \{A,B\} + q^{(\deg A)(\deg B)}\delta_{B\circ A}.$$

Let A and B are in $\mathrm{Der}_F\Omega$, then $\{A,\{B,d\}\}$ is in $\mathrm{Der}_R\Omega$. Applying proposition we get that the difference $\{A,\{B,d\}\} - \{\delta_{A\circ B},d\}$ is in $\mathrm{Der}_F\Omega$, which could be adopted (up to the sign conventions) as the classical definition of the Frölicher-Nijenhuis bracket for q = -1 case

$$N_A B = \{A, \{B, d\}\} - \{\delta_{A\circ B}, d\}. \tag{7}$$

To show that (NA)B is in $\mathrm{Der}_F\Omega$ it is enough to see that (NA)B $\in$ Ann F. In particularly for one-vector fields X and Y, deg X = deg Y = 1, from (7) we get $N_X Y = X\circ Y - Y\circ X$. Definition (6) is more general as valid also for symmetric tensor algebra. Denote $LA = \{d, A\}$ for A in v-$\mathrm{End}_R\Omega$. Also we use $L_T = L((e\circ i)T)$ for T in $\Omega \otimes \mathfrak{F}$. Note the different sign convention from the definition adopted by Michor. As we will see in the next section, our convention has some aesthetic motivation. Clearly $\{d, \{LA, LB\}\} = 0$. If the commutator $\{LA, LB\}$ lay in $\mathrm{Der}_R\Omega$ then above proposition assure the unique existence of $N_A B$ in $\mathrm{Der}_F\Omega$ such that $\{LA, LB\} = = L(N_A B)$. Therefore L composed with the skew-isomorphism $e\circ i$ extent to graded Lie R-algebras isomorphism of the Frölicher and Nijenhuis Lie algebra of a one vector field - valued differential forms and the quotient Lie algebra $\mathrm{Der}_R\Omega/\mathrm{Der}_F\Omega$ with the graded commutator (1).

Schouten and Nijenhuis Lie module

The Schouten and Nijenhuis Lie module of multivector fields is the unique extension, up to the sign, of the Lie algebra structure of the DerF to tensor algebra $\mathfrak{F}$ (cf Michor 1987). Here $\mathfrak{F}$ denote symmetric or Grassmann F-algebra generated by F and F-module DerF. We consider here both q = -1 and q = +1 cases, see (1). The Schouten-Nijenhuis binary Lie R-operation on graded F-module $\mathfrak{F}$ will be denoted either by S or by traditional bracket [,]

$= [\,,\,]_{S-N}$, where $S_X Y = (SX)Y = [X,Y]$. By definition, the Schouten-Nijenhuis Lie structure S is operating as the derivation of the associative algebra structure e:

$$S: \mathfrak{X} \longrightarrow Der_R(\mathfrak{X},\ e), \quad \deg S = -1,$$

$$\{S_X,\ e_Y\} = e(S_X Y)$$

$$S_X Y = (q)^{(\deg X)(\deg Y)}\ S_Y X. \tag{8}$$

Therefore ($\mathfrak{X}$, e, S) looks like graded version of a Poisson algebra. Above conditions (8) specify S up to the values on the generators of the Grassmann algebra $\mathfrak{X}$, i.e. on F and on DerF. For f in F we put $S_f = i_{df}$, then for $X \in DerF$, $[X,f] = = [f,X] = Xf$. For X in DerF we put S_X to be the Lie derivative along X, i.e. $S_X = L_X$. If X and Y are both in DerF then $S_X Y = -S_Y X$ for both $q = +1$ and $q = -1$, and still there are two conventions: either $S_X Y = X{\circ}Y - Y{\circ}X$ or $S_X Y = Y{\circ}X - X{\circ}Y$, cf. with discussion by Michor (1987). The source of this sign ambiguity is clear. On the category of the mutually diffeomorphic manifolds there are two different functors Der. First is covariant, second is contravariant. Here we use these notions as they are used commonly in the category theory. The covariant functor Der can be extended for embeddings. The Lie derivative $L_X Y$ is equal to $X{\circ}Y - Y{\circ}X$ iff Der is a covariant functor and change sign if Der is a contravariant one. In our convention Der is a contra-variant functor.

The (graded) Jacobi identity, i.e. the self-derivation of S by S, follows from the definition (8)

$$\{S_X,\ S_Y\} = S_{(SX)Y} \ \in Der_R(\mathfrak{X},e). \tag{9}$$

This is an equality of the derivations and it is enough to check their values on the generators of the algebra. This is indeed a very tedious calculation. Jacobi identity show that S is a homomorphism of (graded) Lie algebras,

$$S \ \in \ Hom_R((\mathfrak{X},S),(Der_R(\mathfrak{X},e),\{,\})).$$

On the contrary, the classical approach for $q = -1$ goes as follows. It is easy to understand that $\{iY,\ \{d,\ iX\}\} = \{iY,\ L_X\}$ is always the tensor field. (However $\{iY,\{d,iX\}\} \in \text{Ann } F$ iff $\deg X + \deg Y \geq 2$.) This could be adopted as the classical definition of the Schouten-Nijenhuis bracket

$$i_{(SX)Y} = \{i_Y, L_X\}, \tag{10}$$

with Jacobi identity (9), now the very simple consequence of the definition (10). The definition (10) goes back to Schouten (1940) who considered arbitrary contravariant tensor fields. The Jacobi identity (9) for skew-symmetric multivector fields ($q = -1$ case) has been noticed by Nijenhuis (1955) who showed in this way the graded Lie module structure. Coordinate free treatment has been presented late on by Tulczyjew (1974). The important relations of Schouten and Nijenhuis bracket S to Lie algebra cohomology has been pointed by Koszul (1985).

Let's compare both definitions (8) and (10). Evidently (8) is more general, because holds for both $q = +1$ and $q = -1$ cases, whereas (10) holds only for $q = -1$. For $q = -1$ both definitions are equivalent. First, (10) $\Rightarrow$ (8), so S_X is a derivation. To see (8) $\Rightarrow$ (10) it is enough to check (10) on generators. $L_{if} = e_{df} \Rightarrow S_f = i_{df}$. For $\deg X = \deg Y = 1$ from (10) we get $S_X Y = Y \circ X - X \circ Y$. The aesthetic reason for which we choose Der as the contra-variant functor is that the expressions (8) and (10) looks formally like mutually adjoint (pull-back of each other). Let's write them once more together with their formal adjoints:

$$\{ S_X \;, eY\} = e(S_X Y) \quad \Rightarrow \quad \{iY, (S_X)^*\} = i(S_X Y)$$

$$\{(L_X)^*, eY\} = e(S_X Y) \quad \Leftarrow \quad \{iY, \; L_X \;\} = i(S_X Y). \tag{11}$$

This remind the familiar formula for the Lie derivative of, say, one form α, $\deg \alpha = \deg Y = 1$. If $L_X(\alpha Y) = 0$ then $(L_X \alpha)Y = +\alpha(S_X Y)$, therefore formally $(S_X)^* = L_X = \{d,iX\}$. Evidently we should rather consider adjoints wrt $\Omega' = \text{Hom}_R(\Omega,R)$ and $\mathfrak{X}' = \text{HomR}(\mathfrak{X},R)$.

Marmo (these Proceedings) noticed the deep relation of the Schouten Nijenhuis bracket S, for $q = -1$ case, to Chevaley complex.

Lie product on DerF, ∂: $(eX)Y \to X \circ Y - Y \circ X = - S_X Y$, where here e is R-linear Grassmann binary operation, give rise to co-multiplication ∂' in the dual space $(\text{DerF})' = \text{Hom}_R(\text{DerF},R)$ which has unique extension as the degree one, Z-graded, derivation of Grassmann R-algebra $\mathfrak{X}'$ generated by $(\text{DerF})'$, $\partial' \in \text{Der}_R \mathfrak{X}'$. This gives formal expression for ∂'' on $\mathfrak{X}''$, adopted as the extension of ∂ to degree -1 endomorphism of Grassmann R-algebra $\mathfrak{X}$

generated by DerF, $\partial \in \mathrm{End}_R\mathfrak{X}$. Jacobi identity is equivalent to $(\partial')^2 = 0$. Obviously $(\partial')^2 = 0 \Rightarrow \partial^2 = 0$. Therefore $\mathfrak{X}$ denote the functor of the multivector fields in value in the category of the chain R-complexes, i.e. $\mathfrak{X}_M$ is Chevalley complex on M with boundary differential ∂ (Chevaley and Eilenberg 1948, Koszul 1950, see also contribution by Marmo to these Proceedings). $\mathfrak{X}_M$ could be seen as the algebra of alternating R-multilinear derivations on F with values in F, $\mathrm{Alt}_R\mathrm{Der}F = \mathrm{Alt}_R\mathrm{Der}(F,F)$. Explicitly

$$\partial'(\alpha^1\wedge\alpha^2\wedge\ldots) = \Sigma\, (-)^{i+1} \quad \alpha^1\wedge\ldots\wedge(\partial'\alpha^i)\wedge\ldots\ldots\ldots\wedge\alpha_n$$

gives

$$\partial\,(X_1\wedge\ldots\ldots) = \Sigma\,(-)^{i+j+1}\partial(X_i\wedge X_j)\wedge X_1\wedge\ldots\,\hat{x}_i\ldots\,\hat{x}_j\ldots \tag{12}$$

Here $\wedge = e$ and $\partial \notin \mathrm{Der}_R\mathfrak{X}$. However (as noted by Marmo) for X in $\mathfrak{X}_M$, $\{\partial, e_X\} = e_{\partial X} + S_X$, where S_X is in $\mathrm{Der}_R\mathfrak{X}$, i.e. $\{S_X, e_Y\} = e_{(SX)Y}$, which is the definition (8). Therefore the Schouten-Nijenhuis Lie product S has been defined by Marmo in terms of the Chevalley boundary differential ∂, as the measure of the deviation from the derivation,

$$S_X = \{\partial,\ e_X\} - e_{\partial X}. \tag{13}$$

Differential identity involving Schouten and Nijenhuis bracket

Let assume that vectors on forms and vice versa act through the internal product, as defined in the first section, so for example for $\omega \in \Omega$ and X in $\mathfrak{X}$, $(iX)\omega$ is in Ω. We can consider here X as argument for ω, which will be denoted formally as $\omega'X = (iX)\omega$. Therefore differential forms and vector fields can be considered as F-modules homomorphisms,

$$\omega \in \Omega \ \Rightarrow\ \omega' \in \text{m-Hom}_F(\mathfrak{X},\ \Omega), \quad X \in \mathfrak{X}\ \Rightarrow\ X' \in \text{m-Hom}_F(\Omega,\ \mathfrak{X}).$$

Make sense the compositions, $\omega'\circ X' \in \text{m-End}_F\Omega$ and $X'\circ\omega' \in \text{m-End}_F\mathfrak{X}$. We associate also with every form ω in Ω, the F-homomorphism μ_ω of symmetric or Grassmann algebras $\mu_\omega \in \text{a-Hom}_F(\mathfrak{X},\Omega)$, defined by the values on one-vectors (generators), so for X in $\mathfrak{X}$, deg X =1, $(\mu\omega)X = (\mu_\omega)X = (iX)\omega$. The same for $\mu_X \in \text{a-Hom}_F(\Omega,\mathfrak{X})$. Therefore ω' and $\mu\omega$ coincide on one-vectors. Sometimes we will need also F-pull-back (transposition) of ω' and of X'. Evidently $\omega'^* = (\omega')^* \in \text{m-Hom}_F(\mathfrak{X},\Omega)$, so make sense the composition say $X'\circ(\omega'^*)$, etc. For example if deg $\omega = 2$, then $(\mu_\omega)^* = \mu_{\omega*}$ and restriction of $(\mu_\omega)^*$ to k-vectors is the same as $(-)^k\mu_\omega$.

Analogously with every F-module endomorphism, say R in $m\text{-}End_F\mathfrak{X}$, one can associate the unique Grassmann algebra derivation, denoted by $\delta R = \delta_R \in Der_F\mathfrak{X}$. For example for $\omega \in \Omega$ and $P \in \mathfrak{X}$, deg ω = deg P = 2 we have

$$(\delta_{\omega' \circ P'})^* = \delta_{P' \circ \omega'}, \quad (\omega'^* = -\omega',\ P'^* = -P') \quad \text{and} \quad \delta_{id}|(k\text{-vectors}) = k \cdot id.$$

So δ_{id} behaves like a number operator.

PROPOSITION

Let $\omega \in \Omega$ and $P \in \mathfrak{X}$, and deg ω = deg P = 2. Then we have

$$2 \cdot \mu_P d\omega - \delta_{P' \circ \omega'} S_P P + 2 \cdot S_{P \circ \omega \circ P} P = 0 \quad \in \quad \Omega.$$

It is important to stress that ω and bivector field P are not related here at all. The proof of proposition is based on simple lemma, proof of which is omitted. Lemma follows from the definition (10) of the Schouten and Nijenhuis bracket S.

LEMMA. Let P is a bivector field, α, β and γ are one-forms in Ω, and let $R \in End_F\Omega$. Then we have

$$\text{I.} \quad (\alpha\wedge\beta\wedge\gamma) S_{P' \circ R} P = -(P'R\gamma)(\beta P'\alpha) - (P'\gamma)(\beta P'R\alpha) + (\delta_R^* \mu_P d\gamma)(\alpha\wedge\beta) + \text{permutations}.$$

$$\text{II.} \quad \{\delta_R(\alpha\wedge\beta\wedge\gamma)\} S_P P/2 = -(P'R\alpha)(\gamma P'\beta) - 2 \cdot (P'\alpha)(\gamma P'R\beta) + (\mu_P dR\alpha)(\beta\wedge\gamma) + (\delta_R^* \mu P d\alpha)(\beta\wedge\gamma) + \text{permutations}.$$

Remark: no relations between P and R has been assumed in lemma.

Sketch of proof of proposition. The starting point is the well known identity (Henri Cartan identity)

$$(d\omega)(X\wedge Y\wedge Z) = -X\{\omega(X\wedge Z)\} - \{(iX)d\omega\}(Y\wedge Z) + \text{permutations}. \tag{14}$$

Inserting in (14) $X\wedge Y\wedge Z = \mu_P(\alpha\wedge\beta\wedge\gamma)$ we obtain

$$(\alpha\wedge\beta\wedge\gamma)(\mu_P d\omega) = (P'\alpha)\{(\mu_P\omega)(\beta\wedge\gamma)\} + (\mu_P dR\alpha)(\beta\wedge\gamma) + \text{permutations}. \tag{15}$$

Moreover $\quad (\mu_P\omega)(\beta\wedge\gamma) = \omega(P'\beta\wedge P'\gamma) = (\omega' P'\beta)(P'\gamma) = (P'\gamma)(\omega' P'\beta)$.
Let's now relate R with ω and P, so R is going to be the "recursion" operator,

$$R = \omega' \circ P' \qquad \varepsilon \ \ \mathrm{End}_F\Omega.$$

Therefore $(\mu_P\omega)(\beta\wedge\gamma) = P(\gamma\wedge R\beta) = -\gamma P' R\beta = (P'\gamma)(R\beta) = (R^{*}P'\gamma)\beta = \beta P' R\gamma$.
From (15) we arrive at identity

$$(\alpha\wedge\beta\wedge\gamma)(\mu_P d\omega) = (P'\alpha)(\gamma P' R\beta) + (\mu_P dR\alpha)(\beta\wedge\gamma) + \text{permutations}. \tag{16}$$

Joining together both formulas in lemma with last identity (16) for $R = \omega' \circ P'$ gives the statement of the proposition. qed

Corollary. Let R = id, i.e. $\omega' \circ P' = \mathrm{id}_\Omega$ ($\Leftrightarrow$ $P' \circ \omega' = \mathrm{id}$ on $\mathfrak{X}$), then we have equivalence: $d\omega = 0 \Leftrightarrow S_P P = 0$. In fact δ_{id} on three-vectors $= 3\cdot\mathrm{id}$, therefore the proposition gives

$$2\cdot\mu_P d\omega = S_P P, \qquad \text{where } \mu_P \text{ is invertible.}$$

This is the case of the usual symplectic and Poisson structures when $d\omega = 0$ and $S_P P = 0$ (cf Lichnerowicz 1977). It should be evident that the identity in the proposition hold true beyond the symplectic mechanics and could be applied in presymplectic and almost symplectic cases.

The couple of another new identities follow if one use the compositions of the differential forms and multivector fields on both sides with the endomorphisms. Say for R in m-$\mathrm{End}_F\mathfrak{X}$ and Q in m-$\mathrm{End}_F\Omega$, then $\omega \in \Omega \Rightarrow Q\circ\omega'\circ R \in \Omega$, and $X \in \mathfrak{X} \Rightarrow R\circ X'\circ Q \in \mathfrak{X}$. New identities involve the expressions like $d(Q\circ\omega'\circ R)$, $S_{R\circ X'\circ Q}$, and compositions $\delta_Q\circ d$, $\delta_R\circ S$, $\mu_X\circ d$ and $\mu_\omega\circ S$, where S is the Schouten-Nijenhuis bracket. We are not presenting these identities here because all this matter has been not yet systemized enough.

However, as we mention at the begining, the most interesting question which is worthy to investigate is how much analogous relations survive in the noncommutative differential geometry (see Dubois-Violette 1990 and these Proceedings).

REFERENCES

CHEVALEY C. and S. EILENBERG (1948) Cohomology theory of Lie groups and Lie algebras. Trans. Am. Math. Soc. **63** 85

DIRAC P.A.M. (1931) Quantised singularities in the electromagnetic field. Proc. R. Soc. (London) **A 133** 60

DUBOIS-VIOLETTE Michel (1990) Noncommutative differential geometry, quantum mechanics and gauge theory. These Proceedings.

DUBOIS-VIOLETTE Michel, Richard KERNER and John MADORE (1990) Noncommutative differential geometry of matrix algebras. J.Math.Phys. **31** (2) 316-322

FROLICHER A. and A. NIJENHUIS (1956) Theory of vector valued differential forms, part I. Indagationes Math. **18** 338-359

KOLAR I. and P.W. MICHOR (1987) Determination of all natural bilinear operators of the type of the Frölicher-Nijenhuis bracket. Suppl. Rendiconti Circolo Mat. Palermo, Serie II, Nr **16** 101-108

KOSZUL J.L. (1950) Homologie et cohomologie des algèbres de Lie. PH.D. thesis, Gauthier-Villars, Paris

KOSZUL J.L. (1985) Crochet de Schouten-Nijenhuis et cohomologie. Asterisque 257-271

LICHNEROWICZ André (1977) Les variétée de Poisson et leurs algèbres de Lie associées. J. of Diff. Geom. **12** 253-300

LICHNEROWICZ André (1977) New geometrical dynamics. In "Differential Geometrical Methods in Mathematical Physics", ed. by K. Bleuler and A. Reetz. Lecture Notes in Mathematics, vol. 570, Springer-Verlag Berlin 377-394

LICHNEROWICZ André (1978) Les variétés de Jacobi et leurs algèbres de Lie associées. J. Math. pures et appl., **51** 453-488

LICHNEROWICZ André (1988) Applications of the deformations of the algebraic structures to geometry and mathematical physics. In "Deformation Theory of Algebras and Structures and Applications", ed. by M. Hazewinkel and M. Gerstenhaber. Kluver Academic Publishers 855-896

MARMO G. (1990) An algebraic characterization of complete integrability for Hamiltonian systems. These Proceedings.

MICHOR P.W. (1985) A generalization of Hamiltonian mechanics. JGP **2** 67-82

MICHOR P.W. (1987) Remarks on the Frölicher-Nijenhuis bracket. Proceedings of the Conference on Differential Geometry and its Applications, Brno, Czechoslovakia. D. Reidel

MICHOR P.W. (1987) Remarks on the Schouten-Nijenhuis bracket. Proceedings of the Winter School on Differential Geometry and Physics, Srni,

Czechoslovakia. Suppl. Rendiconti Circolo Mat. Palermo, Serie II, No 16

MICHOR P.W.(1988) Graded derivations of the algebra of differential forms associated with a connection. Proceedings of Symposium on Differential Geometry, Peniscola, Spain. Lecture Notes in Mathematics, vol. , Springer-Verlag Berlin

MICHOR P.W. (1989) Multigraded Lie algebras. Proceedings of the Winter School on Differential Geometry and Physics, Srni, Czechoslovakia. Suppl. Rendiconti Circolo Mat. Palermo,

NIJENHUIS A. (1955) Jacobi-type identities for bilinear differential concomitants of certain tensor fields I. Indagationes Math. **17** 390-403

OZIEWICZ Z. and W. GRUHN (1983) On Jacobi's theorem (from a year 1838). Hadronic Journal **6** No 6 1579-1605

OZIEWICZ Z. (1986) From Grassmann to Clifford. In "Clifford Algebras and their Applications in Mathematical Physics", edited by J.S.R. Chisholm and A.K. Common, R. Reidel Publishing Company, Dordrecht, 245-255

SCHOUTEN J.A. (1940) Über Differentialkonkomitanten zweier kontravariater Grössen. Indagationes Math. **2** 449-452

TULCZYJEW W. (1974) The graded Lie algebra of multivector fields and the generalized Lie derivative of forms. Bull. Ached. Pylon. Sci. **22** 937-942

GENERIC IRREDUCIBLE REPRESENTATIONS OF CLASSICAL LIE SUPERALGEBRAS

Ivan Penkov
Department of Mathematics
University of California
Berkeley, CA 94720

Vera Serganova
Department of Mathematics
Yale University
New Haven, CT 06520

Introduction. V. Kac observed in 1977, [K2],[K3] that finite dimensional irreducible representations of classical complex Lie superalgebras (for an explicit description of these superalgebras, see for instance [Kl], [Schl],[P]) split naturally into two classes: typical and atypical. Typical representations behave similarly to irreducible representations of simple finite dimensional Lie algebras and admit a natural and non-difficult generalization of H. Weyl's character formula, while atypical representations behave differently and the problem of finding an explicit character formula for an arbitrary such representation is still open. Since 1977 a large number of articles studying irreducible representations has appeared; (incomplete) lists of references, see for instance in [Lel],[P],[PS2],[JHKT]. In particular there have been several attempts to write down a general character formula for any irreducible finite dimensional representation of the complex general linear Lie superalgebra $gl(m+n\varepsilon)$. The first such attempt known to us is the article [BL]. However the formula proposed in [BL] (which we will call below the Bernstein-Leites formula) turned out to be true for all representations only if $m=1$ or $n=1$, or $mn=0$ (for instance the trivial representation of $gl(2+2\varepsilon)$ provides a counter-example). The most recent attempt known to us to propose a character formula for all finite-dimensional irreducible $gl(m+n\varepsilon)$-modules is the conjecture in [JHKT], which (as far as we know) is still open. The objective of this talk is to announce a structure theorem for a large class of (possibly infinite dimensional) irreducible representations over the Lie superalgebras $\mathfrak{g} = gl(m+n\varepsilon)$, $osp(2+n\varepsilon)$, $\mathfrak{p}(m)$. More precisely we describe explicitly the irreducible composition factors of a generic (see section 1 below) irreducible representation considered as a representation of the even part $\mathfrak{g}_0$ of $\mathfrak{g}$, in terms of a suitably chosen composition factor. In the finite dimensional case our description reduces simply to the Bernstein-Leites formula, showing in particular that this formula is still true for an atypical representation of $gl(m+n\varepsilon)$ "in general position". The proofs will appear in [PS3].

Acknowledgment. I.P. thanks the organizers of the XIX DGMTP meeting in Rapallo for the kind invitation to deliver this talk and for their hospitality.

1. The problem

The ground field will be $\mathbb{C}$. By G we denote one of the complex Lie supergroups $GL(m+n\varepsilon)$, $OSP^{0}(m+n\varepsilon)$,* $Q(m)$, $P(m)$ in the notation of [P], [PS1], or sometimes also a connected reductive complex algebraic group. Here ε is a formal odd variable with $\varepsilon^2=1$, and $m+n\varepsilon$ stands for the dimension of a $\mathbb{Z}_2$-graded vector space isomorphic to $\mathbb{C}^m \oplus \mathbb{C}^n$, $\mathbb{C}^m$ being considered as even and $\mathbb{C}^n$ being considered as odd, see [P],[PS1]. $\mathfrak{g} = \text{Lie } G = \mathfrak{g}_0 \oplus \mathfrak{g}_1$ will be one of the Lie superalgebras $gl(m+n\varepsilon)$, $osp(m+n\varepsilon)$, $\mathfrak{p}(m)$, $q(m)$ (in the notation of [Kl], the first three of them are respectively $\ell(m,n)$, $osp(m,n)$, $P(m-1)\oplus\mathbb{C}$, while $q(m)$ is an extension of $\mathbb{C}^{\varepsilon}$ by the Lie superalgebra which Kac denotes $\tilde{Q}(m-1)$), or a reductive complex Lie algebra. The result announced in this talk will concern only the cases $\mathfrak{g} = gl(m+n\varepsilon)$, $osp(2+n\varepsilon)$, $\mathfrak{p}(m)$.

In [P] one can find all necessary preliminaries about Cartan and Borel subsupergroups of G (respectively Cartan and Borel subsuperalgebras of $\mathfrak{g}$), the center Z_U of the enveloping algebra $U = U(\mathfrak{g})$, regular, typical, and atypical weights, the character formula for typical irreducible representations of finite dimension, etc. For a more compact presentation of the preliminaries needed in this talk, see [PS1]. Here we proceed directly to the problem.

Fix a Cartan subsuperalgebra $\mathfrak{h} = \mathfrak{h}_0 \oplus \mathfrak{h}_1$ of $\mathfrak{g}$ (for $\mathfrak{g} = q(m)$, $\mathfrak{h} = \mathfrak{h}_0$ and for $\mathfrak{g} = q(m)$, $\dim \mathfrak{h}_1 = \dim \mathfrak{h}_0 \cdot \varepsilon = m\varepsilon$). A weight $\eta \in \mathfrak{h}_0^*$ will be called *generic* iff it is sufficiently far from the walls of the Weyl chambers. The latter means more precisely that if η belongs to a certain Weyl chamber, the set $\eta+\Xi$, Ξ being a suitable bounded set of weights, should be a subset of the same chamber. All statements made below about generic weights will formally mean the existence of an appropriate bounded $\Xi \subset \mathfrak{h}_0^*$ such that the corresponding statement is true for all η lying in one and the same chamber together with the set $\eta+\Xi$. Consider in more detail generic weights for $\mathfrak{g} = gl(m+n\varepsilon)$, $osp(2+n\varepsilon)$, $\mathfrak{p}(m)$. Assume $\mathfrak{g} = gl(m+n\varepsilon)$. If $\eta_1,\eta_2,\dots,\eta_m$, $\eta_{m+1},\dots,\eta_{m+n}$ are the standard coordinates of a weight $\eta \in \mathfrak{h}_0^*$ (in this coordinate system roots have coordinates $\pm(0,\dots,0,1,0,\dots,0,-1,0,\dots0)$), η is generic iff $|\text{Re}(\eta_i-\eta_j)| \gg 0$ ($\gg$ denotes much greater) $\forall\, i,j$ with $i,j \le m$, $i\ne j$, or with $i,j > m$, $i\ne j$. Moreover, since η is atypical iff $\eta_i + \eta_j = 0$ for some $i\le m$, $j>m$, the sets of generic and atypical weights intersect essentially. The latter is true also for $\mathfrak{g} = osp(2+n\varepsilon)$, but is false for $\mathfrak{g} = \mathfrak{p}(m)$.

---

*By definition, $OSP^{0}(m+n\varepsilon)$ is the identity component in the orthosymplectic Lie supergroup $OSP(m+n\varepsilon)$.

Indeed if the standard coordinates of a weight $\eta \in \mathfrak{h}_0^*$ are $\eta_1,\ldots,\eta_m$ (here roots have coordinates $\pm(0,\ldots,0,1,0,\ldots,0,-1,0,\ldots 0)$, $\pm(0,\ldots,0,1,0,\ldots,0,1,0,\ldots,0)$, $(0,\ldots,0,\ -2,\ 0,\ldots,0))$, then genericness means $|\mathrm{Re}(\eta_i-\eta_j)| \gg 0\ \ \forall\ i\neq j$, while typicality means $\eta_i-\eta_j \neq 1\ \ \forall\ i\neq j$, and therefore genericness implies typicality.

If V is a representation of $\mathfrak{g}$ (below, for short, we say $\mathfrak{g}$-module), denote by ${}^0V$, V considered to be a representation (module) of $\mathfrak{g}_0$. For any weight $\eta \in \mathfrak{h}_0^*$ we can set

$$^{+}{}_i{}^{\eta}V := {}^0V/(\ker \theta^{\eta}_{red})\cdot{}^0V \quad ,$$

$\theta^{\eta}_{red}\colon Z_{red} \longrightarrow \mathbb{C}$ being by definition the composition

$Z_{red} \xrightarrow{HC_{red}} S^{\bullet}(\mathfrak{h}_0)^W \xrightarrow{\eta_W} \mathbb{C}$, where Z_{red} is the center of $U(\mathfrak{g}_0)$, HC_{red} is the Harish-Chandra homomorphism (see for instance [B] or [Dix]) for the Lie algebra $\mathfrak{g}_0$, W is the Weyl group of $\mathfrak{g}_0$, and η_W is the natural extension of the map $\eta\colon \mathfrak{h}_0 \longrightarrow \mathbb{C}$ to a homomorphism of the algebra of W-invariants $S^{\bullet}(\mathfrak{h}_0)^W$ in the symmetric algebra $S^{\bullet}(\mathfrak{h}_0)$ into $\mathbb{C}$. ${}^{+}{}_i{}^{\eta}$ itself is a functor from the category of $\mathfrak{g}$-modules $\mathfrak{g}$-mod into the category of $\mathbb{Z}_2$-graded $\mathfrak{g}_0$-modules $\mathfrak{g}_0$- mod. Furthermore, assuming that V is irreducible, we shall call V *generic* iff ${}^{+}{}_i{}^{\eta}V \neq 0$ for some generic $\eta \in \mathfrak{h}_0^*$. Below we restrict ourselves to consideration of generic $\mathfrak{g}$-modules. It is not difficult to show that for a generic irreducible V, ${}^0V$ has a finite Jordan-Hölder series. By $\mathrm{gr}\,{}^0V$ we will denote the corresponding $\mathbb{Z}\times\mathbb{Z}_2$-graded $\mathfrak{g}_0$-module.

The problem of reconstructing $\mathrm{gr}\,{}^0V$ in terms of ${}^{+}{}_i{}^{\eta}V$ for an appropriate η is quite natural and generalizes the character formula problem. Indeed, assume that V is a generic irreducible highest weight module with respect to a Borel subsuperalgebra $\mathfrak{b}$ with highest weight λ, and choose η to be $\lambda + \rho_{\mathfrak{b}_0}$, $\rho = \rho_{\mathfrak{b}_0}$ denoting the half sum of roots of $\mathfrak{b}_0$. Then it is not difficult to show that ${}^{+}{}_i{}^{\lambda+\rho_0}V$ is an irreducible highest weight module of $\mathfrak{g}_0$ with $\mathfrak{b}_0$-highest weight λ (such a module with trivial odd or trivial even part shall be denoted below respectively by $V_{\mathfrak{b}_0}(\lambda)$ or $\varepsilon\cdot V_{\mathfrak{b}_0}(\lambda)$) for $\mathfrak{g} \neq q(m)$, and is a direct sum of several such modules for $\mathfrak{g} = q(m)$. Reconstructing $\mathrm{gr}\,{}^0V$ in terms of ${}^{+}{}_i{}^{\lambda+\rho_0}V$ means now presenting a finite set of weights $\lambda = \lambda^0,\lambda^1,\ldots,\lambda^t$ and multiplicities $k^0 = k^0_0 + k^0_1\varepsilon$, $k^1 = k^1_0 + k^1_1\varepsilon,\ldots,k^t = k^t_0 + k^t_1\varepsilon$, $k^i_j \in \mathbb{N}$ (k^0 being determined by the condition ${}^{+}{}_{i}^{\lambda+\rho_0}V = k^0 V_{\mathfrak{b}_0}(\lambda)$*) so that

$$\mathrm{gr}^0 V = \bigoplus_{i=0}^{t} k^i V_{\mathfrak{b}_0}(\lambda^i) \quad .$$

Then also

$$\mathrm{ch}\, V = \sum_{i=0}^{t} k^i \mathrm{ch}\, V_{\mathfrak{b}_0}(\lambda^i) \quad ,$$

(ch denoting the character of a $\mathfrak{g}$- or $\mathfrak{g}_0$-module; see [P] or [PS1]), and if $\dim V < \infty$, $\mathrm{ch}\, V_{\mathfrak{b}_0}(\lambda_i)$ will be given by Weyl's formula.

Our result will be an explicit description of $\mathrm{gr}^0 V$ in terms of ${}^{+}{}_{i}^{\eta}V$ (for a suitable $\eta \in \mathfrak{h}_0^*$) for a generic irreducible representation V of $\mathfrak{g}$ = $\mathfrak{gl}(m+n\varepsilon)$, $\mathfrak{osp}(2+n\varepsilon)$, $\mathfrak{q}(m)$. Before formulating the statement, we need to recall

2. The Beilinson-Bernstein localization theorem

In this section G is a connected reductive complex Lie group. Consider the flag manifold G/B (B being a fixed Borel subgroup) and denote by $\mathcal{O}_{G/B}$ the structure sheaf of G/B (as an algebraic variety). Let $\tilde{\mathfrak{g}} = \mathcal{O}_{G/B} \otimes \mathfrak{g}$ be the sheaf of Lie algebras with commutator

$$[f_1\otimes g_1, f_2\otimes g_2] = f_1\varkappa(g_1)(f_2) \otimes g_2 + f_2\varkappa(g_2)(f_1) \otimes g_1 + f_1 f_2 \otimes [g_1,g_2]$$

f_1, f_2 being local sections of $\mathcal{O}_{G/B}$, $g_1, g_2 \in \mathfrak{g}$, and $\varkappa\colon \mathfrak{g} \longrightarrow \mathcal{T}_{G/B}$ being the canonical morphism of $\mathfrak{g}$ into the tangent sheaf of G/B induced by the action of G on G/B. Denote by $U(\tilde{\mathfrak{g}})$ the sheaf of enveloping algebras of $\tilde{\mathfrak{g}}$. Set

$$\hat{\mathcal{D}} = U(\tilde{\mathfrak{g}})/\hat{\mathfrak{n}} \quad ,$$

where $\hat{\mathfrak{n}} \subset \tilde{\mathfrak{g}}$ is the tautological nilpotent subsheaf of $\tilde{\mathfrak{g}}$ (at each closed point B' of G/B the geometric fibre of $\hat{\mathfrak{n}}$ is [Lie B', Lie B'], and ( ) denotes (here and below) a two-sided ideal. One has a canonical injective homomorphism of sheaves of algebras

$$S^{\cdot}(\mathfrak{h}) \longrightarrow \hat{\mathcal{D}}$$

(the symmetric algebra $S^{\cdot}(\eta)$ being considered as a constant sheaf) induced by the map

$$\mathfrak{h} \hookrightarrow \mathcal{O}_{G/B}\otimes\mathfrak{h} \hookrightarrow \tilde{\mathfrak{g}}/\hat{\mathfrak{n}} \hookrightarrow U(\tilde{\mathfrak{g}})/(\hat{\mathfrak{n}}) \quad .$$

Therefore we can set for each $\eta \in \mathfrak{h}^*$

$$\mathcal{D}^{\eta} = \hat{\mathcal{D}}/(\ker w_m(\eta-\rho_0)) \quad ,$$

where $\rho_0 = \rho_{\mathfrak{b}} = \rho_{\mathfrak{b}_0}$, $\mathfrak{b}$ = Lie B, and w_m denotes the element of maximal

---

* $k^0 = 1$ except when $\mathfrak{g} = \mathfrak{q}(m)$.

length in W, obtaining in this way a family of sheaves of algebras parametrized by $\mathfrak{h}^*$. One has also a canonical injection of sheaves of rings

$$\mathcal{O}_{G/B} \hookrightarrow \mathcal{D}^\eta$$

for any η, induced by the injection $\mathcal{O}_{G/B} \hookrightarrow U(\tilde{\mathfrak{g}})$. Therefore in particular any sheaf of $\mathcal{D}^\eta$-modules on G/B is endowed automatically with a $\mathcal{O}_{G/B}$-module structure. By $\mathcal{D}^\eta$-mod we denote the category of sheaves of $\mathcal{D}^\eta$-modules quasi-coherent as $\mathcal{O}_{G/B}$-modules.

One says that G/B is $\mathcal{D}^\eta$-*affine* for some $\eta \in \mathfrak{h}^*$ iff for any object $\mathcal{F}$ of the category $\mathcal{D}^\eta$-mod $\Gamma(\mathcal{F})$ (Γ denoting global sections) generates $\mathcal{F}$ over $\mathcal{D}^\eta$ and $H^i(\mathcal{F}) = 0 \quad \forall\, i > 0$ (H^i denotes i^{th} sheaf cohomology). It is easy to show that if G/B is $\mathcal{D}^\eta$-affine, the functor

$$\Gamma : \mathcal{D}^\eta\text{-mod} \rightsquigarrow \Gamma(\mathcal{D}^\eta)\text{-mod}$$

($\Gamma(\mathcal{D}^\eta)$-mod denoting simply the category of modules over the algebra $\Gamma(\mathcal{D}^\eta)$) is an equivalence of categories. The opposite equivalence is then given by the localization functor

$$\Gamma(\mathcal{D}^\eta)\text{-mod} \rightsquigarrow \mathcal{D}^\eta\text{-mod}$$

$$V \longmapsto \mathcal{D}^\eta \otimes_{\Gamma(\mathcal{D}^\eta)} V \ .$$

The following important statements have been proven by A. Beilinson and J. Bernstein in [BB]:

- *for any* $\eta \in \mathfrak{h}_0^*$, $\Gamma(\mathcal{D}^\eta) = V^\eta := U(\mathfrak{g})/(\ker \theta^\eta_{red})$ (in the case we are considering in this section $\mathfrak{g} = \mathfrak{g}_0$, but keeping the notation from section 1, we denote the central character again by θ^η_{red});
- *if* η *is dominant and regular* (regularity means here simply that $(\eta,\alpha) \neq 0$ for any root α, (,) denoting the dual of a non-degenerate invariant form on $\mathfrak{g}$), G/B *is* $\mathcal{D}^\eta$-*affine.*

Our main result uses an extension of the last claim to the case of a classical Lie supergroup. However before passing to this case, we need to introduce one notion: the *twist of a* $\mathfrak{g}$-*module by an integral weight.* Assume that V is a U^η-module for some regular dominant η (this means that Z_{red} acts on V via θ^η_{red}) and let ζ be an integral weight, such that $\eta+\zeta$ is again regular dominant. Then one sets

$$V^\zeta := \Gamma(\mathcal{O}(\zeta) \otimes_{\mathcal{O}_{G/B}} \mathcal{D}^\eta \otimes_{U(\mathfrak{g})} V) \quad ,$$

where $\mathcal{O}(\zeta)$ is the G-linearized invertible sheaf induced by the representation of B corresponding to the character $w_m(\zeta)$. Irreducibility of V implies via Beilinson-Bernstein's result irreducibility of V^ζ.

Now we are able to pass to

3. The result

In [P] we have (partially) extended Beilinson-Bernstein's localization theorem to classical Lie supergroups G. In particular we have introduced corresponding sheaves of algebras $\mathcal{D}^\eta$ on any flag supermanifold G/B (here Borel subsupergroups of G are not necessarily conjugated and the supermanifolds G/B are parametrized by the set of conjugacy classes of B) and have proven

THEOREM 1. *Let* $G = GL(m+n\varepsilon), OSP(m+n\varepsilon), P(m), Q(m)$ *and* $\eta \in \mathfrak{h}_0^*$ *be a regular weight in the sense of* [P] *or* [PS1] *(this implies in particular that* η *is typical). Then*

(a) *G/B is* $\mathcal{D}^\eta$*-affine.*

(b) *for* $G = GL(m+n\varepsilon)$, $OSP(2+n\varepsilon)$,

$$\Gamma(\mathcal{D}^\eta) := U^\eta = U(\mathfrak{g})/(\ker \theta^\eta),$$

$\theta^\eta\colon Z_U \longrightarrow \mathbb{C}$ *denoting now the homomorphism via which the center of* $U(\mathfrak{g})$ *acts on a module with highest weight* $\eta-\rho_b$ *with respect to* $b = \mathrm{Lie}\ B$, *where* $\rho_b := \frac{1}{2}\left(\sum_{\alpha\in\Delta_0(b)} - \sum_{\alpha\in\Delta_1(b)}\right)$ *and* $\Delta_0(b)$ *and* $\Delta_1(b)$ *are respectively the sets of even and odd roots of* b.

In [P] we have conjectured that Theorem 1(b) is true also for $G = OSP(m+n\varepsilon)$, $m\neq 2$, $Q(m)$ (but not for $G = P(m)$ because, as proven by M. Scheunert in [Sch2], $Z_U = \mathbb{C}$ for $\mathfrak{g} = p(m)$).

It is essential that in Theorem 1(a) we claim $\mathcal{D}^\eta$-affinity for an arbitrary regular dominant η. If instead of a regular dominant η one considers simply a generic dominant η (now possibly atypical!), the $\mathcal{D}^\eta$-affinity of G/B is also true and is moreover a rather easy consequence of the original theorem of Beilinson-Bernstein. However for a generic but atypical η the equivalence of the categories $\mathcal{D}^\eta$-mod and U^η-mod (here one naturally considers $\mathbb{Z}_2$-graded $\mathcal{D}^\eta$- and U^η-modules) fails because if $\mathcal{F}$ is an irreducible object of $\mathcal{D}^\eta$-mod, $\Gamma(\mathcal{F})$ is generally a reducible $\mathfrak{g}$-module. Nevertheless one can show that for an irreducible $\mathcal{F}$, $\Gamma(\mathcal{F})$ (is an indecomposable $\mathfrak{g}$-module and even more) has a unique irreducible $\mathfrak{g}$-submodule. It is a natural problem to describe explicitly $gr^0 V_{\mathcal{F}}$ in terms of $\mathcal{F}$, or equivalently in terms of the restriction $\mathcal{F}_{red}$ of $\mathcal{F}$ to the reduced part $(G/B)_{red}$ of G/B (in [P] it is shown that the correspondence $\mathcal{F} \longrightarrow \mathcal{F}_{red}$ gives rise to an equivalence of categories). Furthermore, by Beilinson-Bernstein's theorem, the sheaf $\mathcal{F}_{red}$ is encoded in the ($\mathbb{Z}_2$-graded) $\mathfrak{g}_0$-module $\Gamma(\mathcal{F}_{red})$, which, as one proves easily, is of the form $V_{\mathcal{F}}^{\eta'}$ for a certain η' (connected with η by an explicit but lengthy formula which we omit here). Therefore the problem of computing $gr^0 V_{\mathcal{F}}$ in terms of $\mathcal{F}$ is equivalent to the problem of reconstructing $gr^0 V_{\mathfrak{F}}$ in terms of $V^{\eta'}_{\mathfrak{F}}$ for a suitable η'. Using the fact that each generic irreducible $\mathfrak{g}$-module V occurs as $V_{\mathfrak{F}}$ for some $\mathfrak{F}$, we have proven for $\mathfrak{g} = gl(m+n\varepsilon)$, $osp(2+n\varepsilon)$, $\mathfrak{p}(m)$ the following more general purely representation-theoretic theorem, which is the main result of this talk.

THEOREM 2. *Let* $\mathfrak{g} = gl(m+n\varepsilon)$, $osp(2+n\varepsilon)$, $\mathfrak{p}(m)$ *and* V *be a generic irreducible* $\mathfrak{g}$*-module. Then for any Borel subsuperalgebra* $\mathfrak{b} \subset \mathfrak{g}$ *there exists a (generic) dominant weight* λ, *such that*

$$gr^0 V = \bigoplus_{\underline{\ell}=\{\ell_1,\dots,\ell_r\}\subset\{0,1\}^r} \varepsilon^{|\underline{\ell}|}\left({}^{+}_{i}{}^{\lambda+\rho_0}V\right)^{-\sum_{i=1}^{r}\ell_i\alpha_i}, \qquad (1)$$

where α_i *runs over the set*

$$R\begin{cases} = \left\{\begin{array}{l} \textit{all odd roots of } \mathfrak{b} \textit{ such that} \\ (\lambda+\rho_{\mathfrak{b}},\alpha) \neq 0, \quad (\ ,\ ) \textit{ denoting a scalar} \\ \textit{product on } \mathfrak{h}_0^* \textit{ induced by an invariant} \\ \textit{non-degenerate form on } \mathfrak{g} \end{array}\right\} & \textit{for } \mathfrak{g} = gl(m+n\varepsilon),\ osp(2+n\varepsilon) \\ = \left\{\begin{array}{l} \textit{all odd roots } \alpha \textit{ of } \mathfrak{b}, \textit{ such that} \\ -\alpha \textit{ is also a root of } \mathfrak{g} \end{array}\right\} & \textit{for } \mathfrak{g} = \mathfrak{p}(m), \end{cases}$$

$\{0,1\}^r := \underbrace{\{0,1\}\times\{0,1\}\times\dots\times\{0,1\}}_{r \text{ times}}$, *and* $|\underline{\ell}| := \#\{i \mid \ell_i \neq 0\}$.

The proof is geometric and will appear in [PS3]. The restriction $\mathfrak{g} = gl(m+n\varepsilon)$, $osp(2+n\varepsilon)$, $\mathfrak{p}(m)$ comes from the circumstance that we have been unable to avoid the use of a Borel subsuperalgebra $\mathfrak{b}_d$ (called a *distinguished* Borel subsuperalgebra) with the property that its odd part, together with the odd part of the opposite Borel subsuperalgebra, is commutative. Such a Borel subsuperalgebra does not exist for $\mathfrak{g} = osp(m+n\varepsilon)$ with $m\neq 2$, $q(m)$ with $m \geq 2$. (The same effect arises already in the regular case, [P].) It is an open question if the claim of Theorem 2 is true for $osp(m+n\varepsilon)$, $m\neq 2$, or $q(m)$, $m\geq 2$ (in the latter case one has to define the set R using an odd invariant form on $\mathfrak{g}$).

For $\mathfrak{g} = gl(m+n\varepsilon)$, $osp(2+n\varepsilon)$, $\mathfrak{p}(m)$, the result of Theorem 2 is new only for an atypical generic η (for a typical generic η it is an immediate consequence of the main theorem of Chapter 5 in [P]). For $\mathfrak{g} = \mathfrak{p}(m)$, although generic weights are regular, the claim of Theorem 2 is not an obvious consequence of the results of [P] since an analogue of Theorem 1(b) for $G = \mathbb{P}(m)$ is still not known. If V is a highest weight module,

Theorem 2 is a direct generalization of the Bernstein-Leites formula. Indeed if λ is the b-highest weight of V, (1) takes the form

$$gr^{0}V = \bigoplus_{\underline{\ell}\in\{0,1\}^{r}} \varepsilon^{|\underline{\ell}|}\cdot V_{b_0}(\lambda - \sum_{i=1}^{r} \ell_i\alpha_i) \quad , \tag{2}$$

and for a generic V with dim V < ∞, (2) is equivalent to the formula

$$\text{ch } V = \frac{\sum_{w\in W} \text{sgn } w\cdot w(e^{\lambda+\rho_0}\cdot \prod_{\alpha_i\in R}(1+\varepsilon e^{-\alpha_i}))}{\sum_{w\in W} \text{sgn } w\, e^{w(\rho_0)}} \quad . \tag{3}$$

For g = gl(m+nε), osp(2+nε), (3) is exactly the formula proposed by J. Bernstein and D. Leites in [BL]. This formula has been known for all cases considered in Theorem 2 except the case of gl(m+nε) with m>1, n>1 (for g = gl(1+nε), gl(m+ε), osp(2+nε) and b = b_d (3) is moreover true for *any* irreducible finite dimensional representation, see [BL],[Le2]).

REFERENCES

[B] N. Bourbaki, *Groupes et Algèbres de Lie, chapitres VII-VIII*, (Hermann, Paris, 1975).

[BB] A. Beilinson and J. Bernstein, "Localisation de g-modules," *C. R. Acad. Sci. Paris* 292 (1981), 15-18.

[BL] J. Bernstein and D. Leites, "A character formula for irreducible finite dimensional modules over the Lie superalgebras of series gl and sl," *C. R. Acad. Sci. Bulg.* (in Russian) 33 (1980), 1049-1051.

[Dix] J. Dixmier, *Algèbres Enveloppantes*, (Gauthier-Villars, Paris, 1974).

[JHKT] J. van der Jeugt, J.W.B. Hughes, R. C. King, and J. Thierry-Mieg, "A character formula for singly atypical modules of the Lie superalgebra sl(m/n)," preprint 1989.

[K1] V. Kac, "Lie superalgebras," *Advances in Math.* 26 (1977), 8-96.

[K2] V. Kac, "Characters of typical representation of classical Lie superalgebras," *Comm. Alg.* 5 (1977), 889-897.

[K3] V. Kac, "Representations of classical Lie superalgebras," *Lecture Notes in Math.* 676 (Springer-Verlag, Berlin, 1978), 597-626.

[Le1] D. Leites, "Lie superalgebras," *JOSMAR* 30, no. 6 (1984), 2482-2513.

[Le2] D. Leites, "A character formula for irreducible finite dimensional representations of the Lie superalgebras of series C," *C. R. Acad. Sci. Bulg.* (in Russian) 30 (1980), 1053-1055.

[P] I. Penkov, "Geometric representation theory of complex classical Lie supergroups," to appear in *Asterisque.*

[PS1] I. Penkov and V. Serganova, "Cohomology of G/P for classical Lie supergroups and characters of some atypical G-modules," *Ann. Inst. Fourier* 39 (1989), 845-873.

[PS2] I. Penkov and V. Serganova, "Character formulas for some classes of atypical gl(m+nε) and p(m)-modules," *Lett. Math. Phys.* 16 (1988), 251-261.

[PS3] I. Penkov and V. Serganova, "Irreducible modules over classical Lie superalgebras," in preparation.

[Sch1] M. Scheunert, "The theory of Lie superalgebras," *Lecture Notes in Math.* 716, (Springer-Verlag, Berlin, 1976).

[Sch2] M. Scheunert, "Invariant supersymmetric multilinear forms and the Casimir elements of P-type Lie superalgebras," *J. Math. Phys.* 28 (1987), 1180-1191.

Krichever Construction of Solutions to the Super KP Hierarchies

Jeffrey M. Rabin

Dept. of Mathematics, University of California at San Diego, La Jolla, CA 92093-0112 USA (jrabin@ucsd.edu)

Abstract: A super Krichever construction is used to produce solutions to the various super Kadomtsev-Petviashvili (SKP) hierarchies from geometric data consisting primarily of a suitable algebraic supercurve of genus g (generally *not* a super Riemann surface) and a line bundle of degree $g-1$ on it. The known SKP hierarchies deform both the supercurve and the bundle, in contrast to ordinary KP which deforms bundles but not curves, and are distinguished by the specific deformations they implement. A new SKP hierarchy is introduced which deforms the bundle only.

Introduction

This lecture concerns the geometric interpretation of the flows described by various supersymmetric versions of the KP hierarchy. The KP hierarchy itself is of physical interest in connection with the operator formalism for conformal field theory and string theory [1], and, recently, the matrix model approach to two-dimensional quantum gravity [2]. Geometrically, solutions to the KP hierarchy are constructed from "Krichever data" consisting primarily of an algebraic curve (Riemann surface) of genus g and a line bundle of degree $g-1$ on it [3]. The KP equations describe deformations of the line bundle on a fixed curve, that is, they are flows on the Picard variety or Jacobian of the curve. The "Krichever map" assigns a point of an infinite-dimensional Grassmannian to the Krichever data and gives an alternative interpretation of KP as flows on this Grassmannian.

There are at least two supersymmetric versions of the KP hierarchy: that of Manin–Radul (MRSKP hierarchy), expressed in terms of Lax equations for a pseudodifferential operator [4], and that of Kac–van de Leur (KVSKP hierarchy), given as Hirota bilinear equations for a tau function [5], and part of our problem is to understand the relation between them geometrically. LeClair's version [6] is quite similar to KVSKP and will not be discussed separately. After a review of the standard KP theory, we will begin with the Krichever construction of solutions to MRSKP. The necessary geometric data involve line bundles of degree $g-1$ on genus-g algebraic supercurves, or $(1|1)$-dimensional supermanifolds. Because of the appearance of the supersymmetric derivative operator D throughout the MRSKP theory, it has been generally assumed that the relevant supercurves would be super Riemann surfaces, but we will see that this is *not* the case. The true role of D in the theory is to generate coupled deformations of both the supercurve and the line bundle on it, in contrast to the ordinary KP theory. Thus, the flows take place in the universal Jacobian over the moduli space of all appropriate supercurves.

Next, we introduce two new SKP hierarchies. The "Jacobian" hierarchy is the best geometric analogue of ordinary KP in that it deforms line bundles on a fixed curve. It has been discussed independently by Mulase [7], who showed that it solves the Schottky-type problem of characterizing the Jacobians of supercurves. The "maximal" SKP hierarchy includes the equations of the Jacobian hierarchy, as well as additional flows deforming the supercurve while fixing the bundle. Because there is no canonical notion of a fixed bundle on a family of curves, this requires additional choices which amount to a connection on the universal Jacobian. Neither of these SKP hierarchies admits a Lax formulation as does MRSKP, indicating that the latter should be the hierarchy arising in two-dimensional quantum supergravity. The maximal hierarchy will be argued to be equivalent to KVSKP, providing the geometric interpretation of that system. Because deformations of curve and bundle do not generally commute, this hierarchy is not integrable. Several open problems, including that of demonstrating this equivalence explicitly, will be posed. A more detailed exposition of these results can be found in [8].

Review of KP Theory

We begin with Krichever data consisting of: M, a compact Riemann surface of genus g; p, a point of M; z, a local coordinate vanishing at p; $\mathcal{L}$, a line bundle on M satisfying the cohomology conditions $H^0(M,\mathcal{L}) = H^1(M,\mathcal{L}) = 0$; and ϕ, a local trivialization of $\mathcal{L}$ in a neighborhood $U = \{|z| < 1\}$ of p (an identification of sections of $\mathcal{L}$ with ordinary functions, in U). The significance of the cohomology conditions follows from the Riemann–Roch theorem,

$$\dim H^0(M,\mathcal{L}) - \dim H^1(M,\mathcal{L}) = \deg \mathcal{L} + 1 - g, \tag{1}$$

which implies that $\deg \mathcal{L} = g - 1$. An example of such an $\mathcal{L}$ would be one of the spin bundles, or square roots of the canonical bundle of M. By assumption $\mathcal{L}$ has no holomorphic sections, but meromorphic sections with poles at p only can be studied by tensoring $\mathcal{L}$ in the Riemann–Roch theorem with $\mathcal{O}(np)$ (a standard bundle with transition function z^n) to obtain

$$\dim H^0(M,\mathcal{L}\otimes\mathcal{O}(np)) - \dim H^1(M,\mathcal{L}\otimes\mathcal{O}(np)) = n, \tag{2}$$

which implies that $\dim H^0(M,\mathcal{L}\otimes\mathcal{O}(np)) = n$. That is, the space $W \equiv H^0(M\backslash p,\mathcal{L})$ of sections of $\mathcal{L}$ which are holomorphic except for possible poles at p has a basis consisting of sections with the behaviors z^{-n} near p for each $n > 0$. A fancier statement of the same fact is that W is a point in the Grassmannian of all subspaces of $\text{span}\{z^n, n \in \mathbf{Z}\}$ which project isomorphically onto the subspace $H_+ = \text{span}\{z^{-n}, n > 0\}$ via the obvious projection.

Now we wish to describe deformations of $\mathcal{L}$ via their effect on W. M is covered by two regions, U and $M\backslash p$, intersecting in $U\backslash p$. Each is a Stein manifold, so $\mathcal{L}$ is trivial on each region and is completely described by its transition function $h(z)$ in the overlap $U\backslash p$ or on the circle ∂U. It can be deformed by simply changing $h(z)$. Convenient deformation parameters x, t_n are introduced by the change

$$h(z) \to G(z)h(z), \quad G(z) = \exp(xz^{-1} + \sum_{n=1}^{\infty} t_n z^{-n}). \tag{3}$$

The crucial features of this parametrization are the following. $G(z)$ is an eigenfunction of $\partial = d/dx$ with eigenvalue z^{-1}. In the language of quantum field theory, x serves as a source for generating insertions of z^{-1}, allowing functions of z to be replaced by pseudodifferential operators in x. Similarly, t_n is a source for insertions of z^{-n}, or equivalently of the composite operator ∂^n.

The semicontinuity theorem of algebraic geometry guarantees that our results on the structure of the vector space W generically continue to hold as we deform $\mathcal{L}$. So consider the unique section of the deformed bundle with behavior z^{-1} + holomorphic near p; represent it by a function $s(z, x, t)$ in U and a function $w(z, x, t)$ in the other chart $M\backslash p$, so that $w = Ghs$. [Because $h(z)$ is independent of x, t it is customary in the literature to remove this factor and work with the Baker-Akhiezer function $h^{-1}w = Gs$. We retain it to emphasize that w is a section of $\mathcal{L}$.] Note that the derivatives $\partial^k w, k \geq 0$ behave as z^{-k-1} near p and so are representatives in the chart $M\backslash p$ of a complete basis of sections in W. With respect to this basis, any element of W is represented by a differential operator acting on w.

We now use the source x to replace the z-dependence of s by a pseudodifferential operator $S = 1 + a_1(x,t)\partial^{-1} + a_2(x,t)\partial^{-2} + \cdots$ such that

$$s \exp xz^{-1} = z^{-1} S \exp xz^{-1}, \tag{4}$$

or equivalently

$$w = z^{-1} hSG. \tag{5}$$

From this we compute

$$\begin{aligned} \frac{\partial w}{\partial t_n} &= z^{-1} h(\frac{\partial S}{\partial t_n} G + S z^{-n} G) \\ &= z^{-1} h(\frac{\partial S}{\partial t_n} S^{-1} SG + S\partial^n S^{-1} SG) \\ &= (\frac{\partial S}{\partial t_n} S^{-1} + S\partial^n S^{-1}) w \\ &= B_n w, \end{aligned} \tag{6}$$

where B_n is a pure (not pseudo) differential operator since $\partial w/\partial t_n$ belongs to W. Because $(\partial S/\partial t_n)S^{-1}$ involves negative powers of ∂ only, these must precisely cancel the negative powers in $S\partial^n S^{-1}$, denoted $(S\partial^n S^{-1})_-$, which determines B_n as its differential operator part, $(S\partial^n S^{-1})_+$. Thus we find that S solves the KP hierarchy in the form

$$\frac{\partial S}{\partial t_n} = -(S\partial^n S^{-1})_- S. \tag{7}$$

The more familiar Lax form is obtained by defining a first-order pseudodifferential operator $L = \partial + u_1\partial^{-1} + u_2\partial^{-2} + \cdots$ by $L = S\partial S^{-1}$ and verifying that

$$\frac{\partial L}{\partial t_n} = [L_+^n, L]. \tag{8}$$

In these equations, the pseudodifferential operators are manipulated using the generalized Leibniz rule

$$\partial^{-1}(fg) = \sum_{i=0}^{\infty}(-1)^i f^{(i)}\partial^{-1-i}g. \tag{9}$$

The conclusion is that the Krichever construction produces orbits of the KP flows which can be identified as the sets $\mathrm{Pic}^{g-1}(M)$ of line bundles of degree $g-1$ on a fixed Riemann surface M.

The MRSKP Hierarchy

The philosophy behind the MRSKP hierarchy is to repeat the classical KP theory as closely as possible, replacing $\partial = d/dx$ by the supersymmetric derivative operator $D = \partial_\xi + \xi\partial_x$. Thus we begin with the following super Krichever data: M, a compact complex supermanifold of dimension $(1|1)$ and genus g but *not* necessarily a super Riemann surface; p, a point of M; (z,θ), local coordinates near p; $\mathcal{L}$, a line bundle on M satisfying $H^0(M,\mathcal{L}) = H^1(M,\mathcal{L}) = 0$; and ϕ, a local trivialization of $\mathcal{L}$ in $U = \{|z| < 1\}$. We initially assume that M is a single (i.e., split, having no supermoduli) supermanifold; shortly we will allow it to be a family of (nonsplit) supermanifolds with transition functions parametrized by finitely many odd parameters β_i. Then the super Riemann–Roch theorem [9]

$$\dim H^0(M,\mathcal{L}) - \dim H^1(M,\mathcal{L}) = (\deg\mathcal{L} + 1 - g|\deg\mathcal{L} + \deg\mathcal{E} + 1 - g) \tag{10}$$

holds, where $\mathcal{E}$ is the line bundle on the Riemann surface M_{red} whose sections $g(z)$ define odd functions $\theta g(z)$ on M. For M a super Riemann surface, $\mathcal{E}$ would be a spin bundle, of degree $g-1$, but here our cohomology assumptions clearly imply that $\deg\mathcal{L} = g-1$ and $\deg\mathcal{E} = 0$, so that M cannot be a super Riemann surface except in the special case $g = 1$. Tensoring with $\mathcal{O}(np)$ gives (here p really means the divisor $z = 0$)

$$\dim H^0(M,\mathcal{L}\otimes\mathcal{O}(np)) - \dim H^1(M,\mathcal{L}\otimes\mathcal{O}(np)) = (n|n), \tag{11}$$

implying that the space $W \equiv H^0(M\backslash p,\mathcal{L})$ of sections of $\mathcal{L}$ holomorphic except for poles at p has a basis of even and odd sections with behavior z^{-n} and θz^{-n} for every $n > 0$. More pompously, W is a point in the Grassmannian of all subspaces of span$\{z^n, \theta z^n, n \in \mathbf{Z}\}$ which project isomorphically onto the subspace $H_+ =$ span$\{z^{-n}, \theta z^{-n}, n > 0\}$ via the obvious projection [10].

In the nonsplit case, the super Riemann–Roch theorem does not always hold, and spaces of sections such as W do not always have a free basis. The typical pathology is that W contains no section with behavior, say, θz^{-1}, but does contain one with behavior $\beta_1 \theta z^{-1}$. However, one can show that the our cohomology assumptions are strong enough to prevent this pathology, so the results on the structure of W do extend to the nonsplit case [8,11]. This is actually necessary for the consistency of our theory, since we will see that the MRSKP flows can deform a split supermanifold to a nonsplit one.

Now we wish to deform $\mathcal{L}$ by multiplying its transition function $h(z,\theta)$ by some $G(z,\theta)$ containing a factor $\exp(xz^{-1} + \xi\theta)$ to provide sources x and ξ for z^{-1} and θ. However, this time such a factor is *not* an eigenfunction of D:

$$D \exp(xz^{-1} + \xi\theta) = (\theta + \xi z^{-1}) \exp(xz^{-1} + \xi\theta), \tag{12}$$

where $\theta + \xi z^{-1}$ is not an eigenvalue because it depends on the variable ξ on which D acts. In fact D is a nonintegrable vector field having *no* nontrivial eigenfunctions: $D\psi = \lambda\psi$ has only trivial solutions because $D^2 \neq 0$ whereas $\lambda^2 = 0$ for an odd constant. A somewhat unconventional resolution of this problem is to allow *operator-valued* eigenvalues:

$$D \exp(xz^{-1} + \xi\theta) = (\theta - z^{-1}\partial_\theta) \exp(xz^{-1} + \xi\theta), \tag{13}$$

where the "eigenvalue" is indeed independent of the variables on which D acts. Then $G(z,\theta)$ should also include sources t_n (even or odd according as n is) for insertions of the eigenvalues of D^n, so we take

$$G(z,\theta) = \exp \sum_{n=1}^{\infty} [t_{2n} z^{-n} + t_{2n-1}(\theta z^{-n+1} - z^{-n}\partial_\theta)] \exp(xz^{-1} + \xi\theta) \tag{14}$$

and deform $h \to hG$. The resulting relationship between the functions $s(z,\theta,x,\xi,t)$ and $w(z,\theta,x,\xi,t)$ representing the basic section with behavior z^{-1} in the two charts U and $M \backslash p$ will be

$$\begin{aligned} w(z,\theta,x,\xi,t) = {} & \exp \sum_{n=1}^{\infty} (t_{2n} z^{-n} + t_{2n-1}\theta z^{-n+1}) \exp[xz^{-1} + \xi(\theta - \sum_{k=1}^{\infty} t_{2k-1} z^{-k})] \\ & \times h(z,\theta) s(z, \theta - \sum_{k=1}^{\infty} t_{2k-1} z^{-k}, x, \xi, t). \end{aligned} \tag{15}$$

This equation makes it clear that two distinct types of deformation have been performed: a multiplicative change in the transition function of $\mathcal{L}$, and an additive shift in the patching of the θ coordinate across the boundary of U. Because the shift of θ is singular at $z = 0$, it cannot be absorbed in a holomorphic redefinition of the θ coordinate and represents a genuine change in the supermoduli of the curve M. Such changes are familiar in the operator formalism of (super)conformal field theory, where they are generated by the stress tensor (Virasoro algebra) of a conformal field theory on M [1,12]. Thus we have simultaneously deformed the supercurve and the line bundle on it, and these deformations are coupled through the use of the t_{2n-1} to parametrize both. Note that the shift of the θ coordinate is absent in two factors in (15). In the first exponential this shift would actually produce a quadratic term in the odd parameters t_{2n-1} which vanishes by symmetry: these special deformations of curve and bundle parametrized by the same set of odd parameters actually commute with one another, in contrast to such deformations with independent parameters. The θ shift is also absent from the factor $h(z,\theta)$; this was arranged by fiat to lead to the correct MRSKP equations, but has a geometric interpretation which will be given later.

Once again we replace the dependence of s on z and θ by pseudodifferential operators with respect to the sources x and ξ, defining the pseudodifferential operator $S = 1 + s_1 D^{-1} + s_2 D^{-2} + \cdots$ by

$$s \exp(xz^{-1} + \xi\theta) = z^{-1} S \exp(xz^{-1} + \xi\theta). \tag{16}$$

Combining this with (15) yields

$$\begin{aligned} w(z,\theta,x,\xi,t) =& z^{-1} h(z,\theta) S \exp \sum_{n=1}^{\infty} (t_{2n} z^{-n} + t_{2n-1}\theta z^{-n+1}) \\ & \times \exp[xz^{-1} + \xi(\theta - \sum_{k=1}^{\infty} t_{2k-1} z^{-k})]. \end{aligned} \tag{17}$$

The same computation as before shows that S, or equivalently $L \equiv SDS^{-1} = D + u_1 + u_2 D^{-1} + \cdots$ satisfies the MRSKP hierarchy [13],

$$\begin{aligned} \frac{\partial S}{\partial t_{2n}} &= -L_-^{2n} S, \\ \frac{\partial S}{\partial t_{2n-1}} &= -(L_-^{2n-1} + \sum_{k=1}^{\infty} t_{2k-1} L_-^{2n+2k-2}) S, \end{aligned} \tag{18}$$

or, in Lax form,

$$\begin{aligned} \frac{\partial L}{\partial t_{2n}} &= [L_+^{2n}, L], \\ \frac{\partial L}{\partial t_{2n-1}} &= [L_+^{2n-1}, L] - 2L^{2n} + \sum_{k=1}^{\infty} t_{2k-1} [L_+^{2n+2k-2}, L]. \end{aligned} \tag{19}$$

The brackets in these equations are supercommutators, and the pseudodifferential operators are manipulated according to $D^{-1} = D\partial_x^{-1}$ and the generalized Leibniz rule (9). These solutions to the MRSKP hierarchy are thus interpreted as flows in the universal Picard bundle $\mathcal{P}$ over the moduli space of $(1|1)$-dimensional supermanifolds having $\deg \mathcal{E} = 0$, whose fiber over any particular supermanifold M

is the set $\mathrm{Pic}^{g-1}(M)$ of line bundles of degree $g-1$ on that supermanifold. The even flows are purely vertical, deforming $\mathcal{L}$ but not M, while the odd flows are in a specific diagonal direction, deforming both.

One may ask how this diagonal direction is defined geometrically in view of the fact that there is no canonical horizontal direction in a bundle such as $\mathcal{P}$; concretely, there is no canonical notion of deforming M while holding $\mathcal{L}$ fixed. The answer is that by construction we hold the transition function $h(z,\theta)$ of $\mathcal{L}$ fixed while deforming M, and the diagonal MRSKP flow is defined relative to this horizontal direction. The choice of the initial transition function for $\mathcal{L}$ thus defines a connection on $\mathcal{P}$, and different choices distinguish different solutions of (18) with the same initial data. This is in contrast to the ordinary KP theory, which requires only a choice of a local trivialization ϕ, which is less data than the choice of an initial transition function for $\mathcal{L}$. The dependence on this choice does, however, disappear in the split case, where in the absence of additional odd parameters the even function $h(z)$ cannot depend on θ and so can never change as we deform M. This is consistent with the unique solvability of the initial-value problem proven in [13] in the split case.

New SKP Hierarchies

Once the geometric interpretation of the MRSKP flows is understood, it is easy to construct new SKP hierarchies implementing alternative deformations of the geometric data by altering the choice of $G(z,\theta)$. We can obtain one which is a true geometric analog of ordinary KP in the sense that it deforms $\mathcal{L}$ but not M simply by omitting the shift of θ from (15). The resulting "Jacobian" SKP hierarchy is

$$\begin{aligned} \frac{\partial S}{\partial t_{2n}} &= -(SD^{2n}S^{-1})_{-}S, \\ \frac{\partial S}{\partial t_{2n-1}} &= -(S\partial_\xi \partial_x^{n-1} S^{-1})_{-}S \\ &= -[S(D^{2n-1} - \xi D^{2n})S^{-1}]_{-}S. \end{aligned} \tag{20}$$

Mulase [7] has also introduced and studied this new integrable system, showing (in the split case) that it indeed solves the Schottky-type problem of characterizing the Jacobians of supercurves as the finite-dimensional orbits of this system.

Note that even if we introduce $L \equiv SDS^{-1}$, so that $L^n = SD^nS^{-1}$, it is not true that $\xi L^n = S\xi D^n S^{-1}$, because S contains ∂_ξ and will not commute with ξ, so that there is no way to rewrite this system in a Lax form involving only L. This is easily related to the fact that these flows deform only $\mathcal{L}$. The MRSKP hierarchy has a Lax formulation because all flows are expressed in terms of powers of D, which become powers of L after conjugation by S. The even flows can be expressed as simple powers of the odd ones precisely because D is a nonintegrable vector field, which was the property which led to coupled deformations of both M and $\mathcal{L}$. Thus, MRSKP is distinguished as the only SKP system with a simple

Lax form. Since the connection of KP with 2d quantum gravity is made via the Lax form [2], it is natural to expect that the SKP system relevant to 2d quantum supergravity should be that of Manin–Radul [14].

We can enlarge the Jacobian SKP system to a "maximal" SKP hierarchy by including also deformations of M while keeping $\mathcal{L}$ fixed (as explained previously, this requires a choice of a horizontal direction on $\mathcal{P}$). Performing the deformation $\theta \to \theta - \sum_{n=1}^{\infty} \hat{t}_{2n-1} z^{-n}$ alone leads to the flow equation

$$\frac{\partial S}{\partial \hat{t}_{2n-1}} = -(S\xi D^{2n} S^{-1})_{-} S. \tag{21}$$

However, M can also be deformed by a multiplicative change of θ, say $\theta \to \theta \exp \sum_{n=1}^{\infty} \hat{t}_{2n} z^{-n}$, which amounts to a deformation of the bundle $\mathcal{E}$ characterizing M. This leads to

$$\frac{\partial S}{\partial \hat{t}_{2n}} = -(S\xi D^{2n+1} S^{-1})_{-} S. \tag{22}$$

The maximal hierarchy is not integrable, since deformations of M and $\mathcal{L}$ having independent parameters do not commute, and admits no Lax form. It describes all deformations of M and $\mathcal{L}$ which do not alter the patching of the even coordinate z. Technically, if M was initially a "projected" supermanifold, this property will be preserved under such deformations.

Like the maximal system, the KVSKP hierarchy has two infinite sets of even and odd flow parameters. Instead of a flow on a pseudodifferential operator S, it is given as a flow on a super τ function which is just the superdeterminant of the natural projection map from W to H_+. Using the work of Kac and van de Leur [5] and of Bergvelt [15], one can give a general argument for the equivalence of these hierarchies. The key is to view deformations of M or of $\mathcal{L}$ as acting directly on the space of sections W of $\mathcal{L}$ inside U (hold each section fixed outside U as the deformation proceeds, forcing it to change inside), and thereby on τ. In the notation of Kac and van de Leur, the deformations of our maximal hierarchy have infinitesimal generators denoted as:

$\lambda(-n) + \mu(-n)$, acting as z^{-n}, generating the t_{2n} deformation.

$\mu(-n)$, acting as $z^{-n}\theta\partial_\theta$, generating the $\hat{t}_{2n}$ deformation.

$e(-n)$, acting as θz^{-n+1}, generating the t_{2n-1} deformation.

$f(-n+1)$, acting as $z^{-n}\partial_\theta$, generating the $\hat{t}_{2n-1}$ deformation.

The KVSKP hierarchy is

$$\sum_{i \in \frac{1}{2}\mathbf{Z}} (-1)^{2i} \psi_i \otimes \psi_i^* (\tau \otimes \tau) = 0, \tag{23}$$

where the ψ_i are superfermion operators related to the superbosonic λ, μ, e, f by a super boson-fermion correspondence. This makes it clear that this hierarchy describes the same deformations as our maximal one, but the parametrization of τ is determined by the details of the superbosonization, so that these parameters do not have immediate geometric interpretations in terms of specific deformations. In principle the maximal hierarchy could be converted directly to the KVSKP form, and the relation between their parameters determined, via the equations relating τ to w given by Dolgikh and Schwarz [16], but it seems very difficult to carry this out explicitly.

Conclusions and Open Problems

All the known and new SKP systems discussed in this lecture describe deformations of a $(1|1)$-dimensional complex supermanifold M with $\deg \mathcal{E} = 0$ and a line bundle $\mathcal{L}$ of degree $g-1$ on it. This supermanifold cannot be a super Riemann surface unless $g = 1$. The various SKP hierarchies are distinguished by their integrability properties, the existence of a Lax form, and the specific deformations they implement. The Jacobian hierarchy is the true geometric analogue of ordinary KP in that it deforms $\mathcal{L}$ only, is integrable, and solves the super Schottky problem, but it has no Lax formulation. The hierarchy of Manin–Radul does have a Lax form and describes coupled deformations of M and $\mathcal{L}$ such that integrability is preserved. The maximal hierarchy is equivalent to that of Kac and van de Leur and describes all deformations of M and $\mathcal{L}$ preserving only the patching of the even coordinates of M; it is not integrable and has no Lax form.

Many open problems remain to be solved before our understanding of the SKP hierarchies matches that achieved for ordinary KP.

(1) An infinite-parameter system of flows on an infinite-dimensional Grassmannian will generically produce infinite-dimensional orbits. Those obtained by our super Krichever construction are finite-dimensional, however, being isomorphic to subsets of the universal Picard bundle $\mathcal{P}$. Prove, in analogy with ordinary KP, that in fact all the finite-dimensional orbits of the SKP hierarchies are obtained by our construction.

(2) Explicitly demonstrate the equivalence between the maximal, Kac–van de Leur, and LeClair SKP hierarchies, obtaining in particular the exact relation between our flow parameters $t_n, \hat{t}_n$ and those parametrizing the Kac–van de Leur and LeClair super τ functions. This might lead to an expression for the super τ function in terms of an appropriate super theta function.

(3) Investigate further the interpretation of the MRSKP hierarchy in the context of 2d quantum supergravity. In particular, only the even flows have been interpreted thus far. The odd flows seem incompatible with any reasonable string equation, which must be related to our observation that they deform M as well as $\mathcal{L}$.

(4) The SKP deformations make sense even for Krichever data not satisfying the cohomology conditions $H^0 = H^1 = 0$. However, they cannot be described in terms of a flow on a conventional super Grassmannian, which is defined as a set of freely generated subspaces W. Give a generalized notion of super Grassmannian whose points may be non-freely generated modules over the set of odd parameters β_i and use it to describe the general SKP flows.

(5) A significiant part of the (S)KP theory not touched on here is its relation to the classification of (super)commutative rings of (super)differential operators. The ring $H^0(M\backslash p, \mathcal{O})$ of functions on M, holomorphic except at p, can be mapped onto an isomorphic supercommutative ring of superdifferential operators by observing that for any such function f, fs is a meromorphic section of $\mathcal{L}$. Send f to the differential operator $P(f)$ associated to this section via its expansion in the basis of derivatives $D^i w$. The affine curve $M\backslash p$ can be recovered as Spec of such a ring. In ordinary KP theory it is shown that all commutative rings of differential operators are obtained by this or similar constructions. Give the corresponding classification of supercommutative rings.

(6) This is a generalization of problem 5. The ring $H^0(M\backslash p, \mathcal{O})$ can be recovered from the space of sections W as its maximal stabilizer: the largest set A_W of functions of z and θ such that $A_W W \subset W$. The idea is that the functions on M are all the objects which convert sections to sections by multiplication. Our discussion of operator-valued eigenvalues, and of the infinitesimal generators of the maximal SKP hierarchy, motivates the study of more general objects having this property; consider the maximal set $\tilde{A}_W$ of "functions" of z, θ, and ∂_θ such that $\tilde{A}_W W \subset W$. This is a nonsupercommutative ring containing A_W. Does it have an interpretation as the set of functions on some noncommutative geometric object? This may serve as a model for an analogous question which arises in 2d quantum gravity. Here, instead of commutative rings, one is interested in determining all rings generated by a pair of differential operators satisfying $[P, Q] = 1$. What geometric object, analogous to the pair $(M, \mathcal{L})$, classifies such rings [17]?

References

1. L. Alvarez-Gaumé, C. Gomez, G. Moore, C. Vafa: Nucl. Phys. **B303** 455 (1988)
2. M.R. Douglas: Phys. Lett. **B238** 176 (1990)
3. G. Segal, G. Wilson: Publ. Math. IHES **61** 5 (1985)
4. Yu. I. Manin, A.O. Radul: Commun. Math. Phys. **98** 65 (1985)
5. V.G. Kac, J.W. van de Leur: Ann. Inst. Fourier Grenoble **37** 99 (1987); idem: Adv. Ser. in Math. Phys. **7** World Scientific 1989
6. A. LeClair: "Supersymmetric KP hierarchy: free field construction," Princeton preprint PUPT-1107 (1988)
7. M. Mulase: "A new super KP system and a characterization of the Jacobians of arbitrary algebraic supercurves," U.C. Davis preprint ITD 89/90-9 (1990)
8. J.M. Rabin: "The geometry of the super KP flows," U.C.S.D. preprint (1990)
9. A.A. Rosly, A.S. Schwarz, A.A. Voronov: Commun. Math. Phys. **119** 129 (1988)
10. A.S. Schwarz: Nucl. Phys. **B317** 323 (1989)
11. L. Hodgkin: "Problems of fields on super Riemann surfaces," to appear in J. Geom. and Phys.
12. L. Alvarez-Gaumé, C. Gomez, P. Nelson, G. Sierra, C. Vafa: Nucl. Phys. **B311** 333 (1988)
13. M. Mulase: Inv. Math. **92** 1 (1988)

14. P. Di Francesco, J. Distler, D. Kutasov: "Superdiscrete series coupled to 2d supergravity," Princeton preprint PUPT-1189 (1990);
M. Kreuzer, R. Schimmrigk: "Nonperturbative string equations from the generalized super-KdV hierarchy," U.C.S.B. preprint NSF-ITP-90-119 (1990)
15. M.J. Bergvelt: "Infinite super Grassmannians and super Plücker equations," preprint (1988)
16. S.N. Dolgikh, A.S. Schwarz: "Supergrassmannians, super τ-functions and strings," preprint IC/89/48 (1989)
17. G. Moore: "Geometry of the string equations," Yale preprint YCTP-P4-90 (1990)

The structure of supersymplectic supermanifolds

Mitchell Rothstein
Department of Mathematics
University of Georgia
Athens, GA 30602

The purpose of this lecture is to give an explicit description of supersymplectic supermanifolds in the C^∞ case. There will also be a few remarks about the holomorphic case, and about geometric quantization in the super setting. The main results may be viewed as a kind of "symplectic Batchelor's theorem." Batchelor's theorem states that any C^∞ supermanifold is isomorphic to the exterior sheaf of the sheaf of sections of a vector bundle [B] . The isomorphism is far from canonical, but useful nevertheless. Indeed, one often deals in practice with a supermanifold which is given explicitly as the exterior sheaf of a vector bundle, and in such a case it is natural to ask for the meaning of supergeometric objects in terms of the geometry of the original manifold and bundle. In the case of a supersymplectic supermanifold, i.e. a supermanifold equipped with a closed non-degenerate even two-form, the result is that, up to a suitable notion of equivalence, supersymplectic structures correspond in a natural way to the data

$$(M, \omega, E, g, \nabla), \tag{1}$$

where (M, ω) is a symplectic manifold, E is a smooth vector bundle over M, g is a nondegenerate metric on E, and ∇ is a connection on E, compatible with the metric.

First let us recall that a C^∞ *supermanifold* is by definition a ringed space $(M, \mathcal{A})$, $\mathcal{A}$ being a sheaf of **R**-algebras, such that

1. $\mathcal{A}$ is supercommutative.
2. $\mathcal{A}/\mathcal{N}$ gives M the structure of a C^∞ manifold. Here $\mathcal{N}$ is the sheaf of nilpotents.
3. $\mathcal{N}/\mathcal{N}^2$ is a locally free sheaf of C^∞-modules, and $\mathcal{A}$ is locally isomorphic to $\Lambda(\mathcal{N}/\mathcal{N}^2)$.

One inherits the elementary definitions from differential calculus, such as the tangent sheaf,

$$\mathcal{X}(\mathcal{A}) = \text{ the sheaf of graded derivations of } \mathcal{A}, \tag{2}$$

the cotangent sheaf,

$$\Omega^1(\mathcal{A}) = \text{ dual of } \mathcal{X}(\mathcal{A}), \tag{3}$$

and the exterior derivative,

$$\Omega^n(\mathcal{A}) \xrightarrow{d} \Omega^{n+1}(\mathcal{A}). \tag{4}$$

As usual, $d^2 = 0$, and the Poincaré lemma holds [K]. A *supersymplectic supermanifold* is then a triple $(M, \mathcal{A}, \omega)$, where $(M, \mathcal{A})$ is a C^∞ supermanifold, and ω is a section of $\Omega^2(\mathcal{A})_{\text{even}}$, such that ω is nondegenerate, and $d\omega = 0$. One then has the usual apparatus of classical mechanics (Poisson brackets, Hamiltonian vector fields, ...).

Now let

$$\mathcal{E} = \mathcal{N}/\mathcal{N}^2. \tag{5}$$

Batchelor's theorem states that

$$\mathcal{A} \approx \Lambda\mathcal{E}. \tag{6}$$

The statement that this isomorphism is not canonical means that there are morphisms in the category or supermanifolds which are not bundle morphisms. Nevertheless, if one considers only those automorphisms of $\Lambda\mathcal{E}$ which are induced from bundle automorphisms of $\mathcal{E}$, then one can define a subsheaf of the sheaf of differential forms on $\Lambda\mathcal{E}$ as follows: Given a local set of generators for $\mathcal{E}$, say $\theta^1, ...\theta^r$, define

$$\Omega_2^n(\Lambda\mathcal{E}) = \Lambda^2\mathcal{E} \cdot \Omega^n M + \Lambda^1\mathcal{E} \cdot (d\theta) \cdot \Omega^{n-1} M + (d\theta d\theta) \cdot \Omega^{n-2} M. \tag{7}$$

Here $(d\theta)$ and $(d\theta d\theta)$ refer to the $C^\infty(M)$ span of $\{d\theta^a\}$ and $\{d\theta^a d\theta^b\}$ respectively. It is clear that these sheaves are invariantly defined.

The main points are

1. There is a natural one-to-one correspondence between the data (1) and supersymplectic structures of the form

$$\tilde{\omega} = \pi^*(\omega) + d\alpha, \tag{8}$$

where π is the projection corresponding to the inclusion

$$C^\infty \xrightarrow{\pi^*} \Lambda\mathcal{E} \tag{9}$$

and α is a section of $\Omega_2^1(\Lambda\mathcal{E})$.

2. Given any supersymplectic form $\tilde{\omega}$ on the supermanifold $(M, \Lambda\mathcal{E})$, there is a superdiffeomorphism

$$\tau : (M, \Lambda\mathcal{E}) \to (M, \Lambda\mathcal{E}) \tag{10}$$

such that

a. $\tau \equiv \text{id} \bmod \Lambda^2\mathcal{E}$,

b. $\tau^*(\tilde{\omega})$ is of the form stated in item 1.

Condition a. means that τ induces the identity on $\Lambda\mathcal{E}/(\Sigma_{j\geq 2}\Lambda^j\mathcal{E})$. Incidentally, it happens that given any supersymplectic form $\tilde{\omega}$, there is always a one-form α, such that

$$\tilde{\omega} = \pi^*(\omega) + d\alpha, \tag{11}$$

where ω is the induced symplectic form on M.

For general background about supermanifolds, see [K], [M] or [L]. Also, for an earlier investigation into the existence of supersymplectic structures, see [Gi, *et al*].

Let us see how to construct a supersymplectic form from the data (1). Take $\mathcal{E}$ to be the sheaf of linear functionals on E. The first step is to understand the tangent sheaf, $\mathcal{X}(\Lambda\mathcal{E})$. $\mathcal{X}(\Lambda\mathcal{E})$ contains $\mathcal{E}^*$, which acts on $\Lambda\mathcal{E}$ by contraction. Moreover, if $x^1,...,x^n,\theta^1,....,\theta^r$ are a set of supercoordinates, then $\mathcal{X}(\Lambda\mathcal{E})$ is freely generated over $\Lambda\mathcal{E}$ by $\frac{\partial}{\partial x^1},...,\frac{\partial}{\partial x^n},\frac{\partial}{\partial\theta^1},...,\frac{\partial}{\partial\theta^r}$. This means that there is a short exact sequence

$$0\to\Lambda\mathcal{E}\otimes\mathcal{E}^*\to\mathcal{X}(\Lambda\mathcal{E})\to\Lambda\mathcal{E}\otimes\mathcal{X}(M)\to 0, \tag{12}$$

where $\mathcal{X}(M)$ is the sheaf of vector fields on M. So the first thing to do is to use ∇ to split sequence (12):

$$\mathcal{X}(\Lambda\mathcal{E})=\Lambda\mathcal{E}\otimes(\mathcal{E}^*\oplus\mathcal{X}M),\ \text{via }\nabla. \tag{13}$$

Now to define a supersymplectic structure $\tilde{\omega}$, it is enough to define $\tilde{\omega}(\nabla_X,\nabla_Y)$, $\tilde{\omega}(\nabla_X,\phi)$ and $\tilde{\omega}(\phi,\psi)$, where X and Y are vector fields on M and ϕ and ψ are sections of $\mathcal{E}^*$. With our conventions, $\mathcal{E}^*$ is the sheaf of sections of E.

Definition 1 *Given the data (M,ω,E,∇,g), let $R\in\Gamma(M,End\mathcal{E}\otimes\Omega^2M)$ denote the curvature of ∇, and let $\tilde{R}\in\Gamma(M,\Lambda^2\mathcal{E}\otimes\Omega^2M)$ denote the contraction of R with g:*

$$\tilde{R}=g_{bc}R^c_{ija}\theta^a\theta^b dx^i dx^j. \tag{14}$$

Define the two-form $\tilde{\omega}\in\Gamma(M,\Omega^2(\Lambda\mathcal{E}))$ by

$$\begin{aligned}\tilde{\omega}(\nabla_X,\nabla_Y) &= \omega(X,Y)+\frac{1}{2}\tilde{R}(X,Y), &(15)\\ \tilde{\omega}(\phi,\psi) &= g(\phi,\psi), &(16)\\ \tilde{\omega}(\nabla_X,\phi) &= 0. &(17)\end{aligned}$$

Theorem 1 *$\tilde{\omega}$ is a supersymplectic form. In other words, $\tilde{\omega}$ is nondegenerate and $d\tilde{\omega}=0$.*

1st Proof It is clear that $\tilde{\omega}$ is nondegenerate, for modulo nilpotents, its matrix is

$$\begin{pmatrix}\omega & 0\\ 0 & g\end{pmatrix}. \tag{18}$$

Now recall the formula for $d\tilde{\omega}$. If A, B, and C are supervectorfields, then

$$d\tilde{\omega}(A,B,C)=\sum_{A,B,C\,\text{cyclic}}(-1)^{AC}(A\tilde{\omega}(B,C)-\tilde{\omega}([A,B],C)). \tag{19}$$

$[\,,]$ is the supercommutator. One must check that this is 0 for

$$(A,B,C)=\begin{cases}(\phi,\psi,\chi)\\(\nabla_X,\psi,\chi)\\(\nabla_X,\nabla_Y,\chi)\\(\nabla_X,\nabla_Y,\nabla_Z)\end{cases}. \tag{20}$$

Now ϕ, ψ and χ are simply acting on $\Lambda\mathcal{E}$ by contraction, so in particular they annihilate $C^\infty(M)$ and they supercommute. So for the case (ϕ,ψ,χ), all six terms are 0. For the case (∇_X,ψ,χ), one has

$$\begin{aligned} d\tilde{\omega}(\nabla_X,\psi,\chi) &= Xg(\psi,\chi) - g([\nabla_X,\psi],\chi) + g([\chi,\nabla_X],\psi), \\ &= Xg(\psi,\chi) - g(\nabla_X(\psi),\chi) - g(\nabla_X(\chi),\psi), \end{aligned} \tag{21}$$

whose vanishing is the statement that the connection is compatible with the metric. For (∇_X,∇_Y,χ), one has

$$\begin{aligned} d\tilde{\omega}(\nabla_X,\nabla_Y,\chi) &= \chi\tilde{\omega}(\nabla_X,\nabla_Y) - \tilde{\omega}([\nabla_X,\nabla_Y],\chi) \\ &\quad -\tilde{\omega}([\nabla_Y,\chi],\nabla_X) - \tilde{\omega}([\chi,\nabla_X],\nabla_Y) \\ &= \frac{1}{2}\chi\cdot\tilde{R}(X,Y) - \tilde{\omega}(R(X,Y),\chi). \end{aligned} \tag{22}$$

So if $\chi = \frac{\partial}{\partial\theta^a}$ for instance, one has

$$\begin{aligned} \frac{1}{2}\frac{\partial}{\partial\theta^a}(g_{bc}R(X,Y)^c_d\theta^b\theta^d) - \tilde{\omega}(R(X,Y)^c_d\theta^d\frac{\partial}{\partial\theta^c},\frac{\partial}{\partial\theta^a}) &= \\ R(X,Y)^c_b g_{ac}\theta^b - R(X,Y)^c_b g_{ca}\theta^b &= 0. \end{aligned} \tag{23}$$

Finally,

$$d\tilde{\omega}(\nabla_X,\nabla_Y,\nabla_Z) = \sum_{cyclic}\nabla_X\tilde{R}(Y,Z) - \tilde{R}([X,Y],Z), \tag{24}$$

whose vanishing is the Bianchi identity. This completes the proof.

A less explicit but perhaps better motivated proof is the following.

2nd Proof Because we have the decomposition

$$\mathcal{X}(\Lambda\mathcal{E}) = \Lambda\mathcal{E}\otimes\mathcal{E}^*\oplus\Lambda\mathcal{E}\otimes\mathcal{X}M, \text{ via } \nabla, \tag{25}$$

we can define an operator

$$\Lambda\mathcal{E}\xrightarrow{D}\Omega^1(\Lambda\mathcal{E}) \tag{26}$$

by declaring, for $F\in\Lambda\mathcal{E}$, $\phi\in E$ and $X\in\mathcal{X}M$,

$$DF(\phi) = dF(\phi) \tag{27}$$

$$DF(\nabla_x) = 0. \tag{28}$$

Then for $f\in C^\infty$,

$$D(fF) = f\,D(F). \tag{29}$$

In terms of supercoordinates $x^1,\ldots,x^n,\theta^1,\ldots,\theta^r$, let

$$\nabla_{\frac{\partial}{\partial x^i}}(\theta^a) = A^a_{ib}\theta^b. \tag{30}$$

Then

$$Dx^i = 0 \tag{31}$$

$$D\theta^a = d\theta^a - A^a_{ib}\theta^b dx^i. \tag{32}$$

Because D is tensorial, we can define a one-form

$$\alpha_{(g,\nabla)} = g_{ab}\theta^a D\theta^b \in \Omega^1(\Lambda\mathcal{E}). \tag{33}$$

Then one may check directly that $\tilde{\omega}$ given in definition 1 is also given by

$$\tilde{\omega} = \pi^*(\omega) + d\alpha_{(g,\nabla)}. \tag{34}$$

This shows immediately that $\tilde{\omega}$ is closed, and completes the second proof.

Remarks

1. Note that $\alpha_{(g,\nabla)}$ can be constructed from any g and ∇, even if $d^\nabla g \neq 0$. But there is a map

$$\nabla \to \nabla + \frac{1}{2}g^{-1}d^\nabla g = P_g(\nabla) \tag{35}$$

which projects

$$\text{connections} \longrightarrow \text{ g compatible connections.} \tag{36}$$

Then

$$d\alpha_{(g,\nabla)} = d\alpha_{(g,P_g(\nabla))}. \tag{37}$$

2. The coordinate expression for $\tilde{\omega}$ is

$$\tilde{\omega} = (\omega_{ij} + \frac{1}{2}R^c_{ija}g_{cb}\theta^a\theta^b)dx^i dx^j + g_{ab}D\theta^a D\theta^b. \tag{38}$$

3. The Poisson structure associated to $\tilde{\omega}$ involves inverting (38). Write

$$\omega_{ij} + \frac{1}{2}g_{bd}R^d_{ija}\theta^a\theta^b = (\delta^k_i + \frac{1}{2}g_{bd}\omega^{lk}R^d_{ila}\theta^a\theta^b)\omega_{kj}, \tag{39}$$

where $\omega^{lk}\omega_{kj} = \delta^l_j$. Then we have something of the general form

$$(I+B)\omega, \tag{40}$$

where the matrix

$$B^k_i = \frac{1}{2}g_{bd}\omega^{lk}R^d_{ila}\theta^a\theta^b \tag{41}$$

has nilpotent entries. Therefore,

$$\tilde{\omega}^{-1} = [\omega^{-1}(I - B + B^2 - B^3 + \ldots)]^{ij}\nabla_i \wedge \nabla_j + g^{ab}\frac{\partial}{\partial\theta^a} \wedge \frac{\partial}{\partial\theta^b}. \tag{42}$$

In particular, if $f,g \in C^\infty(M)$, the Poisson bracket $\{f,g\}$ is determined by the first term on the right hand side of (42). One see that

$$\{f,g\}^\sim = \{f,g\} + o(R,\theta^2) + o(R^2,\theta^4) + \cdots \tag{43}$$

where $\{\ ,\ \}^{\sim}$ and $\{\ ,\ \}$ are the Poisson brackets corresponding to $\tilde{\omega}$ and ω respectively. In particular, $C^{\infty}(M)$ is closed under $\{\ ,\ \}^{\sim}$ if and only if the curvature of $\nabla = 0$.

Now we don't have all the supersymplectic forms yet. Indeed, because of the formula

$$d(f\theta) = df \cdot \theta + f d\theta, \tag{44}$$

it makes sense to speak of m-forms homogeneous of degree n in θ and $d\theta$ in a coordinate invariant way. In other words, $(\Omega^{*}(\Lambda\mathcal{E}), d)$ splits as a direct sum of complexes, $\oplus(\Omega^{*}_{n}(\Lambda,\mathcal{E}), d)$, where

$$\Omega^{m}_{n}(\Lambda\mathcal{E}) = \sum_{j=0}^{m} \Lambda^{n-j}\mathcal{E} \cdot (d\theta)^{j} \cdot \Omega^{m-j}(M). \tag{45}$$

Here, $(d\theta)^{j}$ denotes the space of j-forms which are polynomials in the $d\theta$'s with $C^{\infty}(M)$ coefficients. Moreover, from the Poincaré lemma for supermanifolds and the usual deRham theorem, it follows that the cohomology of d on global sections of $\Omega^{*}_{0}(\Lambda\mathcal{E})$ is $H^{*}(M, \mathbf{R})$, and the cohomology of d on global sections of $\Omega^{*}_{n}(\Lambda\mathcal{E})$ is 0 for $n \geq 1$. If $\tilde{\omega}$ is any supersymplectic form, set

$$\tilde{\omega} = \omega_0 + \omega_2, + \omega_4 + \cdots, \tag{46}$$

according to the decomposition (45) . Then $\omega_0 + \omega_2$ is already a supersymplectic form. If $\omega_n = 0$ for $n > 2$, let us say $\tilde{\omega}$ is of *quadratic type.* So far we have constructed only supersymplectic forms of quadratic type. The next theorem says that we have constructed all of them.

Theorem 2 *Given the manifold M and vector bundle E, there is a 1-1 correspondence between*

1. *data (ω, g, ∇), where ω is a symplectic form on M, g is a metric on E and ∇ is a g-compatible connection, and*

2. *Supersymplectic forms of quadratic type on $(M, \Lambda\mathcal{E})$.*

Proof This is a good time to emphasize the role of the projection

$$(M, \Lambda\mathcal{E}) \xrightarrow{\pi} (M, C^{\infty}). \tag{47}$$

In the category of supermanifolds, π is noncanonical. What is canonical is the inclusion

$$(M, C^{\infty}) \xrightarrow{\iota} (M, \Lambda\mathcal{E}) \tag{48}$$

corresponding to the short exact sequence

$$0 \to \mathcal{N} \to \Lambda\mathcal{E} \to C^{\infty} \to 0 \tag{49}$$

If $\tilde{\omega}$ is any 2-form on the supermanifold $(M, \Lambda\mathcal{E})$ one always obtains a two-form ω on M and a symmetric 2-tensor g on E. ω is simply the pullback of $\tilde{\omega}$ via ι,

$$\omega = \iota^{*}(\tilde{\omega}). \tag{50}$$

g is defined by

$$g(\phi,\psi) \equiv \tilde{\omega}(\phi,\psi) \text{ mod nilpotents.} \tag{51}$$

The non-canonical projection π comes into play in recovering the connection on E. We ask for a map

$$TM \xrightarrow{\nabla} T(\Lambda\mathcal{E}) \tag{52}$$

such that

1. For all $v \in TM$, $d\pi(\nabla_v) = v$.

2. For all $v \in TM$ and $e \in E$,
$$\tilde{\omega}(\nabla_v, e) = 0. \tag{53}$$

In local coordinates, condition 1. says that

$$\nabla_{\frac{\partial}{\partial x^i}} = \frac{\partial}{\partial x^i} + \lambda_i^a \frac{\partial}{\partial \theta^a}, \tag{54}$$

where λ_i^a are odd sections of $\Lambda\mathcal{E}$. Then condition 2. says that for all i and a,

$$\tilde{\omega}(\frac{\partial}{\partial x^i}, \frac{\partial}{\partial \theta^a}) + \lambda_i^b \tilde{\omega}(\frac{\partial}{\partial \theta^b}, \frac{\partial}{\partial \theta^a}) = 0. \tag{55}$$

So there is a unique solution if and only if g is nondegenerate. To summarize, one recovers from any 2-form $\tilde{\omega}$ on $(M, \Lambda\mathcal{E})$ a set of data (ω, g, ∇). The projection π is used to construct ∇, as is the assumption that g was nondegenerate. It was neither assumed that $d\tilde{\omega} = 0$, nor that $\tilde{\omega}$ was of quadratic type. It is a simple matter to check however that if $d\tilde{\omega} = 0$, then $d\omega = 0$ and $d^\nabla g = 0$. So from any supersymplectic form $\tilde{\omega}$ we obtain the data (1). If we apply definition 1 to this data, we recover $\tilde{\omega}$ precisely in the case that $\tilde{\omega}$ is of quadratic type. This completes the proof of theorem 2.

Remark
The sections λ_i^a defined in equation (54) are in general a combination of terms in all the odd exterior powers of $\mathcal{E}$. Thus in general, the connection ∇ is a connection in $\Lambda\mathcal{E}$, which does not however preserve the subsheaf $\mathcal{E}$. Nevertheless, there is a connection induced on $\mathcal{E}$:

$$\tilde{\nabla}(\theta) \equiv \nabla(\theta) \text{ mod } \Lambda^{(3)}\mathcal{E}. \tag{56}$$

To be convinced that the connection ∇ is not canonical in the category of supermanifolds, consider the following example. Associated to the supermanifold $(M, \Lambda\mathcal{E})$ is another supermanifold, namely its cotangent bundle, $T^*(\Lambda\mathcal{E})$. $T^*(\Lambda\mathcal{E})$ is characterized by the property that its fiberwise polynomial superfunctions coincide with the symmetric algebra of the tangent sheaf of $\Lambda\mathcal{E}$. It follows that $T^*(\Lambda\mathcal{E})$ has a canonical supersymplectic structure, completely analogous to the symplectic structure defined on ordinary cotangent bundles. Indeed, the Poisson structure on the cotangent bundle arises naturally from the fact that its structure sheaf is the associated **Z**-graded sheaf of the filtered sheaf of linear differential operators. Now it is not hard to see that

$$T^*(\Lambda\mathcal{E}) \simeq (T^*(M), \sigma^*\Lambda(\mathcal{E}\oplus\mathcal{E}^*)), \tag{57}$$

where σ is the projection from $T^*(M)$ to M. A choice of the isomorphism (57) will therefore induce a connection on $\sigma^*\mathcal{E}$. The locally free sheaf $\mathcal{E}$ is completely arbitrary, so we cannot expect to have such a connection defined canonically. To put the matter more clearly, consider the analogue of the preceding constructions in the holomorphic setting. For holomorphic supermanifolds, one has the *splitting problem*, which is to determine whether a given supermanifold $(M,\mathcal{A})$ is isomorphic to $(M,\Lambda\mathcal{E})$, where $\mathcal{E}$ is the conormal sheaf of $(M,\mathcal{O})$ in $(M,\mathcal{A})$. For a given locally free sheaf $\mathcal{E}$ of $\mathcal{O}$-modules on a complex manifold M, there is a class

$$a(\mathcal{E}) \in H^1(M, End\mathcal{E}\otimes\Omega^1 M), \tag{58}$$

which measures whether $\mathcal{E}$ admits a holomorphic connection. This class is typically non-zero. See [**A**]. We have

Proposition 3 *Let $\mathcal{E}$ be a locally free sheaf of $\mathcal{O}$-modules on a complex manifold M. Then the obstruction to splitting the cotangent bundle of the supermanifold $(M,\Lambda\mathcal{E})$ is precisely the obstruction to finding a holomorphic connection on $\mathcal{E}$.*

Proof Recall that from a connection on $\mathcal{E}$ one gets an identification

$$\mathcal{X}(\Lambda\mathcal{E}) = \Lambda\mathcal{E}\otimes(\mathcal{E}^*\oplus\mathcal{X}M). \tag{59}$$

This gives us an identification

$$\begin{aligned} Sym_{\Lambda\mathcal{E}}(\mathcal{X}(\Lambda\mathcal{E})) &= Sym_{\Lambda\mathcal{E}}(\Lambda\mathcal{E}\otimes(\mathcal{E}^*\oplus\mathcal{X}M)) \\ &= \Lambda\mathcal{E}\otimes\Lambda\mathcal{E}^*\otimes Sym_{\mathcal{O}}(\mathcal{X}M) = \Lambda(\mathcal{E}\oplus\mathcal{E}^*)\otimes Sym_{\mathcal{O}}(\mathcal{X}M). \end{aligned} \tag{60}$$

Equation (60) gives an isomorphism between the structure sheaf of $T^*(\Lambda\mathcal{E})$ and the sheaf $\sigma^*(\Lambda(\mathcal{E}\oplus\mathcal{E}^*))$, i.e. a splitting of the supermanifold $T^*(M,\Lambda\mathcal{E})$. Conversely, if such an splitting exists, then we get a connection on $\sigma^*\mathcal{E}$, by our previous considerations. This in turns produces a connection on $\mathcal{E}$. Indeed, if we denote by $M \xrightarrow{\iota} T^*M$ the 0-section of T^*M, and if ∇ is the connection on $\sigma^*\mathcal{E}$, then we define a connection on $\mathcal{E}$ by pullback:

$$\nabla'_v(\phi) = \nabla_{d\iota(v)}(1\otimes\phi). \tag{61}$$

This completes the proof of the proposition.

Thus, holomorphic supercotangent bundles give us a large supply of nonsplit holomorphic supermanifolds.

Returning to the main thread of the talk, we now wish to show that if $\tilde{\omega}$ is an arbitrary supersymplectic form on $\Lambda\mathcal{E}$, then it can be brought into the form described in definition 1 by a superdiffeomorphism.

Letting $\Omega_{(j)}$ denote the sum $\sum_{k\geq j}\Omega_k(\Lambda\mathcal{E})$, we set

$$\tilde{\omega} = \tilde{\omega}_2 + \rho, \tag{62}$$

where $\tilde{\omega}_2 \in \Omega_0\oplus\Omega_2$ and $\rho\in\Omega_{(4)}$.

Theorem 4 *There exists a superdiffeomorphism* $\tau : \Lambda\mathcal{E} \to \Lambda\mathcal{E}$ *such that*

$$\tau^*(\tilde{\omega}) = \tilde{\omega}_2. \tag{63}$$

Moreover, τ *is of the form*

$$\tau = e^Y, \tag{64}$$

where Y *is a derivation of* $\Lambda\mathcal{E}$ *taking* $\Lambda\mathcal{E}$ *to* $\Lambda^{(2)}\mathcal{E}$.

(The latter statement is equivalent to the requirement that τ act trivially on M and also on the bundle E, regarded as the dual bundle of the sheaf $\Lambda^{(1)}\mathcal{E}/\Lambda^{(2)}\mathcal{E}$. See [R].)

Proof We have $d\rho = 0$. Since $\rho \in \Omega_4$, we know that

$$\rho = d\alpha, \tag{65}$$

where α is a one-form lying in Ω_4. Since $\tilde{\omega}_2$ is nondegenerate, there is a unique vector field Y, such that

$$-\alpha = Y \rfloor \tilde{\omega}_2. \tag{66}$$

As with the differential forms, we may introduce subsheaves of the sheaf of vector fields:

$$\mathcal{X}_j = \Lambda^j\mathcal{E} \cdot (\frac{\partial}{\partial x} + \mathcal{E}\frac{\partial}{\partial \theta}) \tag{67}$$

$$\mathcal{X}_{(j)} = \oplus_{k \geq j}\mathcal{X}_k. \tag{68}$$

It is easy to verify that

$$\mathcal{X}_j \rfloor \Omega_k \subset \Omega_{k+j}. \tag{69}$$

In particular Y must belong to $\mathcal{X}_{(2)}$. A statement similar to (69) holds for the Lie derivative:

$$\mathcal{L}_{\mathcal{X}_j}(\Omega_k) \subset \Omega_{k+j}. \tag{70}$$

Indeed, the Lie derivative is defined in the usual way,

$$\mathcal{L}_X(\mu) = d(X \rfloor \mu) + X \rfloor d\mu, \tag{71}$$

so (70) follows from a combination of (69) and the fact that d preserves each Ω_n. Now consider the one parameter family of two-forms

$$\gamma(t) = (e^{tY})^*(\tilde{\omega}_2 + t\rho). \tag{72}$$

We compute $\gamma'(t)$:

$$\gamma'(t) = (e^{tY})^*\mathcal{L}_Y(\tilde{\omega}_2 + t\rho) + (e^{tY})^*(\rho). \tag{73}$$

But

$$\mathcal{L}_Y\tilde{\omega}_2 = d(Y \rfloor \tilde{\omega}_2) = -d\alpha = -\rho. \tag{74}$$

So

$$\gamma'(t) = t(e^{tY})^* \mathcal{L}_Y(\rho). \tag{75}$$

By the inclusion (70), $\mathcal{L}_Y(\rho) \in \Omega_{(6)}$. So $\gamma' \equiv 0 \mod \Omega_{(6)}$. Thus $\gamma(0) \equiv \gamma(1) \mod \Omega_{(6)}$, which is to say that

$$(e^Y)^*(\tilde{\omega}_2 + \rho) = \tilde{\omega}_2 + \hat{\rho}, \tag{76}$$

where $\hat{\rho} \in \Omega_{(6)}$. By continuing in this way we can push the order of ρ up to $4 + r$ where r is the rank of $\mathcal{E}$. At that point ρ is 0. The theorem is proved.

Note that the graded Darboux theorem follows quickly from the results presented here. (See **[K]** for a statement of the theorem and an indication of the proof.) Indeed, suppose one has a section

$$\alpha \in H^0(\Sigma, \Lambda^2\mathcal{E} \otimes \Omega^1 M). \tag{77}$$

By means of the given supersymplectic form $\tilde{\omega}$, α may be identified with a section

$$Y_\alpha \in H^0(\Sigma, \mathcal{X}_2(\Lambda\mathcal{E})). \tag{78}$$

Then $(e^{Y_\alpha})^*\tilde{\omega}$ produces the same metric as does $\tilde{\omega}$, but the connection ∇ is changed to $\nabla + \alpha$. In this way one may obtain any connection compatible with the metric on E. Now suppose $\theta^1, \ldots, \theta^r$ is a basis of orthonormal sections of $\mathcal{E}$, defined locally. With respect to this basis, and with a suitable choice of α, $(e^{Y_\alpha})^*\tilde{\omega}$ will produce a connection for which all the connection symbols are 0. Now choose Y as in theorem 4, so that $(e^Y)^*(e^{Y_\alpha})^*\tilde{\omega}$ is of quadratic type and agrees with $(e^{Y_\alpha})^*\tilde{\omega}$ to second order. Then, if $q^1, \ldots, q^n, p_1, \ldots, p_n$ are Darboux coordinates on M, a super Darboux coordinate system is obtained by setting

$$(\tilde{q}^i, \tilde{p}_i, \tilde{\theta}^a) = (e^Y)^*(e^{Y_\alpha})^*(q^i, p_i, \theta^a). \tag{79}$$

Finally, let's look at the geometric (pre)quantization of $(M, \tilde{\omega})$. See **[W]** for background in the non-super case. To prequantize (M, ω), we need a line bundle $\mathcal{L}$ on M, and a connection $\overset{\circ}{\nabla}$ on $\mathcal{L}$ such that

$$\text{curv } \overset{\circ}{\nabla} = \omega. \tag{80}$$

Given such, pull $\mathcal{L}$ back to $\Lambda\mathcal{E}$ (via π!):

$$\tilde{\mathcal{L}} \equiv \mathcal{L} \otimes_{C^\infty_M} \Lambda\mathcal{E}. \tag{81}$$

Then it is easy to see there is a connection $\tilde{\nabla}$ on $\tilde{\mathcal{L}}$, defined canonically in terms of $\overset{\circ}{\nabla}$, such that

$$\text{curv}(\tilde{\nabla}) = \tilde{\omega}. \tag{82}$$

One may then proceed with the usual prequantization procedure.

Example Let M be any manifold (not necessarily symplectic,) and let $(\mathcal{E}, g, \nabla)$ be a vector bundle on M with a metric and a compatible connection. Pull $(\mathcal{E}, g, \nabla)$ back to T^*M. Since T^*M has a symplectic structure, we have all the data needed to construct a supersymplectic form $\tilde{\omega}$ on $(T^*M, \Lambda\mathcal{E})$. By (38), $\tilde{\omega}$ works out to

$$\tilde{\omega} = dp_i dq^i + \frac{1}{2} R^c_{ija} g_{cb} \theta^a \theta^b dq^i dq^j + g_{ab} D\theta^a D\theta^b. \tag{83}$$

Schematically,

$$\tilde{\omega} = \left(\begin{array}{cc|c} Rg\theta^2 & I & \\ -I & 0 & 0 \\ \hline & & \\ \multicolumn{2}{c|}{0} & g \end{array} \right).$$

In this case, $\tilde{\omega}^{-1}$ is easy to calculate. We find the following hamiltonian vector fields:

$$X_{q^i} = -\frac{\partial}{\partial p_i} \tag{84}$$

$$X_{p_i} = \nabla_{\frac{\partial}{\partial q^i}} - R^c_{ija} g_{cb} \theta^a \theta^b \frac{\partial}{\partial p_j} \tag{85}$$

$$X_{\theta^a} = A^a_{ib} \theta^b \frac{\partial}{\partial p_i} + \frac{1}{2} g^{ab} \frac{\partial}{\partial \theta^b}. \tag{86}$$

This leads to Poisson brackets:

$$\{q^i, q^j\} = 0 \qquad \{p_i, p_j\} = -R^c_{ija} g_{cb} \theta^a \theta^b \tag{87}$$

$$\{p_i, q^j\} = \delta^j_i \qquad \{\theta^a, p_j\} = A^a_{jb} \theta^b \tag{88}$$

$$\{\theta^a, q^j\} = 0 \qquad \{\theta^a, \theta^b\} = \frac{1}{2} g^{ab} \tag{89}$$

In this case, $\tilde{\omega}$ is exact:

$$\alpha = p_i dq^i + g_{ab} \theta^a D\theta^b, \tag{90}$$

$$\tilde{\omega} = d\alpha. \tag{91}$$

Thus the prequantization line bundle is trivial, and we get a prequantization prescription:

$$\hat{L}_f = iX_f - \alpha(X_f) + f,$$

which works out to

$$\hat{L}_{q^j} = -i\frac{\partial}{\partial p_j} + q^j \tag{92}$$

$$\hat{L}_{p_j} = i(\nabla_{\frac{\partial}{\partial q_i}} - R^c_{kja} g_{cb} \theta^a \theta^b \frac{\partial}{\partial p_k}) \tag{93}$$

$$\hat{L}_{\theta^a} = iA^a_{jb} \theta^a \frac{\partial}{\partial p_j} + \frac{1}{2}(ig^{ab} \frac{\partial}{\partial \theta_b} + \theta^a). \tag{94}$$

One has a natural polarization, $\{\frac{\partial}{\partial p}=0\}$, giving us the quantum operators

$$L_{q^j} = q^j \quad (95)$$

$$L_{p_j} = i\nabla_{\frac{\partial}{\partial q^j}} \quad (96)$$

$$L_{\theta^a} = \frac{1}{2}(ig^{ab}\frac{\partial}{\partial\theta^b}+\theta^a) \quad (97)$$

acting on functions of q and θ.

It is interesting to note that, at least when $M=\mathbf{R}^n$, this representation of the Poisson brackets is unitarizable in the non-super sense. Simply make $\Gamma(M,\Lambda\mathcal{E})$ a normed space by

$$\alpha\cdot\beta=\int g(\alpha,\beta)d\vec{q},$$

where g is extended from $\mathcal{E}$ to $\Lambda\mathcal{E}$ in the usual (non-super) way. Then our quantum operators satisfy, for even operators, X,

$$X^*=-X$$

and for odd operators, ζ,

$$\zeta^*=i\zeta.$$

Acknowledgements: The author thanks SISSA for the hospitable environment in which this work was done.

References

[A] Atiyah, M.F., *Complex analytic connections on fibre bundles*, Trans. AMS **85** (1957),181-207.

[B] Batchelor, M, *The structure of supermanifolds*, Trans. AMS **253** (1979), 1218-1221.

[Ga] Gawedzki, K., *Supersymmetries – mathematics of supergeometry*, Ann. Inst. H. Poincaré Sect A (N.S.) **27** (1977), 335-366.

[Gi, *et al*] Giachetti, R., Ragionieri, R., Ricci, R., *Symplectic structures on graded manifolds*, J. Diff. Geo. **16** (1981), 247-253.

[K] Kostant, B., *Graded manifolds, graded Lie theory, and prequantization*, Differential Geometric Methods in Mathematical Physics (Proc. Sympos. Univ. Bonn, Bonn, 1975), Lecture Notes in Math., vol. 570, Springer-Verlag, Berlin, 1977, pp.177-306.

[L] Leites, D., *Introduction to the theory of supermanifolds*, Russian Math. Surveys **35** (1980), 1-64.

[M] Manin, Yu. I., Gauge Field Theory and Complex Geometry. Grund. der math. Wissen., vol. 289, Springer-Verlag, Berlin, 1988.

[R] Rothstein, M., *Deformations of complex supermanifolds*, Proc. AMS **95** (1985), 255-260.

[W] Woodhouse, N., Geometric Quantization. Clarendon Press, Oxford, 1980

5. Problems in quantum field theory

GAUGE FIXING : GEOMETRIC AND PROBABILISTIC ASPECTS OF YANG-MILLS GAUGE THEORIES

Gianfausto Dell'Antonio

Dipartimento di Matematica
Università di Roma , La Sapienza
C.N.R. Gruppo di Geometria e Fisica

In this lecture I will discuss briefly some recent results , obtained in part in collaboration with D.Zwanziger, about gauge fixing in non-abelian Yang-Mills theory. I will place these results in the context of the problem of providing a constructive version of (Euclidean) Yang-Mills theory in d space-time dimensions, d ≥1, by constructing a measure on a suitable function space .
Recall first that, given a Hamiltonian system ,there are essentially two ways to construct the corresponding Quantum Theory.
The first approach stresses the algebraic aspects, i.e. the Poisson structure which is given on (a subset of) continuous functions on phase space. Basic objects of the "quantized" version are elements of a C*-algebra *A* ; the canonical structure is introduced through the choice of canonical objects, which satisfy the Heisenberg commutation relations (more precisely, of objects which give a realization of the Weyl algebra in *A* or in the weak closure of a representation of *A*).Dynamics is introduced giving a (weakly continuous) one-parameter group of automorphisms of *A* (or of the weak closure of a representation of *A*).
This is the Heisenberg, or algebraic approach, which has been described in this Conference in particular by D.Kastler.
The other approach (equivalent to the first one for unconstrained Classical Mechanics) is originally due to Schroedinger and consists in introducing a concrete Hilbert space of L^2 functions on a configuration space M, relative to a given measure μ . To the classical Hamiltonian corresponds here a self-adjoint operator H constructed by analogy with its classical counterpart, making use of functions on M and of the generators of the one-parameter unitary groups of translations in M (their role is the same as that of moments in Classical Mechanics) .

For a given dynamical theory with Hamiltonian H ,the observables are the (real part of) the smallest algebra of operators on $L^2(M, \mu)$ which contains all bounded functions on M (acting by pointwise multiplication) and the spectral projections of the Hamiltonian H .
By Von Neumann's uniqueness theorem, for Classical Mechanics and in general for linear systems with a finite number of degrees of freedom, the two methods are essentially equivalent. There may be differences (and difficulties) for non-linear systems or for systems with an infinte number of degrees of freedom.
We shall be concerned here with the Schroedinger point of view.
In this context, measure theory, and more specifically probability theory enter in the following way ("Euclidean approach").
If H is non-negative (or bounded below) instead of regarding H as generator of a unitary one-parameter group, one can regard it as generator of the contraction semigroup $\exp\{-Ht\}$, $t \geq 0$ on $L^2(M,\mu)$; in cases of interest, it can happen that $\exp\{-Ht\}$ has a representative kernel $\exp\{-Ht\}(q,q')$, $q,q' \in M$, which is non-negative and defines the transition probability for a process $\{\xi_t\}$. In many cases, one can find a realization of ξ_t on a subspace K of continuous maps from R to M , endowed with a probability measure υ in such a way that ξ_t is the evaluation process (i.e. if w(t) is a continuous function, then $\xi_t(w(.))=w(t)$) . In this case, H can be reconstructed from the measure υ and therefore the entire theory is described in terms of a family of random variables (measurable functions) and of a measure υ on the space of continuous functions of time with values in M.
Quantum Mechanics and some Scalar Field Theories in $d \geq 2$ space-time dimensions are examples of a realization of this scheme.
In Quantum Mechanics (under mild conditions on the potential) M is configuration space (e.g. R^n, n=number of degrees of freedom) , K can be chosen to be the set of paths on M which are Holder-continuous of order $s<1/2$, and the measure υ is a suitable Radon-Nykodim modification of Wiener measure .
In Scalar Field Theory for space-time dimension d, M is a suitable space of distributions on R^{d-1}, ξ_t is again the evaluation process , K is a subspace of continuous functions from R to M and υ is a modification (but in general not of Radon-Nykodim type) of a natural generalization of Wiener measure (or more precisely of a measure associated to an Ohrnstein-Uhlembeck process) [1] .
In each of these cases the measure υ is constructed by a limit procedure starting from a Radon-Nykodim modification of a simpler (free) measure, along lines suggested by perturbation theory.
There is a modification of this scheme which is worth emphasizing. Instead of considering Markov Processes on can use Markov Fields and choose for K a space of distributions on R^d (e.g. a Sobolev space H_{-k} for some $k>0$) . This wider context may be preferable (or even necessary) for space-time dimensions $d \geq 3$ [2]
In this case one considers as basic measurable functions (coordinates) the elements

of the dual space H_k (if $w \in K$ and $f \in K^*$, the function $f(w)$ is continous and therefore measurable for any Borel probability measure) .
If the measure υ is constructed by some limit procedure from a sequence of measures υ_n , one may have to use a sequence of "rescaled" ("renormalized") coordinates $c_n f$, with $c_n \to 0$ or $+\infty$ for $n \to \infty$, in order to obtain in the limit for each f in K* a well defined measurable function $\phi(f)$.
The question then naturally arises: if F is a measurable function with respect to the measure υ , can it always be written as a function of the basic coordinates $\phi(f)$?
Experience suggests that for scalar field theories with polynomial interactions this is indeed the case, at least in dimensions not greater than three .
When one tries to apply a similar procedure to the construction of a Quantized Gauge Theory, one is faced with severe difficulties which are at least in part connected with the choice of the space K; indeed, if one chooses for K a space on which acts a group GT of Gauge Tranformations , it is in general impossible to construct a σ-additive measure which is gauge invariant (the group GT is in general not locally compact).
If one measures only gauge-invariant functions, one may try to construct the measure on a suitable quotient space. This is difficult to achieve, since taking the quotient requires that one works with spaces of "sufficiently smooth" functions, while experience from Field Theory suggests that for systems with infinitely many degrees of freedom the set of"sufficiently smooth functions" has zero measure for any "reasonable" measure which describes a local theory.
This suggests that from the point of view of constructive field theory the geometric aspects of a Gauge Theory are at odds with the measure-theoretic requirements.
In spite of these difficulties, for gauge theories , in particular of Yang-Mills type, it is widley accepted that the informal starting point in the choice for K be the space of distribution-valued one-forms with values in the Lie algebra of some compact Lie group, or at least a subset of this space.
In this case there are some indications [3] (but so far no proof) that, contrary to the case of scalar field theories, the gauge-invariant measurable functions are not expressible as (limits of) regular functions of the renormalized coordinates $\phi(f)$, but rather as limits of functions of the "unrenormalized" coordinates f . We shall not discuss further here this important point, and we shall concentrate the discussion on the choice to be made for K and on the properties of the set we choose.
In order to measure gauge-invariant quantities it would be sufficient to choose for K a subset Ω (not necessarily a linear subspace) of connections which has the property to have non-empty intersection with each gauge orbit . A choice of such set is called "gauge fixing" . Of course the choice of Ω must be made in such a way that a suitable measure can be defined on it.
I shall describe here one possible choice and some of its properties.
To illustrate the method I will start with a toy example which is very simple but sufficiently instructive .

<u>Toy example</u>:

Consider on C the Gauss measure with covariance one i.e

$$\upsilon(dx) = \pi^{-1}\exp\{-|x|^2\}dx \ , \ x = x_1 + ix_2$$

For group of gauge transformations we take the rotations around the origin $x \to \tau_\theta(x) = e^{i\theta}x$; the observables are then the functions of $|x|$.

The subset $R^+ = \{x \mid x_2 = 0, x_1 \geq 0\}$ is intersected once by every orbit.

The observables are in one-to-one correspondence with functions on R^+ through $f(|x|) \to f(x_1)$. With this correspondence, the measure υ takes the form

$$\upsilon_0(dx_1) = 2\, x_1 \exp[-x_1^2]dx_1.$$

Notice that R^+ is the natural space and υ_0 the natural measure when dealing only with gauge-invariant functions ; indeed , using polar coordinates, it is obvious that in this simple example υ_0 can be derived from the fact that $2\pi x_1$ is the "volume" of the gauge orbit through x_1 (and the same time the curvature of the orbit at x_1 and the Jacobian of the map from cartesian to polar coordinates) .

We give two alternative ways of constructing υ_0 , which extend to more complex cases, and in particular to models of gauge theories.

I) On continuous functions one has $\upsilon_0(f^\#) = \lim_{\varepsilon\to 0} \upsilon_\varepsilon(f)$, where $f^\#$ is the restriction of f to R^+ and $\upsilon_\varepsilon(dx) = c(\varepsilon)f(\varepsilon,x)\exp\{-F(x)\}dx$, where F is any function of class C^2 which takes its minimum value precisely where the orbit intersects R^+ , $f(\varepsilon,x) \equiv \exp\{-\varepsilon^{-1}\int F(\tau_\theta x)d\theta\}$ and $c(\varepsilon)$ is a normalizing factor . Alternatively, one can take $\upsilon_\varepsilon(dx) = c(\varepsilon)\exp\{-\varepsilon^{-1}[F(x)-F(x_1)]\}\, dx$ where F is as before and moreover if $y \geq 0$, $d^2/d^2\theta\, F(\tau_\theta(y)|_{\theta=0}$ is independent of y .

II) Remark first that υ is the unique invariant measure for the Ohrnstein-Uhlembeck process with generator $\Delta + x.\nabla$. Consider the family of processes with generators $\Delta + x.\nabla + \varepsilon^{-1}\Phi(x)$ where $\Phi(x)$ is the image of the vector $\partial/\partial\theta$ under the map $\tau_\theta{}^*$ at x (so that $\Phi(x)$ is tangent to the gauge orbit at x, and vanishes on R^+). Denote by ξ_ε the corresponding processes. One can verify that each ξ_ε has a unique invariant measure υ_ε and that the processes ξ_ε converge when $\varepsilon \to 0$ to a unique process ξ_0, which is supported by R^+ and has υ_0 as unique invariant measure. The proof can be given using methods of stochastic stability, through the construction of a suitable family of Lyapunov functions [4].

We attempt now to repeat in the more complicated case of the (Euclidean) Yang-Mills theory the analysis made for the toy example.Let N be a smooth

Riemannian manifold of dimension d equipped with a Borel measure υ, absolutely continuous relative to Lebesgue measure..

Let G be a compact Lie group concretely represented by unitary MxM matrices and let G be its Lie algebra.

Denote by X the Hilbert space of G-valued connections on N, equipped with the L^2-topology induced on T*M by the Riemann metric and by the trace scalar product on MxM matrices.

Denote by X_0 the subset of smooth connections, and let GT_0 be the group of smooth Gauge Transformations (maps $x \to g(x)$ from N to G); if N is not compact, add the further condition that g(x)=I outside a compact set.

The action of GT_o on X is given by

$$A^g = g^*A\,g + g^*dg \quad , \; A \in X$$

(A is realized in the adjoint representation).

Denote by GT the closure of GT_0 in $L^2_{loc}(N,\upsilon)^{M\times M}$ and by GT_1 the closure of GT_o in the topology induced on smooth maps from N to M×M matrices by the seminorms $\|.\|_K$ defined for every compact K by

$$\|g\|_K^2 = \int |dg|^2 \upsilon(dx) + \int_K |g|^2 \upsilon(dx) \qquad , \qquad |B|^2 \equiv \mathrm{Tr} B^*B$$

Denote by $\sigma_0(A)$ the orbit of GT_o through $A \in X$, and by $\sigma(A)$ its closure in X.

One can then prove [5]:

<u>Proposition 1</u> : GT_1 is a topological group which acts continuously on X. Moreover the orbit of GT_1 through A coincides with $\sigma(A)$. ⌋

<u>Remark</u> 1 One should be aware that many topological properties (e.g. winding numbers) do not extend in general from GT_0 to GT_1 since the defining functionals are in general not continuous in $\|.\|_K$. ⌋

We shall now cosider a functional which plays the same role as F(x) in the toy example and which will allow us to find a region Ω (the analogue of R^+) which is intersected by each gauge orbit (at least) once.

For each $A \in X$, define on GT_1 a functional $S_A(g)$ by

$$S_A(g) \equiv \|A^g\|^2$$

where $\|.\|$ is the L^2-norm on X .

Let W_o be the set of C^∞ maps from N to G with compact support .

We denote by T_A the symmetric form on W_o defined by

$$T_A(w_1,w_2)=(dw_1,dw_2)+[(w_1,A \wedge dw_2) +1/2(w_1, \partial A \wedge w_2)]$$

Here (. ,.) is the scalar product which defines the L^2-norm on X , ∂=*d* is the

Hodge divergence and $\wedge$ denotes wedge product for forms and Lie product in G . One can then verify that $S_A(g)$ attains a local minimum at the identity of G if and only if

1) $\partial A = 0$ weakly (1)

2) $T_A \geq 0$ (2)

Remark 2 If A has some smoothness properties, T_A is closable and bounded below. Correspondingly one has a self-adjoint operator K_A (the Faddeev-Popov operator) formally given by $-\Delta+[A \wedge \partial + 1/2\, \partial A \wedge]$. ⌋

Define Ω by

$$\Omega \equiv \{A \in X \mid \partial A = 0 \,,\, T_A \geq 0 \} \tag{3}$$

and Λ by

$$\Lambda \equiv \{A \in X \mid S_A(g) \geq \|A\|^2 \quad \forall g \in GT_1\} \tag{4}$$

One has $\Omega \supset \Lambda$ and moreover

Proposition 2 [6], [7] Ω and Λ are convex and bounded in every direction ⌋

Proposition 3 [5] Every orbit intersects Λ at least once. ⌋

Proposition 4 [7] ,[8] Some orbits intersect Ω in at least two points. ⌋

Proposition 2 is a simple consequence of the definition of Ω and of the convexity of $S_A(I)$ as a function of A.

We prove Proposition 3 by proving that for every $A \in X$ the function $S_A(g)$ on GT_1 attains its minimum value. This is in turn a consequence of the remarkable fact that every minimizing sequence $\{g_n\}$ for $S_A(g)$ has a subsequence which converges in the topology of GT_1 (convergence in the topology of GT is a simple consequence of compact embedding theorems).

We prove Proposition 4 by determining an open subset P of Ω , $\partial P \cap \partial\Omega \neq \{\phi\}$ such that if $A \in P$ one can find $g_o \in GT_1$ for which $S_A(g_o) < \|A\|^2$.

More precisely, one can prove the following: if $A \in \partial\Omega$ and is sufficiently smooth, then the operator K_A exists and has a kernel, which we denote by N_A. Then the tangent space τ_A to $\partial\Omega$ at A can be defined, and if $N_A \cap \tau_A = \{\phi\}$, there exists an open subset P_A of $\Omega - \Lambda$ such that $A \in \partial P_A$.

Barring degeneracies, one can expect that Λ be a modular set , i.e. that "almost every" orbit intersects Λ in precisely one point ("almost every " with respect to the measure μ to be constructed on Λ).

One can now attempt to construct a measure on Λ along the lines described above for the toy model using Procedure I .
At a formal level, one possible candidate for the informal measure $\mu_\varepsilon(dA)$ is

$$\mu_\varepsilon(dA) \equiv \exp\{-\varepsilon^{-1}\|A\|^2 c^{-1}(A)\} \mu_o(dA) \qquad (5)$$

where $c(A)=\inf_{g\in GT1} S_A(g)$.
The informal expression for the Yang-Mills measure μ_o is

$$\mu_o(dA) = Z^{-1} \exp\{-1/2\, g_o^{-2} \int_N \Sigma_{i<j} Tr(F_{ij}^A(x))^2 \upsilon(dx)\}\, DA \quad (6)$$

where $F^A=dA+A\wedge A$ is the curvature of A , $A=\Sigma_1^N S_i A_i$ for a basis $\{S_i\}$ of G , $DA=\Pi_{i=1}^N \Pi_{x\in N}\, d(A_i(x))$ is "infinite-dimensional Lebesgue measure", g_o is a coupling constant and Z is a "normalization constant".

One can attempt a construction by defining first a regularized version of the theory (for exmple on a finite lattice) so that μ_o is well defined. One can then introduce a measure μ_{reg} on Λ_{reg} and remove the regularization to obtain the measure μ on Λ . Interesting results have been obtained along these lines at a formal level working directly with the non-regularized theory [9]. A rigorous analysis has not been given so far; one of the difficulties is that one lacks an explicit analytic expression for c(A) and an analytic characterization of the subset Λ .

Consider now the region Ω; in view of Proposition 4 , it is not a modular region, and therefore not every function on Ω represents an observable; observables correspond to those functions which take the same value at conjugate points (A and B are conjugate if one can find $g\in GT_1$ such that $A=B^g$).
On the other hand Ω has the explicit characterization as the set of those A for which $S_A(g)$ has a (local) minimum at the identity of GT_1 .
One can then attempt to construct a measure on Ω using the Procedure II described for the toy model. Indeed, the existence of a function which on each orbit has its local minima at Ω strongly suggests the use of Lyapunov techniques to implement Procedure II .
In fact, let φ_t be the gradient flow of $S_A(g)$ on $\cup_A \sigma_A$, and let φ_t^* be the dual flow on measures. Then, if υ is a measure which does not charge the union over A of the set of critical points of $S_A(g)$ (as a function of g) , one has at least formally

$$\Omega \supset \text{supp}\,(\lim_{t\to\infty}\varphi_t{}^*\upsilon)\ .$$

If $\zeta(A)$ is the vector field of φ_t then $\varphi_{t/\varepsilon} = \varphi^\varepsilon{}_t$ where $\varphi^\varepsilon{}_t$ is the flow of $\varepsilon^{-1}\zeta(A)$. Therefore for every $\tau>0$ one has

$$\Omega \supset \text{supp}\,(\lim_{\varepsilon\to 0}\varphi^\varepsilon{}_\tau{}^*\upsilon)\,.$$

It was observed by D.Zwanziger some time ago [6] that, if one considers a family of diffusions wtih generator L_ε given, at least formally, by

$$L_\varepsilon = L + \varepsilon^{-1}\zeta(A) \tag{7}$$

with L strictly elliptic, a similar phenomenon should occur. Roughly speaking, in a Trotter-Kato analysis of the stochastic flow generated by L_ε (which we denote by $\xi^\varepsilon{}_t$) the role of L should be to guarantee that at each time t the measure $\xi^\varepsilon{}_t{}^*\upsilon$ does not charge the critical points of $S_A(g)$ while the role of $\varepsilon^{-1}\zeta(A)$ is the same as in the deterministic case.

The function $S_A(g)$ itself is (locally) a good candidate as Lyapunov function for its gradient flow. One can expect therefore that a suitable modification of $S_A(g)$ can be used as stochastic Lyapunov function for the flow $\xi^\varepsilon{}_t$. [10]

This program has been carried out in [4] for a class of stochastic flows in R^d , $d<+\infty$, including many approximations of the Yang-Mills system on a finite lattice.

One can choose for L an operator of the form $L = \mathcal{L} -1/2\ \nabla_A \|F^A\|^2$, where $\mathcal{L}$ generates Brownian motion in R^d and $\|F^A\|^2$ is a finite-dimensional approximation of the Yang-Mills action.In this case, the unique invariant measure for L is absolutely continuous with respect to Lebesgue measure with density $\exp\{-\|F^A\|^2\}$.

Using methods of stochastic stability, in particular a family of Lyapunov functions, one can prove for these "approximate Yang-Mills theories on a finite lattice" :

<u>Proposition</u> 5 [4] For every ε the process $\xi^\varepsilon{}_t$, with initial distribution υ , has a unique invariant measure $\mu(\varepsilon,\upsilon)$. The processes $\{\xi^\varepsilon{}_t\}$ form a tight family and every limit process for $\varepsilon\to 0$ is supported by Ω . Moreover, if f is gauge-invariant and υ is absolutely continuous with respect to Lebesgue mesure , $\mu(\varepsilon,\upsilon)(f)$ does not depend on ε and coincides with $\mu(f)$ if the latter is well defined (otherwise formally $\mu(\varepsilon,\upsilon)(f)=\mu(f)/\mu(1)$) . ┘

From now on we shall omit the explicit reference to the initial measure υ .

Let ξ^o be a limit point of the family $\{\xi^\varepsilon{}_t\}$ and let L_o be its generator.

Let Ω^o be the interior of Ω (in the finite-dimensional case, Ω is closed with regular Jordan boundary $\partial\Omega$)

Proposition 6 [4] L_o is a self-adjoint extension of a strictly elliptic operator L_o defined on $C_o^2(\Omega^o)$. The coefficients of L_o are given explicitely at $A\in\Omega^o$ as functions of the tangent plane and of the curvature of σ_A at A . ⌋

Notice that in general there may be several self-adjoint extensions of L_o and to each of them may correspond a Markov process which may have one or more invariant measures. Each of these measures is a limit point of the set $\{\mu^\varepsilon\}$, which can be proved to be tight.
The coefficients of L_o can become singular at $\partial\Omega$; in particular the drift term becomes singular (this has a geometrical interpretation : at $\partial\Omega$, σ_A and Ω do not intersect transversally).
Let $\eta(A)$, $A\in\partial\Omega$, be the leading singular term in the drift when $\lambda_n\uparrow 1$, $A_n=\lambda_n A$ (so that $A_n \in\Omega^o\ \forall n$).
One has then $p(A)\equiv(\eta(A),n(A))\leq 0$, where n(A) is the outer normal to $\partial\Omega$ in Ω ($\partial\Omega$ is generically of codimension 1 in Ω).
Using the distance of A from $\partial\Omega$ as Lyapunov function, one can prove

Proposition 7 [4] The process generated by L_o with a stopping condition at $\partial\Omega$ is a diffusion which cannot reach those points of $\partial\Omega$ at which $p(A)<0$. Let

$$\partial_o\Omega\equiv\{A\in\partial\Omega\ ,\ p(A)=0\}$$

If $\partial_o\Omega$ has zero capacity relative to Brownian motion, with probability one the process ξ^o generated by L_o in Ω^o cannot reach the boundary. L_o has then a unique self-adjoint extension $\underline{L}_o$ and ξ^o has a unique invariant measure μ^o. In this case ξ^o is the unique limit point of $\{\xi^\varepsilon\}$ and μ^o is the unique limit point of $\{\mu^\varepsilon\}$. ⌋

Remark 3 The condition $p(A)<0$ coincides with the condition that one can find an open subset P(A) of Ω^o , with $A\in\partial P(A)$ and such that $P(A)\in\Omega^o-\Lambda$. Equivalently, $p(A)<0$ if one can find a subsequence $A_n\in\Omega^o-\Lambda$ with $A_n\to A$. It is however important to notice that Λ and $\partial\Lambda$ seem to play no role in the construction and in the properties of the limit processes ξ^o and of the limit measure μ^o . ⌋

So far, procedure II has not been applied directly to the continuum limit (the corresponding Lyapunov techniques have not been worked out) nor has one studied in this context the continuum limit of the finite dimensional approximations.

Remark 4 It may be of interest to notice that T_A is defined on W_o for every distribution-valued connection A, so that Ω can be defined in this wider context.. On the other hand , the procedure leading to the definition of the set Ω can also be defined in the lattice approximations. It is reasonable to expect that if one starts with a lattice approximation and takes the continuum limit , one obtains a measure supported by Ω' , where Ω' is the subset of distribution-valued connections which satisfy (1) and (2). ⌋

It is then encouraging that one can find a bound on the singular behaviour (as distributions) of the elements of Ω'. This bound may prove important in convergence estimates for the construction of an Euclidean Yang-Mills Theory. In particular one can prove

Proposition 8 [12] Let $N=T_V{}^d$, a torus of volume V in R^d, υ=Lebesgue measure, G = SU(2) in the adjoint representation.. Then $E_V \supset \Omega'$ where the ellipsoid E_V is defined by

$$E_V \equiv \{A \in X \ , \Sigma_k \Sigma_{\mu=1}{}^d \Sigma_{b=1}{}^3 |a_\mu{}^b(k)|^2 |k|^{-2} \leq 60 \, d \, V \} \qquad (9)$$

where $A(x) = \Sigma_{\mu,b} S_b A_\mu{}^b(x) dx_\mu$, S_b is a basis for the adjoint representation of su(2), k takes values in the dual lattice and a(k) is the Fourier Transform of A(x). ⌋

The bound $E_V \supset \Omega'$ leads directly to bounds on the second moments of any probability measure on Ω' and in particular of the ground state measure. This gives the following bound on the two-point function of the <u>unrenormalized</u> Euclidean connection field.

Let E(.) be expectation for the ground state masure, so that

$$E(A_\mu{}^b(x) A_\upsilon{}^c(y)) = V^{-1} \Sigma_k (\delta_{\mu\nu} - k_\mu k_\upsilon / k^2) \delta_{bc} g(k) \exp\{ik(x-y)\}$$
$$\mu,\nu = 1..d \ , \ x,y \in T_V{}^d$$

is the Kallen-Lehmann representation for the two-point function. Then (9) implies

$$\Sigma_{k \neq 0} \, g(k) \, k^{-2} \leq 20 \, V \, d/(d-1) \qquad (10)$$

Ellipsoidal bounds of the type (9) hold for any semisimple Lie group g and for any compact manifold N . Corresponding results hold also for non-compact manifolds.

In particular one has

Proposition 9 [11] Let $N = R^d$, υ and G as in Proposition 8. Then $E' \supset \Omega'$ where

$$E' \equiv \{ A \in X , \Sigma_{\mu,b} \int_{R^d} |a_\mu{}^b(k)|^2 k^{-2} dk \leq 5d/4(d-1)\} \qquad (11)$$ ⌋

The corresponding bound for the weight in the Kallen-Lehmann represenation of the two-point function is

$$\int g(k) k^{-2} dk \leq 5d/4(d-1) \qquad (12)$$

or equivalently

$$\int dx\, \Sigma_{\mu,b} G_0(x) E(A_\mu{}^b(x) A_\mu{}^b(0)) \leq 15d/4$$

where $G_0(x)$ is the Green Function of the Laplace operator in R^d.

Remark 5 In the geometric context described before, in which for all A in X one has $\Sigma_{\mu,b} \int |a_\mu{}^b(k)|^2 dk < +\infty$, (11) provides a qualitative bound only for the infrared content of the connections in Ω (i.e. the behaviour when $k \rightarrow 0$). However, as pointed out in Remark 4, the region Ω' can be defined through (1) and (2) as a subset of all distribution-valued connections. In this wider setting, which should be important in the study of convergence of lattice approximations, (9) and (11) represent a qualitative bound also on the short-scale roughness of $A \in \Omega'$. In particular, when $d \geq 4$, the connections which satisfy (11) are contained in a set of zero measure for the Gaussian field Ξ_0 with covariance G_0. This can be taken as an indication that if Ω' is the support of the Euclidean measure for the Yang-Mills theory, then in $d \geq 4$ dimensions Ξ_0 is not appropriate as starting point for an approximation scheme. ⌋

Remark 6 For $d \leq 4$ the bound (12) excludes a pole at $k=0$ in the Kallen-Lehmann representation. As already indicated by Gribov, this may be relevant for the phenomenon of confinement. One must keep in mind however that the (Euclidean) vector potential is not an observable ; the bound (12) provides no information about the existence of a pole in the Kallen-Lehmann representation for observable fields. ⌋

The proof of propositions 8 and 9 is obtained through a variational inequality. Since $A \in \Omega$ one has $T_A(w) \geq 0$ for $w \in W_0$.

We use a one-parameter family of trial elements $\{w_\alpha\}$ suggested by (formal) first-order perturbation theory for the operator K_A ; if A is not smooth enough, we use a regularized version of the formal expression .
The condition $T_A(w_\alpha)\geq 0$ leads to the inequality

$$X - 2\alpha Y + \alpha^2(Z_1+Z_2) \leq 0 \qquad (13)$$

where X is independent of A , Y and Z_1 are quadratic in A and Z_2 is cubic.In fact Y is essentially the quadratic expression for which (9) and (11) provide bounds.
One must then have

$$Y^2 \leq X(Z_1+Z_2) \qquad (14)$$

It is not dfifficult to provide bounds for Z_1Y^{-1} and also a lower bound for Z_2Y^{-1} . An upper bound for Z_2Y^{-1} concludes the proof of Propositions 8 and 9 ; it is obtained exploiting the specific structure of Z_2 and the properties of the maximal weights in the Dynkin diagram for su(2) .

REFERENCES

[1] S.Albeverio, M.Roeckner Stochastic Differential equations in infinite dimensions University of Bielefeld Preprint 402/89

[2] J.Glimm, A.Jaffe Quantum Physics , 2nd edition, Springer Verlag, 1986

[3] R.Brandt et al. Loop Space Preprint NYU TR-1 January 82.

[4] G.F.Dell'Antonio Proceedings of the Ascona Conference 1988 , S.Albeverio and D.Merlini Eds. Springer Verlag Lecture Notes in Physics.

[5] G.F.Dell'Antonio, D.Zwanziger Every Gauge Orbit passes inside the Gribov Horizon, preprint , submitted to Comm. Math. Phys.

[6] D.Zwanziger, Nuclear Physics B 209 (1982) 336

[7] G.F.Dell'Antonio, D.Zwanziger Proceedings of the Research Workshop on Probabilistic Methods in Q.F.T.,Cargèse 1989 ,Damgaard and Hueffel Eds , Plenum Press

[8] M.A.Semenov-Tyan-Shanskii , V.A.Franke Proceedings of the Seminars of the Leningrad Math. Inst. 1982, English translation 1986, Plenum Publ. Corp.

[9] D.Zwanziger Quantization of Gauge Fields, Classical gauge invariance and Gluon Confinement , NYU preprint

[10] H.Kushner Approximation and weak convergence methods , MIT Press, 1984

[11] G.F.Dell'Antonio , D.Zwanziger Nuclear Physics B 326 (1989) 333

A RENORMALIZABLE THEORY OF QUANTUM GRAVITY

Yuval Ne'eman

Raymond and Beverly Sackler Faculty of Exact Sciences+¶#
Tel-Aviv University, Tel-Aviv, Israel 69978
and
Center for Particle Theory*,
University of Texas, Austin, Texas, USA 78712

and

Chang-Yeong Lee
Center for Particle Theory*
University of Texas, Austin, Texas, USA 78712

Abstract

We present a proof of renormalizability for a non-Riemannian model, based on gauging GL(4,R), in which Einstein's Gravity dominates the low-energy region through a Goldstone-Higgs spontaneous symmetry breakdown mechanism. In other renormalizble models of gravity, this is the result of $1/p^4$ unitarity-violating propagators, in this case it follows only from the Yang-Mills-like features of the theory.

1. INTRODUCTION

At the Como conference in 1987, the XVIth in this Series, one of us (YN) presented a paper entitled "Towards a Renormalizable Theory of Quantum Gravity" [1]. In that paper, he sketched a program leading to a renormalizable theory of quantum gravity, starting from a Yang-Mills like construction with Goldstone-Higgs spontaneous symmetry breakdown. He conjectured that the Yang-Mills structure would ensure renormalizability and help us evade most of the difficulties of Einstein's theory - without involving the $1/p^4$ propagators that make it non-unitary [2]. The present paper represents the realization of this program [3,4,5].

Why should we want to quantize gravity at all? The late Leon Rosenfeld did suggest that gravity might be purely classical, operationally fulfilling the measurement act, classical in the Copenhagen interpretation of Quantum Mechanics. It was shown, however, that such a situation would enable us to violate the Uncertainty Principle, and could therefore not be tolerated by the Quantum Axioms. Thus, the consistency of physical theory requires gravity to possess a quatum version, like all other fundamental interactions.

The issue became pressing with the advent of "GUTs", i.e. theories in which both QCD and the weak-electromagnetic $SU(2)\hat{x}U(1)$ are embedded in one large gauge group. The symmetry breaking yielding the Standard Model $SU(3)x[SU(2)\hat{x}U(1)]$ occurs around 10^{15} Gev. This is rather close to the region where gravity enters the quantum regime, i.e. when the Schwarzschild radius of a particle $R_S = 2Gm/c^2$ (G is Newton's constant) overtakes its Compton wave length $R_C = h/2\pi mc$ (h is Planck's constant). This is the so-called Planck mass,

$$2GM/c^2 = h/2\pi Mc, \rightarrow M = \sqrt{(hc/4\pi G)} \sim 10^{19} \text{ GeV}/c^2. \quad (1)$$

A fact that is less well known is that Quantum Gravity is a <u>strong</u> force. Indeed, from (1) we get (F_g is the gravitational attraction between two Planck mass particles at a distance r from each other)

$$GM^2 = h/4\pi c = F_g.r^2, \quad (2)$$

whereas the electromagnetic force between two particles carrying charges equal to the charge on the electron (e) is

$$F_e.r^2 = e^2/4\pi\epsilon_0 = [e^2/4\pi\epsilon_0(hc/2\pi)].(hc/2\pi) = (hc/2\pi)/137,$$

so that

$$F_g/F_e = (hc/2\pi)/2 : 137/(hc/2\pi) \sim 70 \quad (3)$$

i.e. quantum gravity is a strong force and it should be taken into account when dealing with the physics above 10^{15} GeV/c^2.

Einstein's gravity is perturbatively nonrenormalizable [6]. A perturbatively renormalizable relativistic quantum field theory (RQFT) of gravity, if it exists, should therefore differ from Einstein's in the quantum region (i.e. Planck energy); however, it should reduce to Einstein's theory in the low-energy (long range $\Delta x \geq l_{Planck}$) region. It is possible (perhaps probable) that a finite theory of Quantum Gravity may yet emerge from the string; however, the sub-Planck energy sector should be describable by a renormalized relativistic quantum field theory [7] anyhow. In that picture, our renormalizable theory represents a description of the massless sector, a truncation at Planck energy. There is, however, yet another possibility: as our theory starts from a symmetric gauge quantum field theory above Planck energy, it might well correspond to the true physics there too, relegating the string to the role of yet another "regulator" theory, after Pauli-Villars, dimensional regularization and the lattice.

As things stand at the present moment (summer 1990) the string amplitudes do look finite (but there is, as yet no proof of finiteness to all orders). Sums of diagrams do not obey Borel summability [8], but this is a characteristic of theories that possess non-perturbative solutions (the infinities cancel when these are included). The main remaining difficulty with the string resides in the arbitrariness in selecting the vacuum, i.e. in reducing from 26 or 10 dimensions to four. This relates also to the incertitude with respect to a non-perturbative foundation, either the missing Quantum String Theory or some other presentation.

The main obstacle to the renormalizability of Einstein's theory consists in the dimension d = 2 of Newton's constant and thus in the need for new counter-terms in each order of the perturbation series. Alternatively, the difficulty can be blamed on a complementary aspect, namely the d = 2 of the Einstein-Hilbert Lagrangian, linear in the curvature. This should be compared with the dimensionless coupling and d = 4 curvature-quadratic Lagrangian of a Yang-Mills gauge theory, guaranteeing cut-off independence of independence of S-matrix amplitudes and restricting the renormalization counter-terms to a finite set.

In gravity, the addition (to the scalar curvature) of d = 4 terms, quadratic in the curvatures, has been shown [2] to lead to a power-counting renormalizable theory. Renormalizability is due to the $1/p^4$ graviton propagators, a result of the Riemannian constraint $D_\sigma g_{\mu\nu} = 0$. This relates the connection $\Gamma^a{}_{b\mu}$ to the metric $g_{\mu\nu}$, so that the $(\partial\Gamma)$ and $(\Gamma\Gamma)$ in the curvature R become ~ $(\partial^2 g)$ and $(\partial g)^2$, making the R^2-type terms yield $1/p^4$ propagators. However, quartic propagators can be shown to contain ghosts, and this type of renormalizable theory ends up being non-unitary.

Other models, involving, in addition to Einstein's, all possible quadratic Lorentz-curvature and torsion terms, have been analyzed [4,10,11]. Several combinations were shown to be unitary; however, in every such case, the $1/p^4$ propagators cancel and the resulting power-counting renormalizability is lost.

2. $\overline{GL}(4,R)$ GAUGE THEORY WITH SPONTANEOUS SYMMETRY BREAKDOWN

In our model [12], the Lagrangian has d = 4 and the coupling is dimensionless. The theory is based on gauging the linear group GL(4,R), i.e. a Lagrangian consisting of some linear combination of the irreducible components of the SKY [13] Lagrangian, quadratic in the GL(4,R) curvatures,

$$\mathcal{L}_{inv} = \sqrt{g}\, R^a{}_{b\mu\nu}\, R^c{}_{d\rho\sigma}\, \alpha_i\, \Delta^i{}_a{}^b{}_c{}^{d\mu\nu\rho\sigma}(g_{\gamma\eta}, g^{\zeta\xi}, e^m{}_\lambda, \hat{e}_n{}^\kappa, \delta^u{}_v) \tag{4}$$

where Δ^i is a reduction tensor, i.e. a product of non-invariant pre-metric tensors $g_{\gamma\eta}(x)$ and their inverses $g^{\zeta\xi}(x)$, tetrads $e^m{}_\lambda(x)$ and frames $\hat{e}_n{}^\kappa(x)$ and Kronecker symbols $\delta^u{}_v$; α_i are parameters defining the linear combination; $g = \det g_{\gamma\eta}$. The only physical field for gravity is the connection $\Gamma^a{}_{b\mu}$. The tetrads have no kinetic energy term; the pre-metric is an auxiliary SL(4,R) symmetric tensor field g_{ab}, also with no kinetic energy, coupled to the tetrads, $g_{\mu\nu} = g_{ab}\, e^a{}_\mu\, e^b{}_\nu$. Now $D_\lambda e^a{}_{\mu(x)} \cong 0$ identically, and $K_{\lambda\mu\nu} = D_\lambda g_{\mu\nu} = D_\lambda\{g_{ab}(x).e^a{}_\mu.e^b{}_\nu\} = e^a{}_\mu\, e^b{}_\nu\, D_\lambda g_{ab}(x) \neq 0$, due to the "pre-metric" gab(x) which is an SL(4,R) gauge-dependent symmetric tensor but not a Riemannian metric. The high energy regime (HE) is thus non-Riemannian.

Spontaneous symmetry breakdown (SSB) is then triggered by two Goldstone-Higgs fields: (1) a dilaton $\sigma(x)$, an SL(4,R) scalar; (2) a manifield $\Phi(x)$, i.e. an infinite-component field carrying an irreducible representation of GL(4,R). This particular manifield's lowest state Φ^{00} behaves as a (0,0) Lorentz scalar.

A non-vanishing v.e.v. for either field, i.e. $\langle\sigma(x)\rangle \neq 0$ or $\langle\Phi(x)^{oo}\rangle \neq 0$, suffices to cause both v.e.v. to occur, through a mixed $\lambda\sigma^2\Phi^2$ term in the potential, in addition to $\lambda'\sigma^4$ and $\lambda''\Phi^4$. The Einstein Lagrangian also emerges from a d = 4 term $\sigma^2 R$, i.e. it induces a "soft" effective Newton constant. This fixes a scale $|\langle\sigma(x)\rangle|^2 = 1/16\pi G_N$, i.e. both SSB occur at Planck energies. The v.e.v. of $\Phi(x)$ therefore gives masses of Planck order to the components of the connection gauging the dilations and shears, i.e. gl(4,R)/so(1,3); those gauging the Lorentz subgroup appropriately remain massless. The Riemannian condition appears as an effective low-energy (LE) result $\langle LE|D_\lambda g_{\mu\nu}|LE\rangle = 0$. The pre-metric $g_{\mu\nu}$ has thus become the LE effective metric, as a result of SSB. The components of $\Phi(x)$ other than Φ^{oo} also acquire Planck-size masses.

Matter fields appear as world-spinor manifields [14-16], i.e. they behave as infinite-dimensional unirreps (here de-unitarized [17]) of the covering group $\widetilde{GL}(4,R)$ [18,19].

In the following, we outline a proof of the renormalizability of this theory [5]. A detailed calculation will appear elsewhere [4]. Unitarity remains to be checked out. Note that the only $1/p^4$ propagators in this theory are found in the gauge-fixing part of the quantum Lagrangian and are therefore harmless. Nevertheless, the non-compact nature of the gauge group could still cause a failure of unitarity. We hope, however, that unitarity can be assured through a proper selection of the GL(4,R)-irreducible pieces of the Lagrangian (4).

3. BRST TRANSFORMATIONS IN THE BACKGROUND FIELD FORMALISM.

An apropriate set of BRST transformations was constructed [3]. For our proof, we adopt the "background field" (BF) formalism [20,21]. The gauge fields ("Y") are split Y = A + Q, i.e. into a "background" part A and a "quantum" part Q. The BRST variations $\delta_B Y = -DC$, $\delta_B C = -[C,C]$, C the ghost field, are then reproduced in two different manners which we denote as types I(δ') and II(δ''): $\delta' A = -\underline{D}C$, $\delta' Q = -CfQ$, $\delta' C = -CfC$; $\delta'' A = 0$, $\delta'' Q = -DC$, $\delta'' C = \delta_B C$. Note that $\delta'(A+Q) = \delta''(A+Q) = \delta_B Y$; $\delta' C = \delta'' C = \delta_B C$; here $D = \partial + Y$, $\underline{D} = \partial + A$; f stands for the gl(4,R) algebra structure constants. Type I variations leave $\mathcal{L}_{gf}$ (the gauge-fixing) and $\mathcal{L}_{gh}$ (the ghost lagrangian) separately invariant, i.e. an invariance preserved by the quantum counterterms. Recapitulating the BRST algebra [3],

$$\delta_B\, w^a{}_{b\mu} = -\,(C^\lambda\, w^a{}_{b\mu,\lambda} + C^\lambda{}_{,\mu}\, w^a{}_{b\lambda} - C^a{}_{b,\mu} + C^a{}_d\, w^d{}_{b\mu} - C^d{}_b\, w^a{}_{d\mu}).$$

$$\delta_B\, C^a{}_b = (-\, C^\lambda\, C^a{}_{b,\lambda} - C^a{}_d\, C^d{}_b\,).$$

$$\delta_B\, e^a{}_\mu = -\,(C^\lambda\, e^a{}_{\mu,\lambda} + C^\lambda{}_{,\mu}\, e^a{}_\lambda + C^a{}_b\, e^b{}_\mu).$$

$$\delta_B\, g^{\mu\nu} = C^\lambda\, g^{\mu\nu}{}_{,\lambda} - C^\mu{}_{,\lambda}\, g^{\lambda\nu} - C^\nu{}_{,\lambda}\, g^{\mu\lambda} + 1/2\, C^\lambda{}_{,\lambda}\, g^{\mu\nu}.$$

$$\delta_B\, C^\mu = -\, C^\lambda\, C^\mu{}_{,\lambda}. \tag{5}$$

$\delta_B{}^2 = 0$ for all fields and ghosts. We add to (4) the gauge-fixing and ghost Lagrangians,

$$\mathcal{L}_{gf} = -\,(1/2\alpha)\,(H^{\mu\nu}\,\underline{D}_\mu\,\omega^a{}_{b\nu})\,(H^{\rho\sigma}\,\underline{D}_\rho\,\omega^b{}_{a\sigma})\, -$$

$$-\,(1/2\beta)\;\; H^{\mu\nu}\,\underline{D}_\mu\,\underline{D}_\nu\,\{(\underline{D}_\kappa\,\eta^{\kappa\lambda})\,(\underline{D}_\rho\,\eta^{\rho\sigma})\,H_{\lambda\sigma}\}\,. \tag{6}$$

$$\mathcal{L}_{gh} = -\,C^b{}_a\;\; H^{\mu\nu}\,\underline{D}_\mu\;\;\delta''\omega^a{}_{b\nu} - H^{\mu\nu}\,\underline{D}_\mu\,\underline{D}_\nu\,C_\sigma\,\underline{D}_\rho\,\delta''\eta^{\rho\sigma}\,. \tag{7}$$

implying the antighost variations,

$$\delta_B\,\overline{C}^b{}_a = -\,(1/\alpha)\,H^{\mu\nu}\,\underline{D}_\mu\,\omega^a{}_{b\nu}\,.$$

$$\delta_B\,\overline{C}^b{}_a = -\,(1/\alpha)\,H^{\mu\nu}\,\underline{D}_\mu\,\omega^a{}_{b\nu}\,.$$

$$\delta_B\,\overline{C}_\nu = -\,(1/\beta)\,H^{\mu\nu}\,\underline{D}_\lambda\,\eta^{\mu\lambda}\,. \tag{8}$$

In the above, we have redefined the $g\mu\nu$ fields and the C_ν so as to incorporate in them the $\sqrt{g}$ density factor. With the A/Q split this yields (Ω,E,H are the A, while ω,ϵ,η are the Q),

$$w^a{}_{b\mu} = \Omega^a{}_{b\mu} + \omega^a{}_{b\mu}\,.$$

$$e^a{}_\mu = E^a{}_\mu + \epsilon^a{}_\mu,\quad \hat{e}_a{}^\mu = \hat{E}_a{}^\mu + \epsilon_a{}^\mu(e^b{}_\nu).$$

$$g^{1/4}\, g^{\mu\nu} \cong G^{\mu\nu} = H^{\mu\nu} + \eta^{\mu\nu},\ g^{-1/4}\, g_{\mu\nu} \cong G_{\mu\nu} = H_{\mu\nu} + \eta_{\mu\nu}(\eta^{\rho\sigma}).$$

$$G^{\mu\nu}\, G_{\nu\sigma} = H^{\mu\nu}\, H_{\nu\sigma} = \delta^{\mu}{}_{\sigma}. \tag{9}$$

Equations (5) and (8) also define the type II for ω ϵ η and the ghosts, since $\delta''Q = \delta_B\, Y$. Note that the $\eta^{\rho\sigma}(x)$ had no kinetic term in (4) and only acquires it in $\mathcal{L}_{gf}$. The frame and lower-indices pre-metric are defined by Taylor expansions.

4. EFFECTIVE ACTION AND SLAVNOV-TAYLOR IDENTITIES.

The background field effective action Γ is written as

$$\Gamma(\Omega,E,H;\omega,\epsilon,\eta,\tilde{C}\uparrow,\bar{\tilde{C}}\uparrow,K\uparrow,L\uparrow) = \tilde{W}(\Omega,E,H;J\uparrow,\bar{\zeta}\uparrow,\zeta\uparrow,K\uparrow,L\uparrow) - J\uparrow.\tilde{Q}\uparrow -$$

$$- \tilde{C}\uparrow.\zeta\uparrow - \zeta\uparrow.\tilde{C}\uparrow =$$

$$= -i\ \ln \{\int\Delta(\omega,\epsilon,\eta,C\uparrow,\bar{C}\uparrow)\ \exp\ i\int d^4x[\mathcal{L}(\Omega,E,H;\omega,\epsilon,\eta;C\uparrow,\bar{C}\uparrow;K\uparrow,L\uparrow) +$$

$$+ J_w.\omega + J_e.\epsilon + J_g.\eta + \bar{C}^b{}_a.\zeta^a{}_b + \bar{C}_\nu.\zeta^\nu + \bar{\zeta}^b{}_a.C^a{}_b + \bar{\zeta}_\nu.C^\nu]\} -$$

$$- J_w.\omega - J_e.\epsilon - J_g.\eta - \bar{\tilde{C}}^b{}_a.\zeta^a{}_b - \bar{\tilde{C}}_\nu.\zeta^\nu - \bar{\zeta}^b{}_a.\tilde{C}^a{}_b - \bar{\zeta}_\nu.\tilde{C}^\nu\ . \tag{10}$$

Here $\tilde{Q}\uparrow,\tilde{C}\uparrow,\bar{\tilde{C}}\uparrow$ denote the v.e.v. of the quantum fields, ghosts and antighosts, $\tilde{Q}\uparrow = \delta(-i\ \ln\ \tilde{Z})/\delta J_{Q\uparrow}$, etc.; $\tilde{Z}(A\uparrow,J\uparrow,\zeta\uparrow,\bar{\zeta}\uparrow)$ the argument of the logarithm in (10) is the generating functional; the arrow $\uparrow$ implies a "multiplet" (e.g. $C\uparrow$ denotes the set $C^a{}_b$, C^ν); the $J\uparrow$, $\bar{\zeta}\uparrow,\zeta\uparrow$ are source terms for quantum fields, ghosts and antighosts, $K\uparrow,L\uparrow$ are source terms for $\delta''Q\uparrow$ and $\delta''C\uparrow$. The measure Δ imposes integration over the $Q\uparrow,C\uparrow,\bar{C}\uparrow$. The quantum action in the exponential is given as

$$\Sigma(A\uparrow,Q\uparrow,C\uparrow,C\uparrow,K\uparrow,L\uparrow) = \int d^4x\ \{\mathcal{L}_{inv}(\omega+W,\epsilon+E,\eta+H) + \mathcal{L}_{gf}(W,H,\underline{D}\omega,\underline{D}\eta) +$$

$$+ \mathcal{L}_{gh}(C\uparrow,A\uparrow,\underline{D}\delta''Q\uparrow) + (K_w)^{\nu b}{}_a\ \delta''\omega^a{}_{b\nu} + (K_e)^\mu{}_a\ \delta''\epsilon^a{}_\mu + (K_g)_{\mu\nu}\ \delta''\eta^{\mu\nu} -$$

$$- L^b{}_a\ \delta''C^a{}_b - L_\nu\ \delta''C^\nu\}. \tag{11}$$

Another useful expression is the reduced effective action in which we resubtract $\pounds_{gf}$,

$$\Gamma' = \Gamma - \int d^4x\, \pounds_{gf} \, . \tag{12}$$

The Slavnov-Taylor identities are expressed as

$$\delta' A\!\uparrow . \delta\Gamma/\delta A\!\uparrow + \delta'\tilde{Q}\!\uparrow . \delta\Gamma/\delta\tilde{Q}\!\uparrow + \delta'\tilde{C}\!\uparrow . \delta'\Gamma/\delta\tilde{C}\!\uparrow + \delta'\tilde{C}\!\uparrow . \delta'\Gamma/\delta\tilde{C}\!\uparrow = 0 \tag{13a}$$

$$\delta\Gamma/\delta\tilde{Q}\!\uparrow . \delta\Gamma/\delta K\!\uparrow + \delta\Gamma/\delta\tilde{C}\!\uparrow . \delta\Gamma/\delta L\!\uparrow = 0 \tag{13b}$$

$$H^{\mu\nu}\, \underline{D}_\mu \{\delta\Gamma/\delta (K_w)_\nu{}^b{}_a\} - \delta\Gamma/\delta\bar{\tilde{C}}^b{}_a = 0 \tag{13c}$$

$$H^{\mu\nu}\, \underline{D}_\lambda \{\delta\Gamma/\delta (Kg)_{\mu\lambda}\} - \delta\Gamma/\delta\bar{\tilde{C}}_\nu = 0 \tag{13d}$$

Eq. (13a) results from type I invariance, the other three from type II.

5. PROOF OF RENORMALIZABILITY: REMOVAL OF DIVERGENT CONTRIBUTIONS.

We assume that Γ' has been successfully renormalized up to (n-1) loops, i.e. all divergences have been removed by a redefinition of the fields, parameters α_i and BRST transformations. We then proceed to n-loop order and prove that the divergent parts in $\Gamma'_{(n)}$ can be similarly removed. $\Gamma'_{(n)}$ is split first into finite and divergent parts:

$$\Gamma'_{(n)} = \Gamma'_{(n)}{}^f + \Gamma'_{(n)}{}^d \, . \tag{14}$$

Eqs. (13) have to be satisfied to n-loop order; the new divergent contributions from n-loops have a pole in ϵ in the dimensional reduction regularization method (working in 4-ϵ dim.) and should thus obey (13) separately . We rewrite (13) for $\Gamma'_{(n)}{}^d$:

$$\delta' A\!\uparrow . \delta\Gamma'_{(n)}{}^d/\delta A\!\uparrow + \delta'\tilde{Q}\!\uparrow . \delta\Gamma'_{(n)}{}^d/\delta\tilde{Q}\!\uparrow + \delta'\tilde{C}\!\uparrow . \delta\Gamma'_{(n)}{}^d/\delta\tilde{C}\!\uparrow +$$

$$+ \delta'\bar{\tilde{C}}\!\uparrow . \delta\Gamma'_{(n)}{}^d/\delta\bar{\tilde{C}}\!\uparrow = 0 \tag{15a}$$

$$(\delta\Gamma'_{(n)}{}^d/\delta\tilde{Q}\uparrow)(\delta\Gamma^{o\prime}/\delta K\uparrow) + (\delta\Gamma^{o\prime}/\delta\tilde{Q}\uparrow)(\delta\Gamma'_{(n)}{}^d/\delta K\uparrow)$$

$$+ (\delta\Gamma'_{(n)}{}^d/\delta\tilde{C}\uparrow)\ (\delta\Gamma^{o\prime}/\delta L\uparrow) + (\delta\Gamma^{o\prime}/\delta\tilde{C}\uparrow)(\delta\Gamma'_{(n)}{}^d/\delta L\uparrow) = 0 \qquad (15b)$$

$$H^{\mu\nu}\ \underline{D}_\mu\{\delta\Gamma'_{(n)}{}^d/\delta(K_w)^{\nu b}{}_a\} - \delta\Gamma'_{(n)}{}^d/\delta\bar{\tilde{C}}^b{}_a = 0 \qquad (15c)$$

$$H^{\mu\nu}\ \underline{D}_\lambda\{\delta\Gamma'_{(n)}{}^d/\delta(K_g)^\mu{}_\lambda\} - \delta\Gamma'_{(n)}{}^d/\delta\bar{\tilde{C}}_\nu = 0 \qquad (15d)$$

where $\Gamma^{o\prime}$ coincides with the expression for Γ'.

Eq. (15b) can be reexpressed in the form $X\Gamma'_{(n)}{}^d = 0$, where X contains the variations of $\Gamma^{o\prime}$ as coefficients of the variations of $\Gamma_{(n)}{}^d$. One checks that $X^2 = 0$. As a result we can write,

$$\Gamma'_{(n)}{}^d = G(A\uparrow,\tilde{Q}\uparrow) + X\ P(A\uparrow,\tilde{Q}\uparrow,\tilde{C}\uparrow,\bar{\tilde{C}}\uparrow,K\uparrow,L\uparrow) \qquad (15b')$$

$d(G) = 4$, and to satisfy typeII invariance $G(A\uparrow,\tilde{Q}\uparrow) = G(A\uparrow + \tilde{Q}\uparrow)$. To satisfy type I, it has to coincide with I_{inv} in (4), up to changes in the α_i. $d(X) = d(\Gamma) - d(Q) - d(K) = d(\Gamma) - d(C) - d(L)$ and we see by inspection that $d(X) = 1$ and ghost number $N_{gh} = 1$, so that $d(P) = 3$, $N_{gh}(P) = -1$. From (15c,d), $K\uparrow$ and $\bar{C}\uparrow$ in P are restricted to the combinations $(K_w)^{\nu b}{}_a + \tilde{C}^b{}_a (H^{\mu\nu}\underline{D}_\mu)$ and $(K_g)^\mu{}_\lambda + \tilde{C}_\nu\ (H^{\mu\nu}\underline{D}_\lambda)$. This constrains P to the form (M,N,N' are arbitrary scalar field functionals, F a vector in both tangent and world indices and T a world tensor)

$$P = \kappa_1\{(K_w)^{\nu b}{}_a + \bar{\tilde{C}}^b{}_a.(H^{\mu\nu}D_\mu)\}.(\omega^a{}_{b\nu} + \Omega^a{}_{b\nu})\ M(\epsilon,\eta,E,H)$$

$$+ \kappa_2(K_e)^\mu{}_a\ .\ F^a{}_\mu(\epsilon,\eta,E,H) + \kappa_3\{(K_g)^\mu{}_\lambda + \tilde{C}_\nu.H^{\mu\nu}D_\lambda\}.T_\mu{}^\lambda(\epsilon,\eta,E,H)$$

$$+ \lambda_1\ L^b{}_a\ \tilde{C}^a{}_b.\ N(\epsilon,\eta,E,H) + \lambda_2\ L_\nu\ \tilde{C}^\nu.\ N'(\epsilon,\eta,E,H) \qquad (16)$$

The divergences in XP can all be removed through field renormalizations or by renormalizing the BRST transformations. For example, the κ_2 term yields a divergent contribution (we read out X from eq.(15b) and P in (16)) in $\Gamma'_{(n)}{}^d$ of (15b'):

$$\kappa_2\ \delta''\epsilon^b{}_\nu\ .(K_e)^\mu{}_a\ (\delta F^a{}_\mu\ /\delta\epsilon^b{}_\nu) + \kappa_2\ (\delta\Gamma^o/\delta\epsilon^a{}_\mu)\ F^a{}_\mu\ . \qquad (17a)$$

which can be removed by redefinitions

$$\epsilon^a{}_\mu \rightarrow \epsilon^a{}_\mu - \kappa_2\, F^a{}_\mu(\epsilon,E,\eta,H) \tag{17b}$$

$$\delta''\epsilon^a{}_\mu \rightarrow \delta''\epsilon^a{}_\mu - \kappa_2\{(\delta F^a{}_\mu\, /\delta\epsilon^b{}_\nu)\, \delta''\epsilon^b{}_\nu + (\delta F^a{}_\mu/\delta\eta^{\rho\sigma})\delta''\eta^{\rho\sigma}\} \tag{17c}$$

For the first term in P we use the redefinitions

$$\omega \rightarrow \omega - \kappa_1\, (\omega + \Omega)\, M(\epsilon,\eta,E,H), \tag{17d}$$

$$\delta''\omega \rightarrow \delta''\omega - \kappa_1\{(\omega+\Omega)[(\delta M/\delta\epsilon)\delta''\epsilon + (\delta M/\delta\eta)\delta''\eta] + \delta''\omega\, M\}. \tag{17e}$$

For the third term,

$$\eta^{\mu\nu} \rightarrow \eta^{\mu\nu} - \kappa_3\, T^{\mu\nu}(\epsilon^b{}_\sigma, E^b{}_\sigma, \eta^{\lambda\rho}, H^{\lambda\rho}), \tag{17f}$$

$$\delta''\eta^{\mu\nu} \rightarrow \delta''\eta^{\mu\nu} - \kappa_3\, \{(\delta T/\delta\epsilon)\delta''\epsilon + (\delta T/\delta\eta)\delta''\eta\}. \tag{17g}$$

The λ terms are renormalized through the redefinitions,

$$\tilde{C}^a{}_b \rightarrow \tilde{C}^a{}_b - \lambda_1\, \tilde{C}^a{}_b\, N(\epsilon,\eta,E,H), \tag{18a}$$

$$\delta''\tilde{C}^a{}_b \rightarrow \delta''\tilde{C}^a{}_b - \lambda_1\{\delta''\tilde{C}^a{}_b\, N + \tilde{C}^a{}_b(\delta N/\delta\epsilon)\delta''\epsilon + \tilde{C}^a{}_b(\delta N/\delta\eta)\delta''\eta, \tag{18b}$$

$$\tilde{C}^\nu \rightarrow \tilde{C}^\nu - \lambda_2\, \tilde{C}^\nu\, N'(\epsilon,\eta,E,H), \tag{18c}$$

$$\delta''\tilde{C}^\nu \rightarrow \delta''\tilde{C}^\nu - \lambda_2\{\delta''\tilde{C}^\nu\, N' + \tilde{C}^\nu(\delta N'/\delta\epsilon)\delta''\epsilon + \tilde{C}^\nu(\delta N/\delta\eta)\delta''\eta. \tag{18d}$$

The divergences contributed to $\Gamma'_{(n)}{}^d$ by $I_{inv}(A\uparrow+\tilde{Q}\uparrow)$ of (4) are removed by a redefinition of the α_i in that equation. As a result, the entire divergent piece of $\Gamma'_{(n)}$ has been removed, and our proof by induction from (n-1) to n is complete.

6. PROOF OF RENORMALIZABILITY: FINITE SET OF COUNTER-TERMS.

The degree of divergence of an arbitrary Feynman diagram is

$$D = 4\Lambda + \sum_v \delta_v - \sum_j \Pi_j \theta_j \tag{19}$$

where Λ is the number of independent loops, δ_v is the number of derivatives acting on internal lines at the vertex "v", Π_j the number of internal lines in the diagram corresponding to the field "j" and θ_j is the negative power of the momenta in the relevant propagator for "j". Topologically, with V the number of vertices, we have the identity $\Lambda = \sum_j \Pi_j + 1 - V$. We can thus reexpress D as

$$D = 4 + \sum_j (4 - \theta_j)\, \Pi_j - \sum_v (4 - \delta_v) \tag{20}$$

in the BF method, the 1PI diagrams with external "Q" lines and those with internal "A" lines will not contribute to the effective action. Hence, $\Pi_j = (1/2) \sum_j N^v{}_j$, where $N^v{}_j$ is the number of internal lines j leaving the vertex v. As a result,

$$D = 4 - \sum_v \{4 - \delta_v - (1/2) \sum_j (4 - \theta_j)\, N^v{}_j\} \tag{21}$$

Our effective action yields values of θ_j: 2 for $\omega^a{}_{b\mu}$ and $C^a{}_b$, 4 for $\eta^{\mu\nu}$ and C^μ; $\epsilon^a{}_\mu$ does not propagate, but to make room for torsion-squared terms (though they do not appear essential to the model) we can assume a value 4. If we include terms $(D\sigma)^2$ and $\sigma^2 R$ we find a value 2 for $\sigma(x)$. Inserting these values yields

$$D = 4 + \sum_v d_v$$

$$d_v = \delta_v + N^v(\omega) + N^v(C^a{}_b) + N^v(\sigma) - 4 \tag{22}$$

The theory is renormalizable if all vertices ("v") in the action have non-positive d_v ; this ensures that the divergences can be removed by a finite number of counter-terms, since higher order diagrams will not generate higher degree divergences. In the case of our action, d_v is indeed non-positive. Vertices involving K_g (the κ_3 term in Eq. (16)) have $d_v = -4$ and do not yield divergent contributions at all. As a result, this also ensures that there is no divergent contribution due to the the anti-ghost $\bar{C}_\nu$, i.e.

it does not appear in $\Gamma_{(n)}^{d}$, which implies that any contribution related to the η field in $\Gamma'_{(n)}{}^{d}$ is gauge invariant.

We conclude that our GL(4,R) quadratic Lagrangian with SSB is renormalizable; this is due to its similarity to the Weinberg-Salam model, i.e. a Yang-Mills gauge theory with SSB. This is achieved without quartic propagators. Such propagators appear only for an auxiliary field in the gauge-fixing Lagrangian and do not play a role in achieving renormalizability.

It remains to show that a unitary theory can be extracted from this model.

References

1. Y. Ne'eman and Dj. Sijacki, in Differential Geometrical Methods in Theoretical Physics, K. Bleuler and M. Werner eds., Kluwer Academic Pub., Dordrecht/Boston/London (1988), p. 333.

2. K.S. Stelle, Phys. Rev. D16 (1977) 953 and Gen. Rel. Grav. 9 (1978) 353.

3. C. Lee and Y. Ne'eman, Phys. Lett. B233 (1990) 286.

4. C. Lee, U. of Texas CPT preprint "Renormalization of a Gravity Model with Local GL(4,R) Symmetry", Feb. 1990.

5. C. Lee and Y. Ne'eman, Phys. Lett. B242 (1990) 59.

6. M. Goroff and A. Sagnotti, Phys. Lett. B160 (1985) 81.

7. A. Casher, Phys. Lett. B195 (1987) 50.

8. D.J. Gross and V. Periwal, Phys. Rev. Lett. 60 (1988) 2105.

9. D. Neville, Phys. Rev. D18 (1978) 3535 and D21 (1980) 867.

10. E. Sezgin and P. van Nieuwenhuizen, Phys. Rev. D21 (1980) 3269 and D22 (1980) 301.

11. R. Kuhfuss and J. Nitsch, Gen. Rel. Grav. 18 (1986) 1207.

12. Y. Ne'eman and Dj. Sijacki, Phys. Lett. B200 (1988) 489.

13. G. Stephenson, Nuo. Cim. 9 (1958) 263; C.W. Kilmister and D.J. Newman, Proc. Cam. Phil. Soc. 57 (1961) 851; C.N. Yang, Phys. Rev. Lett. 33 (1974) 445.

14. Y. Ne'eman and Dj. Sijacki, Phys. Lett. B157 (1985) 267, 275.

15. J. Mickelsson, Com. Math. Phys. 88 (1983) 551.

16. A. Cant and Y. Ne'eman, J. Math. Phys. 26 (1985) 3100.

17. Dj. Sijacki and Y. Ne'eman, J. Math. Phys. 26 (1985) 2457.

18. Y. Ne'eman, Ann. Inst. H. Poincaré, 28 (1978) 369.

19. Y. Ne'eman and Dj. Sijacki, Int. J. Mod. Phys. A2 (1987) 1655.

20. B.S. DeWitt, Phys. Rev. 162 (1967) 1195; J. Honerkamp, Nucl. Phys. B36 (1971) 130; G. 't Hooft, ibid, B62 (1973) 444; B.S. DeWitt, in Quantum Gravity 2, C. Isham et al. editors, Clarendon Press, Oxford (1981).

21. L.F. Abbott, Nucl. Phys. B185 (1981) 189 and Acta Phys. Pol. B13 (1982) 33; D.G. Boulware, Phys. Rev. D23 (1983) 389.

6. Short contributions

Third order nonlinear Hamiltonian systems : some remarks on the the action-angle transformation [1]

Sandra Carillo
Dipartimento di Metodi e Modelli Matematici per le Scienze Applicate,
Universtitá *La Sapienza*, Roma, Italy

Abstract: Multi-soliton solutions of third order nonlinear evolution equations admitting a recursion operator as well as a Lax operator are here considered. Specifically, results obtained in [2,8], which give a method to construct the action-angle transformation on the so-called "multi-soliton" manifold [15], are briefly discussed.

Crucial to achieve such a result is the nonlinear link between the eigenvectors of the Lax and the recursion operator. Furthermore, the action-angle transformation can be recognized to be an infinitesimal symmetry generator of the corresponding interacting soliton equation; thus, it can be also obtained via the direct analysis of the structural properties of the underlying dynamics. [2]

Let

$$u_t = K(u) \tag{1}$$

denote a nonlinear evolution equation admitting a hereditary recursion operator $\Phi(u)$ [6], [9] wherein, in general, "formal"integration operators appear [3]. Here we are concerned about solutions vanishing rapidly at infinity. Many examples as well as related detailed computations concerning the construction of action-angle transformations in general are comprised in [8].

As usual [5] we characterize the N-soliton manifold by

$$M_N = \{u \mid there\ are\ c_n\ s.\ t.\ \sum_{n=1}^{N} c_n K_n(u) = 0\} \ . \tag{2}$$

According to the general theory [10], [11], [16] the recursion operator of the given nonlinear system follows to be doubly degenerated when restricted to M_N. Furthermore, when the gradients of canonical action-angle variables are mapped by the hamiltonian formulation onto vector fields, then the eigenvectors of the recursion operator are obtained. In all those cases here considered in which the nonlinear evolution equation admits a hamiltonian formulation, it is well known [12] that it can be written in the form:

$$K(u) = \Theta(u) \ \nabla \ H(u) \tag{3}$$

[1] Work partially supported by the G.N.F.M. of C.N.R.and by the M.U.R.S.T. project "Geometria e Fisica"

[2] In the case of the Korteweg-deVries equation [2] such a symmetry group generator can be proved to generate a group in the kernel of the Miura transformation.

[3] Usually D^{-1} denotes integration from $-\infty$ to x

where $\Theta(u)$ is an implectic operator and $H(u)$ a scalar field. Hence, given an eigenvalue and a related eigenvector, there is a second eigenvector admitting the same eigenvalue.

The first one corresponds to a symmetry generator of the system, the latter represents the result of the application of $\Theta(u)$ to the gradient of an angle variable. Thus, when $\Theta(u)$ coincides with the operator D, i.e., for instance, in the cases of the Korteweg-deVries (KdV) and the modified Korteweg-deVries (mKdV) equation [8], it represents the derivative of the gradient of an angle variable.

Whenever the Lax formulation is given, then the vector-field defined by the nonlinear evolution equation under investigation represents an isospectral flow with respect, not only to its recursion operator, but also the related linear Lax operator [4].

Indeed, one of the many interesting concomitant properties enjoyed by the so-called "integrable" nonlinear evolution equations is that they are, in general, amenable to the Inverse Scattering Transform (I.S.T.) (see for instance [1]). Such is the case of the KdV, mKdV and nonlinear Schrödinger (NLS) equations, only to mention the most well known ones.

Explicit computations show [8] that whenever

$$L\omega = \alpha\omega \tag{4}$$

then

$$\Phi(u)(\omega^2)_x = \lambda(\omega^2)_x \tag{5}$$

where α and λ are the eigenvalues of the two spectral problems. Since this relation is independent on whether or not ω is a genuine eigenvector of L, a nonlinear link between the eigenvectors of these isospectral problems follows.

In fact:

a) linear superposition of different solutions of the linear spectral problem related to the Lax operator delivers other such solutions;

b) thus, the subsequent application of the nonlinear link between the two operators gives the three eigenvectors of the recursion operator since this is a third order eigenvalue problem;

c) Among the three formal solutions obtained, only those with vanishing boundary conditions at infinity belong to the multi-soliton manifold M_N: they are precisely those which correspond to the doubly degenerate spectrum of Φ [5].

Notwithstanding its efficiency, this method is unsatisfactory from the structural viewpoint since always two isospectral problems are needed for the construction of action-angle transformations.

---

[4] the Schrödinger operator: $L = D^2 + u$ if, for instance, we are considering the Korteweg-deVries equation.

[5] details concerning the needed computations are given in [2] in the KdV case and in [8] for further examples

Indeed, a further analysis [8] of the method shows that the action-angle map always is given by a symmetry group generator of the "full" interacting soliton equation [3] not only for the zero-soliton equation which is the analogue of the singularity manifold equation [6].

This characterization, importantly, implies that the action-angle map, as a symmetry group generator of the interacting soliton equation, can be further used to give a systematic approach. This new approach, then, makes the construction of the action-angle map completely independent of the existence of two isospectral operators: only the hereditary recursion operator spectral problem is required [8].

Finally, a further method to compute the action-angle transformation is given in [8]. Indeed, in those cases in which the recursion operator is of higher order, it might be difficult to single out, by an appropriate choice of the integration constants, those eigenvectors satisfying suitable boundary conditions. We can overcome this difficulty concentratings our attention on the links relating various nonlinear evolution equations.

In particular, in the case of the KdV equation, the links connecting such an equations with the mKdV and the KdV interacting soliton equation [13], provide an alternative method to construct the action-angle transformation [3].

Indeed, there are two possible links between the KdV and the related interacting soliton equation (for eigenvalue 0):

1) the direct Bäcklund Transformation [3]

$$u = \frac{1}{4s^2}(\lambda s^2 - 2ss_{xx} + {s_x}^2) \tag{6}$$

2) the composition of the Miura link which connects the KdV and mKdV equations

$$u = i\, v_x + v^2 \tag{7}$$

and the Cole-Hopf transformation relating the latter with the KdV interacting soliton equation.

The analysis of such links, shows that [3] the action-angle transformation, in this case, coincides with a symmetry group generator in the kernel of the Miura Transformation.

It can be furthermore remarked that restriction it is not necessary to consider only the zero-solitons equation since the "full" KdV interacting soliton equation follows on use of the translation group out of the zero soliton equation [3]. Accordingly [3], the same analysis remain valid whenever equations characterized by scaling degree $\alpha = -2$ and $\alpha = -1$ are related via Bäcklund transformation (see [14]).

[6] Indeed, that supports the belief that the singularity manifold equation as a test for complete integrability should be replaced by the equation which corresponds to the the full expansion test introduced in [3].

References

[1] F. Calogero and A. Degasperis: *Spectral Transform and Solitons I*, Studies in Mathematics and its Applications Vol. 13, North Holland Publishing Co., Amsterdam - New York - Oxford, 1982

[2] S. Carillo and B. Fuchssteiner: *The Action-Angle transformation for the Korteweg-deVries Equation*, in: Proceedings of the 5th Workshop on Nonlinear Evolution Equations and Dynamical Systems, Springer Verlag, Berlin-Heidelberg-New York, S. Carillo and O. Ragnisco, eds. , p.127-130, 1990

[3] B. Fuchssteiner and S. Carillo: *Soliton structure versus singularity analysis: Third order completely integrable nonlinear equations in 1+1 dimensions*, Physica, A 152, p.467-510, 1989

[4] P. J. Caudrey and R. K. Dodd and J. D. Gibbon: *A new hierarchy of Korteweg-de Vries equations*, Proc.R.Soc. London, A 351, p.407-422, 1976

[5] B.A. Dubrovin and S.P. Novikov: *Periodic and conditionally periodic analogs of the many-soliton solutions of the Korteweg-De Vries equation*, Sov. Phys.-JETP, 40, p.1058-1063, 1975

[6] B. Fuchssteiner: *Application of Hereditary Symmetries to Nonlinear Evolution equations*, Nonlinear Analysis TMA, 3, p.849-862, 1979

[7] B. Fuchssteiner: *Solitons in Interaction*, Progress of Theoretical Physics, 78, p.1022-1050, 1987

[8] B. Fuchssteiner and S. Carillo: *The Action-Angle transformation for Soliton Equations*, Physica A, 166, p.651-675, 1990

[9] B. Fuchssteiner and A. S. Fokas: *Symplectic Structures, Their, Bäcklund Transformations and Hereditary Symmetries*, Physica, 4 D, p.47-66, 1981

[10] B. Fuchssteiner and G. Oevel: *Geometry and action-angle variables of multisoliton systems*, Reviews in Mathematical Physics, 1990

[11] Gudrun Oevel, B. Fuchssteiner and M. Blaszak: *Action Angle variables and asymptotic data*, in: Proceedings of the 5.th Workshop on Nonlinear Evolution Equationsand Dynamical Systems, Springer Verlag, Berlin-Heidelberg-New York, S. S. DeLillo and O. Ragnisco, eds. , , p., 1990

[12] F. Magri: *A simple model of the integrable Hamiltonian equation*, 19, p.1156-1162, 1978

[13] S. Carillo and B. Fuchssteiner: *Non commutative Symmetries and new solutions of the Harry Dym equation*, in: Proceedings of the Como conference, Manchester University Press, 1989

[14] C. Rogers and S. Carillo: *On Reciprocal Properties of the Caudrey-Dodd-Gibbon and Kaup-Kupershmidt Hierarchies*, Physica Scripta, 36, 1987

[15] S. P. Novikov: *The periodic problem for the Kortewed-de Vries equation*, Functional Analysis and its Applications, 8, p.236-246, 1974

[16] H.M.M. TenEickelder: *On the local structure of Recursion operators for Symmetries*, Indag. Math., 89, p.386-403, 1986

Tensor Products of $q^p = 1$ Quantum Groups and WZW Fusion Rules [1]

Paolo Furlan [2], Alexander Ganchev [3], Valentina Petkova [3]

[2]Dipartimento di Fisica Teorica dell' Universita di Trieste, Italy and INFN, Sezione di Trieste, Italy
[3]Inst. for Nucl. Research and Nucl. Energy, Sofia 1784, Bulgaria

It has been a strange fact of life that the fusion rules for a WZW model based on the affine group $\hat{g}$ [1] resemble the tensor product decomposition rules of the underlying finite dimensional Lie algebra g. The introduction of quantum groups in the RCFT technology raised the hope that the fusion rules are given by some modified tensor product of the quantum deformation of g with a deformation parameter q a root of unity. Below we present a formula for a truncated tensor product multiplicities which is a generilization of a classical formula of Weyl with the role of the Weyl group played by the affine Weyl group (for details see [2]). It turns out to be equivalent to the Verlinde formula [3] for the fusion rules. This and the appearance of the affine Weyl group comes as no surprize having the striking similarity between the embedding patterns of Verma modules for the quantum group with q a root of unity [4-6] and for the affine Lie algebras [7] (see also [8]).

Consider the quantum enveloping algebra $U_q(g)$ of a simple Lie algebra g. The representation theory of $U_q(g)$ for generic q, i.e., q not a root of unity parallels completely [9] the undeformed case $q = 1$. The finite dimensional modules E_Λ for Λ integral dominant are irreducible and are the factor of the Verma module $V(\Lambda)$ by the maximal submodule $\sum_{w\in W} V(w\cdot\Lambda)$, where W is the Weyl group of g and $w\cdot\Lambda = w(\Lambda+\rho)-\rho$, with ρ the sum of the fundamental weights, is the shifted action of W. The multiplicities $m^{\Lambda}_{\alpha\beta}$ in the tensor product decomposition

$$E_\alpha \otimes E_\beta = \sum_{\Lambda \text{ dominant}} m^{\Lambda}_{\alpha\beta}\, E_\Lambda \tag{1}$$

are given by the classical Weyl formula $m^{\Lambda}_{\alpha\beta} = \sum_{w\in W}(-1)^{l(w)}\, n^{\alpha}_{w\cdot\Lambda-\beta}$ (see [10,11]) where n^{Λ}_{λ} is the multiplicity of the weight λ in E_Λ and $l(w)$ is the minimal number of reflections necessary to write w. The derivation is straightforward using the Weyl character formula.

The representation theory of $U_q(g)$ for q a rational phase differs considerably from the generic q case [4-6,12,13]. Let us call regular the integral weights Λ in the interior of the Weyl alcove, i.e., $(\Lambda+\rho,\theta) \le p$, θ - the highest root. (We will consider the case of simply laced g below.) The modules E_Λ for Λ regular are irreducible. Let $w' < w''$ be a partial order (Bruhat order) on the affine Weyl

[1] presented by A.G.

group $\hat{W}$. If Λ is regular, $w' < w''$, and $w' \cdot \Lambda$, $w'' \cdot \Lambda$ are dominant then $E_{w'' \cdot \Lambda}$ contains a submodule isomorphic to $E_{w' \cdot \Lambda}$. There is a quadratic form on E_Λ which is nondegenerate for regular Λ and degenerates if E_Λ is reducible.

Now consider the tensor product $E_\alpha \otimes E_\beta$, for α and β regular. In this case the decomposition (1) holds only as vector space. Since on the lhs we have a nondegenerate form any E_Λ on the rhs with Λ outside the alcove must combine with other $E_{\Lambda'}$, Λ' in the $\hat{W}$ orbit of Λ, producing an indecomposable, non highest weight module $E_{<\Lambda>}$, $<\Lambda> = (\Lambda, \Lambda', \ldots)$, of zero quantum dimension on which the quadratic form is nondegenerate. This grouping into zero q-dimension modules occures in pairs for $sl(2)$ (see [13] for a detailed discussion of this case) and $sl(3)$, in pairs and fours for $sl(4)$ and $sl(5)$, in 2,4, and 8 for $sl(6)$ and $sl(7)$, etc. These $E_{<\Lambda>}$ modules form an ideal in the tensor product ring and factoring out this ideal gives the truncated tensor product $\otimes^T$ [13-15]. Let us denote by $\overline{m}_{\alpha\beta}^{\Lambda}$ the reduced multiplicities:

$$E_\alpha \otimes^T E_\beta = \sum_{\Lambda \text{ regular}} \overline{m}_{\alpha\beta}^{\Lambda} E_\Lambda \tag{2}$$

In order to determine $\overline{m}_{\alpha\beta}^{\Lambda}$ one does not need detailed information about the structure of $E_{<\Lambda>}$. Because the dimensions of weight spaces are rigid under deformations of q and characters simply "count dimensions of weight spaces" ($\chi_\Lambda = \sum_\lambda n_\lambda^\Lambda e_\lambda$, with e_λ a formal exponent) the formula

$$\chi_\alpha \chi_\beta = \sum_{\Lambda \text{ dominant}} m_{\alpha\beta}^{\Lambda} \chi_\Lambda \tag{3}$$

continues to hold for $q^p = 1$ only in this case some of the terms in the rhs will not correspond to a direct summand but rather to a piece of a direct summand, i.e., for $E_{<\Lambda>}$ we have $\chi_{<\Lambda>} = \sum_{\Lambda \in <\Lambda>} \chi_\Lambda$. Now introduce "periodic" characters $\overline{\chi}_\Lambda = \sum n_\lambda^\Lambda \overline{e}_\lambda$, $\overline{e}_\Lambda = \overline{e}_{\Lambda + p\alpha}$ for any root α. Then $\overline{\chi}_{w \cdot \Lambda} = (-1)^{l(w)} \overline{\chi}_\Lambda$, $w \in \hat{W}$ and $\overline{\chi}_{<\Lambda>} = 0$ for any zero q-dimension module $E_{<\Lambda>}$. Thus one obtains

$$\overline{m}_{\alpha\beta}^{\Lambda} = \sum_{w \in \hat{W}} (-1)^{l(w)} m_{\alpha\beta}^{w \cdot \Lambda} = \sum_{w \in \hat{W}} (-1)^{l(w)} n_{w \cdot \Lambda - \beta}^{\alpha} \,. \tag{4}$$

This is a formula giving a simple algorithm for the calculation of the reduced multiplicities $\overline{m}_{\alpha\beta}^{\Lambda}$ (it can be easily implemented into a computer code).

Let us go from the formal exponents e_Λ to functionals on the weight space $e_{\lambda+\rho}(\cdot) = \exp(-(\lambda+\rho), \cdot)$. Their periodic counterparts are $\overline{e}_{\lambda+\rho}(2\pi i(\lambda' + \rho)) = q^{-(\lambda+\rho, \lambda'+\rho)}$, $q = e^{2\pi i/p}$. The characters of the integrable representations of the affine Lie algebra $\hat{g}$ of level $k = p - h$ (which representations are in one to one correspondence with the regular weights) constitute a representation of the modular group $SL(2, \mathbb{Z})$. If $S_{\Lambda\Lambda'}$ are the matrix elements [16] in this representation of the modular transformation of the torus $\tau \to -1/\tau$ one has $\overline{\chi}_\Lambda(2\pi i(\Lambda' + \rho)/p) = S_{\Lambda\Lambda'}/S_{0\Lambda'}$ and $\overline{\chi}(2\pi i\,\rho/p) = S_{\Lambda 0}/S_{00} = dim_q(\Lambda)$. Thus $\overline{\chi}_\alpha \overline{\chi}_\beta = \sum \overline{m}_{\alpha\beta}^{\Lambda} \overline{\chi}_\Lambda$ takes the form

$$\frac{S_{\alpha \Lambda'} S_{\beta \Lambda'}}{S_{0 \Lambda'}} = \sum_{\Lambda} \overline{m}_{\alpha\beta}^{\Lambda} S_{\Lambda' \Lambda} \ . \tag{5}$$

But this is exactly the Verlinde formula [3] for the fusion rules $N_{\alpha\beta}^{\Lambda}$ so we conclude that

$$\overline{m}_{\alpha\beta}^{\Lambda} = N_{\alpha\beta}^{\Lambda}. \tag{6}$$

Thus, in particular, (4) gives an easy algoritm for computing the fusion rules to be contrasted with the impractical depth rule [1]. Let us remark that once we know the final result we may forget everything about quantum groups and simply rearrange the sums in (5) (the ratios of S matrix elements are given by characters which in turn are given by the Weyl character formula as sums over the Weyl group) to obtain (6).

After the appearance of our paper [2] we understood that the new formula for the fusion rules has been reported in [17,18] and various derivations have been given in [19-21].

Acknowledgements. A.G. would like to thank the organizers of this conference for the opportunity to participate and for the financial support making it possible. This research was supported in part by contract No 403 with the Bulgarian Ministry of Science, Culture, and Higher Education.

References

1. D. Gepner and E. Witten, Nucl. Phys. B278 (1986) 493.
2. P. Furlan, A. Ganchev, V. Petkova, "Quantum groups and fusion rules multiplicities", Trieste preprint INFN/AE-89/15, Nucl. Phys. B, to appear.
3. E. Verlinde, Nucl. Phys. B300 [FS22] (1988) 360.
4. G. Lusztig, Contemp. Math. 82 (1989) 59.
5. C. DeConccini and V. Kac, contribution to this volume.
6. V.K. Dobrev, preprint IC/89/142, Proc. of Int. Group Theory Conf., StÀndrews, August 1989, to appear, and Proc. of XIII Johns Hopkins Workshop "Knots, Topology and Field Theory", Florence, June 1989, Ed. L. Lusanna, to appear.
7. V. Kac and D. Kazhdan, Adv. Math. **34** (19779) 97.
8. V. Dobrev, preprint IC/85/9.
9. M. Rosso, Comm. Math. Phys. 117 (1988) 581.
10. H. Weyl, "The Classical Groups" (Princeton Press, Princeton, 1940).
11. D.P. Zhelobenko, "Compact Lie Groups and their Representations" (Amer. Math. Soc., Providence, 1973) (translated from the Russian edition Nauka, Moscow, 1970).
12. P. Roche and D. Arnandon, Lett. Math. Phys. 17 (1989) 235.
13. V. Pasquier and H. Saleur, Nucl. Phys. **B330** (1990) 247.
14. L. Alvarez-Gaumé, C. Gomez, and G. Sierra, Nucl. Phys. **B330** (1990) 347.
15. J. Frohlich, contribution to this volume.
16. V.G. Kac and D. Peterson, Adv. Math. 53 (1984) 125.
17. V. Kac, Montreal conference, 1989.
18. M. Walton, preprint Universite Laval, 1989.
19. J. Fuchs and P. van Driel, preprint Amsterdam (1990).
20. M. Walton, preprint Universite Laval, 1990.
21. F. Goodman and H. Wenzl, preprint 1990, to appear in Adv. Math.

The Modular Group and Super-KMS Functionals[1]

Arthur Jaffe [2] and Orlin Stoytchev [3]

[2]Harvard University, Cambridge, MA 02138, USA
[3]Institute of Nucl. Research and Nucl. Energy, 1784 Sofia, Bulgaria

Supersymmetric generalizations of KMS functionals have been introduced and studied recently [1–4] in connection with their importance to entire cyclic cohomology theory for Banach algebras (noncommutative geometry) [5,6]. Given a $\mathbb{Z}_2$ graded C^*-dynamical system $(\mathcal{A}, \mathbb{R}, \alpha)$ with α commuting with the grading, we call the linear functional ω on $\mathcal{A}$ weakly super-KMS (sKMS) if it satisfies $\omega(ab) = \omega(b^\Gamma \alpha_i(a))$ for any analytic element $a \in \mathcal{A}$, where $b^\Gamma := b_+ - b_-$ with $b_\pm$ respectively the even and odd parts of b and $\alpha_i(a)$ is the element of $\mathcal{A}$, obtained by the action on a of the analitically continued automorphism group α_z at the point $z = \sqrt{-1}$. (In comparison the usual KMS condition is $\omega(ab) = \omega(b\alpha_i(a))$.) Typically such functionals are nonpositive and the most one can require is to be self-adjoint.

The standard KMS condition arises naturally as a property of any normal, faithful state ϕ on a von Neumann algebra $\mathcal{A}$ with respect to the modular group σ^ϕ, as shown in the Tomita–Takesaki theory. The aim of this paper is to prove that the weak sKMS condition also arises naturally as a property of any (not necessarily positive) self-adjoint normal faithful (the exact meaning to be defined later) functional ω on $\mathcal{A}$ with respect to canonically defined $\mathbb{Z}_2$ grading Γ of the algebra and a one-parameter $*$-automorphism group σ^ω. Thus by giving up positivity and keeping just self-adjointness we obtain a "super" Tomita–Takesaki theory, generalizing the usual one. Furthermore a uniqueness property holds in complete analogy with the standard case.

In our construction we use the (unique) Jordan decomposition of ω [7]:

$$\omega = \omega_+ - \omega_- \,, \quad \omega_\pm \text{ positive}, \quad \omega_+ \perp \omega_-$$

and the fact that there exist mutually singular projections $\chi\pm \in \mathcal{A}$ onto the supports of $\omega_\pm$, i.e., the following identities hold [8]:

$$\omega_\pm(a) = \pm\omega(a\chi_\pm) \,, \qquad \forall a \in \mathcal{A} \,,$$

$$\chi_+\chi_- = 0 \,.$$

[1] Presented by O. Stoytchev

We define a positive functional $|\omega|$:

$$|\omega| := \omega_+ + \omega_- \quad .$$

We have proven the following

Theorem . *If* ω *is a normal self-adjoint weakly sKMS functional, then* $|\omega|$ *is a KMS functional.*

In the proof of this theorem the following identities, shown by us, are instrumental:

$$\omega_+(b\chi_- a) = \omega_-(b\chi_+ a) = 0 \ , \qquad a \in \mathcal{A}_+$$
$$\omega_+(b\chi_+ a) = \omega_-(b\chi_- a) = 0 \ , \qquad a \in \mathcal{A}_-$$

The above theorem, apart from being interesting on its own, is essential in the proof of the uniqueness part of our main result.

For the generalization of the Tomita–Takesaki theory we need to extend the notion of a faithful functional to the case of a (nonpositive) self-adjoint functional ω . This justifies the following

Definition . A self-adjoint functional ω will be called *faithful* if the conditions $a \geq 0, \omega(a\chi) = 0$ for any projection χ imply $a = 0$.

It is not hard to prove:

Lemma . *With the above definition the normal self-adjoint functional* ω *is faithful if and only if* $|\omega|$ *is faithful in the usual sense.*

Taking a faithful normal self-adjoint functional ω on the von Neumann algebra $\mathcal{A}$ there is a canonical $\mathbb{Z}_2$ grading of $\mathcal{A}$ defined by:

$$\mathcal{A}_+ := \chi_+\mathcal{A}\chi_+ \oplus \chi_-\mathcal{A}\chi_- \ , \quad \mathcal{A}_- := \chi_+\mathcal{A}\chi_- \oplus \chi_-\mathcal{A}\chi_+ \ .$$

The fact that the direct sum decomposition as a linear space $\mathcal{A} = \mathcal{A}_+ \oplus \mathcal{A}_-$ holds follows from ω being faithful. The algebraic conditions for a $\mathbb{Z}_2$ grading are immediate from the properties of $\chi_\pm$. Also it is not difficult to show that $\mathcal{A}_\pm$ are σ-weakly closed subspaces and invariant under the involution $*$.

We define the one-parameter $*$-automorphism group σ^ω to be just the modular group $\sigma^{|\omega|}$, corresponding to the faithful normal positive functional $|\omega|$. A nontrivial fact is that $\chi_\pm$ are invariant under σ^ω

Finally we state the main result of our paper:

Theorem . *Given a faithful, normal, self-adjoint, linear functional* ω *with norm one on a von Neumann algebra* $\mathcal{A}$ *, there exist a canonically-defined* $\mathbb{Z}_2$ *grading* Γ *on* $\mathcal{A}$ *and a* σ*-weakly continuous one parameter* $*$*-automorphism group* σ^ω *. The grading* Γ *commutes with the action of the automorphism group. The functional*

ω is weakly sKMS with respect to σ^ω and Γ. Furthermore, the canonical $\mathbb{Z}_2$ grading and the group σ^ω are the unique pair of a $\mathbb{Z}_2$ grading and a σ-weakly continuous $*$-automorphism group, commuting with each other, for which ω is weakly sKMS.

Acnowledgements: O.S. would like to thank the organizers of the conference for providing financial support, making his participation possible. The research was supported in part by the U.S.Department of Energy under Grants DE-FG02-88ER25065 and DE-FG05-87ER25033.

References

1. D. Kastler: Commun. Math. Phys. **121** 345 (1989)
2. A. Jaffe, A. Lesniewski, K. Osterwalder: K-Theory **2** 675 (1989)
3. A. Jaffe, A. Lesniewski, M. Wisniowski: Commun. Math. Phys. **121** 527 (1989)
4. O. Stoytchev: A Study of Super-KMS Functionals, Ph.D. Thesis, Virginia Tech. (1989)
5. A. Connes: Publ.Math IHES **62** 257 (1986)
6. A. Connes: K-Theory **1** 519 (1988)
7. G. Pedersen: C^*-Algebras and their Automorphism Groups, Academic press, London, New York, San Francisco (1979).
8. J. Dixmier: Von Neumann Algebras, North-Holland, Amsterdam, New York, Oxford (1981)

New Quantum Representation for Gravity and Yang-Mills Theory

Renate Loll

Physikalisches Institut, Universität Bonn, Nussallee 12, D-5300 Bonn 1

In a recent paper (Rovelli and Smolin 1990), an attempt is made to describe canonical, pure gravity with the help of a new set of loop variables. Its aim is to formulate (quantum) gravity nonperturbatively, i.e. in a regime where general diffeomorphism invariance is unbroken, and where one cannot resort to any perturbative expansion about a fixed background metric. These variables depend on closed curves in three-space and are expressed in terms of Ashtekar's new variables (Ashtekar 1987) (which make the imbedding of the gravity phase space into the phase space of an SU(2,$\mathbb{C}$) Yang-Mills phase space explicit). They are generalizations of Wilson lines (i.e. traces of holonomies), gauge-invariant under SU(2,$\mathbb{C}$)-transformations and nonlocal. The loop variables T_i possess a grading $i \geq 0$, and form an infinite-dimensional closing, graded Poisson-bracket algebra of the form $\{T_0, T_0\} = 0, \{T_m, T_n\} \sim T_{m+n-1}$, for $m + n > 0$, with respect to the canonical symplectic structure on Ashtekar's phase space. The algebra can be described in terms of cutting and glueing of loop diagrams.

The main purpose of the work summarized here is to initiate the construction of a complete and rigorous quantum theory, based on the infinite-dimensional graded algebra of nonlocal loop variables, which is applicable to both canonical gravity and gauge theory (Loll 1990). Although the use of loop variables in Yang-Mills theory is not a new idea (mainly within a path-integral context), the novelty is to have a set of such objects which form a closing Poisson-bracket algebra on phase space. This abstract algebra structure will serve as a starting point for the quantization (in the sense of finding a selfadjoint operator representation of the classical Poisson-bracket relations).

After eliminating certain quadratic constraints among the T-variables (arising from the fact that they are traces of complex 2×2-matrices), I go on working with a smaller set of variables, which I call the L-variables. They form a Poisson-bracket algebra of the same type, but now constitute a set of independent variables. From the knowledge of the structure constants of the L-algebra, I "integrate" the infinite-dimensional subalgebra of the loop variables L_0 and L_1, using a formal group law expansion.

Starting from the formal group law of the semidirect product $\mathcal{L}_0 \circledS \mathcal{L}_1$, I then construct a natural unitary representation of this group, such that the quantum commutators of the corresponding selfadjoint generators reproduce the classical

Poisson-bracket relations of the L-algebra. This is done along the lines of the usual construction of unitary irreducible representations of finite-dimensional semidirect-product groups à la Mackey. Of course, without specifying further the structure of the underlying loop space, one cannot define a rigorous Hilbert space inner product, and also it is unclear whether a concept like "the orbit structure under the $\mathcal{L}_0$-action" is still meaningful in the context of a formal group. Nevertheless, my results are no more formal than those obtained by Rovelli and Smolin, and, in contrast to the representation given there, I find a quantum representation of the full L-algebra (the difference between L- and T-algebra does not matter here) which closes without any additional higher-order terms in the commutation relations.

State vectors arise naturally as functionals $\Psi[\bar{L}]$ on the dual of $\mathcal{L}_0$, which for our purposes we can identify with $\mathcal{L}_0$ itself. One finds that on this representation space the selfadjoint operators $\hat{L}_0$ act by multiplication with L_0, and that the L_1's are essentially represented by terms of the form $L_0 \frac{\delta}{\delta L_0}$. This way the classical Poisson-bracket subalgebra of L_0's and L_1's is represented exactly by quantum commutators, with appropriate factors of $i\hbar$. Moreover, all the higher-order L_n's can be quantized on the same representation space and their quantum commutation relations are isomorphic to those of the classical Poisson-bracket algebra. Even at this level of formality it is remarkable that one can find a consistent, anomaly-free quantization of an infinite-dimensional algebra. The absence of factor-ordering problems is traced back to the special form of the quantum operators.

This so-called L-representation sets the kinematical stage for both gravity and SU(2) Yang-Mills theory. The next step is the incorporation of dynamics, where we have a wide range of possible applications. Lattice gauge models are particularly attractive because the use of L-variables leads to an independent, finite set of quantum states, and the whole quantum theory becomes totally explicit and rigorous. Here the L-representation is genuinely different from other, connection-based quantum representations. Other applications include the continuum theories of pure gravity in both 2+1 and 3+1 dimensions. For the latter one expects new insights into factor-ordering problems of the quantum constraints. Also a generalization of the present framework to SU(3) Yang-Mills theory seems to be feasible.

References

Ashtekar, A. (1987): "New Hamiltonian Formulation of General Relativity", Phys. Rev. D36 (1987) 1587

Loll, R. (1990): "A New Quantum Representation for Canonical Gravity and SU(2) Yang-Mills Theory", Nucl. Phys., in press

Rovelli, C. and Smolin, L. (1990): "Loop Space Representation of Quantum General Relativity", Nucl. Phys. B331 (1990) 80

GEOMETRIC QUANTIZATION OF THE FIVE-DIMENSIONAL KEPLER PROBLEM

Ivailo M. Mladenov

Central Laboratory of Biophysics
Bulgarian Academy of Sciences
Acad. G. Bonchev Str., Bl. 21, 1113 Sofia
Bulgaria

ABSTRACT

An extension of the Hurwitz'transformation to a canonical transformation between the corresponding phase spaces allows conversion of the five-dimensional Kepler problem into that of a constrained Harmonic oscillator in eight-dimensions. Then following Dirac we quantize the extended phase space imposing non-Abelian constraint conditions as superselection rules. In that way the interchangeability of the reduction and the quantization procedures for this problem is proved. As a side result a new regularization of the Kepler problem in the aforementioned dimension is established.

As Hamiltonian system the n-dimensional Kepler problem which is one of the best known and most important problem in Mechanics can be described by the triple (M,ω,H), where

$$M = T^*(R^n\setminus\{0\}) = T^*\dot{R}^n = \{x, p \in R^n;\ x \neq 0\}$$
$$\omega = dp\wedge dx \quad \text{and } H = \frac{|p|^2}{2} - \frac{\alpha}{|x|},\quad \alpha > 0 \tag{1}$$

For negative values of H, Moser [1] has proved the equivalence of the regularized flow of the Kepler problem with that of the geodesic flow on S^n. The orbits of the geodesic flow on S^n are great circles on S^n and can be parametrized by the points of the Grassmanian of the oriented two-planes in R^{n+1}. The Grassmanian (the orbit manifold) is the compact Hermitian symmetric space $SO(n+1)/SO(n-1)\times SO(2)$ which is isomorphic to the nonsingular (n-1)-dimensional complex quadric Q_{n-1} in

Partially supported by contracts 549/90 and 911/90 of the Ministry of Culture, Science and Education.

$\mathbb{C}P^n$. Direct application of geometric quantization scheme [2] to this Kaehler manifold produces the energy spectrum and multiplicities of the geodesic flow on S^n [3]. Combined with Moser's symplectomorphism this result enables us to find also the spectrum and corresponding multiplicities of the n-dimensional Kepler problem [3]. It should be noted that the quantization of the orbit manifold amounts in quantum mechanical terms to transition from Schrödinger to the Heisenberg picture. In addition to reduction, the other possibilities for quantizing any dynamical system is to work directly with the initial phase space or to go over its appropriate extension. The goal of the reducton procedure is to "simplify" as much as possible the equations of motion. However, it is conceivable that quite the reverse can take place - namely, the equations of the quotient system motion appears more complicated. It is one of the purposes of the present work to provide a transperant realization of Dirac's quantization programme in the extended phase space for the five-dimensional Kepler problem case. The Hydrogen atom in one, two and three dimensions and its close connection with Harmonic oscillator in one, two and four dimensions respectively is an object of continual discussion in the literature. Interestingly enough the five-dimensional Hydrogen atom can be placed next in this line. The mathematical basis for these intimate relationships is the Euler identity realized by the so-called non-bijective transformations (extensively studied in [4] and [5]):

$$(x_1^2+x_2^2+\ldots+x_n^2)^{1/2} = u_1^2+u_2^2+\ldots+u_m^2, \quad n < m \tag{2}$$

Here x_i and x_k are the Cartesian coordinates describing the Hydrogen atom and the Harmonic oscillator respectively. It is the beautiful theorem of Hurwitz which tells us that the only possible (n,m) couples are:

$$(1,1),\ (2,2),\ (3,4) \text{ and } (5,8) \tag{3}$$

While the first three pairs have found many applications (see [6] and references therein), we know only one paper [7] dealing with the last case. There the authors have succeeded in obtaining the energy spectrum of the five-dimensional Hydrogen atom from that of the Harmonic oscillator in eight dimensions. The questions about multiplicities, symmetry as well non-invariance algebras remain open. In the present paper these questions have been answered using Geometric Quantization. The starting point is the $R^8\setminus\{0\} \to R^5\setminus\{0\}$ mapping (closely related to the Hopf fibration on spheres $S^7 \to S^4$) and its extension to a canonical transformation between phase spaces $(T^*\dot{R}^8,\ \Omega = dv\wedge du)$ and $(T^*\dot{R}^5,\ \omega = dp\wedge dx)$. Utilizing this symplectic map the five-dimensional Kepler problem is converted into the Hamiltonian system

$$(T^*\dot{R}^8,\ dv\wedge du,\ \tilde{H} = (|v|^2 - 8\alpha)/8|u|^2). \tag{4}$$

Next, we introduce the Hamiltonian of a Harmonic oscillator

$K = (|v|^2 + \lambda|u|^2)/2$, λ - an arbitrary constant

and state the following

Lemma. Let $E < 0$ and $\lambda = \sqrt{-8E}$. Then

$$H^{-1}(R) = K^{-1}(4\alpha) \tag{5}$$

Moreover, the flows defined by the Hamiltonians $\tilde{H}$ and K on energy hypersurfaces (5) coincide up to a monotonic change of parameter. The regularization procedure relies heavily on this lemma as it allows to consider dynamical flow of K in place of $\tilde{H}$. The Hamiltonian SU(2) action of $(T^*\dot{R}^8, \Omega)$ gives rise to a momentum mapping $J: T^*\dot{R}^8 \to su(2)^*$. According Gotay's theorem [8] only the zero level of the momentum J (in non-Abelian case) can be quantized consistently. Taking this into account one can prove (modulo some technicalities) the following results:

Result 1. The energy spectrum (bound states) of the five dimensional Kepler problem (n=5 in (1), α-fixed) consists of the energy levels

$$E = -\alpha^2/2(N+1)^2 \tag{6}$$

with corresponding multiplicities

$$m(E_N) = N(N+1)^2(N+2)/12, \quad N = 1,2,... \tag{7}$$

Result 2. The symmetry (invariance) Lie algebra of the problem under consideration is $su(4) \cong spin(6) \cong so(6)$.

Result 3. The spectrum generating Lie algebra of Hydrogen atom in R^5 is $so(6,2) \cong so^*(8)$ (see Barut & Bracken [9] for the details of this identification).

Combined, the Results (1-3) established the equivalence of extended and reduced phase space quantization of the constrained classical system with symmetry $(\dot{\mathbb{C}}^4, \Omega, \tilde{H}, SU(2), J)$.

REFERENCES

1. Moser, J., Comm. Pure Appl. Math. 23, 609 (1970).
2. Simms, D. and Woodhouse, N., Lectures on Geometric Quantization, L.N.P. vol. 53 (1976).
3. Mladenov, I. and Tsanov, V., J. Geom. Phys. 2, 17 (1985).
4. Lambert, D. and Kibler, M., J. Phys. A, 21, 307 (1988).
5. Kibler, M. and Winternitz, P., J. Phys. A, 21, 1787 (1988).
6. Kibler, M. and Negadi, T., Croatica Chem. Acta 57, 1509 (1984).
7. Davtyan, L., Mardoyan, G., Pogosyan, G., Sissakian, S. and Ter-Antonyan, V., J. Phys. A, 20, 6121 (1987).
8. Gotay, M., J. Math. Phys. 27, 2051 (1986).
9. Barut, A. and Bracken, A., J. Phys. A, 23, 641 (1990).

Structure functions on the usual and exotic symplectic and periplectic supermanifolds

E.Poletaeva

SFB-170, Dept. of Math., Göttingen University; on leave of absence from Dept. of Math., Pennsylvania State University, Mac Allister Build. 305, University Park, PA 16802

Introduction. In this paper I continue calculating structure functions of classical superspaces. The problem, raised in [L3], is also formulated in [LSV], [LP], where there are also reviewed some calculations from [P1] - [P3], and [P4]; certain calculations are interpreted in [LPS1] and [LPS2].

The main result of my paper are theorems describing structure functions (SF) -- for definitions see [LSV] and [LP] -- on the usual and exotic symplectic and periplectic complex supermanifolds.

Recall that a usual symplectic (periplectic) supermaifold is the one with fixed nondegenerate closed even (odd) differential 2-form. The Darboux theorem on supermanifolds [L1], [SH] states that an above described 2-form ω reduces to the following form:

if it is even then $\omega = \omega_0 = \Sigma_i dp_i dq_i + \Sigma_j d\xi_j d\eta_j$ if there is an even number of odd coordinates or $\omega = \omega_0 = \Sigma_i dp_i dq_i + \Sigma_j d\xi_j d\eta_j + d\theta^2$ if there is an odd number of odd coordinates) and,

if it is odd then $\omega = \omega_1 = \Sigma_i dq_i d\xi_i$.

The Lie algebras of vector fields that preserve these forms are denoted, respectively, $\mathfrak{h}(2m|n)$ and $\mathfrak{le}(n)$; their respective central extensions $\mathfrak{po}(2m|n)$ and $\mathfrak{b}(n)$, are called the Poisson and Buttin superalgebra (see [L1], where $\mathfrak{le}(n)$, $\mathfrak{b}(n)$ and their contact cousin were introduced, the review [L2] and [Ko1], where their traceless analogues are described together with deformations, and [Ko2], where the representations of $\mathfrak{le}(n)$ and $\mathfrak{b}(n)$ -- prequantizations -- are described).

The Lie superalgebras of series $\mathfrak{le}(n)$ and $\mathfrak{b}(n)$ and their divergence-free subalgebras have deformations which preserve what I call *exotic periplectic* structures. An exotic structure is, actually, a linear combination of the periplectic form and the volume form. In one case, however, the Lie superalgebra of leitesian fields ($\mathfrak{le}(2)$) is isomorphic to the Lie algebra of hamiltonian fields ($\mathfrak{h}(2|2)$) giving rise to an *exotic symplectic* structure.

This exotic structure was interpreted by G. Shmelev, see [ALSh], [Sh], [L2], in addition to the above interpretation, as preserving (for $\lambda \neq 0, 1$) either of the pseudodifferential forms

$$\omega_+ = d\eta^{(2\lambda-1)/(1-\lambda)}((1-\lambda)dpdq + \lambda d\xi d\eta) \text{ or } \omega_- = d\xi^{1/\lambda-2}(dpdq + d\xi d\eta),\ \lambda \in \mathbb{C}.$$

The "symmetric" counterpart of the structures considered in this paper are metrics. The corresponding structure functions are described in [P1] [P2], [P5], [LP1].

The problems solved here were raised by D.Leites while I was his guest at IAS, Princeton, in 1989. I am thankful to him and IAS for help and hospitality. I am also thankful to SFB-170, the organizers of the DGM-XIX and an NSF grant via Penn. State University for financial support during the wonderful creative summer of 1990.

In what follows $R(\Sigma a_i \pi_i)$ denotes the irreducible $\mathfrak{g}_0$-module with highest weight $\Sigma a_i \pi_i$, where π_i is the i-th fundamental weight; we will denote it sometimes by its numerical labels $R(\Sigma a_i;\, a)$ the highest weight with respect to the center of

$\mathfrak{g}_0$ stands after semicolon, cf.[OV], Reference Chapter. Weights with respect to $\mathfrak{gl}(n)$, however, are most natural to denote with respect to the roots corresponding to coroots realized as matrix units on the main diagonal.

Definitions of the Spencer cohomology of the Cartan prolong of a pair $\mathfrak{g}_* = (\mathfrak{g}_{-1}, \mathfrak{g}_0)_*$, the order of SF, etc. see in [LP], [LSV].

1. An analogue of a theorem by Serre for Lie superalgebras: consequences of involutivity (after V.Serganova, cf. [LPS1]). The theorem ascribed in [LP] and [LSV] to Serre is actually a corollary of his initial statement that $\mathbb{Z}$-graded Lie algebra of the form $\mathfrak{g} = \oplus_{i\geq -1} \mathfrak{g}_i$ is *involutive* if and only if its Spencer cohomology $H^{k,s}\mathfrak{g}_*$ vanish for $s\geq 0$ ([St]). For Lie *super*algebras we only need here a theorem on vanishing of Spencer cohomology. To formulate it we have to superize the notion of involutivity. Let us do so and recall the classical definition of involutivity for Lie algebras.

Let $\mathfrak{g} = \oplus_{i\geq -1} \mathfrak{g}_i$ be a $\mathbb{Z}$-graded Lie superalgebra, $\{a_1, ..., a_n\}$ a (homogeneous) basis of $\mathfrak{g}_{-1}$. Clearly, the map

$a_i: \mathfrak{g} \to \mathfrak{g}$, $x \to [x, a_i]$

is a homomorphism of $\mathfrak{g}_{-1}$-modules. A $\mathbb{Z}$-graded Lie algebra $\mathfrak{g} = \oplus_{i\geq -1} \mathfrak{g}_i$ is called *involutive* if the maps a_i are onto. In the superized version of this notion we have to require the same property of the even maps a_i and to demand that the homology of the odd maps a_i vanish (by Jacoby identity $(a_i)^2 = 0$ for the odd a_i and the homology is well-defined). In scientific terms this is formulated as follows.

For a Lie superalgebra $\mathfrak{g} = \oplus_{i\geq -1} \mathfrak{g}_i$ set

$\mathfrak{g}^r = \ker a_1 \cap \ker a_2 \cap ... \cap \ker a_r.$

Clearly, $\mathfrak{g}^r = \oplus_{i\geq -1} (\mathfrak{g}^r)_i$, where $(\mathfrak{g}^r)_i = \mathfrak{g}^r \cap \mathfrak{g}_i$. Notice that $a_r((\mathfrak{g}^{r-1})_i) \subset (\mathfrak{g}^{r-1})_{i-1}$. The Lie superalgebra $\mathfrak{g} = \oplus_{i\geq -1} \mathfrak{g}_i$ will be called *involutive* if the following conditions are fulfilled:

(1) $\mathfrak{g}^n = \mathfrak{g}_{-1}$;

(2) $a_r(\mathfrak{g}^{r-1}) = \mathfrak{g}^{r-1}$ if a_r is even;

(3) $a_r(\mathfrak{g}^{r-1}) = \mathfrak{g}^r$ if a_r is odd.

Theorem. *If* $\mathfrak{g}$ *is involutive then* $H^{k,s}\mathfrak{g}_* = 0$ *for* $s\geq 0$ *and any* k.

2. Spencer cohomology of vectory Lie superalgebras in the standard grading. Theorem (cf. Theorem 0.2 in [LSV]).

1) For $\mathfrak{g}_* = \mathfrak{vect}(m|n)$ and $\mathfrak{svect}(m|n)$ *SF vanish except for* $\mathfrak{svect}(0|n)$ *when SF are of order n and constitute the* $\mathfrak{g}_0$*-module* $\Pi^n(\mathbb{1})$.

2) For $\mathfrak{g}_* = \mathfrak{h}(2m|n)$, *SF are* $E^3(\mathfrak{g}_{-1}) = \Pi(R(\phi_3)\oplus R(\phi_1))$ *for* $m \neq 0$ *or* $\Pi(R(3\phi_1)\oplus R(\phi_1))$ *for* $m = 0$, $n>3$.

3) For $\mathfrak{g}_* = \mathfrak{sh}(m)$, $m > 3$, *nonzero SF are same as for* $\mathfrak{h}(0/m)$ *and an additional direct summand* $\Pi^n(R(\pi_1))$ *of order* $n-1$.

4) For $\mathfrak{g}_* = \mathfrak{sle}(n)$, n >1, *nonzero SF are* $H^{1,2}\ \mathfrak{spe}(n) = S^3(\mathfrak{g}_{-1}{}^*)$, $H^{2,2}\ \mathfrak{spe}(n) = \Pi(\mathbb{1})$, $H^{n,2}\ \mathfrak{spe}(n) = \Pi^n(\mathbb{1})$.

The conformally symplectic structures were studied in [Le1], [Le2]. (Nobody, however, proved that the above-listed SF exhaust all of the structure functions.) Let us consider the funny equations on integrability of the corresponding G-structures; the existence of their solutions allows one to eliminate, at least partly, the adjective "almost" in the name of these *almost* symplectic manifolds.

Let σ be a nondegenerate (but not closed!) differential 2-form on a 2n-dimensional manifold M. Its differential is the sum of two components for n >2. Allowing the component that belongs to $R(\phi_3)$, the one with the higher highest weight, vanish we get the equation similar in form to the Bianci identity:

$$d\sigma = \lambda\sigma, \qquad \text{(CSy)}$$

where $\lambda \in \Omega^1(M)$. This equation describes a *conformally symplectic* structure. (CSy) implies that if a conformally symplectic structure is integrable then it is symplectic.

For n>1 there is a possibility to vanish the other component of SF which is "divisible" by σ. To nicely describe this possibility, notice that the multiplication by σ^{n-1}

$$\sigma^{n-1} : \Omega^1 \longrightarrow \Omega^{2n-1}$$

is an isomorphism. Therefore, the component divisible by σ vanishes if and only if

$$\sigma^{n-1}d\sigma = 0. \qquad \text{(ASy)}$$

The formula (CSy) is meaningful for supermanifolds as well, whereas (ASy) has to be rewritten. In presence of σ there is a canonical isomorphism (see [L2])

$$* : \Omega^1 \longrightarrow \Sigma_{-1}$$

where Σ_{-1} is the superspace of integrable forms of penultimate degree, cf. [BL]. (very roughly speaking, $*$ is $\sigma_0{}^{n-1}/\sigma_1{}^{(m/2)}$, where σ_0 (resp. σ_1) is the part of σ without differentials of the odd (resp. even) coordinates). In these notations (ASy) on supermanifolds takes the form

$$*d\sigma = 0. \qquad \text{(ASy)}$$

An almost symplectic structure which is not conformal, let me call it *ASy-structure*, i.e. the one described by the formula (ASy) had never been seriosly studied, cf. [Le1], [Le2].

3. SF for the exotic structures. In order to formulate the theorem calculating these SF let us give a finer description of the algebras which preserve exotic structures.

Terminological conventions. 1) The $\mathfrak{g}$-module V with the highest weight ξ and *even* highest vector will be denoted by V_ξ or $R(\xi)$.

2) Let $c\mathfrak{g}$ denote the trivial central "extent", i.e. the result of extension, of the Lie superalgebra $\mathfrak{g}$. Let $z = 1_{2n}$ be the unit matrix and $d = \mathrm{diag}(1_n, -1_n)$.

3.1. Periplectic superalgebras and their Cartan prolongs. Let P be a nondegenerate superskewsymmetric odd bilinear form on a superspace V. Clearly, dim V = (n, n). Define the odd analogue of the symplectic Lie algebra, the periplectic Lie superalgebra $\mathfrak{pe}(n)$, and its special subsuperalgebra setting

$$\mathfrak{pe}(n) = \{X \in \mathfrak{gl}(n, n) : X^{st}P + (-1)^{p(X)}PX = 0\};$$

$$\mathfrak{spe}(n) = \mathfrak{pe}(n) \cap \mathfrak{sl}(n, n).$$

Lemma. 1) *There exists a* $\mathbb{Z}$*-grading of the Lie superalgebra* $\mathfrak{pe}(n+1)$ *of the form*

$$\mathfrak{g}_{-1}\oplus\mathfrak{g}_0\oplus\mathfrak{g}_1\oplus\mathfrak{g}_2$$

where $\mathfrak{g}_{-1} = V$, $\mathfrak{g}_0 = \mathfrak{cpe}(n)$, $\mathfrak{g}_1 = V^* = \Pi(V)$, $\mathfrak{g}_2 = \Pi(<1>)$.

2) *There exists a* $\mathbb{Z}$*-grading of the Lie superalgebra* $\mathfrak{spe}(n+1)$ *of the form*

$$\mathfrak{g}_{-1}\oplus\mathfrak{g}_0\oplus\mathfrak{g}_1\oplus\mathfrak{g}_2$$

where $\mathfrak{g}_{-1} = V$, $\mathfrak{g}_0 = \mathfrak{spe}(n)\oplus<d+nz>$, $\mathfrak{g}_1 = V^* = \Pi(V)$, $\mathfrak{g}_2 = \Pi(<1>)$.

Proof can be deduced from [K].

Theorem. *Let* $\mathfrak{g}_{-1} = V$. *Then*

a) If $\mathfrak{g}_0 = \mathfrak{spe}(n)$, $\mathfrak{pe}(n)$ *or* $\mathfrak{spe}(n)\oplus<ad+bz>$, *where* $a,b \in \mathbb{C}$ *are such that* $ab \neq 0$ *and* $b/a \neq n$, *then* $(\mathfrak{g}_{-1}, \mathfrak{g}_0)_* = \mathfrak{g}_{-1}\oplus\mathfrak{g}_0$.

b) If $\mathfrak{g}_0 = \mathfrak{cpe}(n)$ *or* $\mathfrak{spe}(n)\oplus<d+nz>$ *then* $(\mathfrak{g}_{-1}, \mathfrak{g}_0)_*$ *is either* $\mathfrak{pe}(n+1)$ *or* $\mathfrak{spe}(n+1)$, *respectively, in the* $\mathbb{Z}$*-grading described in Lemma.*

Let us describe now a deformation of $\mathfrak{h}(2|2)$. With respect to the standard $\mathbb{Z}$-grading of $\mathfrak{g}$ we have $\mathfrak{g}_0 = \mathfrak{osp}(2|2) = \mathfrak{vect}(0|2)$.

A transparent description of continuous irreducible $\mathfrak{vect}(m|n)$-modules with highest weight and, in particular, of finite-dimensional $\mathfrak{vect}(0|n)$-modules is given in [BL]. The $\mathfrak{g}_0$-module V with highest weight θ and *even* highest weight vector will be denoted by V_θ or $R(\theta)$. If $\theta = (\lambda, \lambda)$ then I will write just V_λ instead of V_θ. Recall that V_λ and $V_{1-\lambda}$ are dual; V_λ are irreducible for $\lambda \neq 1$ or 0 when there is a 1-dimensional trivial submodule or quotient module, respectively, denoted by $\mathbb{1}$.

Let $\mathfrak{g}_{-1} = \Pi(V_{1/2})$. Recall that $\mathfrak{g}_0 = \mathfrak{osp}(2|2)$; then for this $\mathfrak{g}_{-1}$ we clearly have $(\mathfrak{g}_{-1}, \mathfrak{g}_0)_* = \mathfrak{h}(2|2)$. Let us define the *deform* (as M. Gerstenhaber justly calls the result of a deformation) $\mathfrak{g} = \mathfrak{h}(2|2; \lambda)$ of $\mathfrak{h}(2|2)$ as follows. Set $\mathfrak{g} = (\mathfrak{g}_{-1}, \mathfrak{g}_0)_*$, where $\mathfrak{g}_0$ is the same as for $\mathfrak{h}(2|2)$ whereas $\mathfrak{g}_{-1} = \Pi(V_\lambda)$, $\lambda \in \mathbb{C}$. (An interpretation of $\mathfrak{g}_{-1}$: Recall that $\mathfrak{g}_0 = \mathfrak{vect}(0|2)$; hence, as $\mathfrak{g}_0$-module, $\mathfrak{g}_{-1}$ is, up to the change of parity, the space of λ-densities on the $(0|2)$-dimensional supermanifold.) The superalgebras $\mathfrak{h}(2|2; 0)$ and $\mathfrak{h}(2|2; 1)$ are not simple but contain the Lie superalgebra $\mathfrak{svect}(2|1)$ of divergence-free vector fields on a $(2|1)$-dimensional supermanifold as subsuperalgebra and quotient superalgebra, respectively.

3.2. Theorem. *For* $\mathfrak{g} = \mathfrak{h}(2|2; \lambda)$ *all SF are of order 1, where the structure of the Jordan-Hölder series of the* $\mathfrak{g}_0$*-module SF is as follows (the arrow indicates the submodule of the indecomposable module):*

$V_{\lambda-1}\oplus V_\lambda\oplus V_{\lambda+1}$, *if* $\lambda \neq -1, 0, 1, 2$; in particular, these are SF for $\mathfrak{h}(2|2)$;

$(\mathbb{1}\to V_{(0,-1)})\oplus V_{-1}\oplus V_{-2}$, *if* $\lambda = -1$;

$(V_1\to\mathbb{1})\oplus(\mathbb{1}\to V_{(0,-1)})\oplus V_{-1}$, *if* $\lambda = 0$;

$(\mathbb{1}\to V_{(0,-1)})\oplus(V_1\to\mathbb{1})\oplus V_2$, *if* $\lambda = 1$;

$(V_1\to\mathbb{1})\oplus V_2\oplus V_3$, *if* $\lambda = 2$.

3.3. Theorem. *SFs of the* SpO(m|2n)*-structure on the total superspace of the linear bundle over a symplectic (resp. periplectic) supermanifold with* *connection whose curvature form is proportional to the symplectic (resp. periplectic) form vanish.*

In other words, $H^2(\mathfrak{g}_-; \mathfrak{g}_*) = 0$ *for* $\mathfrak{g}_* = \mathfrak{po}(2m|n)$ *and* $\mathfrak{sm}_\lambda(n)$.

4. Proof. Theorems 3.3 and 2.2 in the part concerning $\mathfrak{vect}$ follow from general statements on cohomology with coeficients in coinduced modules. The short exact sequence that determines the Poisson and Buttin superalgebras as central extensions of $\mathfrak{h}(2m|n)$ and $\mathfrak{le}(n)$, respectively gives rise to a long exact sequence that determines SF on the symplectic and periplectic supermanifolds after we will have recalled that the cohomology of $\mathfrak{g}_-$ with coefficients in $\mathfrak{po}(2m|n)$ or $\mathfrak{b}(n)$ are deRham cohomology and Π of them, respectively. This gives a part of Theorem 2.2 and 3.2 for $\lambda = 1/2$. Serganova's theorem shows that except for $\mathfrak{svect}(0|n)$ SF vanish if the order of SF is not 1; than I directly calculate SF as module over the even part of $\mathfrak{g}_0$; for a generic λ the answer shoul be as above; finally (and this is most difficult part because there are no rules so far to see this immediately) I look which of these modules are glued into indecomposable $\mathfrak{g}_0$-modules.

References

[ALS] D.Alekseevsky, D.Leites, I.Shchepochkina, New examples of simple Lie superalgebras of vector fields, C.r.Acad. bulg. Sci., v. 34, #9, 1980, 1187-1190 (in Russian)

[BL] J.Bernstein, D.Leites, Invariant differential operators and irreducible representations of Lie superalgebras of vector fields. Selecta Math. Sov., v.1, #2, 1981, 143-160

[F] Fuchs D., Cohomology of infinite dimensional Lie algebras, Consultunts Bureau, NY, 1987

[G] A.Goncharov, Infinitesimal structures related to hermitian symmetric spaces, Funct. Anal. Appl, 15, n3 (1981),23-24 (in Russian); a detailed version: id, Generalized conformal structures on manifolds. In: [L3], #11 and Selecta Math. Sov. 1987

[K] Kac V., Classification of simple $\mathbb{Z}$-graded Lie superalgebras and simple Jordan superalgebras. Commun. Alg. 5(13), 1977, 1375-1400

[Ko1] Kotchetkoff Yu. Déformations des superalgèbres de Buttin et quantification. C.R. Acad. Sc. Paris, t.299, série I, n. 14, 1984, 643-645

[Ko2] Kochetkov Yu. Induced representation of Leitesian superalgebras. Queastions of group theory and homological algebra. Yaroslavl Univ. 1983, 120-123; II. id. ibid., 1989, 142-147 (Russian)

[Le1] Lee Hwa-Chung, A kind of even-dimensional differential geometry and its application to exterior calculus. Amer. J. Math. v. 65, 1943, 433-438

[Le2] Lee Hwa-Chung, On even-dimensional skew-metric spaces and their groups of transformations. Amer. J. Math. v. 67, 1945, 321-328.

[L1] D.Leites, New Lie superalgebras and mechanics. Soviet Math. Doklady v.18, n 5, 1977, 1277-1280

[L2] D.Leites, Lie superalgebras. In: Modern problems of mathematics. Recent developments, v. 25, VINITI, Moscow, 1984,3-49 (in Russian) = Engl. transl. J. Sov. Math. (JOSMAR), v. 30, #6, 1985.

[L3] Leites D., Supermanifold theory. Karelia Branch of the USSR Acad. of Sci., Petrozavodsk, 1983, 200pp. (Russian) = in English a still more expanded version (in 7 volumes) is to be published by Kluwer in 1991-92; meanwhile see the preprinted part in: [L3]

[L4] Leites D. (ed.), Seminar on supermanifolds, Reports of Dept. of Math. of Stockholm Univ. n1-34, 2800 pp., 1986-89

[LP] D.Leites, E.Poletaeva, Analogues of the Riemannian structure for classical superspaces. In: Proc. Int. Alg. Conf. Novosibirsk, August 1989. (To appear)

[LPS1] D.Leites, E.Poletaeva, V.Serganova, Einstein equations on manifolds and supermanifolds (to appear)

[LPS2] D.Leites, E.Poletaeva, V.Serganova, Structure functions for contact structures on supermanifolds (to appear)

[LSV] D.Leites, V.Serganova, G.Vinel. Classical superdomains and related structures (to appear)

[OV] Onishchik A.L., Vinberg E.B., Seminar on algebraic groups and Lie groups. Springer, Berlin ea, 1990

[P1] E.Poletaeva, On Spencer cohomology associated with some Lie superalgebras. In: Questions of group theory and homological algebra, Yaroslavl Univ. Press, Yaroslavl, 1988, 162-167 (in Russian)

[P2] E.Poletaeva, On Spencer cohomology associated with certain $\mathbb{Z}$-gradings of simple Lie superalgebras. Proc. of All-Union algebraic conference, Lvov Univ. Press, Lvov, 1987, (in Russian)

[P3] E.Poletaeva, Spencer cohomology of Lie superalgebras of vector fields. In: Questions of group theory and homological algebra.,Yaroslavl Univ. Press, Yaroslavl, 1990, (in Russian)

[P4] E.Poletaeva, Penrose tensors on supergrassmannians. Math. Scand. (to appear)

[P5] E.Poletaeva, Analogues of the Riemann tensor for the odd metrics on supermanifolds. Geom. Dedicata (to appear)

[Sh] G.Shmelev, Differential operators invariant with respect to the Lie superalgebra $h(2|2;\lambda)$ and its irreducible representations. C.R.de l'Acad. bulg des Sci. t. 35, #3, 1982, 287-290 (in Russian)

[SH] Shander V., Analogues of the Frobenius and Darboux theorems for supermanifolds. C. R. de l'Acad. bulg. de Sci., t. 36, #3, 1983, 309-312

[St] Sternberg S. Lectures on differential geometry, 2nd ed, Chelsey, 1985

Symbols alias Generating Functionals — a Supergeometric Point of View

T. Schmitt

Institut für Mathematik, Mohrenstraße 39, Berlin 1086, GDR

Abstract: We exhibit the supergeometric meaning of the symbols or "generating functionals" considered by F. A. Berezin [1] of states and operators in Fock space.

Generating functionals of states and operators in Fock space were invented by F. A. Berezin [1] twenty five years ago. However, while in the purely bosonic case they are analytic functions on the one-particle state space, their geometric interpretation remained obscure on the fermionic side: they are just elements of an "infinite-dimensional Grassmann algebra" the origin of which is not clear.

Here we will indicate how geometry on supermanifolds provides a satisfactory solution of the riddle: the Grassmann algebra mentioned above is nothing but the algebra of superfunctions on the $0|\infty$-dimensional supermanifold deterined by the one-particle state space H. More generally, supergeometry allows to unify bosonic and fermionic case, instead of considering them separately, as Berezin did.

The key for this new approach is the framework of infinite-dimensional real-analytic supermanifolds modelled over locally convex topological vector spaces, which was developed in [2]. Within this framework, a **linear supermanifold** $\mathrm{L}(E)$ is assigned to every locally convex vector space E; it consists of the even part E_0 as underlying topological space, a structure sheaf $\mathcal{O}$ on it, and some additional structure which has to be preserved under morphisms, in order to exclude "nonsense morphisms". In case $E = \mathbb{R}^{m|n}$, one gets essentially the usual $m|n$-dimensional linear superspace of usual supergeometry in the approach of Berezin/Leites.

Our symbols will be real-analytic superfunctions on the linear supermanifold $\mathrm{L}(H)$ determined by the Hilbert space H (or, more exactly, by its underlying real topological vector space). We indicate only the results without proof; a detailed presentation will (probably) appear in the third part of [2].

As one particle state space, let be given a $\mathbb{Z}_2$-graded complex Hilbert space which consists of bosonic and fermionic part:

$$H = H_0 \oplus H_1.$$

The Fock space over H,

$$\mathrm{F}(H) = \mathrm{F}(H_0) \otimes \mathrm{F}(H_1)$$

is the Hilbert tensor product of the bosonic Fock space over H_0 with the fermionic Fock space over H_1.

1. To any state $\Phi \in \mathrm{F}(H)$ we assign an antiholomorphic superfunction

$$\sigma(\Phi) \in \mathcal{O}(\mathrm{L}(H))$$

(**symbol of the state**) on the infinite-dimensional linear superspace determined by the locally convex space H.

2. To any operator from finite vectors to unbounded vectors,

$$A : \mathrm{F}^{\oplus}(H) \to \mathrm{F}^{\Pi}(H)$$

we assign two real-analytic superfunctions

$$\sigma_m(A), \sigma(A) \in \mathcal{O}(\mathrm{L}(H))$$

(**matrix symbol** and **normal symbol** of the operator).

These symbols are connected by

$$\sigma(A) = e^{-<z,z>} \sigma_m(A),$$

where $e^{-<z,z>} \in \mathcal{O}(\mathrm{L}(H))$ is a well-defined superfunction.

In the purely even case $H = H_0$, these symbols are just maps

$$\sigma(\Phi), \sigma_m(A), \sigma(A) : H \to \mathbb{C}$$

while in the purely odd case, they are elements of an infinite-dimensional Grassmann algebra.

3. In terms of Taylor expansion coefficients at the origin, these symbols are characterized by

$$\partial_{g_1} \ldots \partial_{g_k} \sigma(\Phi)(0) =< 0|a(g_1) \ldots a(g_k)|\Phi >,$$

$$\partial_{h_1} \ldots \partial_{h_l} \partial_{g_1} \ldots \partial_{g_k} \sigma_m(A)(0) =$$
$$=< 0|a(h_1) \ldots a(h_l) A a^*(g_1) \ldots a^*(g_k)|0 >$$

where $g_1, \ldots, g_k, h_1, \ldots, h_l \in H$, and $a(.)$, $a^*(.)$ are the usual annihilation and creation operators.

Straightforward estimates show that these formulas define the superfunctions wanted.

4. Characterization in terms of orthonormal bases $(e_i) \in H_0$, $(f_j) \in H_1$: If

$$\Phi = \sum c^{(\mu)}_{(i)(j)} \frac{a^*(e_{i_1})^{\mu_1} \ldots a^*(e_{i_k})^{\mu_k} a^*(f_{j_1}) \ldots a^*(f_{j_l})|0>}{\sqrt{\mu_1! \ldots \mu_k!}}$$

(and hence $\|\Phi\|^2 = \sum |c^{(\mu)}_{(i)(j)}|^2$) then

$$\sigma(\Phi) = \sum c^{(\mu)}_{(i)(j)} \frac{\sigma(e_{i_1})^{\mu_1} \ldots \sigma(e_{i_k})^{\mu_k} \sigma(f_{j_1}) \ldots \sigma(f_{j_l})}{\sqrt{\mu_1! \ldots \mu_k!}}$$

where $\sigma(e_i)$, $\sigma(f_j)$ are the antilinear superfunctions on $L(H)$ determined by e_i, f_j. Also, if

$$A = \sum c^{(\mu)(\nu)}_{(i)(j)(i')(j')} a^*(e_{i_1})^{\mu_1} \dots a^*(e_{i_k})^{\mu_k} a^*(f_{j_1}) \dots$$
$$\dots a^*(f_{j_l}) a(e_{i'_1})^{\nu_1} \dots a(e_{i'_{k'}})^{\nu_{k'}} a(f_{j'_1}) \dots a(f_{j'_{l'}})$$

(with a suitable convergence notion) then

$$\sigma(A) = \sum c^{(\mu)(\nu)}_{(i)(j)(i')(j')} \sigma(e_{i_1})^{\mu_1} \dots \sigma(e_{i_k})^{\mu_k} \sigma(f_{j_1}) \dots$$
$$\dots \sigma(f_{j_l}) \overline{\sigma(e_{i'_1})}^{\nu_1} \dots \overline{\sigma(e_{i'_{k'}})}^{\nu_{k'}} \overline{\sigma(f_{j'_1})} \dots \overline{\sigma(f_{j'_{l'}})}.$$

5. An alternative characterization is given by using coherent states: the usual coherent state map

$$H_0 \to F(H_0), \quad g \mapsto \sum_{i \geq 0} a(g)^i / i! =: Q(g)$$

"superizes" to a (non-linear) morphism of linear supermanifolds

$$Q : L(H) \to L(F(H))$$

which we call the **Poisson morphism**, and which is the same as an $F(H)$-valued superfunction

$$Q \in \mathcal{O}^{F(H)}(L(H)).$$

One has

$$\sigma(\Phi) = \langle Q | \Phi \rangle,$$
$$\sigma_m(A) = \langle Q | A | Q \rangle,$$
$$\sigma(A) = \langle P | A | P \rangle$$

where $P := e^{-\langle z,z \rangle /2} Q$ is the **normalized Poisson morphism.**

6. *Concluding Remark:* More general unbounded operators can be treated, too; their symbols live on smaller supermanifolds $L(D)$ where $D \subseteq H$ is a dense $\mathbb{Z}_2$-graded subspace, equipped with its own locally convex topology.

References

1. Berezin, F. A. (1966): *The method of second quantization*, New York: Academic Press
2. Schmitt, T. (1989): *Infinite-dimensional supermanifolds*, Report des IMath, Berlin

Sheaves of graded Lie algebras over variable Riemann surfaces and a paired Weil-Petersson inner product

Paolo Teofilatto

Dipartimento di Fisica, Università di Roma II
Via E. Carnevale, 00173 Roma, Italy

The bosonic string partition function can be defined by the integrals over the moduli spaces M_g of genus g Riemann surfaces:

$$Z_g = \int_{M_g} (\det \Delta_0)^{-13} \det \Delta_2 \, d\nu$$

where Δ_0,Δ_2 are Laplacians acting on the fibers of the bundles of relative one- and two-forms $\omega \to C_g \to M_g$, $\omega^2 \to C_g \to M_g$, respectively. In the previous formula $d\nu$ is the measure associated with the metric defined by the Weil-Petersson inner product on the fibers of $R^0\omega^2$, the bundle of relative global sections of ω^2. Superstring theory has not such a mathematical sound base: its measure and domain of definition (supermoduli space) are only partially known. Formally superstring measure can be written as:

$$\text{sdet}\,(\Delta_0^s)^{-5}\,\text{sdet}\,\Delta_2^s \, d\nu^s$$

with Δ_0^s, Δ_2^s (super) Laplacians defined on sections of appropriate bundles over supermoduli space, and $d\nu^s$ is some generalization of the Weil-Petersson metric defined by a suitable inner product. Some contributions to the mathematical foundations of superstring theory are presented here:

(a) We identify supermoduli space as the space of deformations of a graded Lie algebras which arises from classical uniformization theory of Riemann surfaces.

(b) Using the Killing form on such graded algebra we define a paired Weil-Petersson inner product, thus introducing the measure $d\nu^s$.

Our starting point is a reformulation of classical uniformization theory in terms of sheaves of graded Lie algebras.

Theorem 1 *Projective uniformizations of a Riemann surface M generate sheaves A of graded Lie algebras over M, with stalks isomorphic to the graded algebra $osp(2,1)$.*

This construction extends to the space of moduli of Riemann surfaces:

Theorem 2 *The process of simultaneous uniformization of variable Riemann surfaces is related to a sheaf $\mathcal{A}$ of graded Lie algebras defined over the universal Teichmuller family.*

Proofs of Theorems 1 and 2 are in [T].

The sheaf $\mathcal{A}$ is a typical model of what we call "graded Riemann surface structure":

Definition 1 A graded Riemann surface structure is a holomorphic family of Riemann surfaces together with an analytic sheaf of $osp(2,1)$ algebras over it.

Remark 1. Graded Riemann surface structures are equivalent to the super Riemann surfaces defined in [M], and the equivalence of the corresponding parameter spaces of local deformations is proved in [T].

We recall that the cotangent bundle of the moduli space of Riemann surfaces is the sheaf $R^0\omega^2$ and the Weil-Petersson inner product defines a metric on its fibers.

Definition 2 The Weil-Petersson inner product on the fibers of $R^0\omega^2$ is defined by

$$(WP)_2\langle\phi_1,\phi_2\rangle = -\int_M Z_2(\phi_1)\wedge_B Z_2(\phi_2)^*$$

where * denotes complex conjugation, Z_2 maps quadratic differentials into $sl(2,C)$ valued one forms, and the exterior product is taken with respect to the Killing form on $sl(2,C)$ [G].

Remark 2. Z_2 is a particular case of the Shimura map [S]

$$Z_{r-1} : H^0(M, K^{r+1/2}) \to S^{r-1}(C^2) + K$$

where K is the cotangent bundle of M and $S^{r-1}(C^2)$ is a flat sheaf with stalks isomorphic to the $r-1$ symmetric tensor product of C^2.

Remark 3. $sl(2,C)$ is the even part of the graded Lie algebra A of Theorem 1.

From Remark 1, it follows that the cotangent bundle of supermoduli space is isomorphic to the sheaf

$$R^0\omega^2 + R^0\omega^{3/2}$$

and we introduce a metric on it, using the following inner product on the fibers.

Definition 3 We define a paired Weil-Petersson inner product by

$$(WP)^s\langle(\phi_1,\psi_1),(\phi_2,\psi_2)\rangle = \int_M Z_s(\phi_1,\psi_1)\wedge_D Z_s(\phi_2,\psi_2)^*$$

where

(i) $Z_s = Z_1 + Z_2$ maps quadratic and 3/2 forms into A-valued one-forms.

(ii) D is the Killing form on A.

The above inner product on the cotangent of supermoduli space can be understood in classical terms.

Theorem 3 *We have*

$$(WP)_s\langle(\phi_1,\psi_1),(\phi_2,\psi_2)\rangle =$$
$$(-3)[(WP)_2\langle\phi_1,\phi_2\rangle\rangle + 2(WP)_{3/2}\langle\psi_1,\psi_2\rangle\rangle]$$

where

$$(WP)_{3/2}\langle\psi_1,\psi_2\rangle = \int_M \psi_1(\psi_2)^*(z-z^*)dz\,dz^*.$$

The proof follows by computation of the Killing form of A.

Acknowledgements. I gratefully thank Dr. W. Harvey for his help and interest in my work.

References

[G] W. Goldmann: "The symplectic nature of fundamental groups of surfaces", Adv. Math. 54 (1984) p.200-245.

[M] Yu.I. Manin: "Critical dimensions in string theories", Funct. Anal. Appl. 20 (1986) p.244-246.

[S] G. Shimura: "Introduction to the arithmetic theory of automorphic functions", Princeton 1971.

[T] P.Teofilatto: "A natural graded Lie algebra sheaf over Riemann surfaces" to appear in Ann. Inst. H. Poincaré Sér. A.

Lecture Notes in Mathematics

Vol. 1236: Stochastic Partial Differential Equations and Applications. Proceedings, 1985. Edited by G. Da Prato and L. Tubaro. V, 257 pages. 1987.

Vol. 1237: Rational Approximation and its Applications in Mathematics and Physics. Proceedings, 1985. Edited by J. Gilewicz, M. Pindor and W. Siemaszko. XII, 350 pages. 1987.

Vol. 1250: Stochastic Processes – Mathematics and Physics II. Proceedings 1985. Edited by S. Albeverio, Ph. Blanchard and L. Streit. VI, 359 pages. 1987.

Vol. 1251: Differential Geometric Methods in Mathematical Physics. Proceedings, 1985. Edited by P.L. García and A. Pérez-Rendón. VII, 300 pages. 1987.

Vol. 1255: Differential Geometry and Differential Equations. Proceedings, 1985. Edited by C. Gu, M. Berger and R.L. Bryant. XII, 243 pages. 1987.

Vol. 1256: Pseudo-Differential Operators. Proceedings, 1986. Edited by H.O. Cordes, B. Gramsch and H. Widom. X, 479 pages. 1987.

Vol. 1258: J. Weidmann, Spectral Theory of Ordinary Differential Operators. VI, 303 pages. 1987.

Vol. 1260: N.H. Pavel, Nonlinear Evolution Operators and Semigroups. VI, 285 pages. 1987.

Vol. 1263: V.L. Hansen (Ed.), Differential Geometry. Proceedings, 1985. XI, 288 pages. 1987.

Vol. 1265: W. Van Assche, Asymptotics for Orthogonal Polynomials. VI, 201 pages. 1987.

Vol. 1267: J. Lindenstrauss, V.D. Milman (Eds.), Geometrical Aspects of Functional Analysis. Seminar. VII, 212 pages. 1987.

Vol. 1269: M. Shiota, Nash Manifolds. VI, 223 pages. 1987.

Vol. 1270: C. Carasso, P.-A. Raviart, D. Serre (Eds.), Nonlinear Hyperbolic Problems. Proceedings, 1986. XV, 341 pages. 1987.

Vol. 1272: M.S. Livšic, L.L. Waksman, Commuting Nonselfadjoint Operators in Hilbert Space. III, 115 pages. 1987.

Vol. 1273: G.-M. Greuel, G. Trautmann (Eds.), Singularities, Representation of Algebras, and Vector Bundles. Proceedings, 1985. XIV, 383 pages. 1987.

Vol. 1275: C.A. Berenstein (Ed.), Complex Analysis I. Proceedings, 1985–86. XV, 331 pages. 1987.

Vol. 1276: C.A. Berenstein (Ed.), Complex Analysis II. Proceedings, 1985–86. IX, 320 pages. 1987.

Vol. 1277: C.A. Berenstein (Ed.), Complex Analysis III. Proceedings, 1985–86. X, 350 pages. 1987.

Vol. 1283: S. Mardešić, J. Segal (Eds.), Geometric Topology and Shape Theory. Proceedings, 1986. V, 261 pages. 1987.

Vol. 1285: I.W. Knowles, Y. Saitō (Eds.), Differential Equations and Mathematical Physics. Proceedings, 1986. XVI, 499 pages. 1987.

Vol. 1287: E.B. Saff (Ed.), Approximation Theory, Tampa. Proceedings, 1985–1986. V, 228 pages. 1987.

Vol. 1288: Yu. L. Rodin, Generalized Analytic Functions on Riemann Surfaces. V, 128 pages, 1987.

Vol. 1294: M. Queffélec, Substitution Dynamical Systems – Spectral Analysis. XIII, 240 pages. 1987.

Vol. 1299: S. Watanabe, Yu.V. Prokhorov (Eds.), Probability Theory and Mathematical Statistics. Proceedings, 1986. VIII, 589 pages. 1988.

Vol. 1300: G.B. Seligman, Constructions of Lie Algebras and their Modules. VI, 190 pages. 1988.

Vol. 1302: M. Cwikel, J. Peetre, Y. Sagher, H. Wallin (Eds.), Function Spaces and Applications. Proceedings, 1986. VI, 445 pages. 1988.

Vol. 1303: L. Accardi, W. von Waldenfels (Eds.), Quantum Probability and Applications III. Proceedings, 1987. VI, 373 pages. 1988.

Lecture Notes in Physics

Vol. 351: A. Dervieux, B. Larrouturou (Eds.), Numerical Combustion. Proceedings, 1989. VII, 481 pages. 1989.

Vol. 352: K.B. Wolf (Ed.), Lie Methods in Optics II. Proceedings, 1988. XII, 197 pages. 1989.

Vol. 353: M. Barthes, J. Léon (Eds.), Nonlinear Coherent Structures. Proceedings, 1989. X, 280 pages. 1990.

Vol. 354: J. De Coninck, F. Dunlop (Eds.), Wetting Phenomena. Proceedings, 1988. VII, 112 pages. 1990.

Vol. 355: R. Lima, L. Streit, R. Vilela Mendes (Eds.), Dynamics and Stochastic Processes. Proceedings, 1988. V, 298 pages. 1990.

Vol. 356: R. Wehrse (Ed.), Accuracy of Element Abundances from Stellar Atmospheres. Proceedings, 1988. V, 109 pages. 1990.

Vol. 357: J.L. Lumley (Ed.), Whither Turbulence? Turbulence at the Crossroads. Proceedings, 1989. V, 535 pages. 1990.

Vol. 358: J. Kovalevsky, Astrométrie moderne. VI, 294 pages. 1990.

Vol. 359: K. Kirchgässner (Ed.), Problems Involving Change of Type. Proceedings, 1988. XII, 207 pages. 1990.

Vol. 360: Y. Mellier, B. Fort, G. Soucail (Eds.), Gravitational Lensing. Proceedings 1989. XV, 315 pages. 1990.

Vol. 361: P. Gaigg, W. Kummer, M. Schweda (Eds.), Physical and Nonstandard Ganges. Proceedings, 1989. IX, 310 pages. 1990.

Vol. 362: N.E. Kassim, K.W. Weiler (Eds.), Low Frequency Astrophysics from Space. Proceedings. XII, 280 pages. 1990.

Vol. 363: V. Ruždjak, E. Tandberg-Hanssen (Eds.), Dynamics of Quiescent Prominences. Proceedings, 1989. XI, 304 pages. 1990.

Vol. 364: T.T.S. Kuo, E. Osnes, Folded-Diagram Theory of the Effective Interaction in Nuclei, Atoms and Molecules. V, 175 pages. 1990.

Vol. 365: M. Schumacher, G. Tamas (Eds.), Perspectives of Photon Interactions with Hadrons and Nuclei. Proceedings. IX, 251 pages. 1990.

Vol. 366: M.-J. Goupil, J.-P. Zahn (Eds.), Rotation and Mixing in Stellar Interiors. Proceedings, 1989. XIII, 183 pages. 1990.

Vol. 367: Y. Osaki, H. Shibahashi (Eds.), Progress of Seismology of the Sun and Stars. Proceedings, 1989. XIII, 467 pages. 1990.

Vol. 368: L. Garrido (Ed.), Statistical Mechanics of Neural Networks. Proceedings. VI, 477 pages. 1990.

Vol. 369: A. Cassatella, R. Viotti (Eds.), Physics of Classical Novae. Proceedings, 1989. XII, 462 pages. 1990.

Vol. 370: H.-D. Doebner, J.-D. Hennig (Eds.), Quantum Groups. Proceedings, 1989. X, 434 pages. 1990.

Vol. 371: K.W. Morton (Ed.), Twelfth International Conference on Numerical Methods in Fluid Dynamics. Proceedings, 1990. XIV, 562 pages. 1990.

Vol. 372: F. Dobran, Theory of Structured Multiphase Mixtures. IX, 223 pages. 1991.

Vol. 374: L. Ting, R. Klein, Viscous Nortical Flows. V, 222 pages. 1991.

Vol. 375: C. Bartocci, U. Bruzzo, R. Cianci (Eds.), Differential Geometric Methods in Theoretical Physics. Proceedings, 1990. XIX, 401 pages. 1991.